Essentials of Hydraulics

Written for a one-semester course in hydraulics, this concise textbook is rooted in the fundamental principles of fluid mechanics and aims to promote sound hydraulic engineering practice. Basic methods are presented to underline the theory and engineering applications, and examples and problems build in complexity as students work their way through the textbook. Abundant worked examples and calculations, real-world case studies and revision exercises, as well as precisely crafted end-of-chapter exercises, ensure students learn exactly what they need in order to consolidate their knowledge and progress in their career. Students learn to solve pipe networks, optimize pumping systems, design pumps and turbines, solve differential equations for gradually varied flow and unsteady flow, and gain knowledge of hydraulic structures such as spillways, gates, valves, and culverts. An essential textbook for intermediate to advanced undergraduate and graduate students in civil and environmental engineering.

Pierre Y. Julien is Professor of Civil and Environmental Engineering at Colorado State University. He has 40 years of experience in the fields of hydraulics, sedimentation, and river engineering. He has authored over 600 scientific contributions including two textbooks, 35 book chapters and manuals, 195 refereed articles, and over 250 conference papers and presentations. He has delivered 25 keynote addresses world-wide, taught 20 short courses, and has guided 40 Ph.D. and over 100 Master's students to graduation. His other two textbooks are *Erosion and Sedimentation* (second edition, 2010, Cambridge University Press) and *River Mechanics* (second edition, 2018, Cambridge University Press).

"I was lucky to be Pierre's Ph.D. student at Colorado State University many years ago. I took several courses from him and was deeply influenced by his teaching style and methods.
I am more than happy to see the publication of his *Essentials of Hydraulics* so that the rest of the world of civil engineering students have a chance to learn from this great teacher and scholar."

Junke Guo, University of Nebraska–Lincoln

"*Essential of Hydraulics* by Professor P. Y. Julien is an excellent and well-needed addition to the literature on hydraulic engineering. The textbook encompasses all subject areas of hydraulics with clarity, and provides an in-depth understanding of the theoretical aspects by using detailed step-by-step worked examples. In addition, the plethora of exercises and problems provide a solid pedagogical tool for mastering the material. The textbook is suitable for undergraduate and graduate students, but also for engineers practicing in the general area of hydraulics. Based on my 30 years of academic experience in hydraulic engineering, I fully appreciate and unequivocally endorse this textbook."

Panagiotis (Pete) D. Scarlatos, Florida Atlantic University

"This handily focused and lucidly written textbook presents the indispensable information needed for a course on civil engineering hydraulics. The textbook's author writes from his extensive experience teaching hydraulics, and draws on his considerable insights into the practical hydraulics issues often faced by civil engineers."

Robert Ettema, Colorado State University

"An excellent reference for a course in hydraulics covering fundamental principles in pipe flow, pumps, and open channel flow. With numerous examples, this textbook will support learning very effectively in an undergraduate course, or serve as review of hydraulics for a graduate course with exposure to more advanced topics."

Paola Passalacqua, University of Texas at Austin

"This is an excellent textbook for learning and teaching the fundamentals of hydraulics and their applications in the fields of civil and environmental engineering. The topics covered in the book are comprehensive. The examples of numerical calculation help undergraduate and graduate students to better understand the fundamental concepts, and the problems are well designed with different levels of challenge and importance."

Ming Ye, Florida State University

Essentials of Hydraulics

Pierre Y. Julien
Colorado State University

CAMBRIDGE
UNIVERSITY PRESS

CAMBRIDGE
UNIVERSITY PRESS

University Printing House, Cambridge CB2 8BS, United Kingdom

One Liberty Plaza, 20th Floor, New York, NY 10006, USA

477 Williamstown Road, Port Melbourne, VIC 3207, Australia

314–321, 3rd Floor, Plot 3, Splendor Forum, Jasola District Centre, New Delhi – 110025, India

103 Penang Road, #05–06/07, Visioncrest Commercial, Singapore 238467

Cambridge University Press is part of the University of Cambridge.

It furthers the University's mission by disseminating knowledge in the pursuit of education, learning, and research at the highest international levels of excellence.

www.cambridge.org
Information on this title: www.cambridge.org/highereducation/isbn/9781316513095
DOI: 10.1017/9781108907446

First published 2022

Printed in the United Kingdom by TJ Books Limited, Padstow, Cornwall, 2022

A catalogue record for this publication is available from the British Library.

ISBN 978-1-316-51309-5 Hardback
ISBN 978-1-108-81630-4 Paperback

Additional resources for this publication at www.cambridge.org/julien-hydraulics

To my father Guy,
and
to all engineers working in developing countries

Contents

Preface

Water sustains life and engineers supply drinking water to a growing world population. Meeting this daunting challenge requires a proper understanding of hydraulic engineering. Hydraulics is an integral part of the curriculum for all civil and environmental engineers. Water flows naturally in river systems and the propagation of large floods can be devastating. In controlling floods through dams and multiple hydraulic structures, engineers have developed techniques to generate hydropower and distribute water through open channels, culverts and pipe networks. The principles governing the motion of water include conservation of mass, momentum and energy, as well as resistance to flow. These governing equations constitute essential knowledge for the understanding of the motion of fluids in pipes and open channels. This textbook demonstrates the benefits of mastering these governing principles. Fundamentally, hydraulic engineering requires an understanding of hydrodynamic forces governing the motion of water in closed conduits and open channels. Engineers apply these fundamental concepts to meet the practical needs of society. These needs include flood control, water distribution systems for agriculture and water supply, hydropower production and sustainable development of water resources.

This textbook prepares engineers through understanding fundamental concepts and problem solving. The essential complementarity of theory and practice cannot be overemphasized. Theory can best enhance engineering applications when the fundamental understanding has been grounded in practical observations. Rather than a voluminous encyclopedia, this textbook scrutinizes selected and proven methods meeting pedagogical objectives. The prerequisites required of the reader include basic knowledge of engineering dynamics, fluid mechanics and differential equations. Besides basic theory and lecture material, the chapters of this book contain numerous examples and solved problems, several data sets, computer problems and case studies. They illustrate specific aspects of the profession from theoretical derivations to practical solutions with the analysis of typical problems. The key to successfully mastering the material from this book is to solve problems. Most problems can be solved with algebraic equations while a few require the use of computers. No specific computer code or language is required. Instead of promoting the use of commercial software packages, student creativity and originality is stimulated by developing their own spreadsheets and computer programs. Throughout the book, a solid diamond (♦) denotes examples and problems of particular significance, double diamond (♦♦) denotes the most important and/or difficult.

This textbook proceeds from closed conduits to open channels as follows:
Chapter 1 describes the physical properties of water and hydrostatics;
Chapter 2 reviews the motion of water in pipes;
Chapter 3 derives the governing equations describing the motion of water;
Chapter 4 guides the design of pumping systems;
Chapter 5 treats the topic of hydropower generation from turbines;

Chapter 6 delves into the compressibility effects of water in pipes;
Chapter 7 presents fundamental methods for unsteady flow in pipes;
Chapter 8 focuses on steady open-channel flow;
Chapter 9 deals with rapidly varied open-channel flow;
Chapter 10 delineates backwater curves and gradually varied open-channel flows;
Chapter 11 broaches the advanced topic of unsteady flow in open channels;
Chapter 12 copes with more complex flows through culverts;
Chapter 13 introduces different types of spillways and gates;
Chapter 14 outlines broad-based concepts in hydrology;
Chapter 15 summarizes the fundamentals of geohydrology; and
Chapter 16 covers essential knowledge of groundwater.

There is more than sufficient material for a 45-hour undergraduate-level course. On a three-hour per week basis, the core material can be subdivided into four parts (four weeks each) with the possibility for monthly mid-term exams. The first part covers hydrostatics and flow in pipes (Chapters 1–3). The unique aspect of this section is to solve pipe networks. The second part focuses on hydro-machinery with pumps and turbines (Chapters 4–6). The unique skill being developed is the optimization of a pipe and pump system. The third part covers open-channel flows (Chapters 8–10). This section uniquely presents methods to obtain synthetic knowledge and analytical solutions, besides the numerical solution of differential equations for backwater profiles. The last section deals with hydraulic structures (Chapters 12–13). This shorter and more descriptive section leads to a comprehensive final exam. One unique aspect of this last section is to introduce practical engineering issues and professional ethics. The end of the semester is also convenient for broader guest lectures from practitioners.

Graduate and honor students will gain advanced knowledge of unsteady flow in pipes (Chapter 7) and in open channels (Chapter 11). In preparation for graduate studies, additional knowledge of hydraulics is provided along complementary concepts in hydrology (Chapter 14), geohydrology (Chapter 15) and groundwater (Chapter 16). All these chapters can support teaching in a quarter system. My preference for presenting pipes before open channels is rooted in the fact that pipes have constant cross sections which makes the geometry simpler to handle. It is also possible to reverse the teaching order and follow the sequence: hydrology, open channels, pipes and hydro-machinery.

This book benefitted from my teaching experience at university level since the late 1970s. The suggestions of a couple of generations of undergraduate students helped tailor the content to meet their needs under the constraints of quality, concision and affordability. I am grateful to Drs. Jai Hong Lee, Joon Hak Lee and Seong Joon Byeon who helped me prepare earlier versions of lecture notes for the students of hydraulic engineering at Colorado State University. Jean Parent deserves all the credit for patiently drafting all the figures. It has been a renewed pleasure to collaborate with Matt Lloyd, Jane Adams, Rachel Norridge, Zoë Lewin and the Cambridge University Press production staff.

Notation

Symbols

a	acceleration
a_x, a_y, a_z	Cartesian acceleration
a_s	seismic acceleration
A	surface area
A	activity of a clay
A_T	surge tank area
A_t	watershed drainage area
B	(base) channel width
B	wicket-gate height, turbine-blade height
c	wave celerity
c	damping coefficient
c'	wave celerity in an elastic pipe
c_v	consolidation coefficient
C	resistance coefficient
C, C_D	discharge coefficient
C	runoff coefficient
C_c	orifice coefficient
C_C	consolidation coefficient
C_g	volumetric gas concentration in water
C_G	Venturi geometric coefficient
C_H, C_Q	pump head and discharge dimensionless parameters
C_s	seismic coefficient
C_v	coefficient of variation
C_V	Venturi coefficient
C_Δ	triangular weir discharge coefficient
$C_\$$	cost function
CMP	corrugated metal pipe
d, w	opening height of a Tainter gate
d_{10}, d_{50}	particle size distribution, % finer by weight
d_s	particle size
D, d	pipe/culvert diameter
D	scour depth below a flip-bucket spillway
D_1, D_2	upstream and downstream flow depths of a stilling basin
DC	downstream control
e, e_m, e_p	efficiency of pumps
e	pipe thickness
e	void ratio
E	specific energy
E_w	water modulus of elasticity
E_p	pipe modulus of elasticity
EGL	energy grade line, EGL = HGL + $V^2/2g$
ΔE	elevation gain for pumps
ΔE	specific energy lost in a hydraulic jump
f	Darcy–Weisbach friction factor
f_d, f_n	damped and natural frequency of oscillations
F	force
F	network frequency
F	future value
F_h	hydrostatic force
F_H, F_V, W	horizontal, vertical and weight forces
Fr	Froude number
g	gravitational acceleration
G	specific gravity of sediment
h	flow depth
Δh	flow depth increment
h_b	block height of baffled chutes
h_c	critical flow depth
h_c	capillary rise
h_d, h_u	downstream and upstream flow depths
h_f	friction loss in pipes
h_n	normal flow depth
h_t	tailwater depth
h_v	vapor head
h_1 to h_4	various sizes of stilling basins

Δh	local change in flow depth
H	Bernoulli sum, or flow depth
H_D	design head of a spillway
H_O, H_a	initial and added pressure head
HDPE	high-density polyethylene
HGL	hydraulic grade line, also piezometric line
ΔH	energy loss
i, i_r	rainfall intensity
i	annual interest rate
i	gradient of the groundwater surface
i_b	riverbed infiltration rate
i_c	critical groundwater gradient for boiling sands
$\bar{I}$	area moment of inertia about the center of gravity
I_0	area moment of inertia about axis 0
k	pipe conveyance coefficient ($h_f = k\ Q^2$)
k	open-channel conveyance coefficient ($Q = k\ S_f^{1/2}$)
k	constant of a spring–mass system
k	number of occurrences
k	intrinsic permeability coefficient
k_1, h_1/D	culvert inlet control parameters
k_s, ε	surface roughness
K	pipe loss coefficient ($h_f = K\ V^2/2g$)
K	conveyance coefficient
K	permeability coefficient
K	flood diffusion (dispersion) coefficient
K	saturated hydraulic conductivity ($V = K\ i$)
K_b	pipe bend coefficient
K_C	pipe contraction coefficient
K_e	pipe or culvert entrance coefficient
K_E	pipe expansion coefficient
K_f, K_f'	ratio of horizontal to vertical normal stresses
$K_G(T)$	frequency factor of the Gumbel distribution
L	pipe, spillway or weir length
L_e	equivalent pipe length for minor losses
L_F	loading factor for pipe trenches
L_j	length of hydraulic jump
L_{II}, L_{III}, L_{IV}	lengths of stilling basins
L_r	river length
LL	liquid limit
m	Manning equation unit coefficient ($m = 1.49$ in customary units, $m = 1$ in SI units)
M, m	mass
M_s, M_w, M_t	mass of solids, water and total
M	specific momentum
M_0	moment of force about 0
n	Manning roughness coefficient
n	porosity of a soil
n	number of independent trials
n	number of years
N	rotational speed
N	dimensionless parameter in Allievi's formula
N	normal force
N	number of years in a sample
N_p, N_s	number of potential and streamlines
N_{rpm}	number of rotations per minute
N_s	specific speed of a pump, or turbine
N_{sq}	specific speed for turbines in SI units
NPSH	net positive suction head
p	pressure
p_{atm}	atmospheric pressure
p_{rel}	relative pressure
p_{abs}	absolute pressure
p	probability of occurrence
p, q	average and half-difference soil pressure
p_0	number of poles in a generator
p_0	porosity
p_v	vapor pressure
P	power
P	wetted perimeter
P	present value
P	flow depth at the spillway crest
P_k	binomial probability distribution
P_{kW}, P_{hp}	power in kilowatt and horsepower
PI	plasticity index
PL	plastic limit

P	probability
Δp	pressure increase from a water hammer
ΔP	power loss in a hydraulic jump
q	unit discharge
q_l	lateral inflow unit discharge
Q	discharge
Q_{gpm}, Q_{cfs}, Q_{cms}	discharge in gallons per minute, ft^3/s and m^3/s
r, R	radius
R	reaction force
R	manometer reading
R	hydrologic risk
R	repeated uniform annual amount
Re	Reynolds number
R_h	hydraulic radius
RCP	reinforced concrete pipe
s_u	subsidence
S	slope
S	surge-tank water elevation
S	degree of saturation
S_F	safety factor
S_0, S_f, S_w	bed, friction and water-surface slope
S_y	specific yield of an aquifer
S_{XY}	sign of pipe flow (+ clockwise from X to Y)
S_3	three-edge pipe-bearing strength
t	time
t	trapezoidal section parameter (H:V)
Δt, Δx	time and space increment
t_C	time of pipe closure
t_e	watershed time to equilibrium
t_r	rainfall duration
T	torque
T	tension, and surface tension
T	surge-tank oscillation period
T	period of return of extreme events
T	trench width
T	transmissivity of an aquifer
T	dimensionless consolidation time
T°	temperature
T_n, T_d	natural and damped period of oscillations
u	pore pressure in a soil
U, $U_{\%}$	consolidation percentage
UC	upstream control
v_p	plate velocity
v_x, v_y	local velocity components
V	mean flow velocity
V_r, V_t	radial and tangential velocity
$\forall$	volume
$\forall_v$, $\forall_t$	volume of voids and total volume
w, d	opening height of a Tainter gate
w	water content of a soil
W	channel surface width
W	weight
W_T	throat width of a Parshall flume
$W(u)$	Theis integral
x, y, z	coordinates usually x downstream, y lateral, and z upward
x, y	coordinates of an ogee spillway
x_b	flip-bucket launching distance
Δx, Δy, Δz	infinitesimal fluid element
Δx, Δy	infinitesimal distance and flow depth in backwater calculations
X_{max}	maximum elevation without cavitation
y	open-channel flow depth
y_b	flip-bucket launching height
y_c	critical flow depth
y_{cp}	center of pressure
y_n	normal flow depth
z_b	bed elevation
z_{max}	maximum successive flow oscillations
z_t	turbine elevation above the tailrace

Greek Symbols

α	angle of failure in the p-q diagram
α	flip bucket launching angle
α, β	coefficient and exponent of the stage discharge relationship
α_s	ratio of seismic acceleration to gravity
β	turbine-blade angle
γ	specific weight of water
ε	surface roughness
θ	angular coordinate
ζ	dimensionless damping coefficient
$\kappa - \varepsilon, \kappa - \omega$	turbulence models
μ	dynamic viscosity of water
υ	kinematic viscosity of water
ξ, η	orthogonal functions for potential and streamlines in groundwater
ρ, ρ_w	mass density of water
ρ_d	dry mass density of a water–sediment mixture
ρ_g	mass density of gas or air
ρ_m, ρ_t	total mass density of a water–sediment mixture
ρ_s	mass density of solid
ρ_{sat}	mass density of a saturated soil
σ	turbine cavitation index
σ	standard deviation
σ	applied stresses in a soil
σ'	effective stress in a soil
σ_g	gradation coefficient
σ_p	pipe tension
ϕ	turbine speed ratio
ϕ	dimensionless turbulent flow oscillations
ϕ, ϕ'	internal friction angle, and effective internal friction angle
ϕ, φ	groundwater potential and streamlines
φ	oscillation phase angle
τ, σ	tangential and normal shear stresses
ω	turbine (angular) rotational speed
ω_n, ω_d	natural and damped circular frequency of oscillations
$\Omega = 1 - \mathrm{Fr}^2$	coefficient of the diffusive-wave equation

Superscripts and Diacriticals

$\bar{e}$	average value
$x, \dot{x}$ *and* $\ddot{x}$.	position, velocity and acceleration, e.g. spring–mass system
$\forall$	volume

Subscripts

a_x, a_y, a_z	Cartesian components
y_n, y_c	normal and critical flow depths
ρ_s, γ_s	solid properties

1 Hydrostatics

Chapter 1 covers water properties, unit conversions and forces on dams. Fundamental dimensions, units and water properties are reviewed in Section 1.1. The concept of pressure and piezometric head in Section 1.2 is expanded into hydrostatic forces on plane surfaces in Section 1.3 and dams in Section 1.4.

1.1 Units and Water Properties

This section reviews fundamental dimensions and units (Section 1.1.1) followed by water properties (Section 1.1.2) and fluid density (Section 1.1.3).

1.1.1 Dimensions and Units

Physical water properties are expressed in terms of the fundamental dimensions: mass (M), length (L) and time (T). In the Système International (SI), the units are the kilogram (kg), the meter (m) and the second (s). To honor Isaac Newton, a newton (N) is the force to accelerate 1 kg at 1 m/s^2, or 1 N = 1 kg m/s^2. The gravitational acceleration at the Earth's surface is $g = 9.81$ m/s^2. The weight corresponding to a mass of one kilogram is W = mass × g = 1 kg × 9.81 m/s^2 = 9.81 N. The pressure denotes a force per unit area perpendicular to a surface, and in memory of Blaise Pascal, it is given in pascals or 1 Pa = 1 N/m^2. Commemorating James Joule, the unit of work (or energy) is the joule (J) equal to the product 1 J = 1 N × 1 m in Table 1.1. Remembering James Watt, the unit of power is a watt (W), 1 W = 1 J/s.

Engineers must be able to use two systems of units. Besides SI, the US customary units use the slug (slug), the foot (ft) and the second (s). The force to accelerate a mass of one slug at 1 ft/s^2 is a pound (lb), which is a force and not a mass. The temperature T° in degrees Celsius ($T_{\circ C}$) is converted to degrees Fahrenheit ($T_{\circ F}$) from $T_{\circ F}$ = 32.2 °F + 1.8 $T_{\circ C}$. Useful unit conversion factors are found in Table 1.2.

Examples include the weight 1 lb = 4.448 N, and discharge 1 m^3/s = 35.32 ft^3/s. The abbreviation 1 cms = 1m^3/s and 1 cfs = 1ft^3/s. Prefixes indicate multiples of units [e.g. k (kilo) = 10^3, M (mega) = 10^6 or G (giga) = 10^9] or fractions of units [n (nano) = 10^{-9}, μ (micro) = 10^{-6}, or m (milli) = 10^{-3}]. For example, one megawatt (MW) equals one million watts (1,000,000 W or 10^6 W), and a micron 1 μm = 10^{-6} m.

1.1.2 Properties of Water

Mass density of water ρ. The mass of water per unit volume defines the mass density ρ. The maximum mass density of water at 4 °C is 1,000 kg/m^3 and decreases with temperature, as shown in Table 1.3 (SI) and Table 1.4 (customary units). The mass

Table 1.1. Geometric, kinematic, dynamic and dimensionless variables

Variable	Symbol	Fundamental dimensions	SI units
Geometric (L)			
Length	L, x, h	L	m
Area	A	L^2	m^2
Volume	$\forall$	L^3	m^3
Kinematic (L, T)			
Velocity	V	LT^{-1}	m/s
Acceleration	a, g	LT^{-2}	m/s^2
Kinematic viscosity	υ	L^2T^{-1}	m^2/s
Unit discharge	q	L^2T^{-1}	m^2/s
Discharge	Q	L^3T^{-1}	m^3/s
Dynamic (M, L, T)			
Mass	m	M	kg
Force	$F = ma, mg$	MLT^{-2}	1 kg m/s^2 = 1 N
Pressure	$p = F/A$	$ML^{-1}T^{-2}$	1 N/m^2 = 1 Pa
Work or energy	$E = Fd$	ML^2T^{-2}	1 Nm = 1 J
Mass density	ρ	ML^{-3}	kg/m^3
Specific weight	$\gamma = \rho g$	$ML^{-2}T^{-2}$	N/m^3
Dynamic viscosity	$\mu = \rho\upsilon$	$ML^{-1}T^{-1}$	1 kg/m s = 1 Pa s
Dimensionless			
Slope	S_0, S_f	–	–
Reynolds number	$\mathrm{Re} = Vh/\upsilon$	–	–
Froude number	$\mathrm{Fr} = V/\sqrt{gh}$	–	–

density of seawater is approximately 1,025 kg/m^3 and the mass density of air at sea level is 1.29 kg/m^3 at 0 °C. The conversion factor for mass density is 1 slug/ft^3 = 515.4 kg/m^3.

Specific weight of water γ. The weight per unit volume is the specific weight γ. At a temperature of 10 °C, water has a specific weight γ = 9,810 N/m^3, or 62.4 lb/ft^3 (1 lb/ft^3 = 157.09 N/m^3). The specific weight varies with temperature in Tables 1.3/1.4. The specific weight γ is the product of mass density ρ and gravitational acceleration g = 32.2 ft/s^2 = 9.81 m/s^2:

$$\gamma = \rho g. \tag{1.1}$$

Dynamic viscosity μ. As a fluid is brought into slow deformation, the fluid velocity at any boundary is the velocity of the boundary. The viscosity is a measure of the fluid resistance to angular deformation. The lower the viscosity, the thinner (or more liquid) a fluid is. The fundamental dimensions of dynamic viscosity μ are M/LT. In Tables 1.3/1.4, the dynamic viscosity of water decreases with temperature. At 20 °C,

Table 1.2. Useful unit conversions from US customary units to SI

Unit	kg, m, s	N, Pa, W
1 acre (1 ac. = 43,560 ft^2)	4,047 m^2	
1 atmosphere (atm)	101.3 kg/m s^2	101.3 kPa
1 cubic foot per second (ft^3/s)	0.0283 m^3/s	
1 degree Fahrenheit (°F) = 32 + 1.8 $T_{°C}$	0.5556 °K	
1 foot (ft)	0.3048 m	
1 gallon (US) (1 US gal = 3.785 liters)	0.003785 m^3	
1 horsepower (hp) = 550 lb ft/s	745.7 kg m^2/s^3	745.7 W
1 inch (in.) (1 ft = 12 in.)	0.0254 m	
1 inch of mercury (in. Hg)	3,386 kg/m s^2	3,386 Pa
1 inch of water	248.8 kg/m s^2	248.8 Pa
1 mile (statute), (1 mile = 5,280 ft)	1,609 m	
1 million gallons/day (1 mgd = 1.55 ft^3/s)	0.04382 m^3/s	
1 pound-force (lb) (1 lb = 1 slug × 1 ft/s^2)	4.448 kg m/s^2	4.448 N
1 pound per square foot (lb/ft^2 or psf)	47.88 kg/m s^2	47.88 Pa
1 pound per square inch (lb/in.^{2} or psi)	6,895 kg/m s^2	6,895 Pa
1 quart (US) (1 qt = 2 pints = 4 cups)	0.0009463 m^3	
1 slug	14.594 kg	
1 ton (SI) (1 metric ton = 1,000 kg = 1 Mg)	1,000 kg	
1 ton (US short) = 2,000 lb	8,900 kg m/s^2	8.9 kN

μ = 0.001 kg/m · s = 0.001 N s/m^2 = 0.001 Pa s. The conversion factor of the dynamic viscosity is 1 lb s/ft^2 = 47.88 N s/m^2 = 47.9 Pa s.

Kinematic viscosity of water υ. The kinematic viscosity $\upsilon = \mu/\rho$ in L^2/T is obtained from the dynamic viscosity of a fluid μ divided by its mass density ρ. Water at 20 °C has υ = 0.01 cm^2/s = 1 × 10^{-6} m^2/s. The conversion factor is 1 ft^2/s = 0.0929 m^2/s. The kinematic viscosity of water υ depends on the temperature T° in degrees Celsius.

$$\upsilon = \frac{\mu}{\rho} \simeq \frac{1.78 \times 10^{-6}\ \mathrm{m^2/s}}{1 + 0.0337 T_{^\circ C} + 0.0002217 T_{^\circ C}^{\,2}}. \tag{1.2}$$

Commonly used values of specific weight and viscosity are:

$$\gamma = \frac{weight}{volume} = \frac{mass}{volume} \times g = \rho \times g = \frac{1{,}000\ \mathrm{kg}}{\mathrm{m^3}} \times \frac{9.81\ \mathrm{m}}{\mathrm{s^2}} = \frac{9.81\ \mathrm{kN}}{\mathrm{m^3}},$$

$$\mu = \rho \times \upsilon = \frac{1{,}000\ \mathrm{kg}}{\mathrm{m^3}} \times \frac{1 \times 10^{-6}\ \mathrm{m^2}}{\mathrm{s}} = \frac{1 \times 10^{-3}\ \mathrm{kg}}{\mathrm{m \cdot s}} = \frac{1 \times 10^{-3}\ \mathrm{N \cdot s}}{\mathrm{m^2}}.$$

Table 1.3. Physical properties of water in SI units at atmospheric pressure

Temperature $T°$ (°C)	Vapor pressure p_v (kN/m^2) absolute	Density ρ (kg/m^3)	Specific weight γ (kN/m^3)	Dynamic viscosity μ (N-s/m^2) $\times 10^{-3}$	Kinematic viscosity υ (m^2/s) $\times 10^{-6}$	Surface tension σ_t (N/m)	Elasticity modulus E_w (kN/m^2) $\times 10^6$
0	0.61	999.8	9.805	1.781	1.785	0.0756	2.02
5	0.87	1000.0	9.807	1.518	1.519	0.0749	2.06
10	1.23	999.7	9.804	1.307	1.306	0.0742	2.10
15	1.70	999.1	9.798	1.139	1.139	0.0735	2.14
20	2.34	998.2	9.789	1.002	1.003	0.0728	2.18
25	3.17	997.0	9.777	0.890	0.893	0.0720	2.22
30	4.24	995.7	9.764	0.798	0.800	0.0712	2.25
40	7.38	992.2	9.730	0.653	0.658	0.0696	2.28
50	12.33	988.0	9.689	0.547	0.553	0.0679	2.29
60	19.92	983.2	9.642	0.466	0.474	0.0662	2.28
70	31.16	977.8	9.589	0.404	0.413	0.0644	2.25
80	47.34	971.8	9.530	0.354	0.364	0.0626	2.20
90	70.10	965.3	9.466	0.315	0.326	0.0608	2.14
100	101.33	958.4	9.399	0.282	0.294	0.0589	2.07

Table 1.4. Properties of water in customary units at atmospheric pressure

Temperature $T°$ (°F)	Vapor pressure p_v (psia)	Density ρ (slug/ft^3)	Specific weight γ (lb/ft^3)	Dynamic viscosity μ (lb · s/ft^2) $\times 10^{-5}$	Kinematic viscosity υ (ft^2/s) $\times 10^{-5}$	Surface tension σ_t (lb/ft) $\times 10^{-3}$	Elasticity modulus E_w (psi) $\times 10^3$
32	0.09	1.940	62.42	3.746	1.931	5.18	293
40	0.12	1.940	62.43	3.229	1.664	5.14	294
50	0.18	1.940	62.41	2.735	1.410	5.09	305
60	0.26	1.938	62.37	2.359	1.217	5.04	311
70	0.36	1.936	62.30	2.050	1.059	4.98	320
80	0.51	1.934	62.22	1.799	0.930	4.92	322
90	0.70	1.931	62.11	1.595	0.826	4.86	323
100	0.95	1.927	62.00	1.424	0.739	4.80	327
120	1.69	1.918	61.71	1.168	0.609	4.67	333
140	2.89	1.908	61.38	0.981	0.514	4.54	330
160	4.74	1.896	61.00	0.838	0.442	4.41	326
180	7.51	1.883	60.58	0.726	0.385	4.27	318
200	11.52	1.868	60.12	0.637	0.341	4.13	308
212	14.70	1.860	59.83	0.593	0.319	4.04	300

Commonly used values of specific weight and viscosity are:

$$\gamma = \frac{weight}{volume} = \frac{mass}{volume} \times g = \rho \times g = \frac{1.94 \text{ slug}}{\text{ft}^3} \times \frac{32.2 \text{ ft}}{\text{s}^2} = \frac{62.4 \text{ lb}}{\text{ft}^3},$$

$$\mu = \rho \times \upsilon = \frac{1.94 \text{ slug}}{\text{ft}^3} \times \frac{1 \times 10^{-5} \text{ ft}^2}{\text{s}} = \frac{2 \times 10^{-5} \text{ slug}}{\text{ft} \cdot \text{s}} = \frac{2 \times 10^{-5} \text{ lb} \cdot \text{s}}{\text{ft}^2}.$$

1.1.3 Fluid Density

The specific gravity G of a fluid is defined as the ratio of the density of the fluid to the density of water. Normally, the temperature is taken as 4 °C (39.2 °F), at which the density of water is 1.94 slug/ft^3 or 1,000 kg/m^3. One can also define the specific gravity as the weight of the liquid divided by the specific weight of water, or

$$G = \frac{\rho_{fluid}}{\rho_{water}} = \frac{\gamma_{fluid}}{\gamma_{water}}. \tag{1.3}$$

The specific gravity is a dimensionless number per Example 1.1 (SI) and Example 1.2 (customary units). Materials with $G < 1$ float in water, e.g., wood, ice, etc.

Example 1.1: Mass density and specific gravity

If a 2 m^3 volume of oil has a mass of 1,720 kg, calculate its specific weight, density and specific gravity.

Solution:

$$\gamma = \frac{weight}{volume} = \frac{1{,}720 \text{ kg}}{2 \text{ m}^3} \times \frac{9.81 \text{ m}}{\text{s}^2} = 8{,}437 \text{ N/m}^3,$$

$$\rho = \frac{\gamma}{g} = \frac{8{,}437}{9.81} = 860 \text{ kg/m}^3$$

and

$$G = \frac{\gamma_{oil}}{\gamma_{water}} = \frac{8{,}437}{9{,}810} = 0.86.$$

♦ **Example 1.2:** Specific weight and density

A container weighs 5,000 lb when filled with 60 ft^3 of a certain liquid. The empty weight of the container is 500 lb. What is the density of the liquid?

Solution:

$$\gamma = \frac{weight}{volume} = \frac{(5{,}000 - 500) \text{ lb}}{60 \text{ ft}^3} = 75 \text{ lb/ft}^3,$$

$$\rho = \frac{\gamma}{g} = \frac{75}{32.2} = 2.3 \text{ slug/ft}^3.$$

In general, calculations should include all decimals but the final answer should have as many significant digits as the input parameters. For instance, it would not be appropriate to write $\rho = 2.329193$ slug/ft^3 in this case because the input parameters only have one significant digit.

1.2 Hydrostatic Pressure

Pressure is the force per unit area perpendicular to a surface. In a fluid, the pressure is the same in all directions. Atmospheric pressure is discussed in Section 1.2.1, followed by hydrostatic pressure in Section 1.2.2. The difference between relative and absolute pressure is explained in Section 1.2.3. Finally, the hydraulic grade line is covered in Section 1.2.4 with vapor pressure (absolute in Section 1.2.5 and relative in Section 1.2.6).

1.2.1 Atmospheric Pressure

The atmospheric pressure was first measured by the French scientist Blaise Pascal, in the seventeenth century. The atmospheric pressure decreases with altitude, as shown in Table 1.5. At sea level, the atmospheric pressure is $p_{atm} = 101.3\,\text{kPa} = 14.7\,\text{psi} = 2{,}116\,\text{psf}$. Note that 1 psi is the pressure generated by a column of 2.31 ft, or 70.4 cm of water.

Table 1.5. Atmospheric pressure vs altitude (1 psi = 6,895 Pa)

Altitude above sea level		Atmospheric pressure			
ft	m	mm Hg	psia	psfa	kPa
0	0	760	14.7	2,116	101.3
1,000	305	733	14.2	2,045	97.7
2,000	610	707	13.7	1,973	94.2
3,000	914	681	13.2	1,901	90.8
4,000	1,219	656	12.7	1,829	87.5
5,000	1,524	632	12.2	1,757	84.3
6,000	1,829	609	11.8	1,700	81.2
7,000	2,134	586	11.3	1,627	78.2
8,000	2,438	564	10.9	1,570	75.3
9,000	2,743	543	10.5	1,512	72.4
10,000	3,048	523	10.1	1,454	69.7
15,000	4,572	429	8.29	1,194	57.2
20,000	6,096	349	6.75	972	46.6
25,000	7,620	282	5.45	785	37.6
30,000	9,144	226	4.36	628	30.1

1.2.2 Hydrostatic Pressure

Hydrostatics refers to fluids at rest. The free surface of a fluid at rest is horizontal, which defines a surface of equal pressure (SEP). A datum is a reference SEP. Commonly used datum elevations are the mean sea level, the floor of a building and a benchmark elevation. Starting from the atmospheric pressure p_0 at the free surface, the hydrostatic pressure distribution can be obtained from the analysis of forces on an infinitesimal element of fluid of volume $\forall$ and weight W shown in Figure 1.1:

$$W = \rho g \forall = \gamma \forall = \gamma \Delta x\, \Delta y\, \Delta z.$$

The sum of upward forces from the pressure p and weight are

$$\sum F_z = p\Delta x\,\Delta y - (p + \Delta p)\Delta x\,\Delta y - \gamma\Delta x\,\Delta y\,\Delta z = 0,$$

such that

$$\Delta p = -\gamma\,\Delta z,$$

or

$$\frac{-dp}{dz} = \gamma,$$

which is integrated,

$$\int_{p_0}^{p} dp = \int_{z_0}^{z} -\gamma dz,$$

to give

$$p = p_0 + \gamma h, \tag{1.4}$$

where $h = z_0 - z$ from the elevations z and z_0 measured above a horizontal SEP, called a datum, as shown in Figure 1.2.

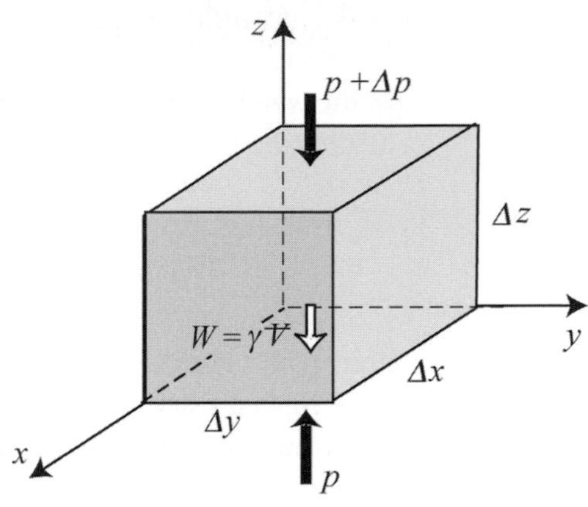

Figure 1.1 Element of fluid at rest

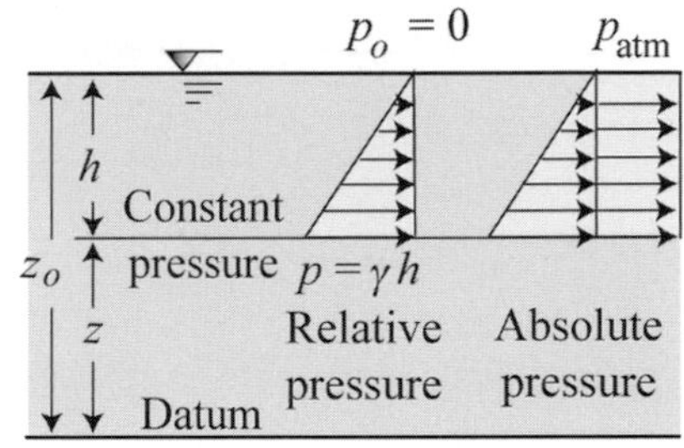

Figure 1.2 Relative and absolute pressure

1.2.3 Relative and Absolute Pressure

As shown in Figure 1.2, the relative pressure p_{rel} (also called the gauge pressure) is relative to the atmospheric pressure. It is obtained by subtracting the atmospheric pressure from the absolute pressure, $p_{rel} = p_{abs} - p_{atm}$. The gauge pressure (in psig) is always less than the absolute pressure (in psia), as shown in Example 1.3:

$$p_{abs} = p_{atm} + p_{rel}. \tag{1.5}$$

♦ **Example 1.3:** Absolute and relative pressure

At an altitude of 10,000 ft and water temperature of 5 °C (or 41 °F), use Tables 1.4 and 1.5 to find the absolute and relative pressure at $h = 10$ ft below the surface of a reservoir (1 psf = 47.9 Pa).

Solution:

$$p_{rel} = \gamma h = \frac{62.4\text{ lb}}{\text{ft}^3} \times 10\text{ ft} = 624\text{ psfg} = 29.9\text{ kPa},$$

$$p_{abs} = p_{atm} + p_{rel} = 10.1\text{ psia} \times 144\frac{\text{in.}^2}{\text{ft}^2} + 624\text{ psfg} = 2{,}078\text{ psfa} = 99.5\text{ kPa}.$$

1.2.4 Hydraulic Grade Line

A piezometer measures the fluid pressure. Dividing the relative pressure by the fluid specific weight defines the pressure head p/γ. The pressure head is a length which can be added to the elevation above a datum to define the piezometric head:

$$\textit{piezometric head} = z + \frac{p}{\gamma}. \tag{1.6}$$

As shown in Figure 1.3, the piezometric head in a fluid at rest is constant everywhere. The piezometric head at different locations defines the hydraulic grade line (HGL). In hydrostatics, the piezometric line and the HGL are simply the free surface of the fluid.

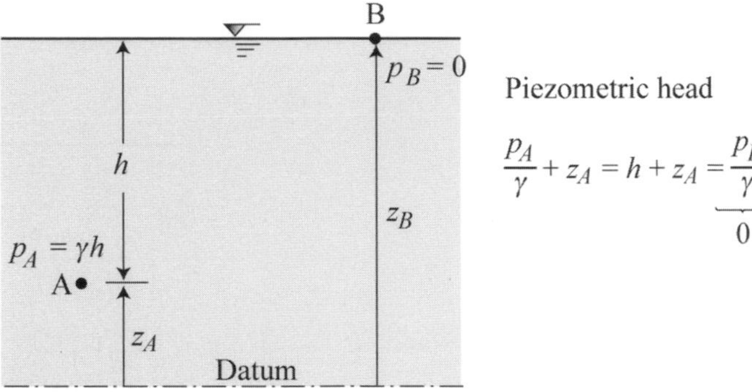

Figure 1.3 Constant piezometric head

A piezometer measures point pressure (Example 1.4). A manometer (Example 1.5) measures the pressure difference between two points.

♦♦ Example 1.4: Piezometer

In Fig. E-1.4, find the absolute and relative pressure at B.

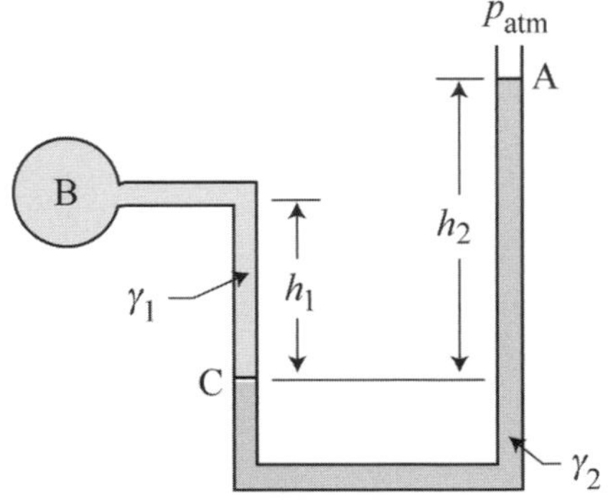

Fig. E-1.4 Piezometer

Solution:

$$p_C = p_{atm} + \gamma_2 h_2, \quad p_C = p_B + \gamma_1 h_1,$$

$$p_B = p_C - \gamma_1 h_1 = p_{atm} + \gamma_2 h_2 - \gamma_1 h_1 \quad \text{(absolute)},$$

$$p_{B\,rel} = p_{B\,abs} - p_{atm} = \gamma_2 h_2 - \gamma_1 h_1 \quad \text{(relative)}.$$

Two points in the same fluid have the same piezometric head. Therefore, two points at the same elevation in the same fluid have the same pressure. The piezometric heads at A and C are the same, $z_C + h_2 = z_A$. However, it is different from the piezometric head at B.

♦ **Example 1.5:** Manometer

In Fig. E-1.5, find the gauge pressure p_A when $G_{oil} = 0.8$ and $p_B = 200$ psf.

$$p_C = p_B + \gamma \times 1 \text{ ft} + 13.6\gamma \times R = p_A + 0.8\gamma \times 6 \text{ ft}.$$

Solution: With $\gamma = 62.4 \text{ lb/ft}^3$,

$$p_A = 200 + (62.4 \times 1) + (13.6 \times 62.4 \times 3) - (0.8 \times 62.4 \times 6) = 2{,}509 \text{ psf}.$$

With 1 psi = 144 psf, and 1 kPa = 0.145038 psi,

$$p_A = 2{,}509 \text{ psf} = 17.4 \text{ psi} = 120 \text{ kPa}.$$

The reader can check that the piezometric head at point A and point C are identical.

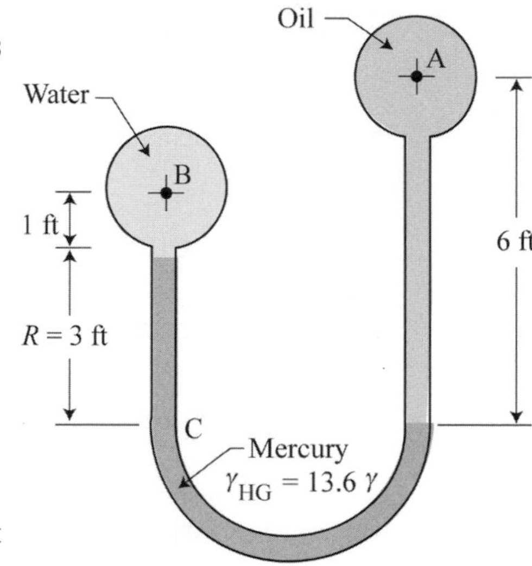

Fig. E-1.5 Manometer

1.2.5 Absolute Vapor Pressure

The pressure at which liquids turn into vapor is called the vapor pressure. The vapor pressure of water varies with temperature as listed in Tables 1.3 and 1.4. There is a wide variation in vapor pressure among liquids. As seen in Table 1.6, the vapor pressure of mercury (Hg) is extremely low) and was therefore used in traditional barometers. Typical calculations are shown for water (Example 1.6) and for mercury (Example 1.7).

Table 1.6. Vapor pressure p_v for mercury and water

Liquid	Pressure (psia)	Pressure head (in.)	Pressure (N/m^2 abs)	Pressure head (mm)
Mercury at 70 °F	0.000025	0.00005	0.17	0.00128
Water at 20 °C (68°F)	0.34	9.4	2,340	238

♦ **Example 1.6:** Head of water vapor

In Fig. E-1.6, find the maximum height of a water column h_v at sea level assuming 20 °C, and p_v = 0.34 psi = 2.34 kPa.

Solution: Consider absolute pressures

$$p_{atm} = 101.3 \text{ kPa}, \quad \gamma = 9.81 \text{ kN/m}^3$$

$$p_B = p_{atm} = p_A + \gamma h_v = p_v + \gamma h_v,$$

$$h_v = \frac{p_{atm} - p_v}{\gamma} = \frac{101.3 - 2.34}{9.81} \text{m}$$
$$= 10.1 \text{ m} = 33.1 \text{ ft}.$$

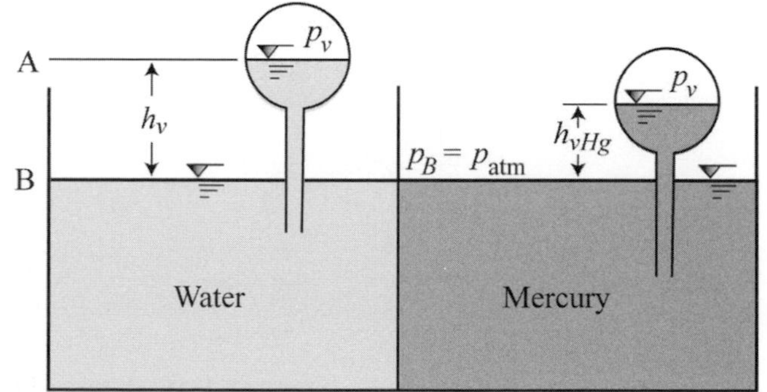

Fig. E-1.6 Head of water vapor

Example 1.7: Barometric head

Mercury is very dense ($G = 13.6$ or $\gamma_{Hg} = 13.6\,\gamma$) with a very low vapor pressure at 20°C, $p_v \cong 0$ kPa. Can you find the maximum height of a column of mercury?

Solution: The maximum height of a column of mercury is given as

$$h_{vHg} = \frac{p_{atm} - p_v}{\gamma_{Hg}} = \frac{101.3 - 0}{13.6 \times 9.81}\,\text{m} = 760 \text{ mm of Hg}.$$

This is the reference value of the atmospheric pressure head in traditional barometers.

1.2.6 Relative Vapor Pressure

As shown in Figure 1.4a, the vapor pressure depends on temperature (Tables 1.3 and 1.4) while the atmospheric pressure depends on altitude (Table 1.5). This knowledge can be combined in Table 1.7 to define the pressure head of water vapor as

$$h_v = \frac{p_{atm} - p_v}{\gamma},$$

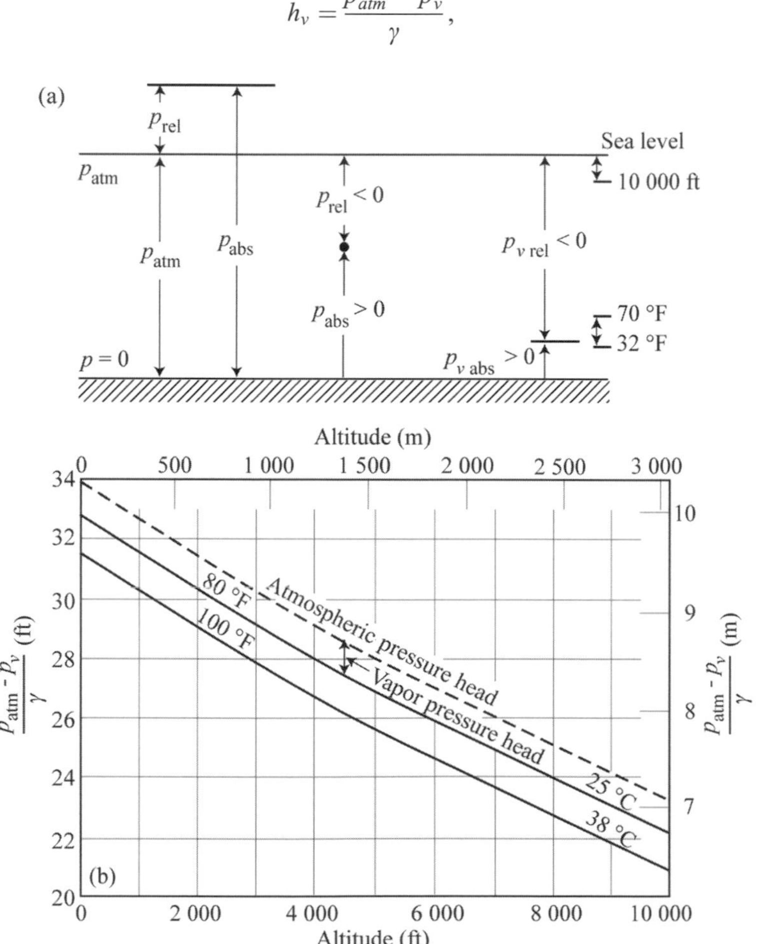

Figure 1.4 Relative pressure head of water vapor

Table 1.7. Atmospheric and water vapor pressure

Temperature $T_{°F}$	Absolute vapor pressure (always positive)		Altitude (ft)	Atmospheric pressure (always absolute)	
	p_v (psia)	p_v (kPa)		p_{atm} (psia)	p_{atm} (kPa)
32	0.09	0.61	0	14.7	101.3
40	0.12	0.87	5,000	12.2	84.0
50	0.18	1.23	10,000	10.1	69.5
60	0.26	1.70	15,000	8.3	57.1
70	0.36	2.34	20,000	6.8	46.5

with the results plotted on Figure 1.4b. A example calculation of the pressure head of water vapor is detailed in Example 1.8.

♦ **Example 1.8:** Relative and absolute vapor pressure

Find the relative vapor pressure of water at sea level and $T_{°F} = 70$ °F.

Solution:

$$p_{v\ rel} = p_v - p_{atm},$$

$$p_{v\ rel} = 0.36 \text{ psia} - 14.7 \text{ psi} = -14.3 \text{ psig} = -2{,}065 \text{ psf}.$$

Water will turn to vapor when the relative pressure decreases below $p_{v\ rel}$.

1.3 Hydrostatic Force

Calculations of hydrostatic forces on plane surfaces require a knowledge of the area moment of inertia in Section 1.3.1. The force magnitude is discussed in Section 1.3.2 and its point of application in Section 1.3.3.

1.3.1 Area Moment of Inertia

The area moment of inertia of a planar surface about its center of gravity is $\bar{I} = \int y^2 dA$. For a rectangular plate of area $A = ab$ sketched in Figure 1.5, this moment of inertia is

$$\bar{I} = \int y^2 dA = \int_{-a/2}^{a/2} y^2 dA = \int_{-a/2}^{a/2} y^2 b dy$$
$$= \frac{b}{3}\left[\left(\frac{a}{2}\right)^3 - \left(\frac{-a}{2}\right)^3\right] = \frac{ba^3}{12}. \quad (1.7)$$

The parallel-axis theorem gives the area moment of inertia about the axis 0 as $I_0 = \bar{I} + AL^2$. Figure 1.6 shows the area moments of inertia of other surfaces.

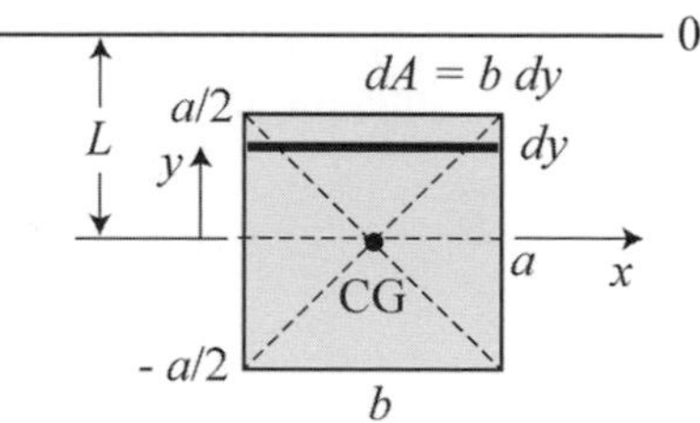

Figure 1.5 Area moment of inertia

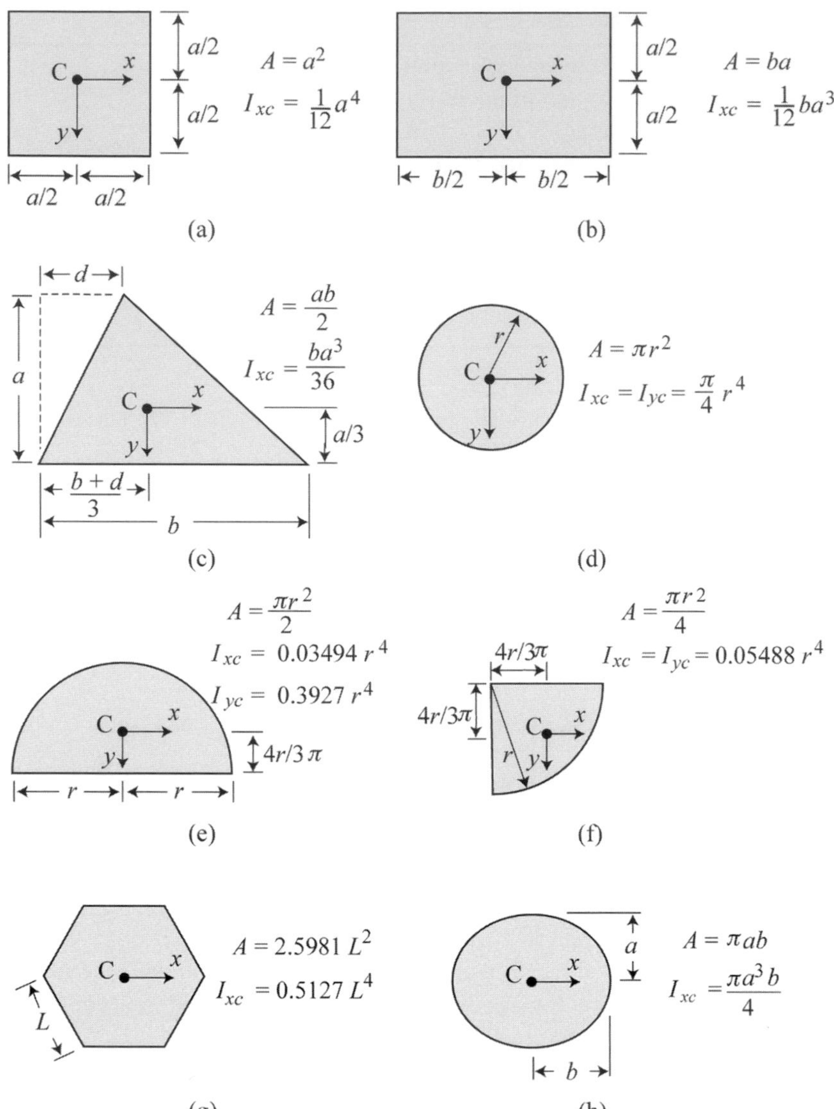

Figure 1.6 Area moments of inertia of various planar surfaces

1.3.2 Force Magnitude on a Plate

The magnitude of the resultant force on a plate in Figure 1.7 is obtained from the integration of pressure:

$$p = \gamma h = \gamma y \sin\theta,$$

$$dF = p dA = \gamma y \sin\theta \, dA,$$

$$F = \int dF = \gamma \sin\theta \int_A y dA = \gamma A \bar{y} \sin\theta,$$

$$F = \gamma A \bar{h}. \tag{1.8}$$

The hydrostatic force equals the surface area times the pressure at the centroid.

1.3.3 Center of Pressure

The point of application of the resultant force F is called the center of pressure y_{cp}. The point of application of the force is different from the centroid because the pressure is not uniformly distributed on the plate. The center of pressure is obtained from the sum of moments about 0, $dM_0 = ydF = ypdA = y(\gamma y \sin\theta dA)$:

$$Fy_{cp} = M_0 = \int dM_0 = \int_A y(\gamma y \sin\theta)dA = \gamma \sin\theta \int_A y^2 dA$$
$$= \gamma \sin\theta I_0.$$

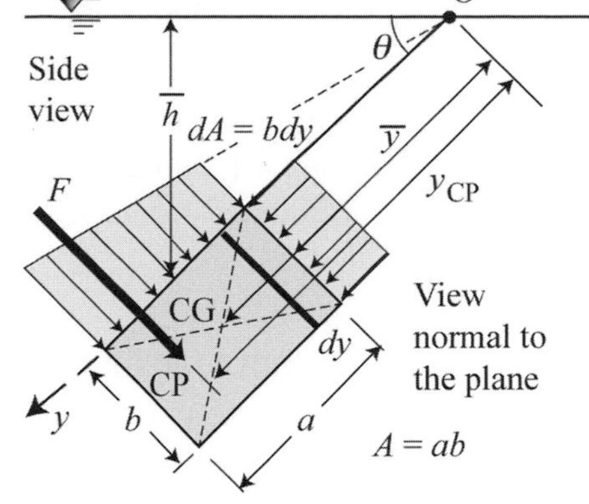

Figure 1.7 Force on a plate

From the definition of the area moment of inertia with the parallel axis theorem, we obtain

$$I_0 = \bar{I} + A\bar{y}^2 = \int y^2 dA = \frac{Fy_{cp}}{\gamma \sin\theta}$$

and

$$y_{cp} = \frac{\gamma I_0 \sin\theta}{F} = \frac{\gamma \sin\theta(\bar{I} + A\bar{y}^2)}{\gamma \sin\theta A\bar{y}} = \frac{(\bar{I} + A\bar{y}^2)}{A\bar{y}},$$

or

$$y_{cp} = \bar{y} + \frac{\bar{I}}{A\bar{y}}. \tag{1.9}$$

The center of pressure y_{cp} is always below the center of gravity $\bar{y}$.

Some calculations are illustrated in Example 1.9.

♦♦ **Example 1.9:** Force on a vertical plate

In Fig. E-1.9, find the force F and its point of the application on a vertical ($y = h$) rectangular plate of height = 24 m and unit width, area $A = 24$ m^2.

Solution:

$$\bar{I} = \frac{ba^3}{12} = \frac{1 \times 24^3}{12} = 1{,}152 \text{ m}^4,$$

$$F = \gamma A\bar{h} = 9{,}810 \times 24 \times 12 = 2{,}825 \text{ kN}$$

and

$$y_{cp} = \bar{y} + \frac{\bar{I}}{A\bar{y}} = 12 + \frac{1{,}152}{24 \times 12} = 16 \text{ m}.$$

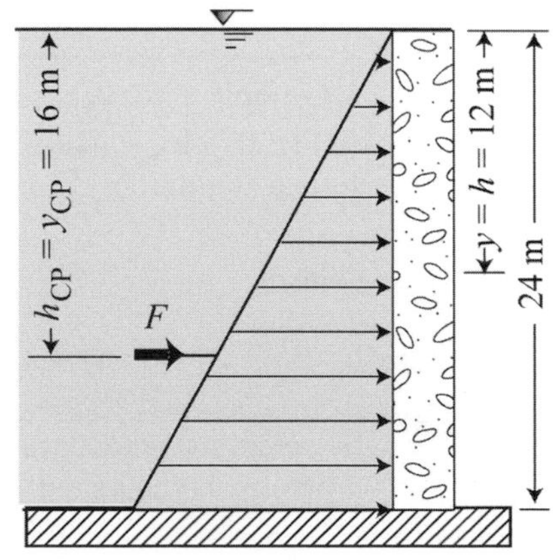

Fig. E-1.9 Vertical plate

Note that the force on the plate per unit width equals the area of the triangular pressure distribution times the specific weight, or $F = \gamma 0.5h^2 = 9{,}810 \times 0.5 \times 24^2 = 2{,}825$ kN per meter of width.

1.4 Forces on Dams

This section covers more detailed applications of hydrostatic forces with illustrations and calculation examples. Basic concepts of concrete dam stability are presented in Section 1.4.1 with a long application example for a curved surface. Additional forces are introduced for gravity dams in Section 1.4.2 and for earth and rock-fill dams in Section 1.4.3.

1.4.1 Dam Stability Concepts

The main concept for the stability of gravity dams involves the hydrostatic force and the weight of concrete. There are at least two main concepts regarding dam stability, and it is imperative to prevent: (1) overturning or toppling of the dam; and (2) tension cracks at the base of concrete dams. In the first case, the resultant force must pass through the base of the dam. The second criterion is more restrictive: to prevent tension in the base of a solid concrete dam without reinforcement, the resultant force must pass through the central third of the base (Figure 1.8). When the force is outside the central third, tension cracks can develop at the base.

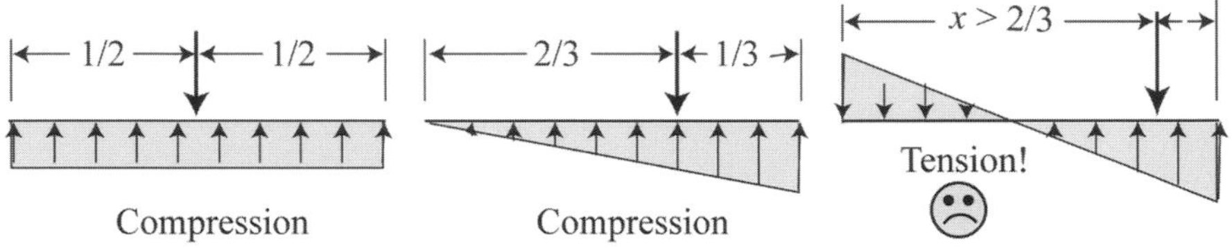

Figure 1.8 Resultant force through the central third of the base

Two basic dam stability cases are considered: (1) a vertical face in Example 1.10; and (2) a curved surface in Example 1.11.

♦ **Example 1.10:** Dam stability

In Fig. E-1.10, will tension develop at the base of the following dam configuration from Example 1.9?

Solution:

$$F_c = \gamma_c W A_c = 24 \frac{\text{kN}}{\text{m}^3}(1\text{ m})\left(\frac{30\text{ m} \times 24\text{ m}}{2}\right) = 8{,}640\text{ kN},$$

where A_c is the area of concrete and W is the unit width of the dam.

Note that $l_h = 24 - h_{cp}$, and the clockwise sum of moments about 0 is

$$\sum M_0 = F_c l_c + F_h l_h$$
$$= 8{,}640 \times 10 + 2{,}825 \times 8$$
$$= 109{,}000 \text{ kNm}.$$

At the base,

$$\sum M_0 = F_c X_R + F_h \times 0,$$

from which

$$X_R = \frac{\sum M_0}{F_c} = \frac{109{,}000}{8{,}640} = 12.6 \text{ m}.$$

The force passes through the central third, i.e. $10 < X_R < 20$ m, and there is no tension.

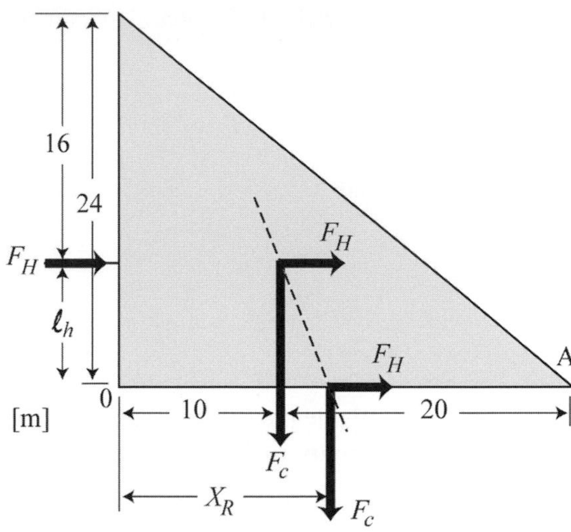

Fig. E-1.10 Gravity dam

Example 1.11: Curved surfaces

For the dam sketched in Fig. E-1.11, consider a unit width and determine the following:

(a) the constant k of the parabolic equation $y = kx^2$, with $24 = k(12)^2$ and $k = 1/6$.

(b) the horizontal hydrostatic force:

$$F_H = \gamma A \bar{h} = \frac{9{,}810 \text{ N}}{\text{m}^3} \times 24 \text{ m}^2 \times 12 \text{ m} = 2{,}825 \text{ kN}.$$

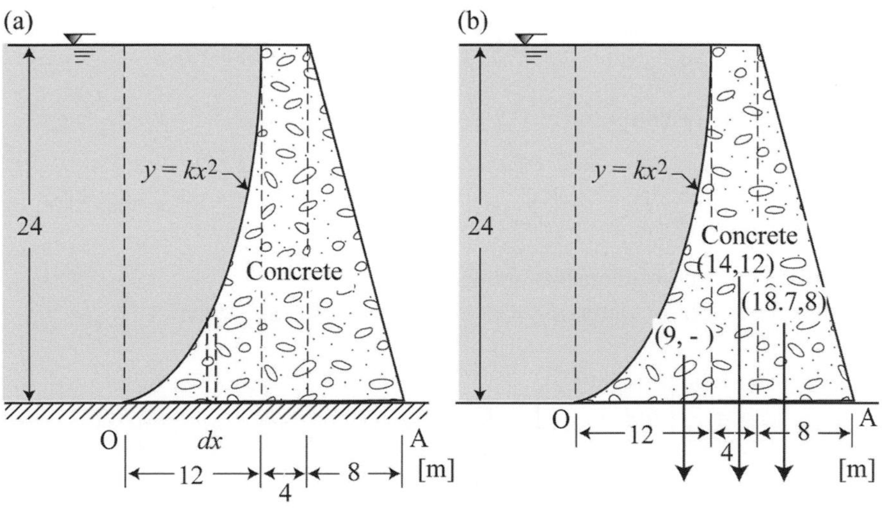

Fig. E-1.11 Stability of a curved gravity dam

(c) the weight of water above the dam:

$$A_{below} = \int_0^{12} y dx = \int_0^{12} kx^2 dx = \frac{1}{6}\left[\frac{x^3}{3}\right]_0^{12} = 96 \text{ m}^2$$

and

$$A_{above} = (12 \times 24) \text{ m}^2 - 96 \text{ m}^2 = 192 \text{ m}^2;$$

$$F_w = \gamma \forall_w = \frac{9{,}81 \text{ kN}}{\text{m}^3} \times 192 \text{ m}^3 = 1{,}883 \text{ kN}.$$

(d) Divide the concrete part into three segments and determine the weight and center of gravity for each segment (see Fig. E-1.11b):

$$W_{c1} = \gamma_c \forall_{c1} = \frac{24 \text{ kN}}{\text{m}^3} \times 96 \text{ m}^3 = 2{,}304 \text{ kN}.$$

$$W_{c2} = \gamma_c \forall_{c2} = \frac{24 \text{ kN}}{\text{m}^3} \times (4 \times 24 \times 1) \text{ m}^3 = 2{,}304 \text{ kN},$$

$$W_{c3} = \gamma_c \forall_{c3} = \frac{24 \text{ kN}}{\text{m}^3} \times 0.5(8 \times 24 \times 1) \text{ m}^3 = 2{,}304 \text{ kN}.$$

(e) Calculate the sum of horizontal and vertical forces:

$$\sum F_H = 2{,}825 \text{ kN}$$

and

$$\sum F_V = 1{,}883 + (3 \times 2{,}304) = 8{,}795 \text{ kN}.$$

(f) Determine the resultant force:

$$\left|\vec{F}_R\right| = \sqrt{\left(\sum F_H\right)^2 + \left(\sum F_V\right)^2} = \sqrt{(2{,}825)^2 + (8{,}795)^2} = 9{,}237 \text{ kN}.$$

(g) Calculate the sum of moments about 0:
For the weight of water, $dA = (24 - y)dx = (24 - kx^2)dx$;

$$\bar{x} = \frac{1}{A}\int x dA = \frac{1}{192}\int_0^{12} x(24 - kx^2)dx = \frac{1}{192}\left[\frac{24x^2}{2} - \frac{kx^4}{4}\right]_0^{12},$$

$$\bar{x} = \frac{1}{192}\left[24\frac{12^2}{2} - \frac{1}{6}\frac{12^4}{4}\right] = \frac{1{,}728 - 864}{192} = 4.5 \text{ m}.$$

The reader can find the area under the curved surface as $A_c = 96\ \text{m}^2$ and the position of the centroid and sum of moments are

$$\bar{x}_c = \left(\int_0^{12} x\,kx^2 dx\right) / \left(\int_0^{12} kx^2 dx\right) = 9\ \text{m}$$

and

$$\sum M_0 = (F_H \times 8) + (F_W \times 4.5) + [2{,}304 \times (9 + 14 + 18.67)] = 127{,}100\ \text{kNm}.$$

(h) Is the resultant force passing through the central third of the base?

$$X_R = \frac{\sum M_0}{\sum F_V} = \frac{127{,}100}{8{,}795} = 14.5\ \text{m}$$

and is without tension at the base since the resultant X_R is passing through the central third of the base, i.e. $8 < (X_R = 14.5\ \text{m}) < 16$.

Note also that the resultant force is passing to the left of point A and therefore the dam is safe against overturning.

1.4.2 Gravity Dams

Besides the hydrostatic fluid force and the weight of concrete, the detailed analysis of gravity dams involves more force components sketched in Figure 1.9 including: (1) uplift pore pressure at the base of the dam; (2) active sediment force at the base of the dam; (3) seismic loading and inertia of the dam; and (4) wave and ice forces near the surface. Accordingly, the design of dams requires a knowledge of hydraulics, structures and geological engineering.

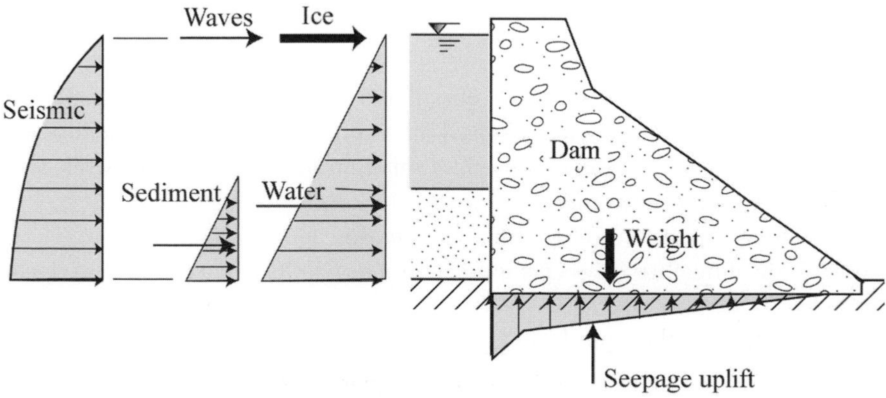

Figure 1.9 Example of various forces applied on a gravity dam

The analysis of each force component can become quite complex. For instance, the seismic force often becomes dominant in deep reservoirs of depth H, because the pressure increases with the square root of the water depth h, from the Westergaard formula

$$p_{seismic} = 0.875\gamma\alpha_s\sqrt{hH} = C_s\rho a_s h, \tag{1.10}$$

where $\alpha_s = a_s/g$ is the ratio of the seismic acceleration a_s to gravity g and C_s is a seismic coefficient which varies with depth and the face angle of the dam (Chwang and Housner 1977, Chwang 1978, USBR 2011). Seepage uplift forces will be discussed in Chapter 16.

In bedrock canyons, it becomes possible to anchor the dam on the side rather than at the base. In wide open areas, buttresses and multiple arch dams can also be favorably designed, as shown in Figure 1.10.

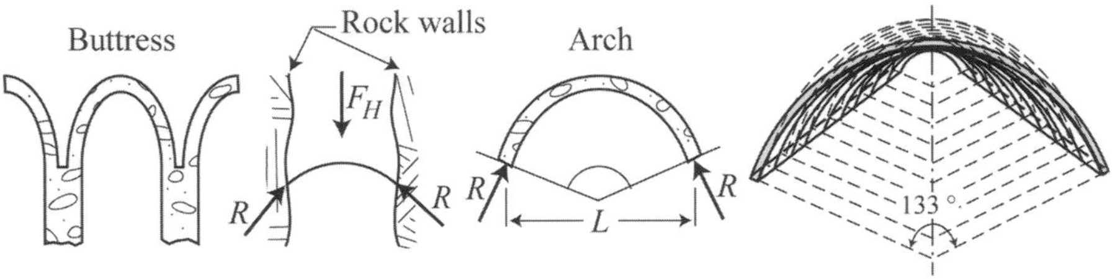

Figure 1.10 Sketches of buttress and arch dams

Bartlet Dam, AZ, in the USA and Manic 5 in Canada represent good examples of buttress and multiple-arch dams, respectively. More details on the design of gravity dams can be found in USBR (1976).

1.4.3 Rock and Earth-Fill Dams

Rock and earth-fill dams are permeable and allow seepage as shown in Figure 1.11. On impervious foundations, this can be done with a rock fill toe or a drainage blanket. Levees and permeable dams are also sensitive to failure from a sudden drawdown in water level. Seepage of rock and earth-fill dams will be discussed further in Chapter 16.

EXERCISES

These exercises provide a few review questions covering the main concepts of the chapter. All the exercises should be solved all at once before moving on to the next chapter. Also, web searches for additional information can be easily done while waiting for the bus ride on the way home!

1. Do we use a pound force or a pound mass in this book?
2. What is the difference between a ton in customary units and a ton in SI? Which one is larger?
3. Do you remember the typical value and units of mass density for water in customary units and in SI?
4. Do you remember the typical value and units of the specific weight for water in customary units and in SI?

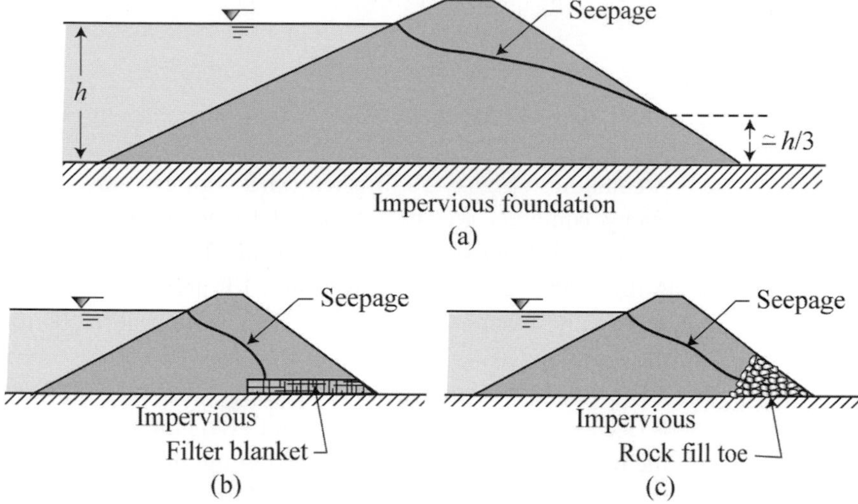

Figure 1.11 Seepage control for rock and earth-fill dams

5. Do you add or subtract the atmospheric pressure to the relative pressure to get the absolute pressure? Which pressure is larger?
6. What is the difference between the pressure head and the piezometric head?
7. Is the force on a submerged plate applied at the centroid? If not, where else?
8. What is the difference between I_0 and $\bar{I}$? Which is larger?
9. What would happen if the resultant force does not pass through the central third of the base of a concrete dam?
10. What is the difference between an arch and a buttress dam?
11. True or false?
    (a) A watt is 1 Nm/s.
    (b) Water viscosity decreases with temperature.
    (c) The mass density of water is maximum at 4 °F.
    (d) The atmospheric pressure is 14.7 psig.
    (e) The pressure in a fluid at rest can be less than the atmospheric pressure.
    (f) The piezometric head is equivalent to the pressure head.
    (g) In Fig. E-1.6, the piezometric heads at A and B are the same.
    (h) The vapor pressure of water decreases with temperature.
    (i) The centroid of a triangular pressure distribution on a rectangular plate is located one-third from the base.
    (j) The seismic pressure distribution increases with the square root of the water depth.

SEARCHING THE WEB

Can you find more information about the following?

1. ♦ Where is Tarbela Dam located? What is its main purpose?
2. ♦♦ Why do we design a clay core in earth-fill dams?
3. ♦♦ What happened to the 21-mile dam in Nevada? What type of dam is it?
4. ♦ Where is Bonneville Dam located? What type of dam is it?

PROBLEMS

Basic Concepts

1. A student weighs 135 lb, find the weight in newtons, the mass in slugs and in kilograms.

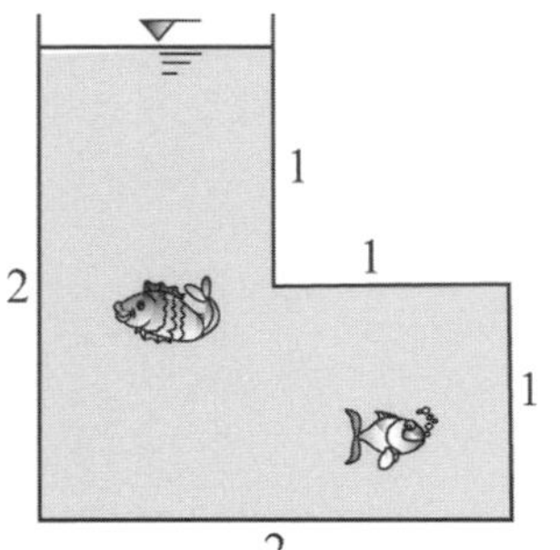

Fig. P-1.4

2. A container weighs 450 lb when filled with 4 ft^3 of fluid. If the empty container weighs 50 lb, determine the specific weight, mass density and specific gravity of this fluid.
3. A fish tank is 2 ft long, 1 ft wide and 1 ft deep, and is filled with water only. What are the pressure and the force at the bottom of the tank?
4. ♦ The 3 ft^3 (2 ft at the base, 2 ft high and 1 ft wide) light fish tank is filled with water, as shown in Fig. P-1.4. Find the pressure and the hydrostatic force applied at the bottom. If the tank is placed on a scale, compare the hydrostatic force at the bottom of the tank with the weight on the scale and explain why the two forces are different. What happened to the force difference?
5. A transducer measures a relative pressure of 65 psi. Determine (1) the value of the pressure in Pascals and (2) the absolute pressure in psi and kPa.
6. What is the absolute vapor pressure of water (1) at 30 °C in kPa and (2) at 180 °F in psia and psfa.
7. A circular plate has a 2 m diameter. What is its area moment of inertia about the center of gravity?

   ♦ What is the area moment of inertia with rotation axis at the top of the plate?
8. ♦ In Fig. P-1.8, a triangular plate has a base of 4 ft and a height of 3 ft. Calculate the following:
   (a) the surface area A;
   (b) the position of the center of gravity;
   (c) the area moment of inertia about the horizontal axis through the center of gravity; and
   (d) the area moment of inertia about the base?

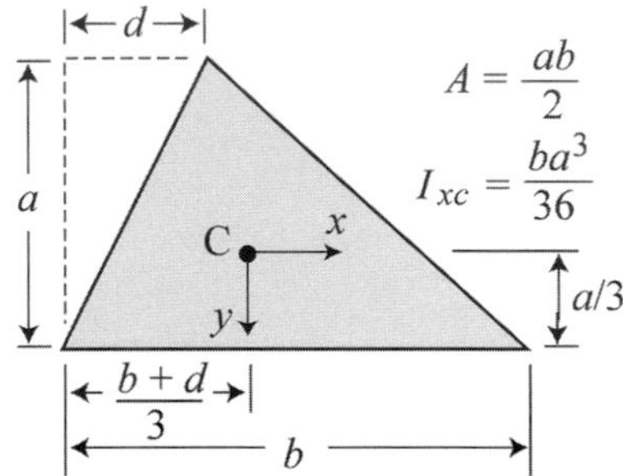

Fig. P-1.8

Pressure, Piezometric Head and Forces

9. ♦ Find the relative and absolute pressure at the pipe centerline at A in kPa, psi and psf. What are the piezometric heads at A and B in Fig. P-1.9.
10. ♦ A circular plate 2 m in diameter serves as a submarine window. Determine the average pressure and force on the vertical plate when the center of gravity of the plate is 500 m below the free surface. Also, calculate the distance separating the center of gravity and the center of pressure.

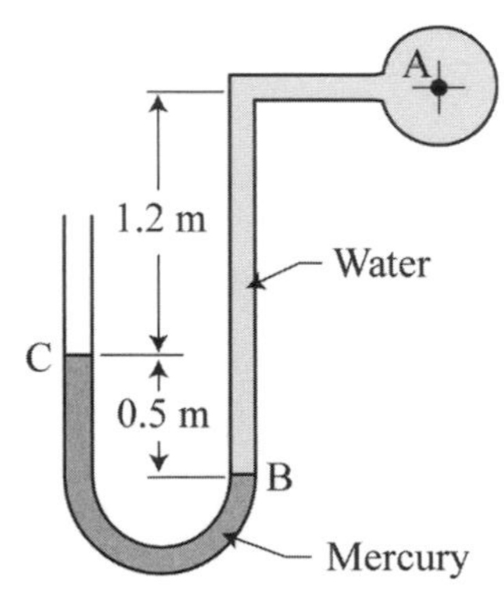

Fig. P-1.9

11. ♦♦ In Fig. P-1.11, the $Z_m = 2$ m circular gate centered at O can only rotate clockwise around a horizontal pivot at C located 5 cm below the center of gravity O of the plate. Determine the range of flow depths h for which the gate opens.

♦♦ As a design problem, can you determine the distance OC that would open the gate when $h >$ 6 m?

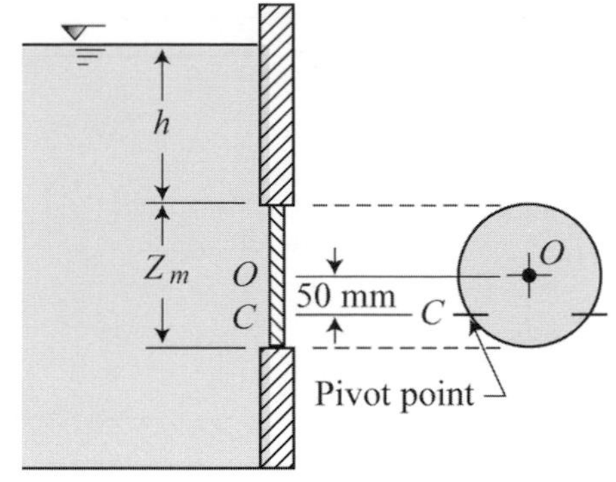

Fig. P-1.11

12. ♦ In Fig. P-1.12, a rectangular plate 4 ft wide and 8 ft long is submerged at an angle of 45° with the vertical. If the center of the plate is located 10 ft vertically below the free water surface, calculate the magnitude of the hydrostatic force on the plate and locate the center of pressure.

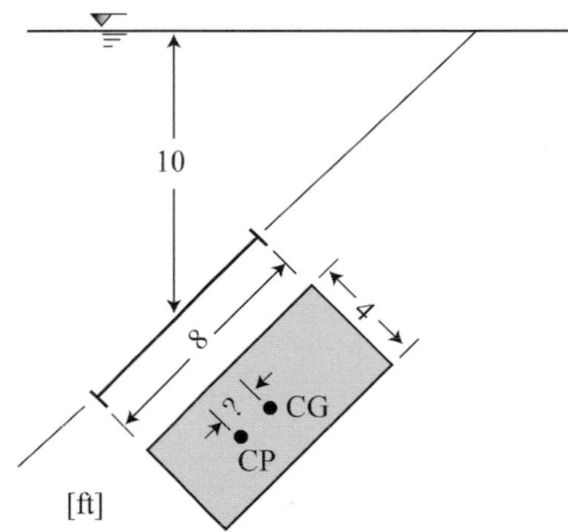

Fig. P-1.12

Dam Stability

13. ♦♦ For the dam sketched in Fig. P-1.13, consider a unit width and determine the following:
    (a) the hydrostatic force on the dam;
    (b) the location of the center of pressure on the face of the dam;
    (c) the weight of concrete of the dam;
    (d) the resultant force;
    (e) the sum of the moments about O; and
    (f) the point of application of the resultant force on the base of the dam OA.
    (g) Finally, is the resultant passing through the central third of the base?

14. ♦♦ For the dam sketched in Fig. P-1.14, consider a unit width and determine the following:
    (a) the horizontal hydrostatic force;
    (b) the weight of water above the dam;
    (c) the weight and center of gravity for each segment, on dividing the concrete part into three segments;
    (d) the sum of horizontal and vertical forces;
    (e) the resultant force; and
    (f) the sum of moments about B.
    (g) Finally, is the resultant passing through the central third of the base?

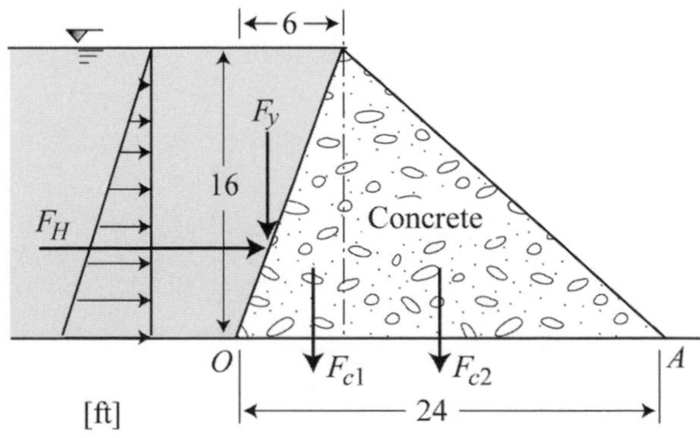

Fig. P-1.13

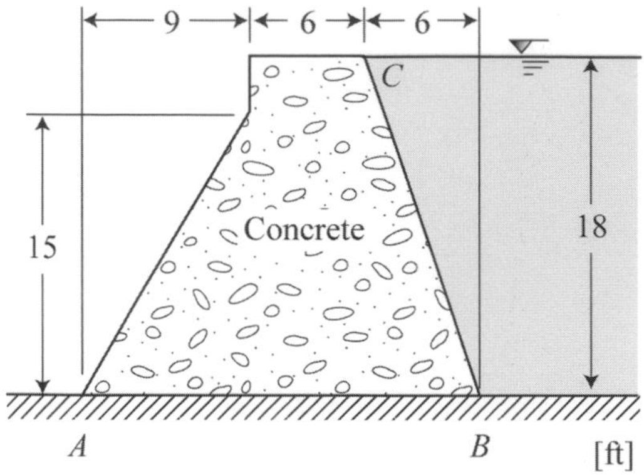

Fig. P-1.14

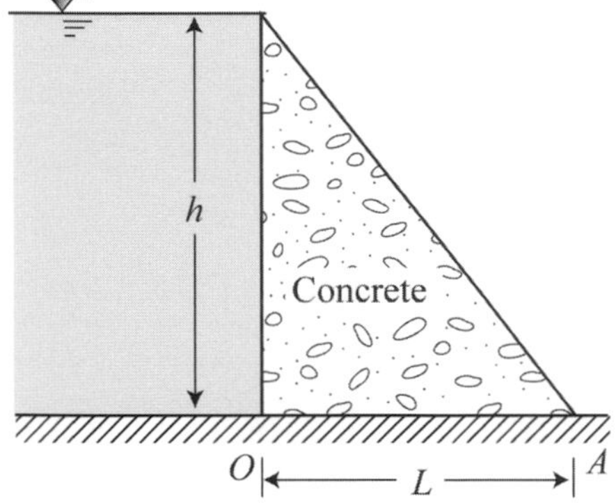

Fig. P-1.15

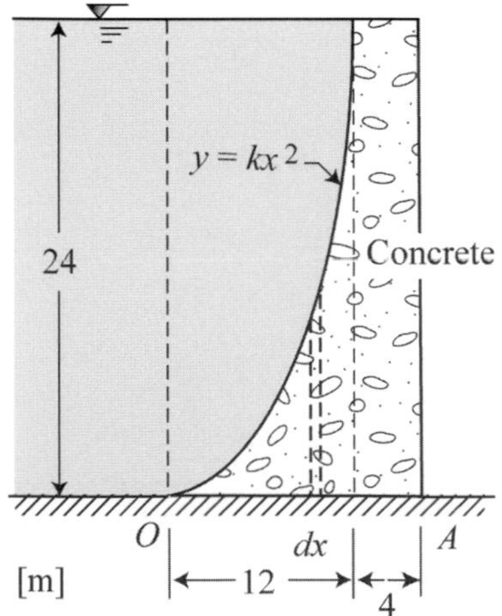

Fig. P-1.16

15. ♦♦ Optimization problem! For the dam sketched in Fig. P-1.15, determine the following per unit width:

(a) the horizontal force;
(b) the vertical force from the weight of concrete;
(c) the expression for the sum of moments about O;
(d) the expression for the point application of the resultant force on the base of the dam; and
(e) the ratio L/h for which the resultant will pass through the midpoint of the base OA.
(f) Finally, what is the range of L/h for which the resultant force would pass through the central portion of the base.

16. ♦♦ For the dam sketched in Fig. P-1.16, determine the following per unit width:

(a) the constant k of the equation $y = kx^2$;
(b) the horizontal hydrostatic force;
(c) the weight of water above the dam;
(d) the weight and center of gravity for each segment, on dividing the concrete part into two segments;
(e) the sum of horizontal and vertical forces;
(f) the resultant force; and
(g) the sum of moments about O.
(h) Finally, is the resultant through the central third of the base?

2 Flow in Pipes

This chapter explains resistance to flow and major friction losses in Section 2.1, and minor head losses in Section 2.2. Head losses are combined with conservation of mass for the analysis of pipe branches and networks in Section 2.3.

2.1 Friction Losses in Pipes

Flow in pipes is very important to the design of water distribution systems. Pipes flowing full (also called closed conduits) have simple geometries and constant velocities. The three main concepts of this section are pipe properties in Section 2.1.1, the energy grade line in Section 2.1.2 and pipe friction losses in Section 2.1.3.

2.1.1 Pipe Properties

The properties of circular pipes in Figure 2.1 include the wetted perimeter $P = \pi D$, cross-section area $A = \pi D^2/4$ and the hydraulic radius $R_h = A/P = D/4$.

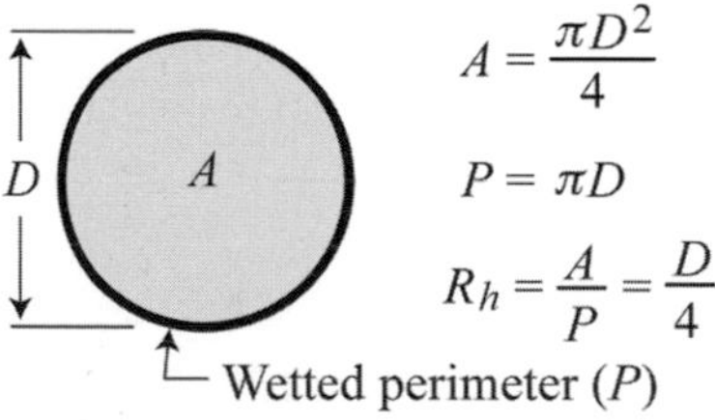

Figure 2.1 Circular pipe

The most common pipe materials include commercial steel, cast iron, concrete, polyvinyl chloride (PVC) and high-density polyethylene (HDPE). The pipe roughness ε measures the surface roughness height of the inner pipe surface, as shown in Figure 2.2 (in the flow direction above and cross section below). Typical values of pipe roughness are listed in Table 2.1, with values in the lower range for new pipes and in the upper range after 20+ years. The relative roughness ε/D is obtained from the roughness height ε and the pipe diameter D. Over time, calcification can also decrease the pipe diameter.

Table 2.1. Pipe roughness

Pipe	Roughness ε	
	ft × 10^{-4}	mm
Concrete	10–200	0.3–5
Wood	6–30	0.18–0.9
Iron and steel	3–50	0.08–1.5
PVC, HDPE, glass	0.7–4	0.02–0.13

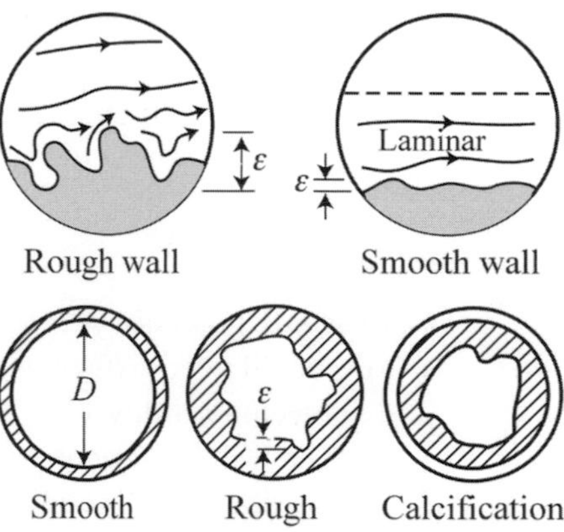

Figure 2.2 Pipe roughness

2.1.2 Energy Grade Line

The concept of flux is a quantity moving per unit time. The flux of water in a pipe is defined as the quantity of water (volume, mass or weight) carried per unit time. The discharge Q is the volumetric flux given as the product of the cross-sectional area A and the mean flow velocity V, or $Q = AV$. In a constant-diameter pipe, the discharge and mean flow velocity also remain constant, $V = Q/A = 4Q/\pi D^2$.

It is remembered from fluid mechanics that the kinetic energy per unit weight of fluid is the velocity head $V^2/2g$, which corresponds to a length

$$\frac{V^2}{2g} = \frac{1}{2g}\left(\frac{4Q}{\pi D^2}\right)^2. \tag{2.1}$$

As shown in Figure 2.3, the pipe centerline is at an elevation z above the datum. The hydraulic grade line (HGL), or piezometric line, is defined as the sum of the pipe centerline elevation z and the pressure head p/γ, as in Chapter 1. Adding the velocity head $V^2/2g$ to the HGL defines the energy grade line (EGL).

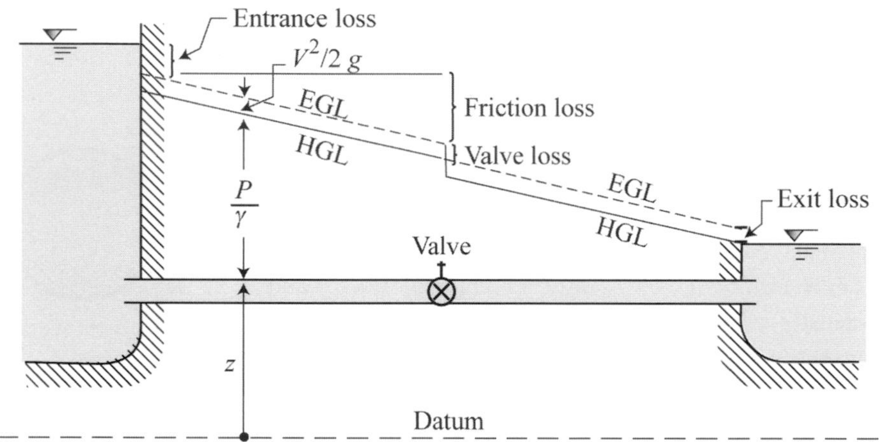

Figure 2.3 Energy and hydraulic grade lines

The Bernoulli sum defines the EGL:

$$H = z + \frac{p}{\gamma} + \frac{V^2}{2g}. \tag{2.2}$$

The EGL graphically represents the energy level in terms of the head of water. The name Bernoulli refers to the family of Swiss mathematicians who developed this concept in the eighteenth century.

The EGL is horizontal in a fluid without friction because the sum is constant. In real fluids like water, we identify gradual friction losses and local minor losses (entrance, exit, valve, etc.), as shown in Figure 2.3. The dissipation of kinetic energy results in a decrease in energy level in the flow direction.

2.1.3 Friction Losses in a Pipe

The head loss ΔH describes the reduction in kinetic energy through friction in a pipe. It corresponds to the decrease in the Bernoulli sum,

$$\Delta H = \left(\frac{p_1}{\gamma} + z_1 + \frac{V_1^2}{2g}\right) - \left(\frac{p_2}{\gamma} + z_2 + \frac{V_2^2}{2g}\right), \tag{2.3}$$

as shown in Figure 2.4. Henri Darcy observed that the energy losses in pipes were proportional to the velocity head. His formula was refined by Julius Weisbach in the nineteenth century by defining the friction factor f. This factor is used to quantify the head loss ΔH in a pipe from the mean flow velocity V and the pipe length L and diameter D according to the Darcy–Weisbach equation:

$$\Delta H = \frac{fL}{D}\frac{V^2}{2g}. \tag{2.4}$$

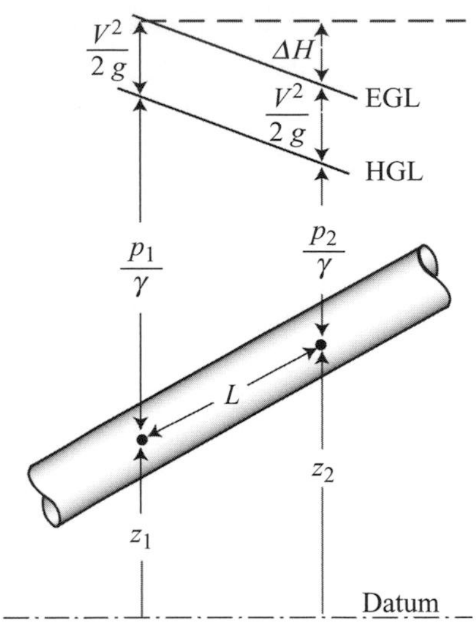

Figure 2.4 Pipe friction losses

In the twentieth century, Lewis Moody developed the diagram in Figure 2.5 describing the Darcy–Weisbach friction factor f vs relative roughness ε/D and the Reynolds number $\mathrm{Re} = VD/\upsilon$. The Reynolds number is named after Osborne Reynolds for his

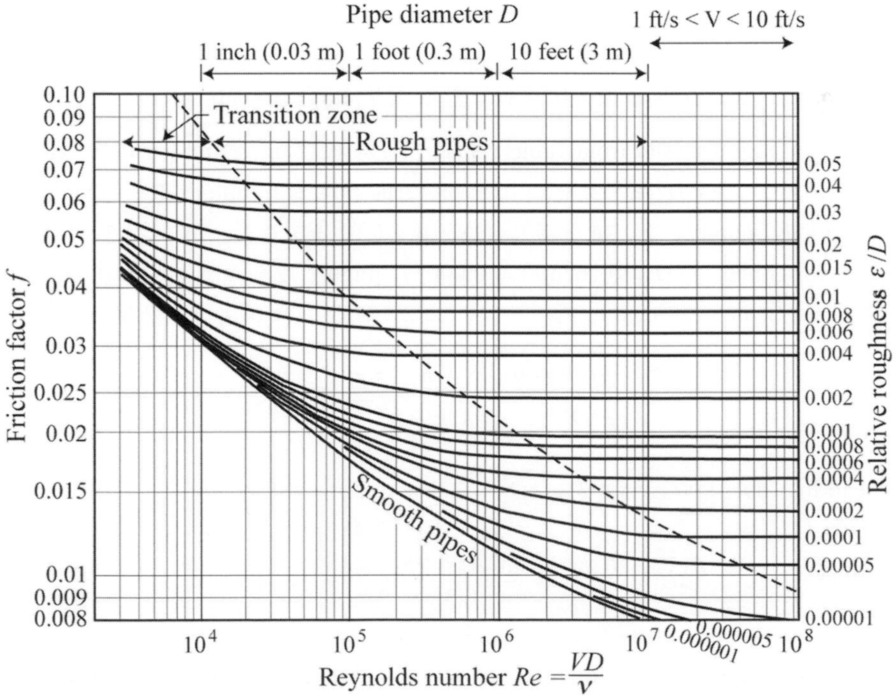

Figure 2.5 Moody diagram for turbulent flow in pipes

seminal experiments on laminar and turbulent flows in the nineteenth century. For values of Re $< 2{,}000$, the flow is laminar and $f = 64/\text{Re}$.

A modified equation of Swamee and Jain (1976) uses $\text{Re} = VD/\upsilon = 4Q/\pi D\upsilon$ to calculate the Darcy–Weisbach friction factor f in terms of discharge Q as

$$f = \frac{0.25}{\left\{ \log \left[\frac{\varepsilon}{3.7D} + \frac{5.74}{(4Q/\pi D\upsilon)^{0.9}} \right] \right\}^2}. \tag{2.5}$$

In practice, constant values of the Darcy–Weisbach friction factor f are often assumed (e.g. $f \cong 0.02$ for problem calculations), and it is always a good practice to double check the assumed f value with the Moody diagram.

Friction losses are usually written as a function of flow discharge Q as

$$\Delta H = h_f = \frac{fL}{D}\frac{V^2}{2g} = \frac{fL}{D}\left(\frac{4Q}{\pi D^2}\right)^2 \frac{1}{2g} = \left(\frac{8fL}{g\pi^2 D^5}\right) Q^2 = kQ^2. \tag{2.6}$$

The constant $k = \frac{8fL}{g\pi^2 D^5}$ has dimensions T^2/L^5. Example 2.1 shows calculation details and Table 2.2 lists typical friction losses for steel and cast-iron pipes.

♦ **Example 2.1:** Friction losses in a single pipe

Calculate the friction loss in a 500-m-long cast-iron pipe with $\varepsilon = 0.26$ mm. The diameter is 30 cm, the flow rate is 0.02 m^3/s, and assume a temperature of 20 °C. Note that 1 m^3/s = 35.32 ft^2/s, and 1 ft^3/s = 450 gpm (gallons per minute).

Solution: The fluid viscosity from Table 1.3 is $\upsilon = 1 \times 10^{-6}$ m^2/s, and the flow velocity is $V = 4Q/\pi D^2 = 4 \times 0.02/\pi \times 0.3^2 = 0.28$ m/s. The flow is turbulent (Re $> 2{,}000$) since the Reynolds number is $\text{Re} = 4Q/\pi D\upsilon = 4 \times 0.02/\pi \times 0.3 \times 10^6 = 8.5 \times 10^4$. The relative roughness $\varepsilon/D = 0.26/300 = 0.00087$. Figure 2.5 gives $f \approx 0.022$. Also, Eq. (2.5) can be programmed to give $f = 0.02225$. The value of k (Eq. (2.6)) is

$$k = \frac{8fL}{g\pi^2 D^5} = \frac{8 \times 0.022 \times 500}{9.81\pi^2 0.3^5} = 374 \text{ s}^2/\text{m}^5.$$

The head loss over 500 m is $\Delta H = kQ^2 = 374 \times 0.02^2 = 0.15$ m, or

$$\Delta H = \frac{fL}{D}\frac{V^2}{2g} = \frac{0.022 \times 500}{0.3}\frac{0.28^2}{2 \times 9.81} = 0.15 \text{ m}.$$

Note that all digits are used for the calculations while only two significant digits are used to display the parameters since all input parameters really have one significant digit.

Table 2.2. Typical flow characteristics in steel and cast-iron pipes

Steel pipe schedule 40				Schedule 80			
Nominal diameter (in.)	Inside diameter (in.)	Wall thickness (in.)	Pressure max. (73 °F) (psi)	Nominal diameter (in.)	Inside diameter (in.)	Min. wall thickness (in.)	Pressure max. (73 °F) (psi)
1	1.033	0.133	140	1	0.935	0.179	630
1.5	1.592	0.145	100	1.5	1.476	0.200	470
2	2.049	0.154	90	2	1.913	0.218	400
3	3.042	0.216	80	3	2.864	0.300	370
4	3.998	0.237	70	4	3.786	0.337	320
6	6.031	0.280	55	6	5.709	0.432	280
8	7.943	0.300	50	8	7.565	0.500	250
10	9.976	0.365	45	10	9.492	0.593	230
12	11.89	0.406	40	12	11.294	0.687	230

	125 psi Cast iron		250 psi Cast iron					
Nominal diameter (in.)	Outside diameter (in.)	Inside diameter (in.)	Outside diameter (in.)	Inside diameter (in.)	Flow velocity (ft/s)	Flow discharge (gpm)	Head loss per 100 ft (ft)	Friction factor f
4	5.00	4.04			5.1	200	3.2	0.026
6	7.10	6.08	7.22	6.00	6.8	600	3.3	0.023
8	9.3	8.18	9.42	8.00	7.7	1,200	2.9	0.021
10	11.4	10.16	11.60	10.00	9.0	2,200	3.0	0.02
12	13.50	12.14	13.78	12.00	9.7	3,400	2.8	0.019
16	17.80	16.20	18.16	16.00	9.6	6,000	1.9	0.018
18	19.92	18.18	20.34	18.00	10	8,000	1.9	0.018
20	22.06	20.22	22.54	20.00	9.2	9,000	1.35	0.017
24	26.32	24.22	26.90	24.00	10	14,000	1.25	0.016
30	32.4	30.00	33.46	30.00	10	22,000	1.0	0.016
36	38.70	35.98	40.04	36.00	9.5	30,000	0.7	0.015
42	45.10	42.02			9.3	40,000	0.55	0.015
48	51.40	47.98			9.8	55,000	0.5	0.014

Note: 1 gpm = 1 gal/min, 1 m^3/s = 35.32 ft^3/s, 1 ft^3/s = 450 gpm, 1 gallon = 3.78 liters, 1 psi = 6.89 kPa, 1 ft = 12 in. = 0.3048 m, 1 in. = 0.0254 m.

2.2 Minor Losses in Pipes

This section describes appurtenances to pipe systems. Different types of valves are identified in Section 2.2.1 followed with pipe couplings in Section 2.2.2. Minor losses and equivalent pipe lengths are calculated in Section 2.2.3. Negative pipe pressure and cavitation are discussed in Section 2.2.4.

2.2.1 Valves

Figure 2.6 sketches various valve types. Some key characteristics (and approximate K values) of several different valve types include:

(1) gate valves that lower a metal gate are used in water supply lines, $K \simeq 0.1 - 0.3$;
(2) needle valves have a cone precisely closing the opening;
(3) globe valves commonly used in faucets dissipate a lot of energy, $K \simeq 1 - 10$;
(4) plug valves rotate a pierced cylindrical plug, $K \simeq 0.5 - 1$;
(5) ball valves are spherical plug valves, $K \approx 0.04$;
(6) butterfly valves have a rotating disk, $K \approx 0.16 - 0.35$;
(7) swing valves or check valves handle one-way flow, $K \approx 1.8 - 3$;
(8) pinch valves contract the opening, $K \approx 0.2 - 0.8$; and
(9) relief valves spray water out once the pressure becomes excessive.

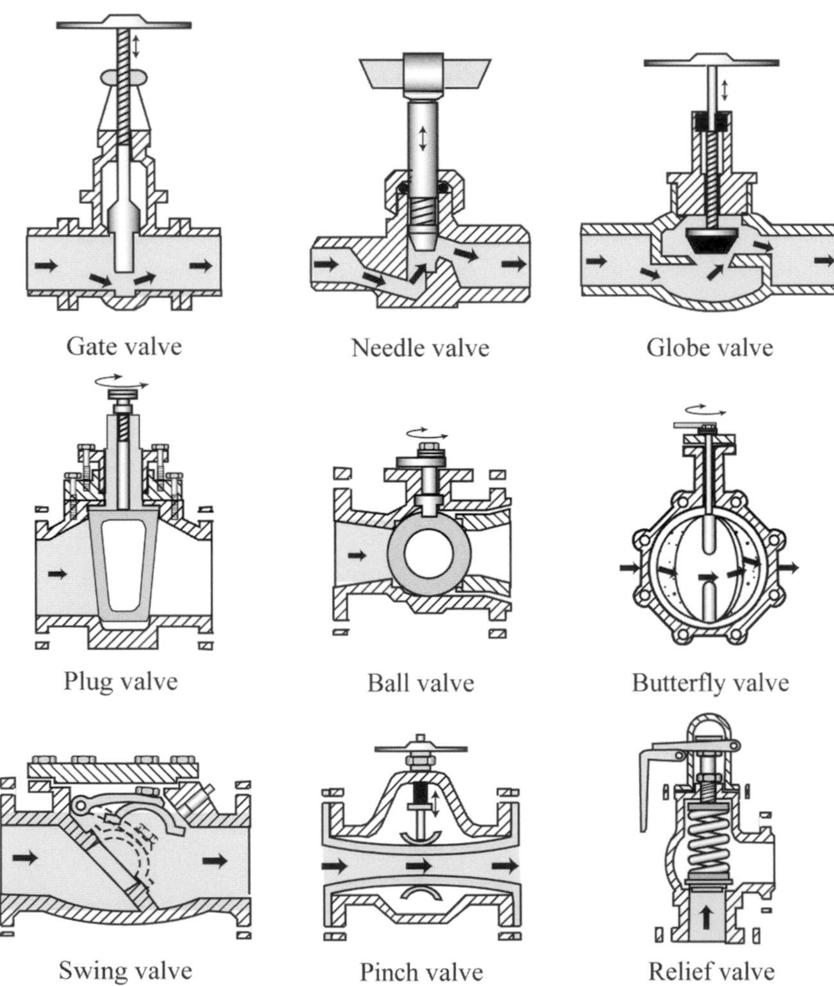

Figure 2.6 Valve types

2.2.2 Pipe Couplings

Figure 2.7 illustrates several ways to connect pipes together including the following.

(1) Threaded joints require screwing cast-iron, copper or PVC pipes 6–300 mm. Threaded joints work best under low pressure but have greater energy losses. They also require maintenance from leaking.

(2) Welded joints require melting a tin-based filler metal between copper pipes. Soldered joints have a low mechanical strength. Butt-welded joints are the most common type of welding for large commercial and industrial pipes. They have smooth internal surfaces and resist high pressure, but are expensive.

(3) Socket joints are connected one into another and welded. They are used where leakage may occur and have a lower fatigue resistance than butt-welded joints.

(4) Gasket joints are often used for PVC pipes.

(5) Grooved joints provide mechanical pipe connection with ductile iron connected by bolts. Dresser couplings are easily removable and ideal for maintenance.

(6) Flanged joints are used for large-diameter pipes under high pressure. Typically used for cast-iron and steel pipes, the flanges are bolted together and components can be easily replaced or repaired.

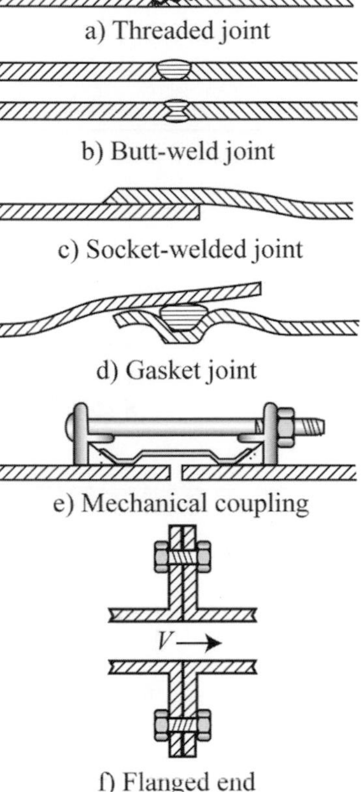

Figure 2.7 Pipe-coupling types

2.2.3 Minor Losses and Equivalent Length

Minor head losses h_L are proportional to the velocity head and a coefficient K:

$$h_L = K\frac{V^2}{2g} = \frac{K}{2g}\left(\frac{4}{\pi D^2}\right)^2 Q^2 = \left(\frac{8K}{g\pi^2 D^4}\right)Q^2. \tag{2.7}$$

Note that the minor-loss coefficient K is different from the friction-loss coefficient k used in Section 2.1. Minor losses are usually treated as an equivalent pipe length L_e obtained from $h_L = KV^2/2g$ where $K = fL_e/D$. The equivalent pipe length of minor losses is

$$L_e = \frac{KD}{f}. \tag{2.8}$$

This equivalent length is added to the nominal pipe length in pipe networks.

Table 2.3 and Figure 2.8 provide practical values of the minor losses in pipes. These standard values are compiled from Albertson et al. (1960), Streeter (1971), Vennard and Street (1975), Roberson and Crowe (1985) and Idelchik (2008).

Table 2.3. Minor-loss coefficient K for threaded and flanged bends and tees

	45° Elbow	90° Elbow	90° Miter bend	180° Bend	Tee flow through	Tee branch flow	Threaded fitting
Threaded	0.4	0.9	1.1 no vanes	1.5	0.9	2.0	0.08
Flanged	–	0.3	0.2 with vanes	0.2	0.2	1.0	–

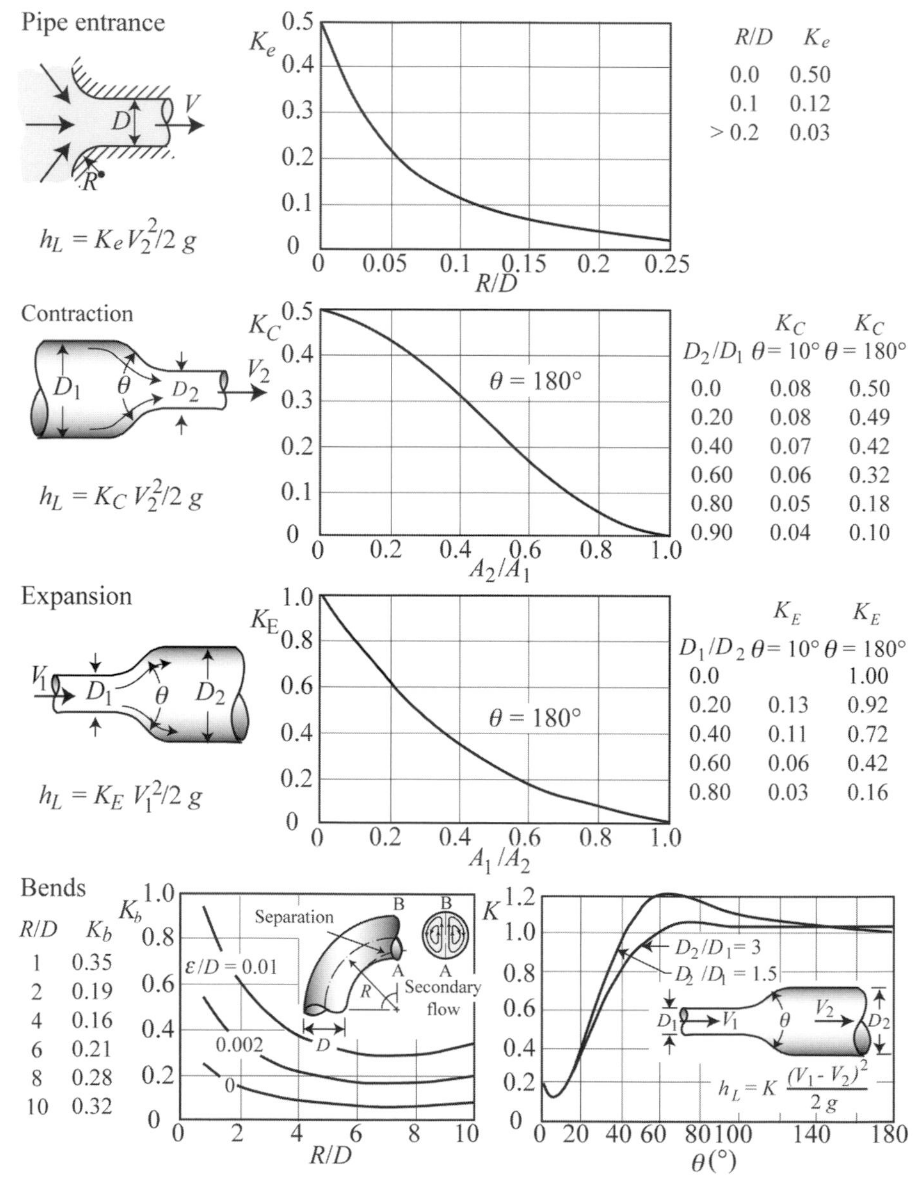

Figure 2.8 Minor losses in pipe entrances, contractions, expansions and bends

The equivalent lengths for typical K values and pipe diameter D are listed in Table 2.4. Example 2.2 shows calculations using the equivalent-length concept.

Table 2.4. Equivalent lengths L_e in meters for various K values and diameters D

Diam. D (in.)	Diam. D (mm)	Friction factor f	Bend $K = 0.2$	45° Bend $K = 0.4$	90° Bend $K = 0.9$	180° Bend $K = 1.5$	Swing valve $K = 3$	Angle valve $K = 5$	Globe valve $K = 10$
1	25	0.023	0.2	0.5	1.0	1.6	3.3	5.4	11
2	50	0.019	0.5	1.1	2.4	3.9	7.9	13	26
4	100	0.017	1.2	2.4	5.3	8.8	18	29	58
6	150	0.015	2.0	4.0	9.0	15	30	50	100
8	200	0.014	2.9	5.7	13	21	43	71	142
12	300	0.013	4.6	9.2	21	35	69	115	230
15	375	0.013	5.8	11	26	43	86	144	288
18	450	0.012	7.5	15	34	56	112	187	375
24	600	0.012	10	20	45	75	150	250	500
30	750	0.011	13.6	27	61	102	204	340	681
36	900	0.011	16.3	32	74	122	245	409	818
42	1,050	0.010	21	42	94	157	315	525	1050

♦ **Example 2.2:** Equivalent length and minor losses

The entrance coefficient for a protruding pipe is $K_e = 1$. Find the equivalent length of entrance/exit losses and the discharge in the pipe, including the minor losses.

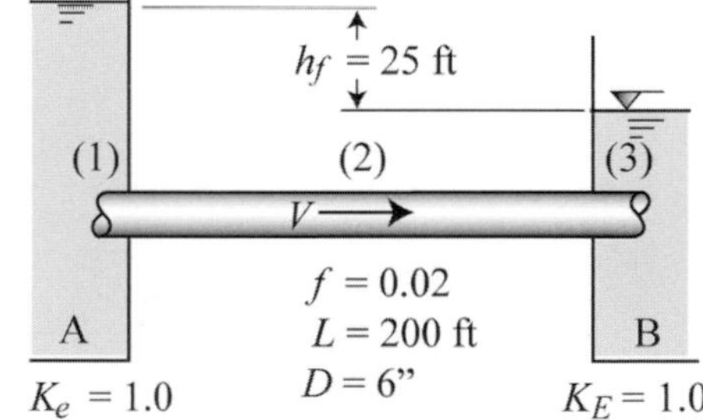

Fig. E-2.2 Pipe losses

Solution: We can solve directly with the equivalent length. The equivalent lengths of the entrance and exit losses in Fig. E-2.2 are in this case equal to

$$L_e = \frac{K_e D}{f} = \frac{1 \times 0.5}{0.02} = 25 \text{ ft.} \quad \text{(E-2.2)}$$

The total equivalent pipe length is $L_t = L + 2 \times L_e = 200 + 2(25) = 250$ ft;

$$k_t = \frac{8f L_t}{g\pi^2 D^5} = \frac{8 \times 0.02 \times 250}{32.2 \times \pi^2 \times 0.5^5} = 4.03;$$

and

$$Q = \sqrt{h_f / k_t} = \sqrt{25/4.03} = 2.49 \text{ cfs.}$$

The head losses at the pipe entrance and exit are 10% of the total losses, or 2.5 ft each. The friction loss in the pipe is 20 ft. Can you plot the HGL and EGL? (See Fig. E-2.3 for the answer.)

2.2.4 Negative Pressure and Cavitation

The relative pressure is negative at any point above the HGL. Cavitation occurs when the pressure is below the vapor pressure. To prevent cavitation, the distance between the pipe elevation and the HGL needs to be less than the relative vapor pressure head. Calculations of negative pipe pressure and cavitation are detailed in Example 2.3.

♦♦♦ **Example 2.3:** Negative pressure

The central part of the pipe in Example 2.2 is lifted above the reservoir level. If the reservoir is near sea level at $T^\circ = 70\ ^\circ$F, how far can you raise the midpoint in the pipe before water turns to vapor?

Solution: From the Bernoulli equation with minor and friction losses in Fig. E-2.3:

$$\frac{p_c}{\gamma} + Z_c + \frac{V^2}{2g} + \Delta H_c = Z_0$$

or

$$\frac{p_c}{\gamma} = (Z_0 - Z_c) - \left(\frac{V^2}{2g} + \Delta H_c\right) = -x - H_c.$$

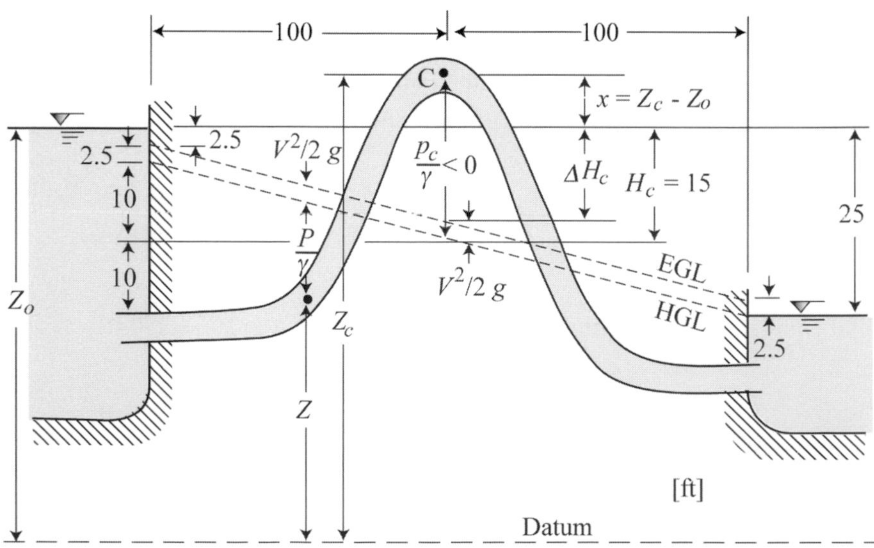

Fig. E-2.3 Negative pressure in a pipe

To prevent cavitation, it is important to check how far up point C can be raised without turning water into vapor. In other words, $x = X_{max}$ defines the maximum possible elevation without causing cavitation. This corresponds to $p_c = p_{v\,rel} = p_v - p_{atm}$.

The governing equation to prevent cavitation becomes

$$\frac{-p_{v\,rel}}{\gamma} = \frac{p_{atm}}{\gamma} - \frac{p_v}{\gamma} = \frac{V^2}{2g} + \Delta H_c + X_{max}. \tag{E-2.3.1}$$

From Examples 1.6 and 1.8, $p_{v\,rel} = -p_{atm} + p_v = -2.065$ psf at sea level and $T^\circ = 70\ ^\circ\text{F}$,

$$X_{max} = \frac{-p_{v\,rel}}{\gamma} - \frac{V^2}{2g} - \Delta H_c = \frac{2{,}065}{62.4} - 2.5 - 12.5 = 33.1 - 15 = 18.1 \text{ ft}.$$

The pipe at point C needs to be below X_{max} to prevent cavitation in the pipe.

2.3 Pipe Branches and Networks

After considering the conservation of mass in Section 2.3.1, we discuss pipe branches in Section 2.3.2 and pipe networks in Section 2.3.3. The Hazen–Williams approach is outlined in Section 2.3.4 and high-pressure valves are discussed in Section 2.3.5.

2.3.1 Continuity or Conservation of Mass

The conservation of mass, or continuity, simply states that the fluid mass cannot be created or destroyed:

$$\textit{Mass flux in} - \textit{Mass flux out} = \textit{Mass change/time}.$$

At low pressure, it is considered that water is incompressible. When connecting two pipes in series per Figure 2.9a, the discharge in both pipes is the same since water cannot be stored in perfectly rigid pipes.

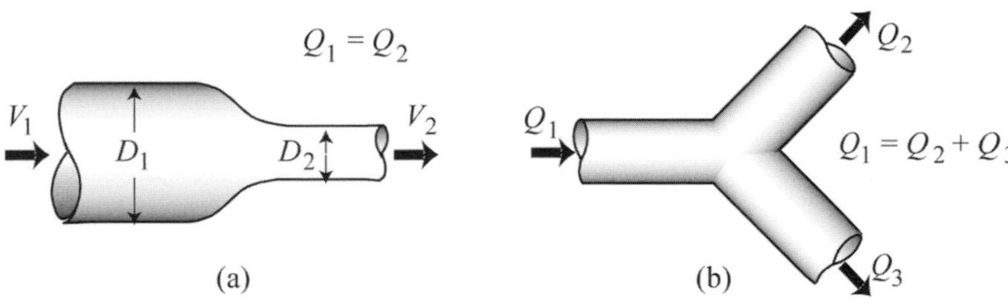

Figure 2.9 Pipes in (a) series and (b) branches

The continuity for pipe branches is also sketched in Figure 2.9b, and the sum of discharges entering the node equals the sum of discharges leaving:

$$\sum Q_i = 0.$$

Example 2.4 provides calculations for two long pipes in series.

♦♦ Example 2.4: Friction losses in two pipes in series

Consider the two pipes in series in Fig. E-2.4. Assume $f = 0.02$ in the two pipes and find the discharge Q when $h_f = 25$ ft.

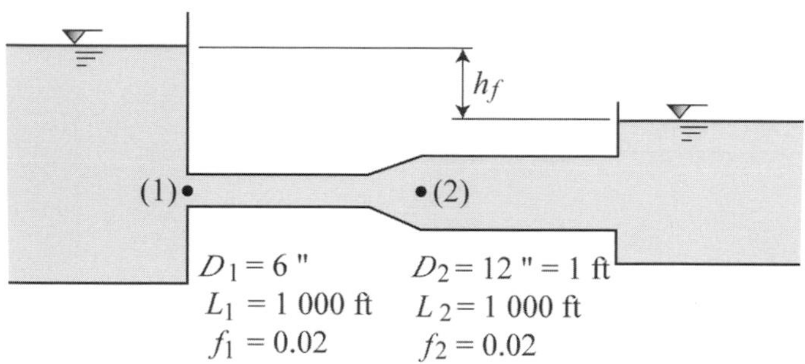

Fig. E-2.4 Expanding pipe flow

Solution: The equivalent lengths for minor losses are often neglected in very long pipes. In this case, $L_e = 0.25/0.02 = 12.5$ ft, which is very small.

$$k_1 = \frac{8fL_1}{g\pi^2 D_1^5} = \frac{8 \times 0.02 \times 1{,}000}{32.2 \times \pi^2 \times 0.5^5} = 16.1$$

and

$$k_2 = \frac{8fL_2}{g\pi^2 D_2^5} = \frac{8 \times 0.02 \times 1{,}000}{32.2 \times \pi^2 \times 1^5} = 0.5;$$

$$h_f = h_{f1} + h_{f2} = (k_1 + k_2)Q^2 = 16.6\,Q^2 = 25 \text{ ft}$$

and

$$Q = \sqrt{\frac{25}{16.6}} = 1.23 \text{ cfs};$$

$$h_{f1} = k_1 Q^2 = 16.1 \times 1.23^2 = 24.2 \text{ ft}$$

and

$$h_{f2} = 0.5 \times 1.23^2 = 0.76 \text{ ft}.$$

The head losses in the small pipe are $(D_2/D_1)^5 = 32$ times greater than for the large pipe.

2.3.2 Pipe Branches

Three-reservoir problems sketched in Figure 2.10 require the calculations of: (1) the head H_0 at the junction of the three pipes; and (2) the flow discharge in each pipe. We use the basic equations:

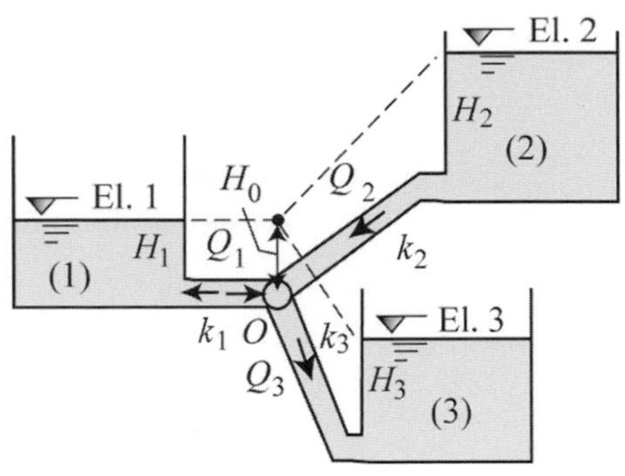

Figure 2.10 Three reservoirs

$$h_f = \frac{fL}{D}\frac{V^2}{2g} = \left(\frac{8fL}{g\pi^2 D^5}\right)Q^2 = kQ^2,$$

$$H_1 - H_0 = k_1 {Q_1}^2,$$

$$H_2 - H_0 = k_2 {Q_2}^2$$

and

$$H_0 - H_3 = k_3 {Q_3}^2.$$

Three-reservoir problems are solved in four steps:

(1) Calculate k_1, k_2 and k_3.

(2) Assume H_0 and calculate Q_1, Q_2 and Q_3 – it is helpful to assume $H_0 = H_1$ for the first iteration to find whether the junction is above or below H_1.

(3) Check the continuity $Q_3 = Q_2 \pm Q_1$ – note here that $Q_3 = Q_2 + Q_1$ when H_1 is above H_0 and $Q_3 = Q_2 - Q_1$ when H_1 is below H_0.

(4) Change H_0 and iterate until convergence.

Example 2.5 illustrates this calculation procedure.

♦ **Example 2.5:** Three reservoirs

Solve the three-reservoir problem in Fig. E-2.5.

Solution:

(1) Calculate k_1, k_2 and k_3.

$$k_1 = \frac{8fL_1}{g\pi^2 D_1^5} = \frac{8 \times 0.02 \times 750}{32.2 \times \pi^2 \times (8/12)^5} = 2.867,$$

$$k_2 = \frac{8fL_2}{g\pi^2 D_2^5} = \frac{8 \times 0.02 \times 1{,}000}{32.2 \times \pi^2 \times 1^5} = 0.503,$$

$$k_3 = \frac{8fL_3}{g\pi^2 D_3^5} = \frac{8 \times 0.02 \times 500}{32.2 \times \pi^2 \times (8/12)^5} = 1.911.$$

(2) Assume H_0 and calculate Q_1, Q_2 and Q_3. First, assume $H_0 = H_1$ such that $Q_1 = 0$,

$$Q_2 = \sqrt{\frac{H_2 - H_0}{k_2}} = \sqrt{\frac{120 - 110}{0.503}} = 4.46 \text{ ft}^3/\text{s},$$

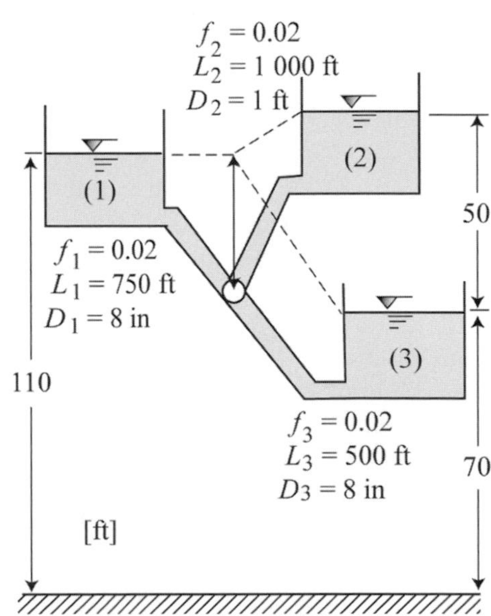

Fig. E-2.5 Three reservoirs

$$Q_3 = \sqrt{\frac{H_0 - H_3}{k_3}} = \sqrt{\frac{110 - 70}{1.911}} = 4.58 \text{ ft}^3/\text{s}.$$

(3) We find that H_0 has to be lowered below H_1 because $Q_3 > Q_2$.

(4) Here, we need to slightly lower H_0 and iterate until $Q_3 = Q_2 + Q_1$.

As a second iteration, let's assume $H_0 = 109$ ft:

$$Q_1 = \sqrt{\frac{H_1 - H_0}{k_1}} = \sqrt{\frac{110 - 109}{2.867}} = 0.59 \text{ ft}^3/\text{s},$$

$$Q_2 = \sqrt{\frac{H_2 - H_0}{k_2}} = \sqrt{\frac{120 - 109}{0.503}} = 4.68 \text{ ft}^3/\text{s},$$

$$Q_3 = \sqrt{\frac{H_0 - H_3}{k_3}} = \sqrt{\frac{109 - 70}{1.911}} = 4.52 \text{ ft}^3/\text{s}.$$

In checking the continuity, $Q_3 < Q_2 + Q_1$, so we need to raise H_0 between 109 and 110 ft.

Third iteration, assume $H_0 = 109.9$ ft:

$$Q_1 = \sqrt{\frac{H_1 - H_0}{k_1}} = \sqrt{\frac{110 - 109.9}{2.867}} = 0.19 \text{ ft}^3/\text{s},$$

$$Q_2 = \sqrt{\frac{H_2 - H_0}{k_2}} = \sqrt{\frac{120 - 109.9}{0.503}} = 4.48 \text{ ft}^3/\text{s},$$

$$Q_3 = \sqrt{\frac{H_0 - H_3}{k_3}} = \sqrt{\frac{109.9 - 70}{1.911}} = 4.57 \text{ ft}^3/\text{s}.$$

Check the continuity and, in this case, we find $Q_3 \cong Q_2 + Q_1$ and we are very close to the answer. This process can be repeated or programmed to obtain more accurate answers.

Note that after the first iteration, we can also use a solver for the following equation:

$$\sqrt{\frac{H_0 - 70}{1.911}} = \sqrt{\frac{110 - H_0}{2.867}} + \sqrt{\frac{120 - H_0}{0.503}},$$

to obtain $H_0 = 109.967$ ft.

Such great precision, typical of computer calculations, is not warranted in practice because the accuracy of the prediction is limited to the two significant digits of the problem statement.

2.3.3 Pipe Networks

A new challenge arises when solving the flow distribution in pipe networks commonly found in urban water distribution systems. Contrary to pipe branches, a network of pipe loops enables the water to reach all nodes even after closing some pipes for maintenance. Our basic equations are: (1) the continuity as the sum of all flows is zero at each node; and (2) the resistance to flow as the head loss in each pipe is calculated from the Darcy–Weisbach equation:

$$h_f = \frac{fL}{D}\frac{V^2}{2g} = \left(\frac{8fL}{g\pi^2 D^5}\right)Q^2 = kQ^2. \quad (2.9)$$

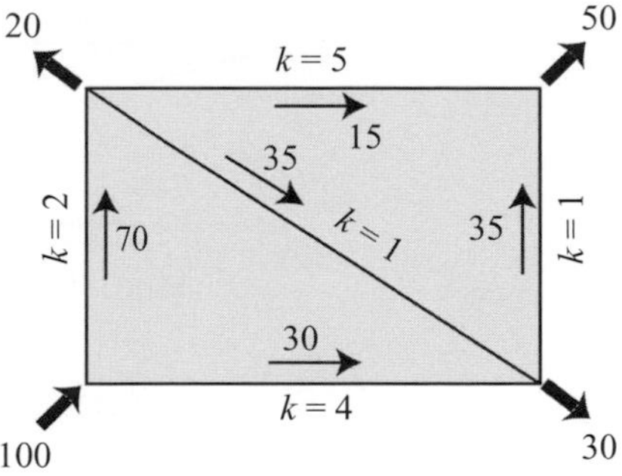

Figure 2.11 Pipe network

The length includes the nominal pipe length plus the equivalent length from minor losses when desirable. Solving this system consists of finding the flow discharge in each pipe and the energy level (energy head) at each node. The Hardy–Cross method solves flow networks in nine steps.

(1) Determine the $k = \dfrac{8fL}{g\pi^2 D^5}$ values for each pipe. A pipe closure is calculated with $k \to \infty$.
(2) Assume a flow distribution (see the arrows shown on Figure 2.11).
(3) Calculate h_f in each pipe from $h_f = kQ^2$.
(4) Sum kQ^2 in each loop in the positive clockwise (CW) direction.
(5) Sum $|2kQ|$ in each loop, where $|x|$ represents the absolute value of x.
(6) Apply a correction

$$\Delta Q = \frac{\sum kQ^2}{\sum |2kQ|} \quad (2.10)$$

for each loop.

If $\Delta Q > 0$, add $|\Delta Q|$ in the counterclockwise (CCW) direction.

If $\Delta Q < 0$, add $|\Delta Q|$ in the CW direction.

(7) Compute the new discharge $Q_{new} = Q_{old} \pm \Delta Q$ in each pipe and each loop – note that pipes adjacent to two loops will have two corrections.
(8) Repeat steps 3–7 until convergence.
(9) The head at each node is obtained from $H_0 - H_1 = k_1 Q_1^2$, with discharge Q_1 from node 0 to node 1.

Figure 2.12 guides the development of a robust pipe network program.

(a) Because of possible changes in the flow direction, $\Delta H = kQ^2$ is replaced with $\Delta H = kQ|Q|$.
(b) For a given loop, there is a sign for the flow direction in each pipe. Per Figure 2.12, the sign from A to B is positive as it is CW around the loop ($S_{AB} = +1$, and $S_{BA} = -1$). Pipe CA is CW and $S_{CA} = +1$, but AC is opposite (CCW), and $S_{AC} = -1$.
(c) For pipes adjacent to two loops, the signs for the two loops are opposite. For instance, $S_{BC1} = +1$ for CW flow in loop 1, but $S_{BC2} = -S_{BC1} = -1$ for CCW flow in loop 2.
(d) The sum of head losses in each loop is calculated as shown in Eq. (2.10).

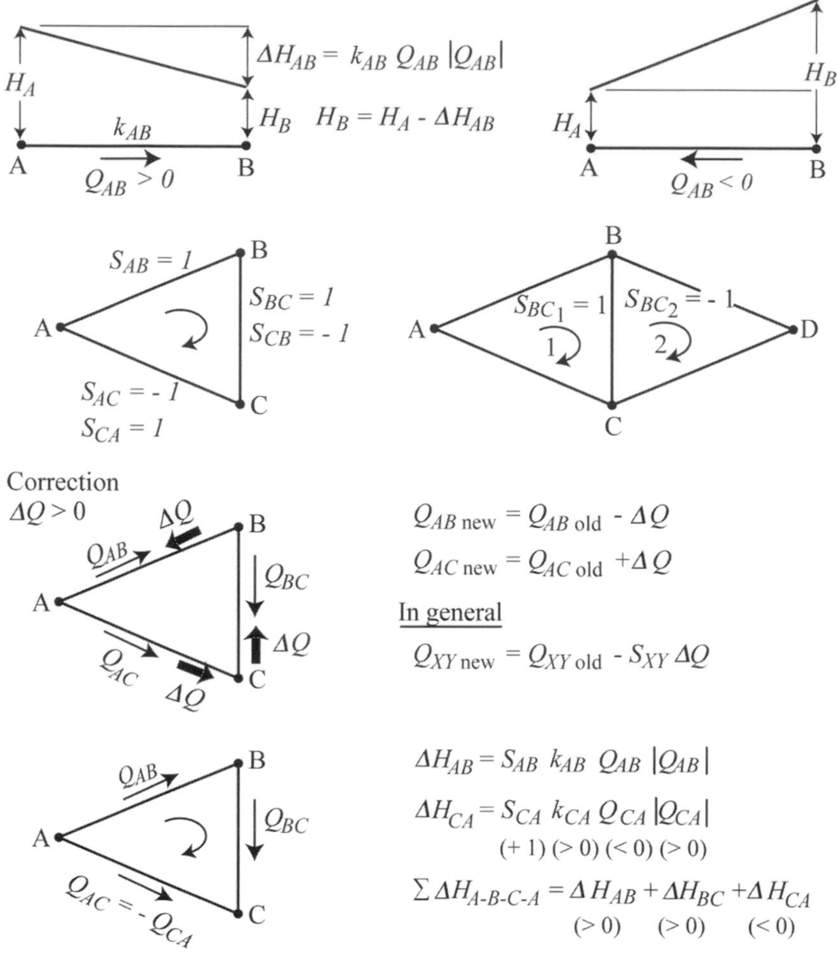

Figure 2.12 Flow net algorithm

(e) The correction ΔQ for each loop (calculated + CW) is in the opposite (CCW) direction. For instance, $Q_{AB\ new} = Q_{AB\ old} - \Delta Q$ and $Q_{AC\ new} = Q_{AC\ old} + \Delta Q$. In general, $Q_{XY\ new} = Q_{XY\ old} - S_{XY}\Delta Q$ and pipes adjacent to two loops have two corrections.

(f) The head (EGL) at each node is calculated from $H_B = H_A - \Delta H_{AB}$.

Example 2.6 shows the detailed calculations of a sample pipe network.

♦♦♦ **Example 2.6:** Solved pipe network

Solve the pipe network of Fig. E-2.6 from Streeter (1971).

Solution: In this study example, the lower-loop calculations are presented on the left-hand side, and the upper loop on the right-hand side. It is important to keep track of the signs. For instance, notice the double discharge corrections for the diagonal pipe. In this example, values of ΔQ are rounded up simply to ease the presentation – computers keep all decimals. Notice how quickly the correction ΔQ decreases from Fig. E-2.6a to c.

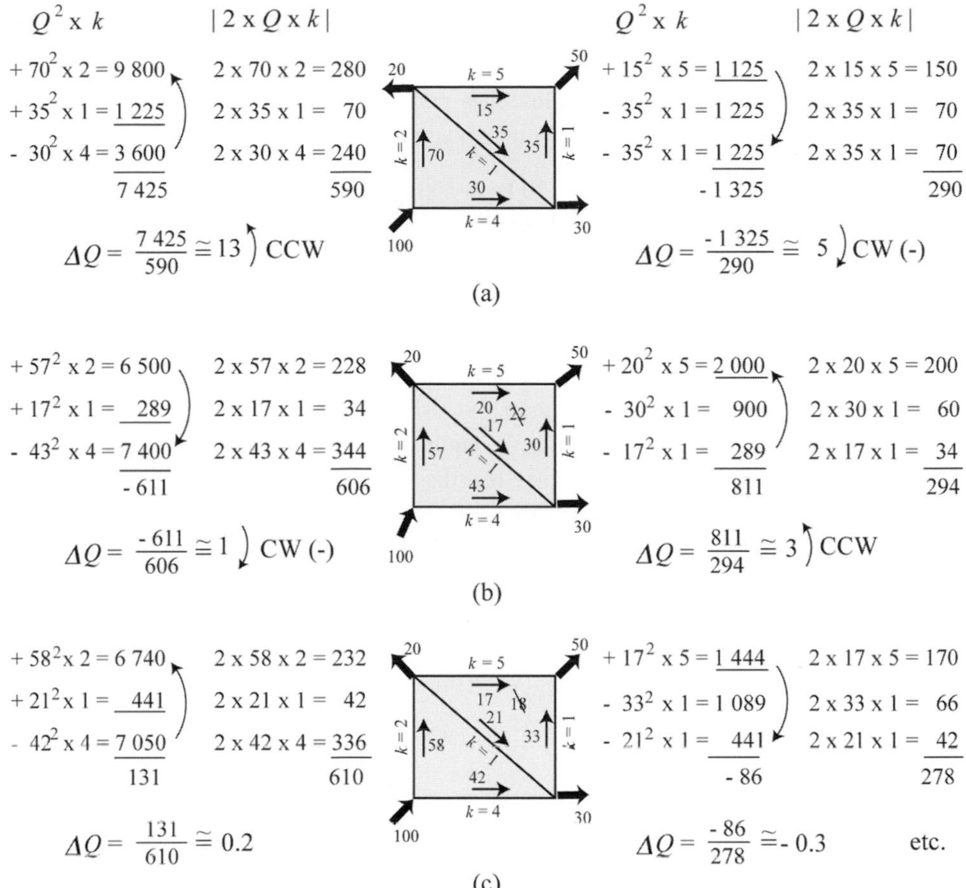

Fig. E-2.6 Pipe-network solution

2.3.4 Hazen–Williams Approach

Friction losses in pipes can also be calculated by the Hazen–Williams formula. Hazen–Williams coefficients C_{HW} describe conveyance rather than flow resistance. Accordingly, smooth pipes have high values of $C_{HW} \simeq 150$, while iron and steel pipes have $130 < C_{HW} < 140$ when new, and typically decreasing to $90 < C_{HW} < 100$ after 20 years of service. The head loss formula is

$$\Delta H = \left[\frac{\phi_{HW} L}{C_{HW}^{1.85} D^{4.87}}\right] Q^{1.85}$$

In SI units, $\phi_{HW} = 10.66$, with $\Delta H, L$ and D in m and Q in m^3/s; and in US customary units, $\phi_{HW} = 4.73$, with $\Delta H, L$ and D in ft and Q in ft^3/s.

For instance, the head loss at a flow rate of 0.02 m^3/s in a 500-m-long 30-cm-diameter cast-iron pipe with $C_{HW} \approx 135$ (see Example 2.1) can be calculated as

$$\Delta H = \left[\frac{10.66 \times 500}{135^{1.85} 0.3^{4.87}}\right] 0.02^{1.85} = 0.155 \text{ m}.$$

Similarly, friction losses in old pipes can be assessed by decreasing $C_{HW} \approx 95$ to estimate losses in these pipes. In this example we would obtain

$$\Delta H = \left[\frac{10.66 \times 500}{95^{1.85} 0.3^{4.87}}\right] 0.02^{1.85} = 0.295 \text{ m}.$$

Simply stated, the losses will nearly double – $(135/95)^{1.85} = 1.91$ – over time.

The calculations for pipe networks using the Hardy–Cross method can be done by replacing the head-loss calculations by $k_{HW}Q|Q|^{0.85}$ instead of $kQ|Q|$. Also, the correction at each iteration becomes $1.85k_{HW}|Q|^{0.85}$ instead of $2k|Q|$. Overall, the Hazen–Williams approach has been used extensively but it is less flexible in the way to incorporate minor losses because the equivalent lengths for the minor losses cannot be directly added to the nominal pipe lengths.

In practice, there are two basic approaches to solve large flow network problems: (1) use the Hardy–Cross method above; or (2) solve a linearized system of equations (matrix inversion) with coefficients that vary at each iteration. EPANET is a public-domain software package to model large water-distribution systems. More examples of pipe networks can be found in other textbooks including Roberson et al. (1997), Houghtalen et al. (1996) and Mays (2019).

2.3.5 High-Pressure Valves

On large infrastructure, there is a need to build valves that can work under very high pressure and control the flow at large flow velocities without causing cavitation. Several types of large valves are sketched in Figure 2.13. Inlet needle valves essentially close the

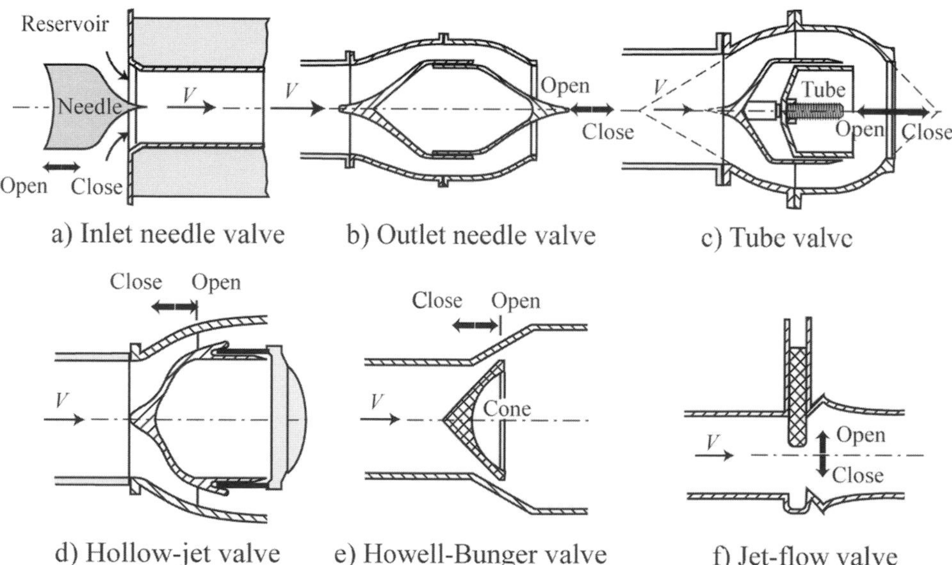

Figure 2.13 Various valves for large infrastructure

flow at the source, i.e. in the reservoir. They function well as long as there is no debris or ice accumulation at the site. Outlet needle valves and tube valves move in the downstream direction to close the outlet of the pipe. Hollow-jet and Howell–Bunger valves are similar in concept and they both move in the upstream direction to close the flow. Jet-flow valves are large gate valves that close vertically. A large steel plate can be raised or lowered to prevent or allow water flow. All valves are designed to withstand large hydrostatic pressures when closed and excessively high velocities when they are open. The very large velocities also imply that they are prone to cavitation when there are slight changes in direction or magnitude of the flow-velocity vectors. A very slight change in cross-sectional area results in local increases in flow velocity that can induce cavitation.

Case Study 2.1 describes the jet-flow gates at Hoover Dam.

♦ **Case Study 2.1:** Jet flow gates at Hoover Dam, USA

Hoover Dam is a concrete arch-gravity dam on the Colorado River between Nevada and Arizona. In operation since 1936, the pool level of Lake Mead was designed at 1,229 ft, and the highest level reached 1,225.4 ft in July 1983. The maximum water elevation drop from Lake Mead to the Colorado River is 590 ft.

In June 1998, 12 obsolete needle valves were replaced with new jet-flow gates operating like gate valves under high pressure. At Hoover Dam, there are four upper gates and eight lower gates excavated in bedrock as shown in Fig. CS-2.1.

The eight lower gates are 68 inches in diameter. When a gate is closed, the hydrostatic pressure is 248 psi and the force exceeds 900,000 pounds per gate. The maximum speed of the water coming out of each gate can reach 175 ft/s! On average, each gate can discharge approximately 3,800 cubic feet per second. At such high velocities, there is always a great risk of cavitation.

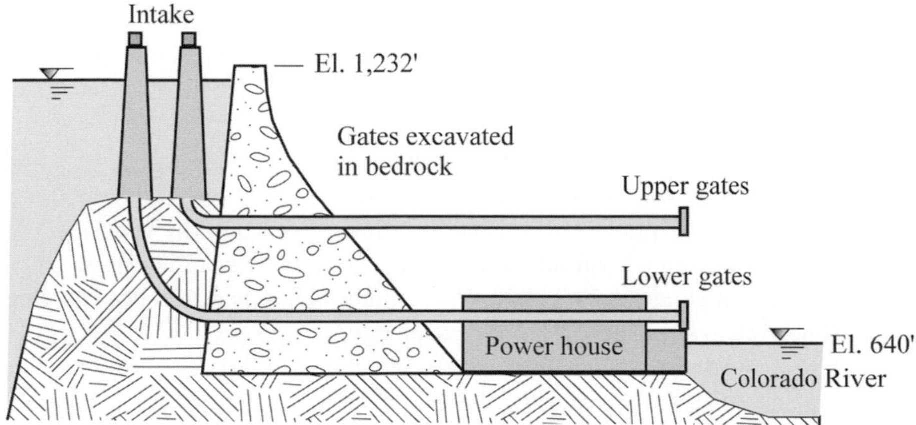

Fig. CS-2.1 Jet Valves at Hoover Dam

The four upper gates are excavated in bedrock and exit on the canyon side 180 feet above the river. Each gate is 90 inches in diameter and discharges 5,400 cfs at a flow velocity of 120 ft/s.

Climate variability with increased evaporation and reduced snowmelt runoff combine with increasing demographics to put pressure on the water demand from Lake Mead and electricity from Hoover Dam. Hoover Dam was retrofitted with wide-head turbines, designed to work efficiently at lower water levels. Since 2002, Las Vegas has cut its water use per capita from 314 to 212 gallons per capita a day. To ensure water supply to the city of Las Vegas, nearly $817 million was spent on a new water intake tunnel named "Third Straw." The 3-mile and 24-ft diameter tunnel has been in operation since late 2015. The water level in May 2021 fell below 1,075 ft, which is below the threshold to declare a water shortage.

Additional Resources

Additional information on pipe coupling can be found in Camp and Lawler (1984) and Novak et al. (2001). More information about valves can be found in Ball and Hebert (1948), Ball (1981) and Kohler and Ball (1984).

EXERCISES

These exercises review essential concepts from this chapter.

1. What is the difference between pipe roughness and relative roughness?
2. What is the velocity head?
3. In Figure 2.4, why can water move up a pipe?
4. What is the difference between HGL and EGL? Which is higher?
5. Why is gravity not a parameter in the Moody diagram?
6. What is the maximum energy-loss coefficient from pipe expansion?
7. What is the purpose of using equivalent lengths for minor losses in pipes?
8. On Fig. E-2.5, can you identify the part of the pipe under negative pressure?
9. Why would you use $kQ|Q|$ instead of kQ^2 to calculate friction losses in pipe networks?
10. At Hoover Dam, what is the pressure head if $p = 248$ psi? What is the velocity head if the flow velocity is 175 ft/s?
11. True or false?
    - (a) Steel pipes have the lowest possible roughness.
    - (b) Friction losses are caused by fluid viscosity.
    - (c) At high Reynolds numbers, resistance to flow is controlled by viscosity.
    - (d) The friction factor f typically decreases as the pipe diameter increases.
    - (e) Flanged couplings typically have lower minor losses than threaded couplings.
    - (f) Globe valves dissipate a lot of kinetic energy.
    - (g) Pipe networks are solved using the conservation of mass.
    - (h) Pipe networks are solved using the conservation of energy.
    - (i) Pipe networks are solved using the conservation of momentum.
    - (j) Pipe networks are solved using resistance to flow.

SEARCHING THE WEB

Can you find more information about the following?

1. ♦ What is the difference between a steel pipe Schedule 40 and 80?
2. ♦ What is the difference between a cast-iron pipe 125 and 250 psi?
3. ♦ What type of coupling is a Dresser coupling?
4. ♦ What type of valve is a Howell–Bunger valve? When is it used?
5. ♦ Where is Hoover Dam located?
6. ♦ Find the EPANET web link.

PROBLEMS

Losses in Single Pipes

1. ♦ For the pipe shown in Fig. P-2.1, neglect all minor losses and calculate the velocity of the jet and the discharge in the pipe. Plot the HGL and EGL. Check f on the Moody diagram.
2. ♦ In Fig. P-2.2, water flows from reservoir A to B at a water temperature of 10 °C. Given the cast-iron pipe length 300 m and diameter 1 m, use the Moody diagram to determine the discharge when $H = 18$ m and $h = 3$ m. Plot the HGL and the EGL and find the pressure at point P halfway between the two reservoirs.

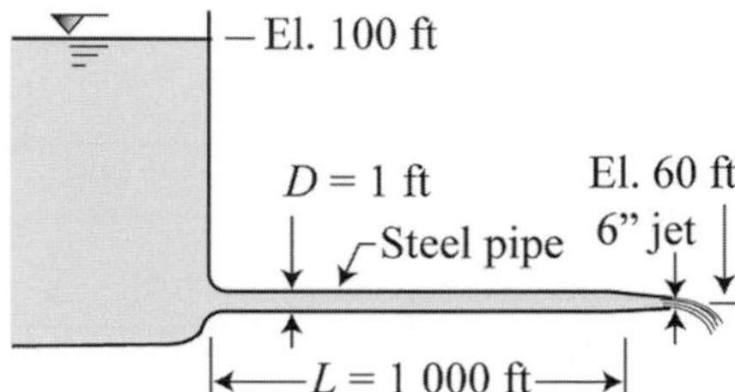

Fig. P-2.1

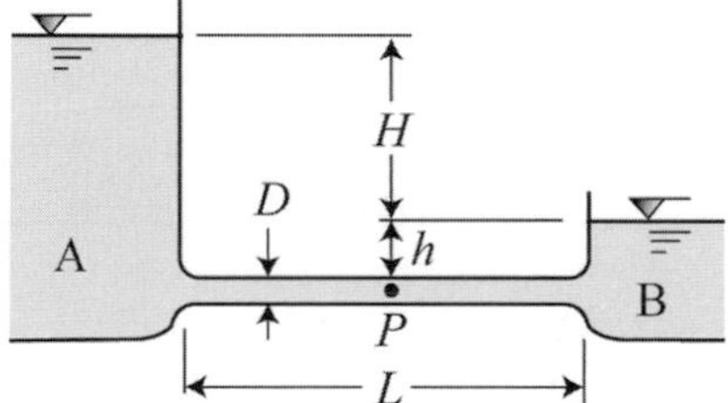

Fig. P-2.2

Three Reservoirs

3. ♦♦ For the system shown in Fig. P-2.3, assume $f = 0.02$ and determine the discharges in the three pipes and the elevation of the EGL at the junction, given

$$k = \frac{8fL}{g\pi^2 D^5}$$

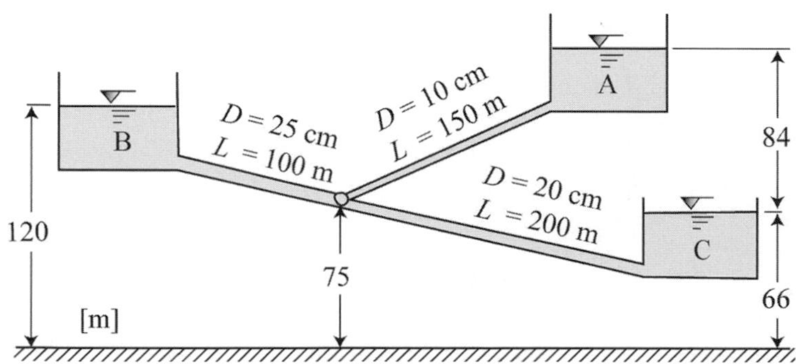

Fig. P-2.3

4. ♦♦ For the three reservoirs in Fig. P-2.4, neglect minor losses and consider $f = 0.02$. Determine the discharge in each pipe and the EGL at node B.

[*Hint:* $k = \frac{8fL}{g\pi^2 D^5}$, $K_{AB} = 12.53$, $K_{CB} = 15.29$.]

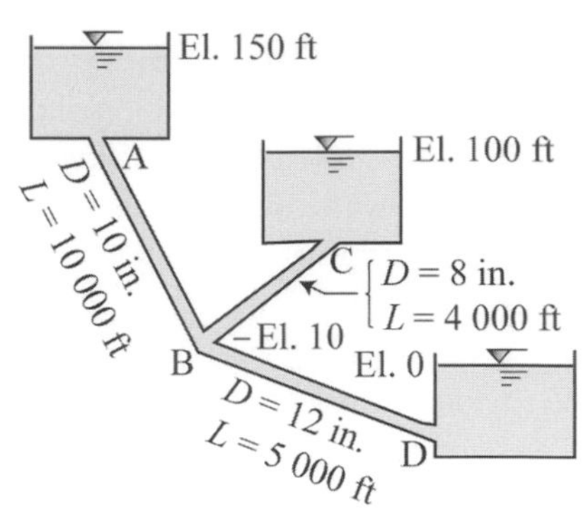

Fig. P-2.4

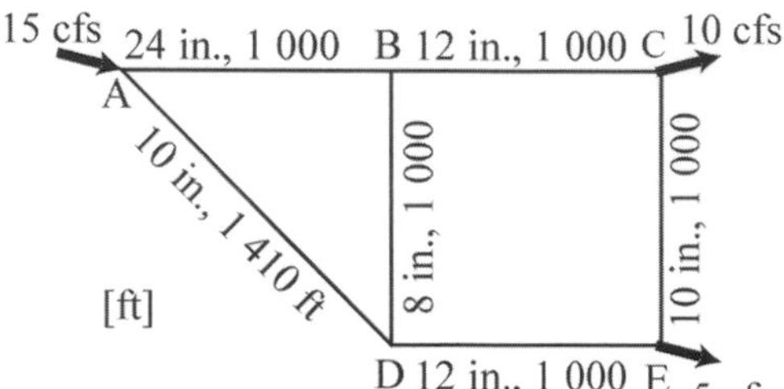

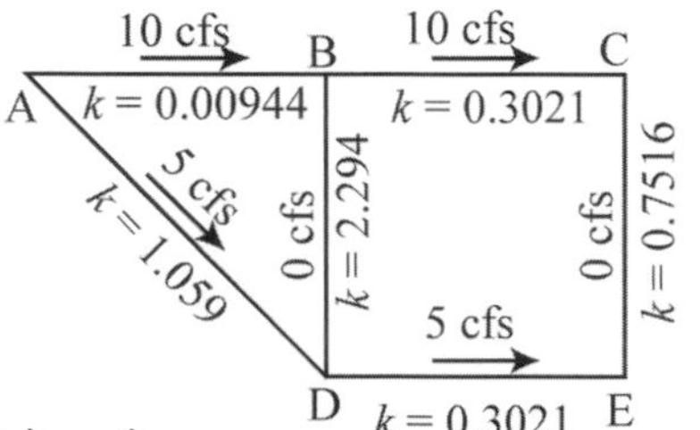

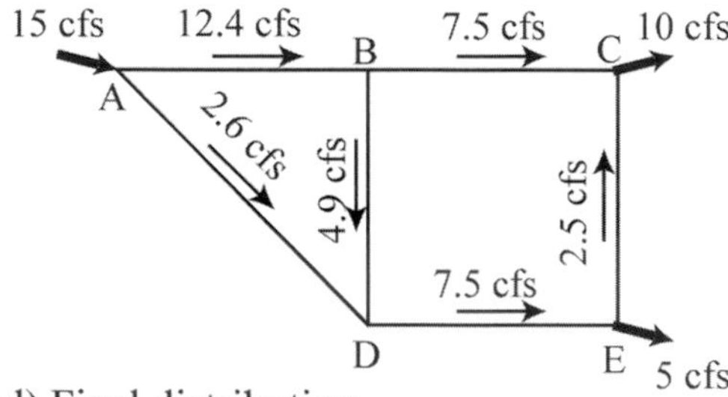

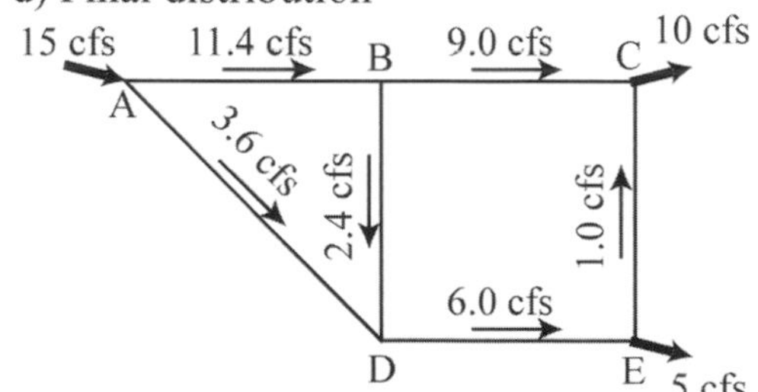

Fig. P-2.5

Pipe Networks

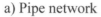

The problems of this chapter are listed with pipe networks before minor losses and negative pressure in this order to allow more time to solve these more challenging pipe-network problems!

5. ♦♦ In the network Fig. P-2.5a, the pressure head at A is 60 ft, all pipes are horizontal, and $f = 0.012$. Part I of this problem is solved in Fig. P-2.5b–d. In writing a program, double check the signs of your calculations (things can be tricky in part I if you consider Q_{CE} negative).

Loop ABD		
	kQ^2	$2kQ$
AB	+0.944	0.189
AD	−26.475	10.59
BD	0	0
Σ	−25.53	10.78

$\Delta Q = (-25.53)/10.78 = -2.4$ cfs

Loop BCDE		
	kQ^2	$2kQ$
BC	30.21	6.042
BD	0	0
CE	0	0
DE	−7.55	3.02
Σ	22.66	9.062

$\Delta Q = (22.66)/9.062 = 2.5$ cfs

Part II is easy to solve once the program works: solve when the outflows are $Q_C = 5$ cfs and $Q_E = 10$ cfs.

6. ♦♦ For the system shown in Fig. P-2.6, assume $f = 0.02$ and a piezometric head of $H_A = 100$ ft at point A. Determine the discharges in all pipes and the elevation of the EGL at all nodes.
7. ♦♦♦ For this example, see Fig. P-2.7. After solving Part I, your program should easily calculate Part II.

   Part I. For the system shown, all pipes are 12 in. in diameter. Assume $f = 0.02$ and a head $H_A = 200$ ft at point A. Write a computer program to solve this pipe network. Find the discharges in all pipes and the elevation of the EGL at all nodes.

   Part II. Once you have solved Part I, replace the outflow of 10 cfs at F with 5 cfs outflows at both nodes C and H. Which of the two cases gives a higher pressure at point E?
8. ♦♦♦ Write a computer program to solve the pipe network in Fig. P-2.8. All pipe diameters are 750 mm and all pipe lengths are indicated. Assume $f = 0.02$ and neglect all minor losses. A head of 40 m is given at point A. Determine the discharges in all pipes and the EGL at all nodes. Keep printed evidence of your program. Once your model works, recalculate the conditions for the following cases: (a) replace the outflow at D with an outflow of 0.5 m^3/s at both B and F; (b) starting from the initial flow condition (i.e. not case a), assume that you have an extra 900 m length of pipe at a diameter of 150 cm; which pipe would you replace to maximize the head at point F?

Minor Losses in Pipes

9. ♦ The pipe system in Fig. P-2.9 is open to the atmosphere at point 2. Calculate the discharge given $f = 0.02$ and $H = 30$ ft.
10. ♦♦ The pipe system in Fig. P-2.10 has a 0.5-in. faucet exit diameter. The friction factor for the 1-in. pipe is 0.025 and 0.03 for the 0.5-in. pipe. Find the time to fill a 20-US gallon bathtub.
11. ♦♦ For the pipe in Fig. P-2.11, assume $f = 0.02$ and determine the discharge in the pipe when $H = 50$ ft and plot the EGL and HGL. Secondly, close the valve and find the value of K required to decrease the discharge by 50%?

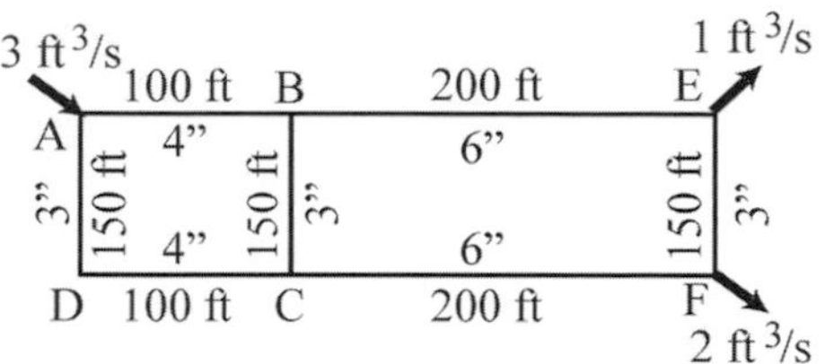

Fig. P-2.6

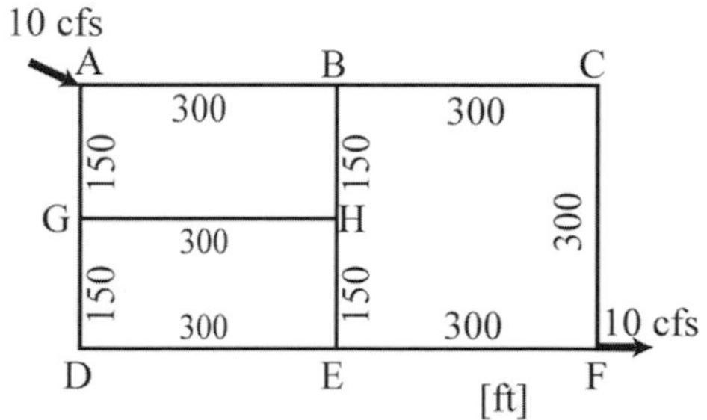

Fig. P-2.7

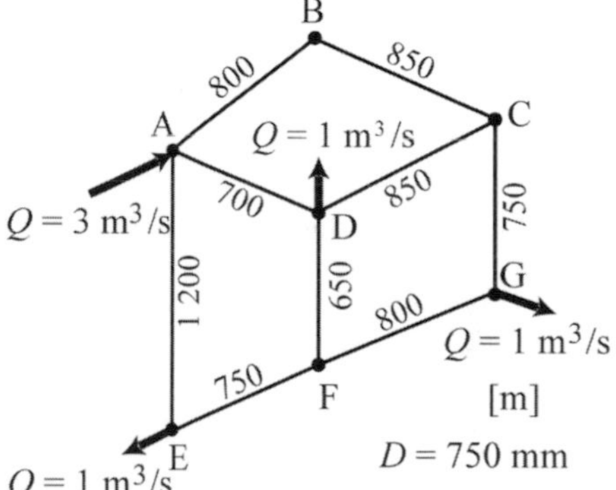

Fig. P-2.8

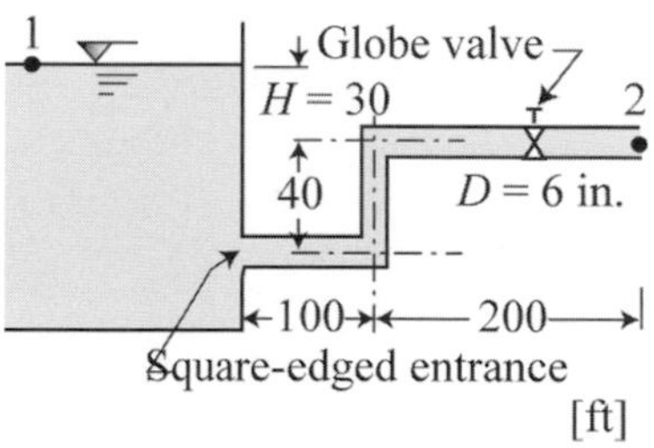

Fig. P-2.9

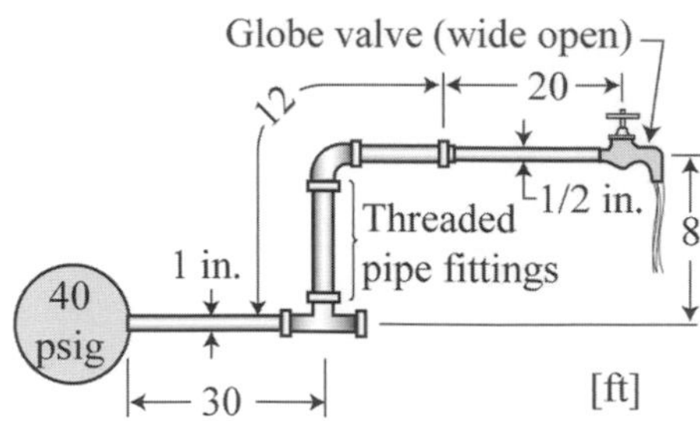

Fig. P-2.10

Fig. P-2.11

12. ♦ For the pipe in Fig. P-2.12, find the pressure in psi halfway between A and B. What is the energy loss in the valve between B and C? Why is the nozzle at the end of the pipe?

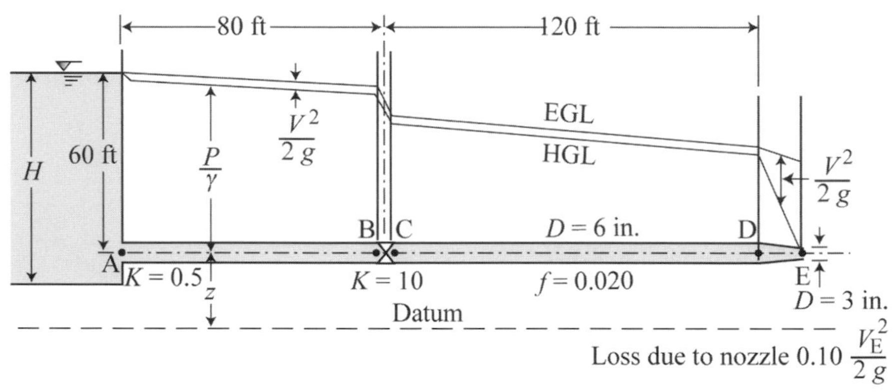

Fig. P-2.12

Siphon and Negative Pressure

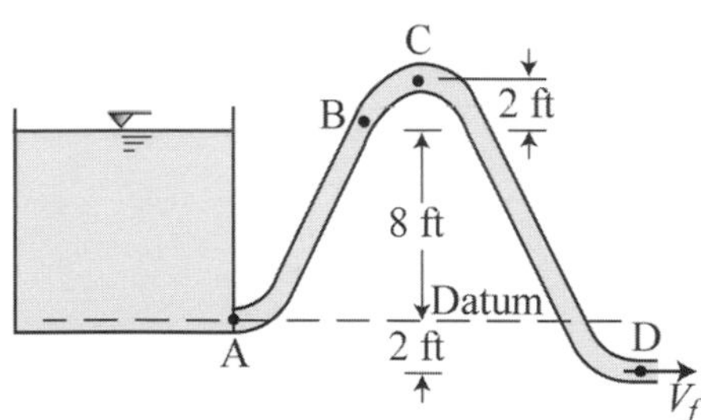

Fig. P-2.13

13. ♦ For the 4-in.-diameter pipe in Fig. P-2.13, point C is the midpoint of the 200-ft-long pipe and neglect all friction losses and determine the following:
    (a) What is the velocity in the pipe?
    (b) Plot the EGL and HGL.
    (c) What is the energy level at point C?
    (d) What is the pressure at point C?
    (e) Assuming the project at sea level and water T° at 65 °F, how far can you elevate point C without cavitation
14. ♦ Redo Problem 2.13 with friction losses $f = 0.02$ and find the discharge.

3 Hydrodynamics

This chapter uses the concept of impulse and momentum to calculate hydrodynamic forces. We examine the forces from water jets in Section 3.1 and forces in pipes in Section 3.2 prior to a review of flow measurement techniques in Section 3.3.

3.1 Hydrodynamic Force on a Plate

The main concepts of this section are the hydrodynamic force from a water jet on a stationary plate in Section 3.1.1 and on a moving plate in Section 3.1.2.

The hydrodynamic force is the force exerted by moving fluids on solid surfaces. There are two components to the analysis of hydrodynamic forces: (1) the pressure force, as in Chapter 1; and (2) the force exerted by the moving fluid. The pressure force is the product of pressure at the centroid and the surface area. In pipes, the pressure is obtained from the position of the hydraulic grade line. The pressure force in pipes pA is exerted towards the control volume of fluid.

A fluid moving at a velocity V exerts an additional force called momentum flux due to the change in flow direction or magnitude of the linear momentum. In preparing a free-body diagram (FBD), we learn in this section that the momentum flux is also exerted towards the fluid control volume. The hydrodynamic force is the sum of the pressure force and the momentum flux, both applied towards the control volume, as sketched Figure 3.1.

Our analysis of hydrodynamic forces starts with the simpler case of forces exerted by jets on plates. In this case, the pressure force can be eliminated under constant ambient pressure. The momentum flux is due to the change in direction of the fluid momentum. Two cases are examined: (1) a fixed plate and (2) a moving plate. The analysis is then extended to hydrodynamic forces in pipes where both the pressure force and the momentum flux are combined.

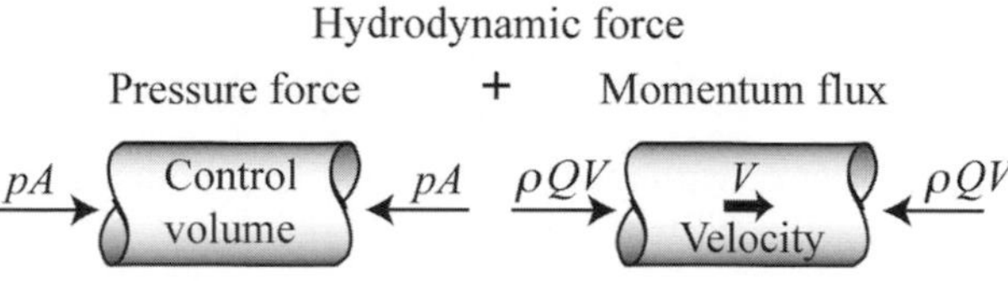

Figure 3.1 Hydrodynamic force in a pipe

3.1.1 Hydrodynamic Force on a Stationary Plate

The hydrodynamic force of a water jet impacting a curved stationary plate is first examined. Figure 3.2, in plan view (horizontal plane), shows a water jet entering at section 1 and leaving the control volume at section 2. The flow velocity is constant at the same elevation (from Bernoulli).

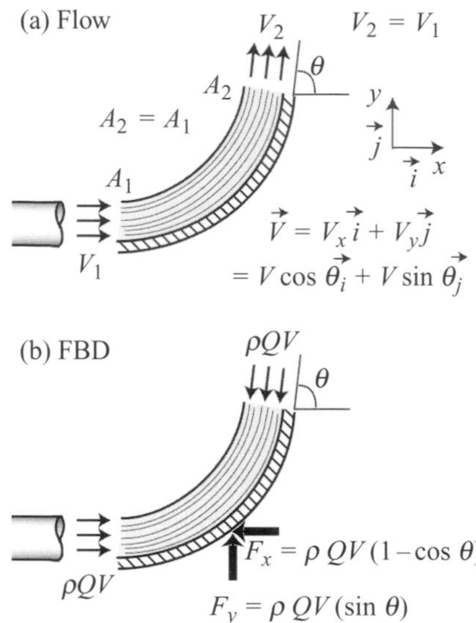

Figure 3.2 Force on a plate

The distance traveled by the fluid over a time interval Δt is $\Delta s = V\Delta t$, and the corresponding mass Δm entering/leaving the control volume is

$$\Delta m = \rho\Delta\forall = \rho A\Delta s = \rho AV\Delta t = \rho Q\Delta t. \tag{3.1}$$

The concept of flux designates the rate of change of a quantity over time. From Eq. (3.1), the mass flux is $\Delta m/\Delta t = \rho Q$ and the volumetric flux of water is Q.

Since the atmospheric pressure is constant, there is no net pressure force on the curved plate. The only force $\vec{F}$ that the plate exerts onto the fluid results from the change in direction of the linear momentum of the fluid. In other words, as learned in dynamics, the impulse $\vec{F}\Delta t$ causes a change in linear momentum of the fluid $\Delta m\vec{V}$. This impulse–momentum relationship is written as

$$\Delta m\vec{V}_{in} + \vec{F}\Delta t = \Delta m\vec{V}_{out}$$

and it can be solved for the force as

$$\vec{F} = \frac{1}{\Delta t}\left(\Delta m\vec{V}_{out} - \Delta m\vec{V}_{in}\right) = \frac{\Delta m}{\Delta t}\left(\vec{V}_2 - \vec{V}_1\right) = \rho Q\left(\vec{V}_2 - \vec{V}_1\right) = \rho Q\left(\vec{V}_{out} - \vec{V}_{in}\right). \tag{3.2}$$

With the use of FBDs, we usually solve $\vec{F} + \rho Q\vec{V}_{in} - \rho Q\vec{V}_{out} = 0$, where the negative momentum flux $-\rho Q\vec{V}_{out}$ leaving the control volume is equivalent to a positive force entering the control volume. Therefore, the momentum flux $\rho Q\vec{V}$ sketched on an FBD is applied towards the control volume as shown in Figure 3.2b. This case is solved for the force $\vec{F} = \rho Q\vec{V}_{out} - \rho Q\vec{V}_{in} = \rho Q\vec{V}_2 - \rho Q\vec{V}_1$, given that $\vec{V}$ is a vector:

$$\vec{V}_{out} = \vec{V}_2 = V\cos\theta\vec{i} + V\sin\theta\vec{j}$$

and

$$\vec{V}_{in} = \vec{V}_1 = V\vec{i} + 0\vec{j}.$$

Force is therefore a vector with two orthogonal components $\vec{F} = F_x\vec{i} + F_y\vec{j} = \rho Q\left[\left(V\cos\theta\vec{i} + V\sin\theta\vec{j}\right) - \left(V\vec{i} + 0\vec{j}\right)\right]$ where $F_x = \rho QV(\cos\theta - 1)$, and $F_y = \rho QV\sin\theta$.

3.1.2 Force and Power on a Moving Plate

Consider the flow of water on a vertical plate with a jet area A and velocity V while the plate is moving at v_p. Considering the axis y horizontal, as sketched in Figure 3.3, the jet splits equally in the lateral direction. As an observer moves at the same velocity as the plate (to make the flow steady), the velocity enters the cross-sectional area at the velocity $V - v_p$ relative to the moving plate. Accordingly, the flow rate entering the cross-sectional area is reduced to

$$Q = A(V - v_p). \tag{3.3}$$

Notice that if the jet and plate move at the same velocity, the water would not exert any force on the plate. The hydrodynamic forces on the moving plate are

$$F_x = \rho Q(V - v_p) = \rho A(V - v_p)^2$$

and

$$F_y = 0. \tag{3.4}$$

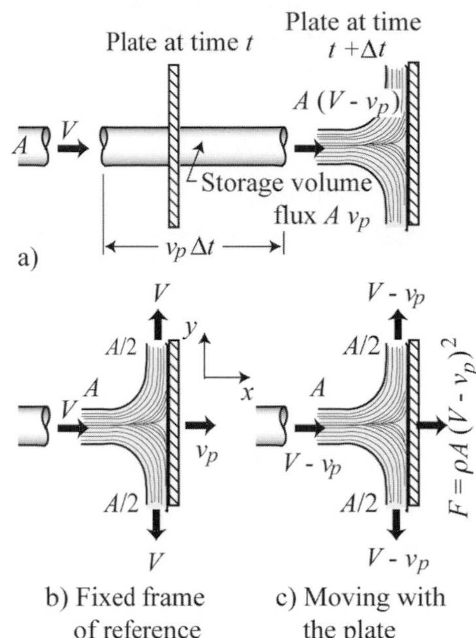

Figure 3.3 Moving vertical plate

The plate will gain energy equal to the force times the displacement of the plate. The power is then simply the rate at which energy is added to the plate. In other words, the power exerted on the plate is the product of the force applied on the plate and the velocity of the plate:

$$P = \rho Q(V - v_p)v_p = \rho A(V - v_p)^2 v_p = \rho A\left(V^2 v_p - 2V{v_p}^2 + {v_p}^3\right). \tag{3.5}$$

It becomes interesting to find out what plate velocity will maximize the power, or

$$\frac{\partial P}{\partial v_p} = \rho A\left(V^2 - 4V v_p + 3{v_p}^2\right) = \rho A\left(V - v_p\right)\left(V - 3v_p\right) = 0.$$

The maximum power is obtained when $v_p = V/3$.

The force and power are calculated for a jet on a moving vertical plate in Example 3.1 and an inclined plate in Example 3.2.

♦ **Example 3.1:** Force on a moving vertical plate

A 10-cfs water jet moving at 100 ft/s hits a vertical plate moving at 10 ft/s. The area of the jet is 0.1 ft^2. Calculate the force and power imparted onto the moving plate.

Solution: $F_x = \rho A(V - v_p)^2 = 1.94 \times 0.1 \times (90)^2 = 1{,}571$ lb, and the power is

$$P = \vec{F} \cdot \vec{V} = \rho A(V - v_p)^2 v_p = 1{,}571 \text{ lb} \times 10 \text{ ft/s} = 15{,}710 \text{ lb} \cdot \text{ft/s}.$$

♦♦ Example 3.2: Force and power on a moving inclined plate

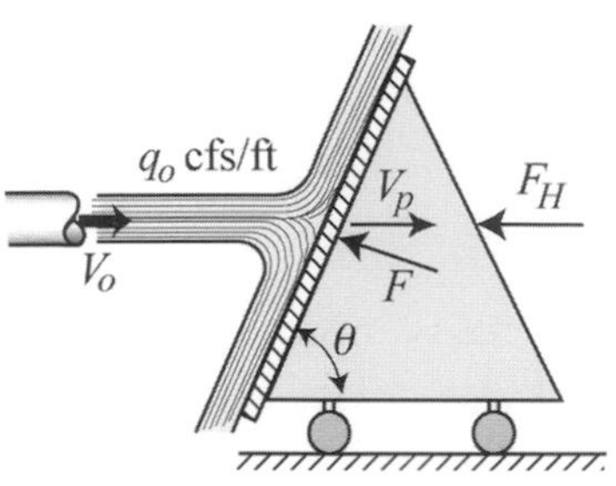

Fig. E-3.2 Inclined plate

A jet is impacting the inclined plate in Fig. E-3.2. Neglect the weight of water and find the following per unit width when $\theta = 60°$, $Q = 2\ \text{ft}^3/\text{s}$, $V = V_0 = 85$ ft/s and $A = Q/V$:

(1) the force F to hold the plate in place at $v_p = 0$;

(2) define the work done on the plate per unit time when the plate is moving to the right at $v_p = 10$ ft/s; and

(3) at what constant plate velocity (v_p to the right) will the work done be maximum?

Solution:

(1) On a stationary plate, the force tangential to the plate is zero because there is no friction on the plate. The force normal to the plate F is now coming at an angle with the horizontal:

$$F = \rho A V_0 V_0 \sin 60° = 1.94 \times 2 \times 85 \sin 60° = 285 \text{ lb},$$

$$F_H = F \sin 60° = 285 \sin 60° = 247 \text{ lb}.$$

(2) When the plate is moving to the right at v_p, not all the fluid will reach the plate because it fills the space left behind the moving plate. Therefore, the flow rate entering the control volume moving with the plate is such that the area is the same, but the velocities both entering and leaving the control volume are reduced to $V - v_p$.

The power P is

$$P = \frac{work}{time} = F_H v_p = \rho A \left(V - v_p\right)^2 \sin^2 60° v_p = 1.94 \times \frac{2}{85}(85 - 10)^2 \sin^2 60° \times 10$$
$$= 1{,}925 \text{ lb} \cdot \text{ft/s}.$$

(3) To find the plate velocity that will yield the maximum power,

$$P = \rho A \left(V - v_p\right)^2 \sin^2 60° v_p = \rho A \sin^2 60° \left(V^2 v_p - 2\, V {v_p}^2 + {v_p}^3\right),$$

$$\frac{\partial P}{\partial v_p} = \rho A \sin^2 60° \left(V^2 - 4\, V v_p + 3{v_p}^2\right) = 0.$$

The equation $\left(V - v_p\right)\left(V - 3\, v_p\right) = 0$ gives the trivial root $v_{p\ min} = V$ with a minimum $P = 0$, and $v_{p\ max} = V/3$ which gives the maximum power

$$P_{max} = \frac{4}{27} \rho A V^3 \sin^2 60° = 3{,}115 \text{ lb} \cdot \text{ft/s when } v_{p\ \max} = V/3 = 28.3 \text{ ft/s};$$

$$P_{max} = \frac{3{,}115 \text{ lb} \cdot \text{ft}}{\text{s}} \times \frac{4.45 \text{ N}}{\text{lb}} \times \frac{\text{m}}{3.28 \text{ ft}} = 4.22 \text{ kW}$$

Note that 1 lb · ft/s = 1.36 Nm/s = 1.36 watts.

In the eighteenth century, the development of steam engines notably by James Watt contributed to the success of the industrial revolution.

3.2 Hydrodynamic Force on a Pipe Bend

The difference between a water jet on a plate and flow in a pipe is twofold: (1) the change in pressure inside the pipe depending on the flow velocity; and (2) a change in direction and/or magnitude of the momentum flux. The pressure change is calculated from the Bernoulli equation. We typically calculate forces in pipe bends in three steps as follows: (1) solve the continuity and calculate areas, flow velocities, discharges and momentum fluxes ρQV for each pipe; (2) apply the Bernoulli equation to determine the pressure p and pressure force pA in each pipe; and (3) sketch an FBD with all forces (pressure force pA, momentum flux ρQV, and weight W) to determine the resultant force R required to anchor the system. As a general case, the flow diagram in Figure 3.4 is in elevation such that it also includes the weight component of the fluid W inside the control volume. The case of a single pipe bend is detailed in Examples 3.3 and 3.4. The more complex case of branching pipes is illustrated step by step in Example 3.5.

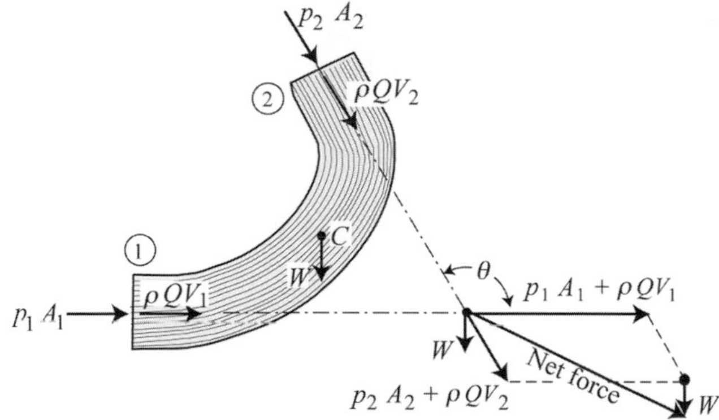

Figure 3.4 Elevation view with pressure, momentum and weight forces

---

♦ **Example 3.3:** Bernoulli application in a pipe

The horizontal pipe in Fig. E-3.3 ($z_2 = z_1$) discharges water into the atmosphere at section 2 ($p_2 = 0$). What is the pressure p_1 inside the pipe?

Solution: From the continuity, $Q = A_1 V_1 = 10 \times \pi \times 2^2/4 = 31.4\ \text{ft}^3/\text{s}$. The Bernoulli equation is solved to find the pressure:

$$\frac{p_1}{\gamma} + z_1 + \frac{V_1^2}{2g} = \frac{p_2}{\gamma} + z_2 + \frac{V_2^2}{2g}$$

and

$$p_1 = p_2 + \gamma\left(\frac{V_2^2 - V_1^2}{2g}\right);$$

$$p_1 = 0 + 62.4\left(\frac{40^2 - 10^2}{2 \times 32.2}\right) = 1{,}453\ \text{lb/ft}^2.$$

$\theta = 30°$

$V_2 = 40$ ft/s
$D_2 = 1$ ft

$V_1 = 10$ ft/s
$D_1 = 2$ ft

Fig. E-3.3 Pipe bend

◆◆ Example 3.4: Force to anchor a pipe bend

The horizontal pipe bend from Example 3.3 has $D_1 = 2$ ft ($V_1 = 10$ ft/s) and $D_2 = 1$ ft ($V_2 = 40$ ft/s). What horizontal force anchors this pipe bend in place?

Solution: The FBD of the pipe is sketched in Fig. E-3.4. Note that the direction of all hydrodynamic force vectors is towards the control volume. This is because the term $-\rho QV_2$ leaving the control volume is equivalent to $+\rho QV_2$ entering the control volume, as shown on Fig. E-3.4. In the x direction, with $p_2 = 0$,

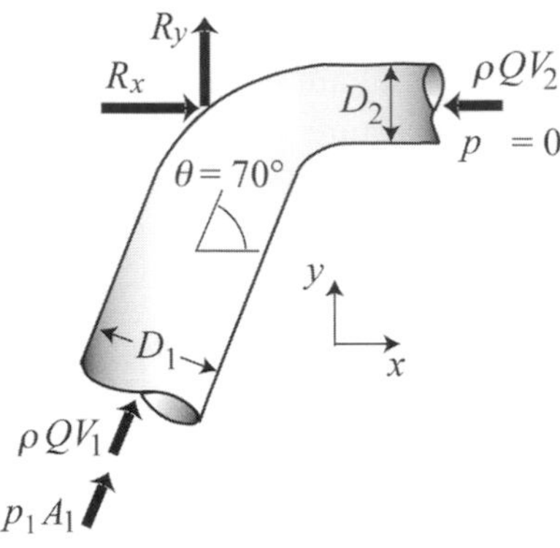

Fig. E-3.4 FBD of a pipe bend

$$\sum F_x = R_x + (p_1 A_1 + \rho QV_1)\cos 70° - \rho QV_2 = 0,$$

$$R_x = \rho QV_2 - (p_1 A_1 + \rho QV_1)\cos 70°,$$

$$\begin{aligned} R_x &= (1.94 \times 31.4 \times 40) \\ &\quad - [(1{,}453 \times 3.14) + (1.94 \times 31.4 \times 10)]\cos 70° \\ &= 668 \text{ lb}; \end{aligned}$$

and in the y direction (note that is + in the upward y direction),

$$\sum F_y = (p_1 A_1 + \rho QV_1)\sin 70° + R_y = 0,$$

$$\begin{aligned} R_y &= -(p_1 A_1 + \rho QV_1)\sin 70° = -[(1{,}453 \times 3.14) + (1.94 \times 31.4 \times 10)]\sin 70° \\ &= -4{,}860 \text{ lb}. \end{aligned}$$

The horizontal force magnitude is

$$\left|\vec{R}\right| = \sqrt{R_x{}^2 + R_y{}^2} = \sqrt{668^2 + 4{,}860^2} = 4{,}906 \text{ lb}.$$

◆◆ Example 3.5: Anchoring a pipe junction

From the pipe junction in a horizontal plane in Fig. E-3.5, what is the flow velocity at section 3 if the diameter is 0.15 m? If the pressure head at section 2 is 10 m, determine the pressure head at section 3 and find the net force required to hold this junction when $\theta = 30°$. Note that this problem can be programmed in parametric form to be solved for different values of D_3 and θ.

Solution:

Step (1): The geometry, continuity and momentum fluxes are

$$D_1 = 0.1 \text{ m}, A_1 = \pi \times 0.1^2/4 = 0.00785 \text{ m}^2,$$

$$D_2 = 0.12 \text{ m}, A_2 = \pi \times 0.12^2/4 = 0.0113 \text{ m}^2$$

and

$$D_3 = 0.15 \text{ m}, A_3 = \pi \times 0.15^2/4 = 0.01767 \text{ m}^2.$$

The discharges are, respectively,

$$Q_1 = A_1 V_1 = 0.00785 \times 4 = 0.0314 \text{ m}^3/\text{s},$$

$$Q_2 = A_2 V_2 = 0.0113 \times 6 = 0.0678 \text{ m}^3/\text{s},$$

$$Q_3 = Q_1 + Q_2 = 0.0314 + 0.0768$$
$$= 0.0993 \text{ m}^3/\text{s}$$

and

$$V_3 = Q_3/A_3 = 0.0992/0.01767 = 5.62 \text{ m/s};$$

and the momentum fluxes are, respectively,

$$\rho Q_1 V_1 = 1{,}000 \times 0.0314 \times 4 = 125.7 \text{ N},$$

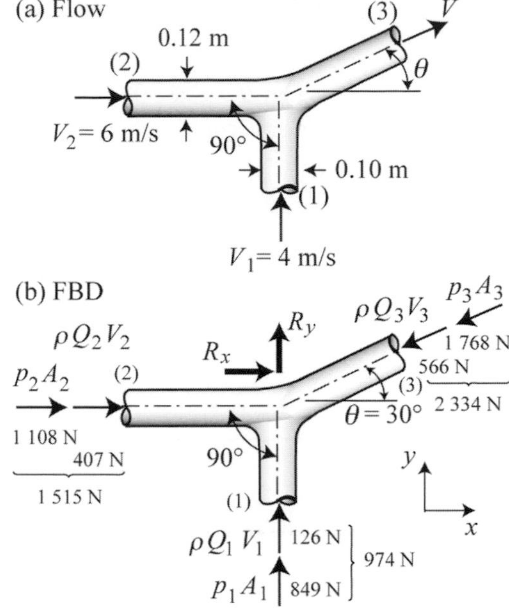

Fig. E-3.5 Pipe branch

$$\rho Q_2 V_2 = 1{,}000 \times 0.0678 \times 6 = 407.2 \text{ N}$$

and

$$\rho Q_3 V_3 = 1{,}000 \times 0.0992 \times 5.66 = 558 \text{ N}.$$

Step (2): Apply the Bernoulli equation with $p_2/\gamma = 10$ m to find the pressure and pressure force:

$p_2 = 10\gamma = 98{,}100$ Pa and $p_2 A_2 = 98{,}100 \times 0.00113 = 1{,}109$ N;

$$p_1 = p_2 + \rho\left(\frac{V_2^2 - V_1^2}{2}\right) = 98{,}100 + 1{,}000\left(\frac{6^2 - 4^2}{2}\right) = 108{,}100 \text{ Pa}$$

and $p_1 A_1 = 108{,}100 \times 0.00784 = 849$ N;

$$p_3 = p_2 + \rho\left(\frac{V_2^2 - V_3^2}{2}\right) = 98{,}100 + 1{,}000\left(\frac{6^2 - 5.62^2}{2}\right) = 100{,}320 \text{ Pa}$$

and $p_3 A_3 = 100{,}320 \times 0.01767 = 1{,}772$ N.

Step (3): From the FBD in Fig. E-3.5, the hydrodynamic forces are

$$p_1 A_1 + \rho Q_1 V_1 = (848.6 + 125.6) = 974.7 \text{ N},$$

$$p_2 A_2 + \rho Q_2 V_2 = (1{,}108 + 406.8) = 1{,}517 \text{ N}$$

and

$$p_3A_3 + \rho Q_3V_3 = (1{,}768 + 566) = 2{,}330 \text{ N}.$$

When $\theta = 30°$, the reaction forces (in the x and y directions) holding the pipe in place become

$$R_x + (p_2A_2 + \rho Q_2V_2) - (p_3A_3 + \rho Q_3V_3)\cos\theta = 0,$$

$$R_x = -1{,}515 + 2{,}334\cos 30° = 502 \text{ N (to the right)}$$

and

$$R_y + (p_1A_1 + \rho Q_1V_1) - (p_3A_3 + \rho Q_3V_3)\sin\theta = 0,$$

$$R_y = -974 + 2{,}334\sin 30° = 191 \text{ N (up)}.$$

3.3 Flow Meters

A flow meter measures the flow velocity, or flow discharge. In pipes, common meters include Venturi meters (Section 3.3.1) and flow nozzles and orifices (Section 3.3.2). Other measuring devices are briefly discussed in Section 3.3.3.

3.3.1 Venturi Meters

In the late eighteenth century, the Italian Giovanni Venturi designed a pipe flow meter. A Venturi meter measures the pressure difference in a manometer connected to a pipe contraction, from which the velocity is determined, as shown in Figure 3.5.

Let's define $G = \gamma_{Hg}/\gamma = 13.6$ for mercury with reading $R = z_B - z_A$, and $\Delta z = z_2 - z_1$. Starting at A, consider the pressure head on both sides of the manometer:

$$\frac{p_A}{\gamma} = z_1 + \frac{p_1}{\gamma} = GR + (z_1 + \Delta z) - R + \frac{p_2}{\gamma},$$

which gives

$$\frac{p_1}{\gamma} - \frac{p_2}{\gamma} = \Delta z + (G-1)R. \tag{3.6}$$

The difference in the piezometric level ΔH_{1-2} is $\Delta H_{1-2} = \left(z_1 + \frac{p_1}{\gamma}\right) - \left(z_2 + \frac{p_2}{\gamma}\right) = \left[\frac{p_1}{\gamma} - \frac{p_2}{\gamma}\right] - \Delta z.$

Combining with Eq. (3.6) gives

$$\Delta H_{1-2} = \left[\frac{p_1}{\gamma} - \frac{p_2}{\gamma}\right] - \Delta z = [\Delta z + (G-1)R] - \Delta z$$
$$= R(G-1). \tag{3.7}$$

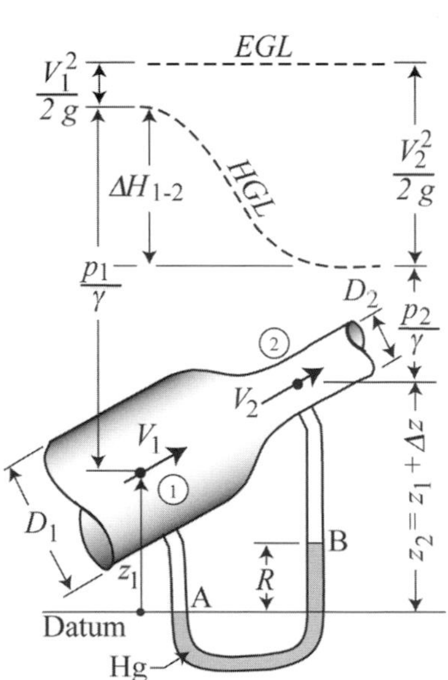

Figure 3.5 Venturi meter

Note that the piezometric head difference is independent of the pipe orientation, or Δz. From the geometry, the ratio of velocities depends on the ratio of pipe diameters as

$$\frac{A_2}{A_1} = \left(\frac{D_2}{D_1}\right)^2$$

and

$$\left(\frac{V_1}{V_2}\right)^2 = \left(\frac{D_2}{D_1}\right)^4.$$

Assuming conservation of the total energy (the same energy grade line [EGL]), the difference in the piezometric head $\Delta H_{1-2} = R(G-1)$ equals the change in the velocity head:

$$\Delta H_{1-2} = \left(\frac{V_2^2}{2g} - \frac{V_1^2}{2g}\right) = \frac{V_2^2}{2g}\left[1 - \left(\frac{V_1}{V_2}\right)^2\right] = \frac{V_2^2}{2g}\left[1 - \left(\frac{D_2}{D_1}\right)^4\right] = R(G-1), \qquad (3.8)$$

and we obtain the velocity $V_2 = C_G\sqrt{2gR(G-1)}$ where $C_G = 1/\sqrt{\left[1-(D_2/D_1)^4\right]}$.

For practical applications, we define an experimental coefficient C_V to account for possible energy losses. The general discharge formula for a Venturi meter is

$$Q = C_V C_G A_2 \sqrt{2gR(G-1)}. \qquad (3.9)$$

At high flows, the piezometric head (or hydraulic grade line [HGL]) can fall below the pipe elevation and cause negative pressure in the contracted section of the pipe as discussed in Section 2.2.4. When reaching the vapor pressure, vapor bubbles form and their collapse can cause cavitation as sketched in Figure 3.6. Cavitation is always an issue when the velocity in the contracted section approaches a critical velocity $V_c = \sqrt{2g[H + (p_{atm} - p_v)/\gamma]}$, where H is the EGL elevation above the pipe. Approximations are $V_{c\,\mathrm{m/s}} \approx \sqrt{20(H_\mathrm{m} + 10)}$ in SI and $V_{c\,\mathrm{ft/s}} \approx \sqrt{64(H_\mathrm{ft} + 33)}$ in customary units. Calculations for a Venturi meter are shown in Example 3.6.

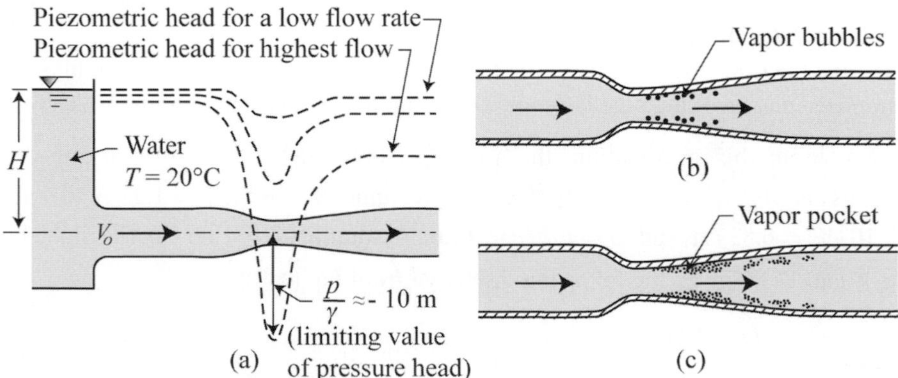

Figure 3.6 Cavitation in a Venturi meter

♦♦ Example 3.6: Venturi meter

Find the flow velocity and discharge in a Venturi meter with diameters $D_1 = 1$ ft and $D_2 = 0.5$ ft with a reading of $R = 1$ ft in a mercury manometer.

Solution: Assuming conservation of energy gives $C_V = 1$, and the geometric coefficient is

$$C_G = 1/\sqrt{\left[1 - (D_2/D_1)^4\right]} = 1/\sqrt{\left[1 - 0.5^4\right]} = 1.0328.$$

The flow velocity is

$$V_2 = C_V C_G \sqrt{2gR(G-1)} = 1.033\sqrt{2 \times 32.2(12.6)} = 29.42 \text{ ft/s}.$$

The area $A_2 = 0.25 \times \pi \times 0.5^2 = 0.196$ ft^2 and the discharge from Eq. (3.9) is

$$Q = C_V C_G A_2 \sqrt{2gR(G-1)} = 1.033 \times 0.196\sqrt{2 \times 32.2 \times 1 \times (13.6 - 1)}$$
$$= 5.77 \text{ ft}^3/\text{s}.$$

3.3.2 Flow Nozzles and Orifices

The analysis of flow nozzles and orifices is very similar to the Venturi meter except that the energy is not conserved. A flow nozzle has a profiled contraction in a pipe compared to the short and sharp-edged orifice. The difference in the piezometric head simply becomes $\Delta H = R(G-1)$, where R is the manometer reading and $G = \rho_m/\rho$ is the density ratio of the two manometer fluids. Expansion losses depend on the Reynolds number, and a discharge coefficient $C = C_G C_V$ can be pre-calibrated for each instrument. such that the discharge relationship becomes

$$Q = CA_2\sqrt{2gR(G-1)}. \tag{3.10}$$

Flow calculations for a nozzle (Example 3.7), an orifice (Example 3.8) and a Pitot tube (Example 3.9) include more details on measurement devices.

♦ Example 3.7: Flow in a nozzle

Determine the flow discharge in a 6-in. line with a 4-in. flow nozzle. The mercury–water manometer measures a gauge difference of 10 in. at a water temperature of 60 °F.

Solution: From this information, the area $A_2 = 0.25\pi(0.333)^2 = 0.0873$ ft^2, the ratio of areas is $A_2/A_1 = (4/6)^2 = 0.444$, the dynamic viscosity $\upsilon = 1.22 \times 10^{-5}$ ft^2/s, $R = 10/12 = 0.833$ ft and $G = \rho_{Hg}/\rho = 13.6$. Assuming $C = 1.06$ on the right-hand side of Fig. E-3.7 gives the following discharge from Eq. (3.10):

$$Q = 1.06 \times 0.0873\sqrt{2 \times 32.2 \times 0.833\,(13.6 - 1)} = 2.4 \text{ cfs}.$$

We then check the coefficient from the velocity $V_1 = 4Q/\left(\pi \times 0.5^2\right) = 12.2$ ft/s and $\text{Re} = V_1 D_1/\upsilon = 12.2 \times 0.5/1.22 \times 10^{-5} = 5 \times 10^5$, such that $C = 1.06$ was correct.

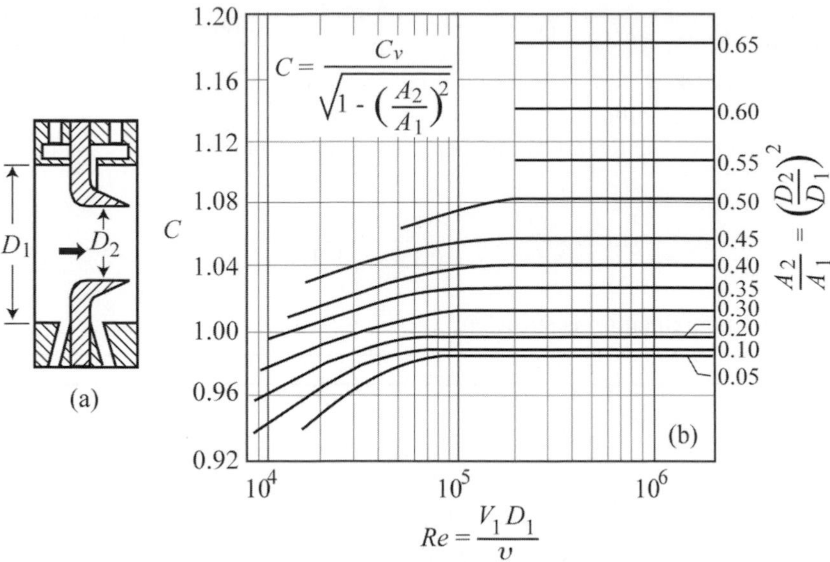

Fig. E-3.7 Discharge coefficient for a flow nozzle

Example 3.8: Orifice flow meter

A water mercury manometer is attached to both sides of a 15-cm orifice placed in a 30-cm pipe. If the manometer reading is 20 cm, what is the discharge in the pipe at room temperature.

Solution: The calibration coefficient for orifices is shown in Fig. E-3.8. The mercury manometer yields a density $G = 13.6$, and the contracted area is $A = 0.25\pi d^2 = 0.0177\ \text{m}^2$. The flow discharge in the pipe is determined from Eq. (3.10) with the coefficient obtained from assuming a high Reynolds number, thus $C \cong 0.63$, and the discharge is

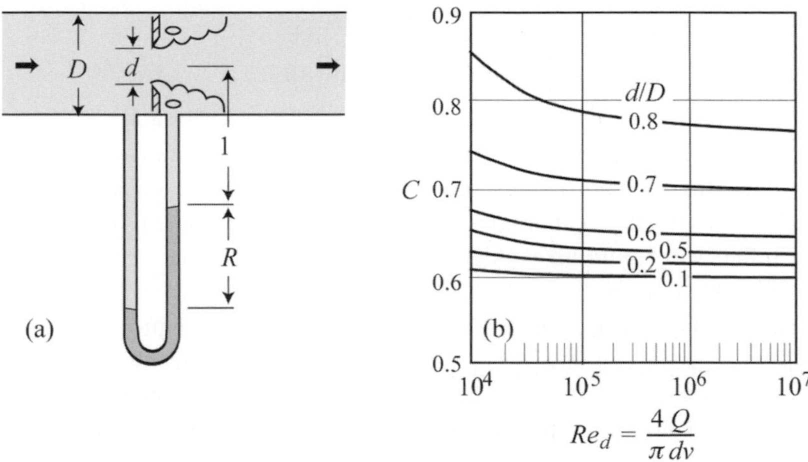

Fig. E-3.8 Discharge coefficient for an orifice

$$Q = CA_2\sqrt{2gR(G-1)} = 0.63\left(\frac{\pi \times 0.15^2}{4}\right)\sqrt{2 \times 9.81 \times 0.2(13.6-1)}$$
$$= 0.078\ \text{m}^3/\text{s}.$$

It is useful to check the Reynolds number even if the variability in C is small:

$$\text{Re} = \frac{4Q}{\pi d \upsilon} = \frac{4 \times 0.078}{\pi \times 0.15 \times 10^{-6}} = 6.6 \times 10^5$$

and the first approximation is sufficient.

♦ **Example 3.9:** Pitot tube

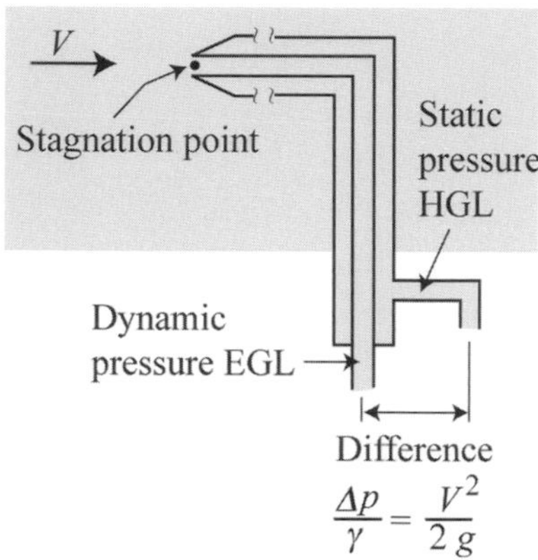

Fig. E-3.9 Pitot tube

The French engineer Henri Pitot developed an instrument to measure the fluid flow velocity in the early eighteenth century. As sketched in Fig. E-3.9, the instrument has two tubes, one inside another. The interior tube is facing the flow direction and measures both the pressure and the velocity head, or EGL. The external tube has openings along the side in the direction perpendicular to the flow. The external tube therefore measures the pressure only, i.e. the HGL. The difference in pressure between the two tubes (EGL – HGL) is the velocity head. For instance, if the pressure difference in water is 500 Pa, what is the velocity head?

Solution:

$\Delta p = 500$ Pa, and $\frac{\Delta p}{\gamma} = \frac{V^2}{2g}$, thus $V = \sqrt{2\Delta p/\rho} = \sqrt{2 \times 500/1000} = 1$ m/s.

3.3.3 Pressure Gauges and Other Flow Meters

This section briefly reviews various devices to measure velocity and pressure. Besides piezometers and manometers, nowadays pressure can be easily measured with pressure sensors, illustrated in Figure 3.7. A simple pivot gauge is sketched in Figure 3.7a. Sensors can be placed on flexible diaphragms to measure the net pressure (Figure 3.7b), or the pressure difference (Figure 3.7c). A Bourdon in Figure 3.7d has a curved pipe, which stiffens under high pressure.

Several other types of flow meters for pipes are sketched in Figure 3.8.

- The Venturi meter and flow nozzles (Figure 3.8a, b) have already been discussed.
- Ultrasonic flow meters (Figure 3.8c) measure the Doppler effect on acoustic waves propagating in the fluid. The acoustic waves are reflected from air bubbles and solid particles moving in the fluid. The frequency shift between the emitted and return signals describes the Doppler effect, which can be electronically measured with great accuracy.

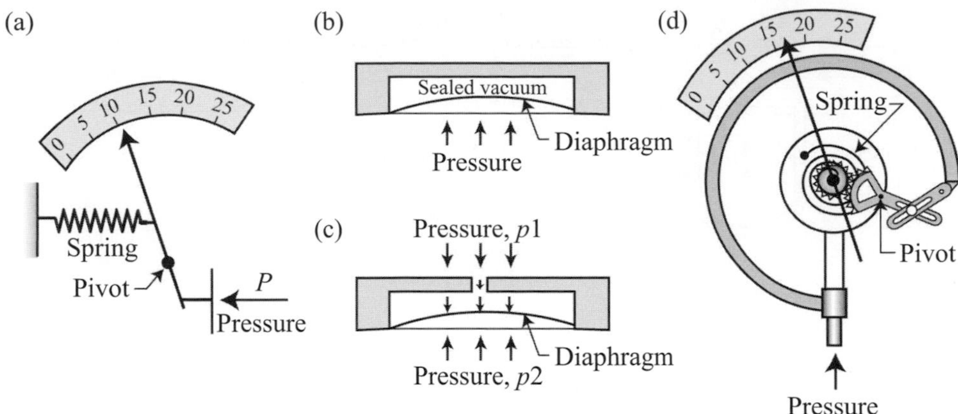

Figure 3.7 Pressure measurement devices

- In large industrial pipes, the Coriolis flow meter (Figure 3.8d) vibrates a pipe loop in the direction orthogonal to the flow direction. The Coriolis force induces a twisting angle of the pipe, which increases with the flow velocity in the pipe.
- Magnetic flow meters (Figure 3.8e) track the current induced by the flow of water under a magnetic field.
- For fluid mechanics research in the hydraulics laboratory, very precise point velocity measurements can be obtained with laser Doppler anemometers (LDAs) (Figure 3.8f) or velocimeters (LDVs). An LDA also uses the Doppler-shift concept and measures the frequency shift between emitted and reflected signals from air bubbles or dust particles moving with the flow. A clean air environment is desirable for LDAs.
- In noncorrosive clear fluids, simple turbines (Figure 3.8g) can also be placed inside the flow to measure the flow rate. The rotational speed of the propeller is calibrated to the flow velocity and thus the flow rate in the pipe.

Figure 3.9 illustrates various flow devices for streams and rivers. Propeller devices Figure 3.9a and Price current meters Figure 3.9b are useful to obtain point flow velocity measurements in turbulent flows over longer periods of time. They are excellent for determining the mean flow velocity. The acoustic Doppler current profiler (ADCP) in Figure 3.9c measures the Doppler shift of acoustic waves propagating with the fluid and reflected from air bubbles and sediment particles. It gives instantaneous values of flow velocities as a function of depth. When coupled with the use of global positioning systems, ADCP measurements can integrate the velocity and depth signals in real time to determine the discharge as the instrument travels the entire cross-sectional area of a river.

Additional Resources

Useful references on hydrodynamic forces and flow meters in pipes include Albertson et al. (1960), Streeter (1971), Vennard and Street (1975) and Roberson et al. (1997). Useful references for flow measuring devices in rivers include Bos (1989) and USBR (1997). A free surface velocimeter was also tested in Lee and Julien (2006).

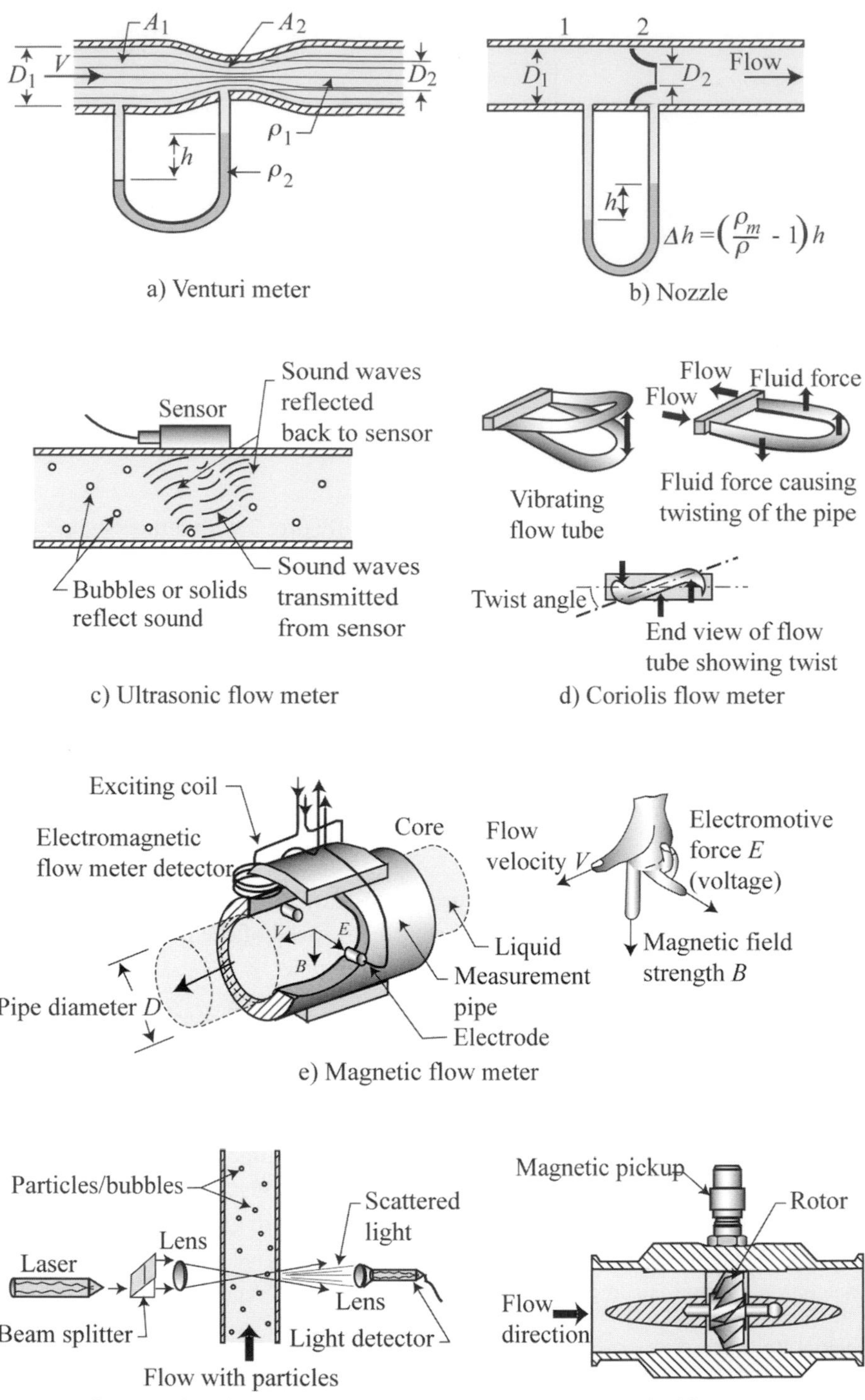

Figure 3.8 Sketches of various pipe flow meters

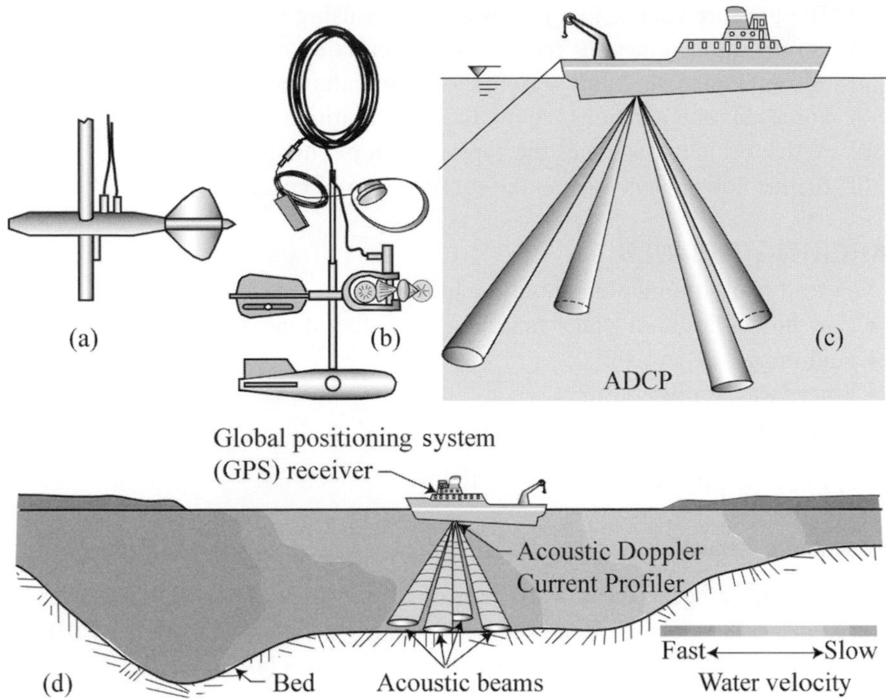

Figure 3.9 Flow measurement devices for streams and rivers

EXERCISES

These exercises review the essential concepts from this chapter.

1. What is the force component that is due to the fluid motion?
2. What are the two components of the hydrodynamic force?
3. What is the pressure force on a curved plate?
4. What is the work done on a stationary plate?
5. At what speed does a moving jet exert the maximum power on a single vertical plate?
6. What are the three force components in a pipe bend?
7. What is the difference between a Venturi meter and a flow nozzle?
8. What is the difference between a flow nozzle and an orifice?
9. What does a Pitot tube measure?
10. What does a Bourdon gauge measure?
11. Would a Venturi meter be appropriate for the pipes in Case Study 2.1?
12. Would a Pitot tube work in the pipes of Case Study 2.1?
13. True or false?
    (a) The pressure force is applied towards the control volume.
    (b) The momentum flux entering and leaving a control volume is applied in the flow direction.
    (c) The power and work done by a jet of water on a moving plate are the same.
    (d) The momentum flux is only important when the velocity changes direction.

(e) The pressure change in a pipe is calculated using the Bernoulli equation.
(f) Forces at pipe junctions require the conservation of mass.
(g) Forces in pipe junctions require the conservation of energy.
(h) Forces in pipe junctions require the conservation of momentum.
(i) A Venturi meter assumes the conservation of energy.
(j) Orifice flow meters assume the conservation of energy.

SEARCHING THE WEB

Can you find more information about the following?

1. ♦ How does a Bourdon gauge work and what does it measure?
2. ♦ Find images of an ADCP.
3. ♦ Where was Venturi born?
4. ♦ Where did Pitot live?

PROBLEMS

Hydrodynamic Forces

1. In Fig. P-3.1, a nozzle ejects water horizontally at 40 mi/h at a volumetric flow rate of 30 m^3/s. The stream is deflected horizontally by a plate in plan view. Determine the force in kN exerted by the plate on the jet in cases (a), (b) and (c).

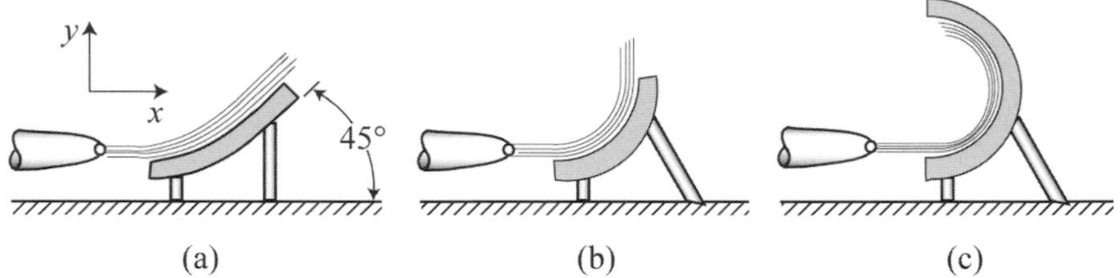

Fig. P-3.1

2. ♦♦ In Fig. P-3.2, water flows through a horizontal contraction at $Q = 25$ cfs. The contraction head loss is 0.2 times that of the higher velocity head. The upstream pipe pressure is 30 psig. What force holds the system in place?

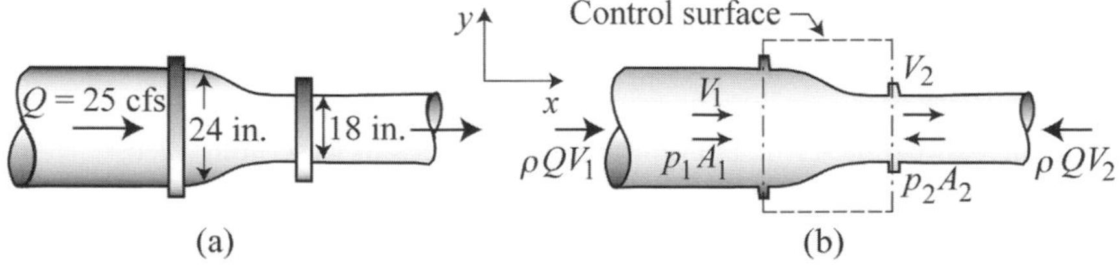

Fig. P-3.2

3. ♦♦ In Fig. P-3.3, the 1-ft-wide vane shown with y horizontal has a discharge $Q_0 = 3$ cfs and a jet velocity of 300 ft/s. Determine the following:

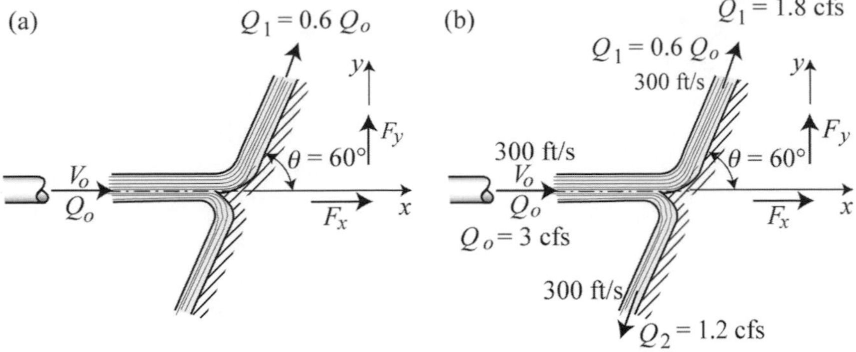

Fig. P-3.3

(a) What force components F_x and F_y can hold the vane in place?

(b) Find the force when the plate moves to the right at $v_p = 100$ ft/s?

(c) What is the work done per unit time on the moving plate?

4. ♦ In Fig. P-3.4, use the continuity and the sum of forces along the inclined plate to determine: (1) the flow distribution Q_1 and Q_2 as a function of the angle θ, and (2) the horizontal force on the plate.

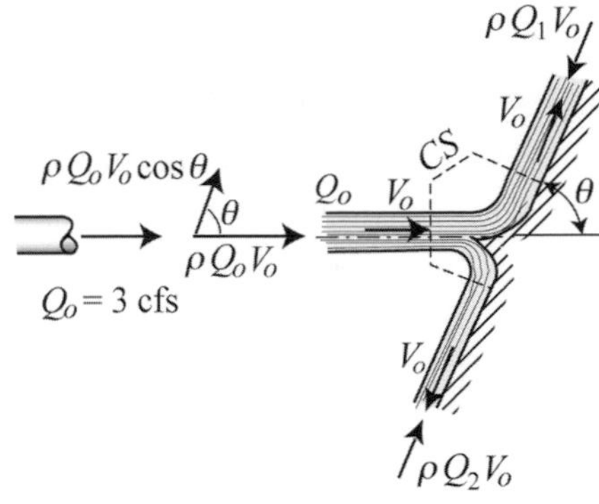

Fig. P-3.4

5. ♦ In Fig. P-3.5, a discharge of 0.1 m^3/s passes through the pipe bend shown in plan view. If the pressure head on the left-hand side is 10 m of water and $d_2 = 0.1$ m, calculate the reaction force to anchor this pipe.

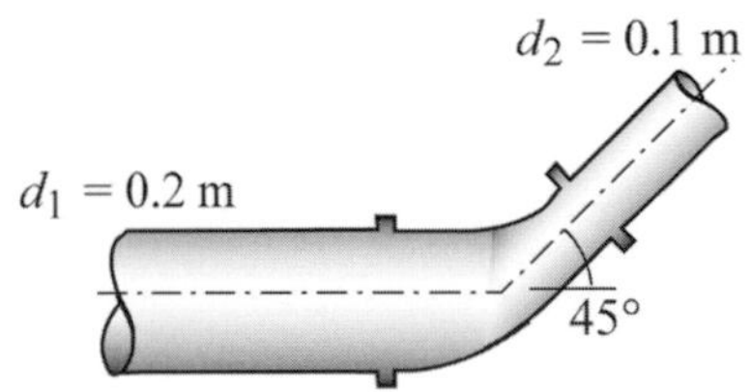

Fig. P-3.5

6. ♦♦ In Fig. P-3.6, a 1-m-diameter pipe carries 3 m^3/s of water at 10 °C under constant pressure at 75 kPa. A 30° horizontal pipe bend weighs 4 kN and contains 1.8 m^3 of water. What is the anchor force to hold the bend in place?

Fig. P-3.6

7. ♦♦ From the pipe junction in a horizontal plane shown in Fig. P-3.7, what is the velocity of section 3 if the diameter is 0.15 m? If the pressure head at section 2 is 10 m, determine the pressure head at section 3 and find the net force required to hold this junction when $\theta = 60°$.

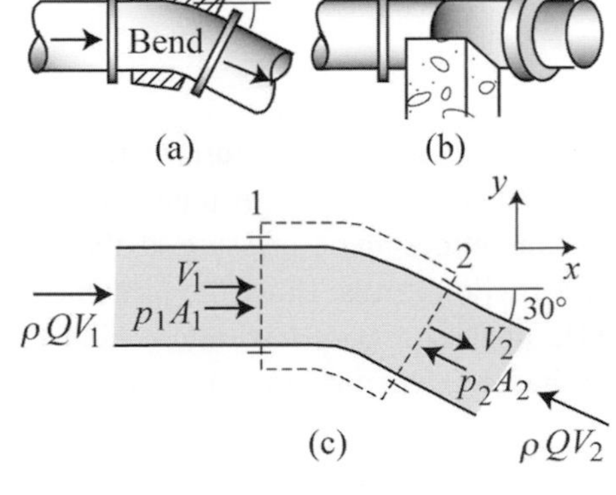

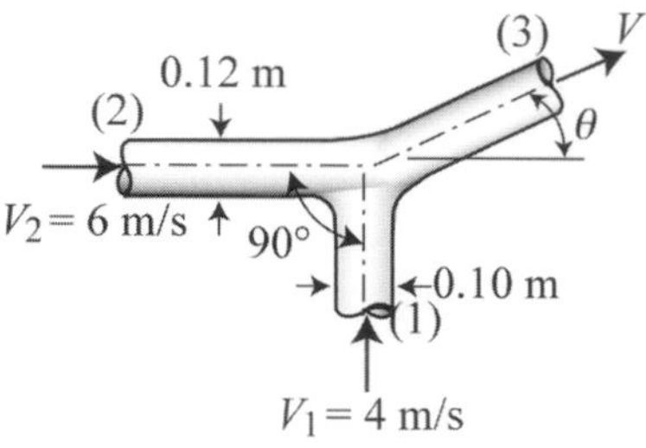

Fig. P-3.7

8. ♦♦ In the horizontal pipe branch of Fig. P-3.8, determine the pressure in each branch and the force required to hold the junction in place.

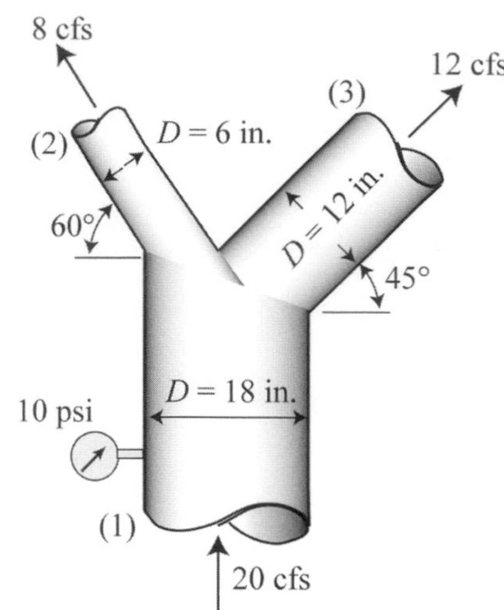

Fig. P-3.8

Flow Meters

9. ♦ First, neglect friction in the Venturi meter shown in Fig. P-3.9 and determine the discharge in the pipe. Then consider C_V from the calibration curve. What is the discharge difference when using the correction coefficient?

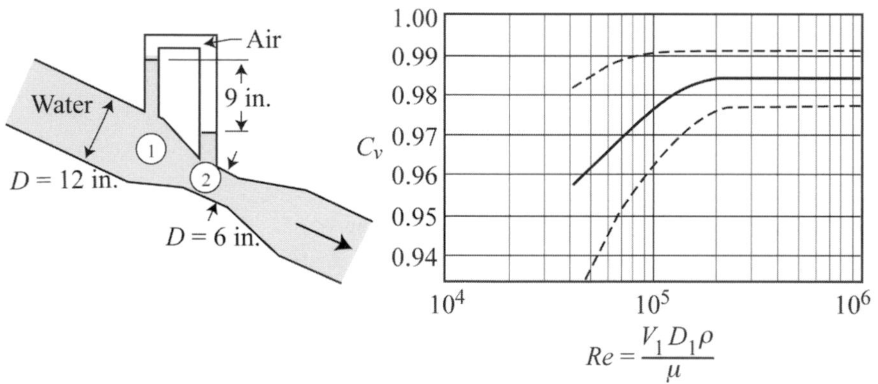

Fig. P-3.9

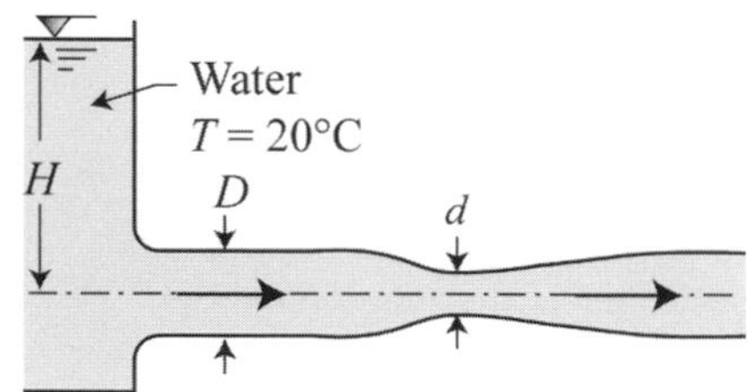

Fig. P-3.10

10. ♦ In Breckenridge Colorado, a Venturi meter with $d = 15$ cm is placed in a $D = 30$-cm pipe shown in Fig. P-3.10. Assume atmospheric pressure at the outflow and neglect friction losses. Draw the piezometric line and the EGL at the onset of cavitation. Find the head H and discharge Q when cavitation occurs.

4 Pumps

Pumps move water through pipes systems. This chapter discusses pump types in Section 4.1, pump performance in Section 4.2 and cavitation in Section 4.3.

4.1 Pump Types

The head and power of pumps are introduced in Section 4.1.1, followed by the specific speed of pumps in Section 4.1.2 and pump types in Section 4.1.3.

4.1.1 Pump Head and Power

In Figure 4.1, the head at the pump H is the difference in the piezometric head at both ends of a pump. In a pipe system, it also corresponds to the elevation gain ΔE plus the energy losses in the pipe $\sum h_f$. The work done by the pump is the applied force times the displacement of water in the direction of the force.

The power is the work done per unit time. The hydraulic power is the product of the force times the displacement over time, or simply the product of the force times the flow velocity in the pipe. Therefore,

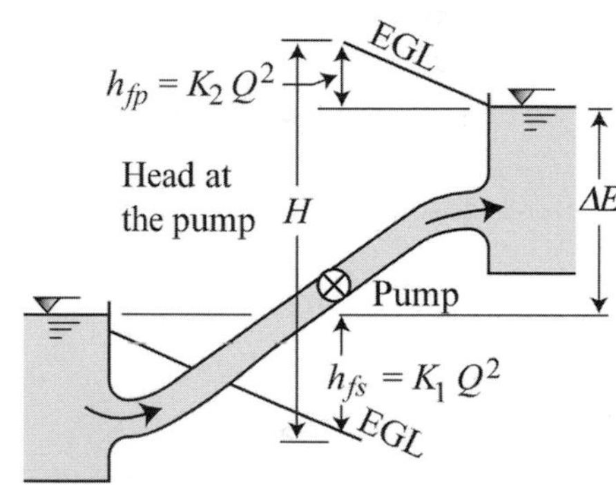

Figure 4.1 Head at the pump

$$power = \frac{work}{time} = \frac{\vec{F} \cdot \vec{d}}{t} = \vec{F} \cdot \vec{V} = pAV = \gamma HAV = \gamma HQ, \tag{4.1}$$

$$P_{\text{hp}} = \frac{\gamma H_{\text{ft}} Q_{\text{cfs}}}{550},$$

with P_{hp} in horsepower from γ in lb/ft^3, H_{ft} in ft and Q_{cfs} in ft^3/s, or

$$P_{\text{kW}} = \frac{\gamma H_{\text{m}} Q_{\text{cms}}}{1{,}000}$$

for power in kW (kilowatt), γ in N/m^3, H_{m} in m and Q_{cms} in m^3/s; and

$$P_{\text{kW}} = \frac{\gamma H_{\text{ft}} Q_{\text{cfs}}}{737}$$

for power in kW (kilowatt), γ in lb/ft^3, H_{ft} in ft and Q_{cfs} in ft^3/s.

Useful unit conversions include 1 cfs = 450 gpm (gallons per minute), 1 hp = 550 lb · ft/s = 746 W = 0.746 kW, and 1 kW = 1,000 N · m/s = 1.34 hp. Example 4.1 illustrates how to calculate the hydraulic power of a pump.

♦♦ Example 4.1: Pump head and power

A pump in Fig. E-4.1 draws water from a reservoir at an elevation of 520 ft, and pushes the water through a 1-ft-diameter pipe, 5,000 ft long. This pipe discharges water into a reservoir at an elevation of 620 ft at a flow rate of 7.85 cfs. Neglect minor losses except the friction loss with $f = 0.01$. Determine the head and hydraulic power at the pump.

Solution: First, solve for the head at the pump H by use of the energy equation at two points at the surface of the reservoirs:

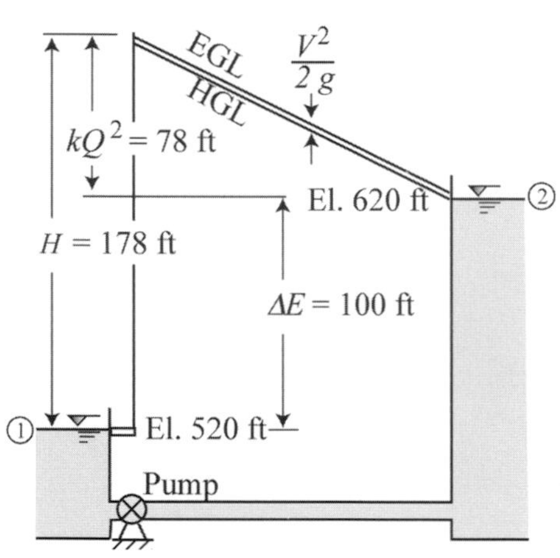

Fig. E-4.1 Pumping system

$$\frac{p_1}{\gamma} + \frac{V_1^2}{2g} + z_1 + H = \frac{p_2}{\gamma} + \frac{V_2^2}{2g} + z_2 + h_L,$$

with $\Delta E = z_2 - z_1 = 100$ ft, $p_1 = p_2 = 0$ and $V_1 = V_2 = 0$, or $\Delta H = \Delta E + kQ^2$.

The velocity is obtained from

$$V = \frac{Q}{A} = \frac{4 \times 7.85}{\pi 1^2} = 10 \text{ ft/s}.$$

The friction loss is

$$h_L = \frac{fL}{D}\frac{V^2}{2g} = \left(\frac{0.01 \times 5{,}000}{1}\right)\frac{10^2}{2 \times 32.2} = 78 \text{ ft}.$$

The velocity head loss at the pipe exit, $V^2/2g = 10^2/(2 \times 32.2) = 1.55$ ft, is typically small compared to the friction losses in long pipes. The head at the pump is $H = z_2 - z_1 + h_L = \Delta E + h_L = 620 - 520 + 78 = 178$ ft.

The hydraulic power at the pump is

$$P = \frac{\gamma H Q}{550} = \frac{62.4 \times 178 \times 7.85}{550} = 158 \text{ hp},$$

or $P = 0.746 \times 158 = 118$ kW. Alternatively, the power in SI units can be calculated from

$$P = \frac{9.81 \text{ kN}}{\text{m}^3} \times \frac{178 \text{ m}}{3.28} \times \frac{7.85 \text{ m}^3}{35.32 \text{ s}} = 118 \text{ kW}.$$

4.1.2 Specific Speed of Pumps

In Figure 4.2, the rotational speed N of a pump is measured in rotations per minute (rpm). We can also define the angular speed ω from $\omega = N\pi/30$ in rad/s. The type of pump that is best suited for a given head and discharge depends on the specific speed of a pump N_s. The specific speed of a pump is determined from a combination of two main dimensionless parameters C_H and C_Q defined for the head at the pump H and flow discharge Q:

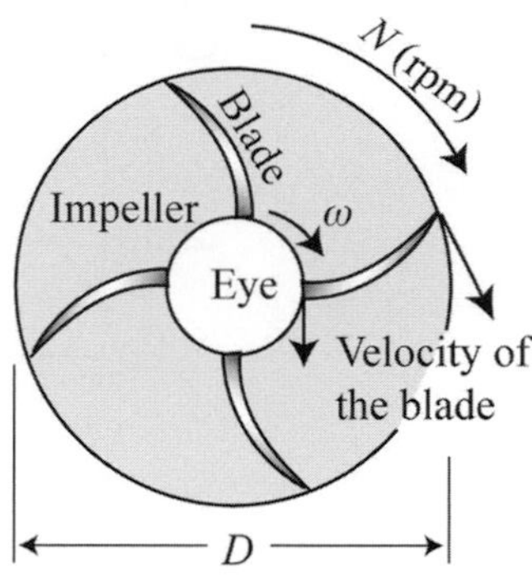

Figure 4.2 Pump rotation

$$\left.\begin{array}{l}\left.\begin{array}{l} V \sim ND \\ H \sim \dfrac{V^2}{2g} \end{array}\right\} C_H = \dfrac{gH}{N^2D^2} \\ \left.\begin{array}{l} Q \sim AV \\ \sim D^2ND \end{array}\right\} C_Q = \dfrac{Q}{ND^3} \end{array}\right\} N_S = \frac{C_Q^{1/2}}{C_H^{3/4}} = \frac{NQ^{1/2}}{H^{3/4}}.$$

Note that C_H defines the changes in the head H of a pump from changes in rotational speed N or diameter D. Similarly, C_Q relates the discharge Q to N and D. To find the specific speed, we eliminate D from C_H and C_Q, and solve for N. Given that g is constant, the result is called the specific speed N_s of a pump:

$$N_s = \frac{N_{\text{rpm}} Q_{\text{gpm}}^{1/2}}{H_{\text{ft}}^{3/4}}, \tag{4.2}$$

given N_{rpm} in rpm, Q_{gpm} in gallons per min, and H_{ft} in ft.

Examples 4.2 and 4.3 calculate power and specific speed.

♦ **Example 4.2:** Pump specific speed

Calculate the specific speed N_s of a pump for $N_{\text{rpm}} = 690$ rpm, $Q = 0.25$ cms $= (0.25 \times 35.32 \times 450) = 3{,}974$ gpm, and $H = 6$ ft.

Solution:

$$N_s = \frac{690 \times 3{,}974^{1/2}}{6^{3/4}} = 11{,}300,$$

and, from Figure 4.3, an axial pump (propeller) is desirable.

♦♦ **Example 4.3:** Pump power

Calculate the power and specific speed given $N = 60$ Hz, or $N_{rpm} = 3{,}600$ rpm. $Q = 2.5$ cfs $= (2.5 \times 450) = 1{,}125$ gpm, and $H = 65$ m $= 65 \times 3.28 = 213$ ft.

Solution:

$$P = \frac{\gamma HQ}{550} = \frac{62.4 \times 213 \times 2.5}{550} = 61 \text{ hp} = 46 \text{ kW},$$

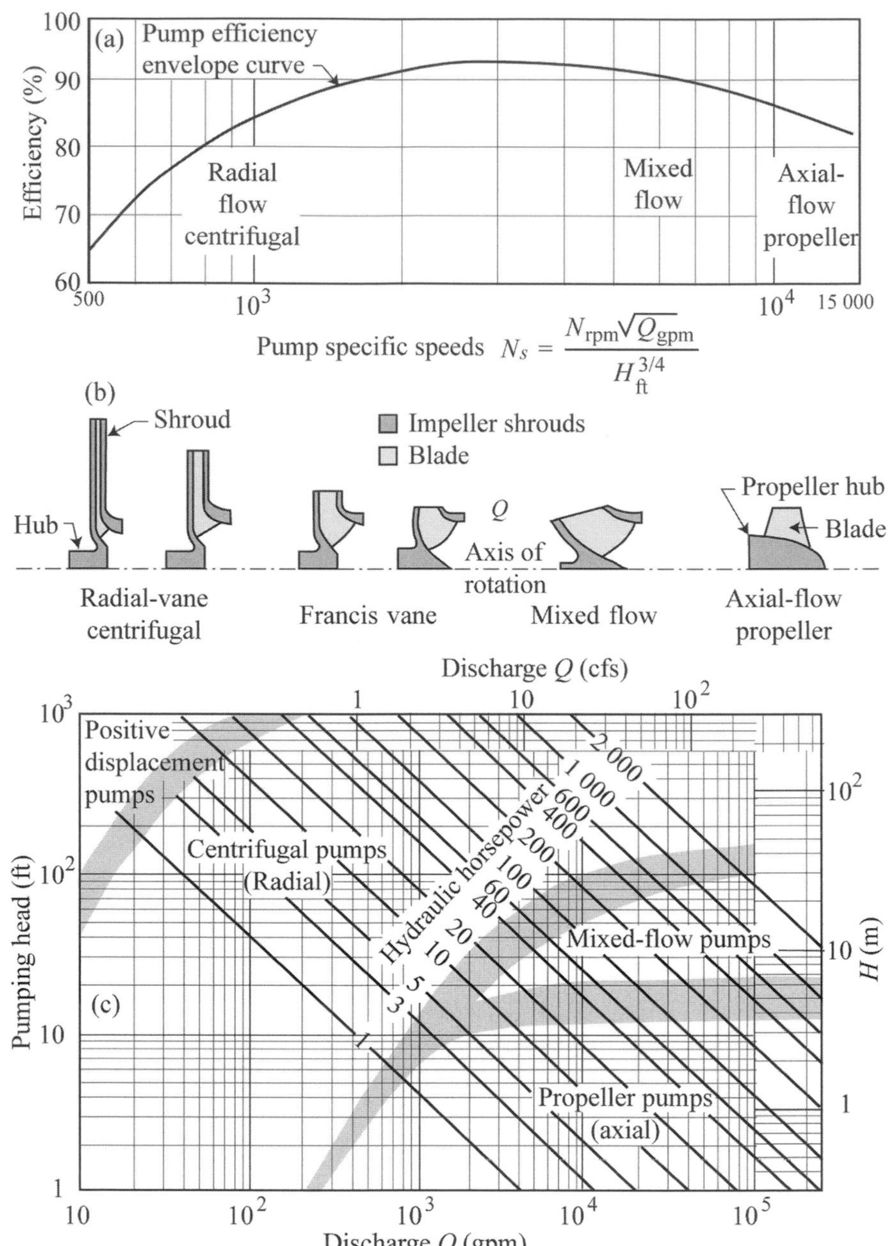

Figure 4.3 Specific speed and pump classification

$$N_s = \frac{3600 \times \sqrt{1{,}125}}{(213)^{3/4}} = 2{,}150$$

and a radial (centrifugal) pump is indicated in this case.

The pump chart in Figure 4.3a is used to identify the best pump type for different heads and discharges. In general, low N_s values require centrifugal pumps and high

N_s values require axial pumps. The maximum efficiency of centrifugal pumps is around $2{,}000 < N_s < 3{,}000$, and drops significantly when $N_s < 1{,}000$. It becomes interesting to consider that Eq. (4.2) can be rewritten as an approximation $N_{\text{rpm}} \approx 0.05 N_s H_{\text{ft}}^{3/4}/Q_{\text{cfs}}^{1/2} \simeq 0.02 N_s H_{\text{m}}^{3/4}/Q_{\text{cms}}^{1/2}$. This relationship indicates that the rotational speed will increase for high heads and low discharges. The effective range of rotational speeds for pumps is $1{,}000 < N_{\text{rpm}} < 6{,}000$, with pumps requiring higher maintenance when $N_{\text{rpm}} > 3{,}000$. Therefore, in Figure 4.3c, positive-displacement pumps and gear pumps will become more effective than centrifugal pumps at high heads and low discharge. Conversely, at high discharges and low head, the rotational speed becomes very slow and the rotational effects become ineffective such that axial pumps become preferable.

Centrifugal pumps are particularly interesting since the centrifugal force raises the water pressure. In a forced vortex, the head generated H is a function of the angular velocity ω as $H = \omega^2 r^2/2g$, and using $\omega = 2\pi N_{\text{rpm}}/60$ gives $N_{\text{rpm}} D = 27\sqrt{gH}$. This results in an important relationship for hydro-machinery, that, at a given head, the size of the impeller is inversely proportional to the rotational speed. This can be rewritten to estimate a pump's size as $D_{\text{ft}} \approx 150\sqrt{H_{\text{ft}}}/N_{\text{rpm}}$ or $D_{\text{m}} \approx 85\sqrt{H_{\text{m}}}/N_{\text{rpm}}$. It is interesting to consider that the size of the impeller depends on the head rather than the discharge and varies inversely with the rotational speed of the pump.

Finally, the cavitation criterion for centrifugal pumps depends on the velocity head on the suction side of the pump. The points of minimal pressure are located near the eye of the pump and on the suction side of the impeller blades. These characteristics of centrifugal pumps will be investigated further in the coming sections.

4.1.3 Pump Types

The positive-displacement, jet and gear pumps sketched in Figure 4.4 are not very efficient at high discharges. However, they are best suited for very low discharges and high heads (low Q and high H). Axial pumps, shown in Figure 4.5, are best for the opposite conditions (high Q and low H). These pumps typically use propellers to displace water in a single axial direction.

Centrifugal pumps, shown in Figure 4.6, are most common and serve a wide range of head and discharge conditions. Centrifugal pumps displace water in a circular motion such that the centrifugal acceleration of the water causes a pressure increase in the radial direction. The water enters in the axial direction in the eye of the pump. The rotation of the impeller brings water into a spiral motion outward in the radial direction and a circular motion outside of the impeller. The water flows in a spiral casing and exits the pump in a circular pipe.

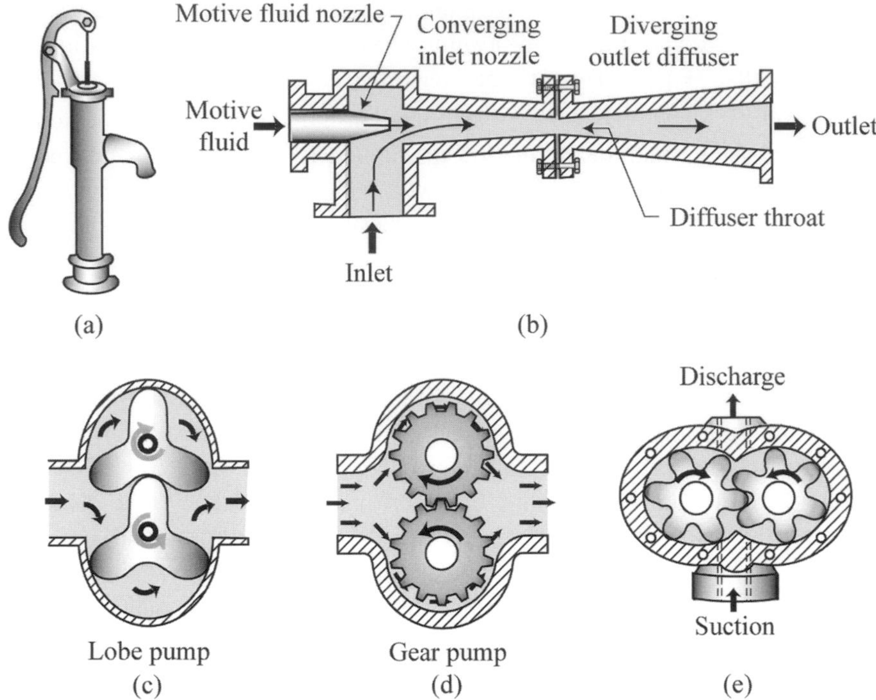

Figure 4.4 Positive-displacement, jet and gear pumps

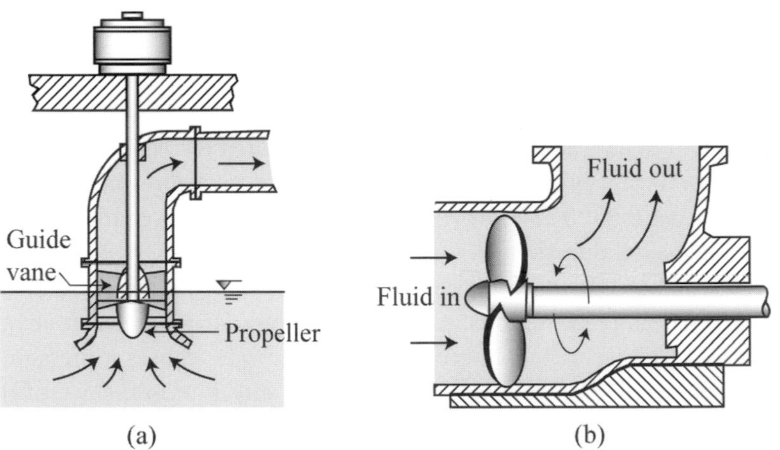

Figure 4.5 Axial pumps

We can recognize three velocity vectors in centrifugal pumps. First, the velocity of the blade $\vec{U}$ is simply the circular motion of the impeller given by the product of the rotational speed ω and the distance r from the axis of rotation. The vector $\vec{W}$ describes the velocity of the fluid relative to the blade. Obviously, the fluid cannot flow across the blade and the direction of the $\vec{W}$ vector is always in the direction of the blade. Finally, the third vector is the velocity of the fluid $\vec{V}$, which is the vectorial sum of the other two vectors $\vec{V} = \vec{U} + \vec{W}$. Therefore, $\vec{U}$ and $\vec{V}$ are two absolute velocity vectors in the fixed frame of reference while $\vec{W}$ is a relative velocity vector.

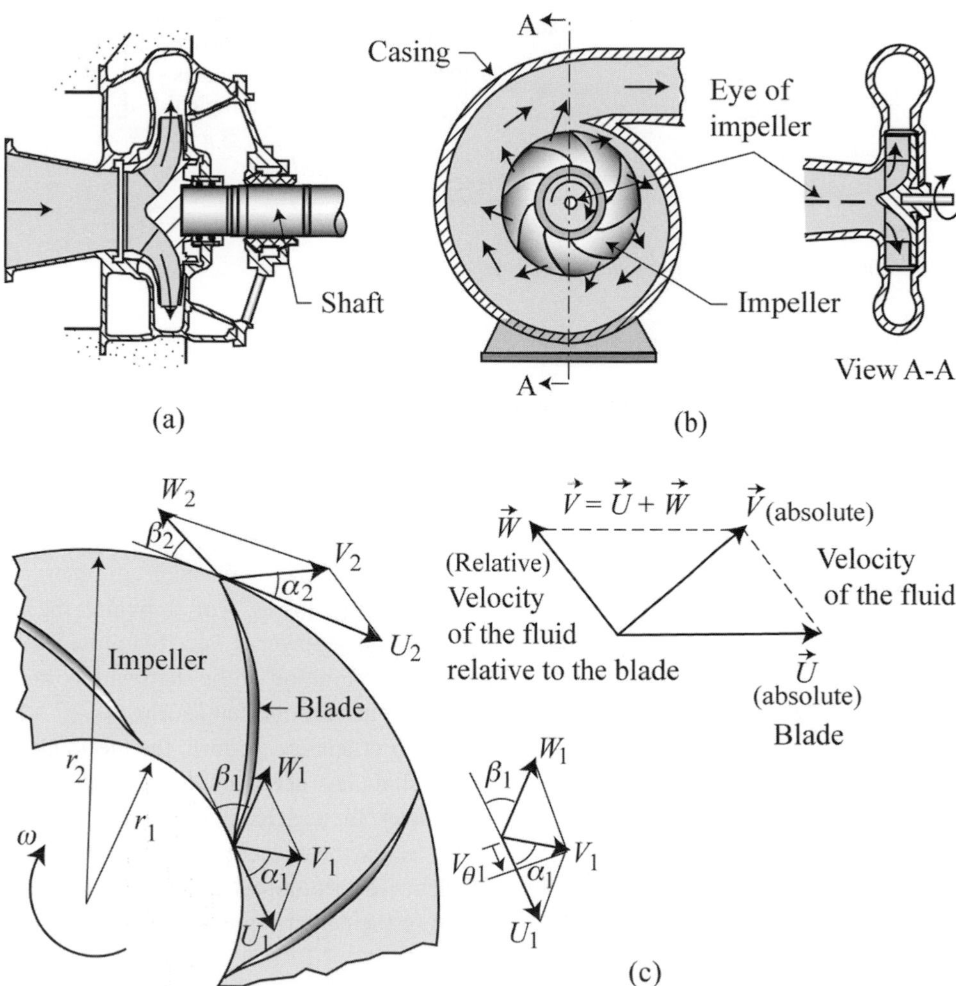

Figure 4.6 Centrifugal pumps

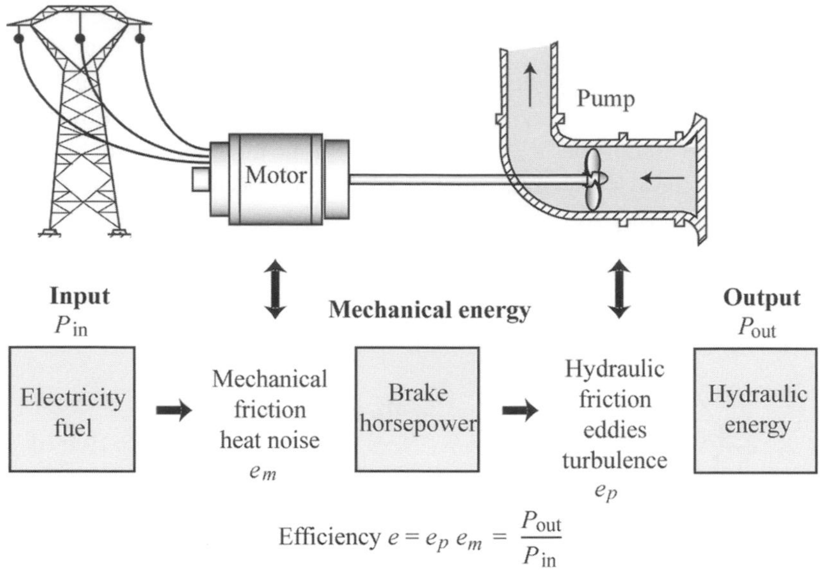

Figure 4.7 Energy conversion and efficiency

4.2 Pump Performance

Energy conversions and pump efficiency are covered in Section 4.2.1, followed by pump performance curves (Section 4.2.2), operating points (Section 4.2.3) and system design (Section 4.2.4).

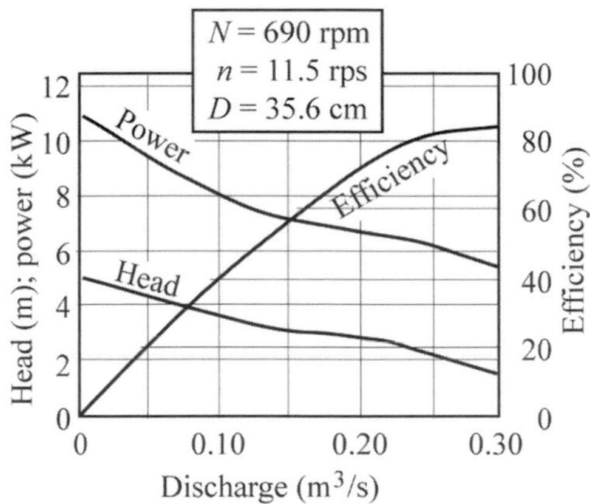

Figure 4.8 Pump 1 performance curve

4.2.1 Energy Losses

In Figure 4.7, the conversion of energy always implies energy losses. For a pump, the desirable hydraulic horsepower is P_{hp}. It is provided mechanically by the rotation of the shaft with corresponding brake horsepower (mechanical) as P_{hp}/e_p, where e_p is the pump efficiency. In turn, the mechanical energy of the shaft has been converted by the motor from the power input (electricity or fuel) as $P_{\text{hp}}/e_p e_m$, where e_m is the mechanical efficiency. The overall efficiency of a pump is the ratio of the power produced to the power input to the system, or $e = e_p e_m$.

4.2.2 Pump Performance Curves

Pump performance curves are provided by the manufacturer with the head H at the pump as a function of the discharge Q. Most pump curves will indicate the rotational speed N_{rpm} and the pump impeller diameter D. Note that the power P on the chart refers to the mechanical energy (brake horsepower), and thus the hydraulic efficiency e_p represents only the conversion of mechanical energy into hydraulic energy. A simple axial pump curve for sample pump 1 is shown in Figure 4.8.

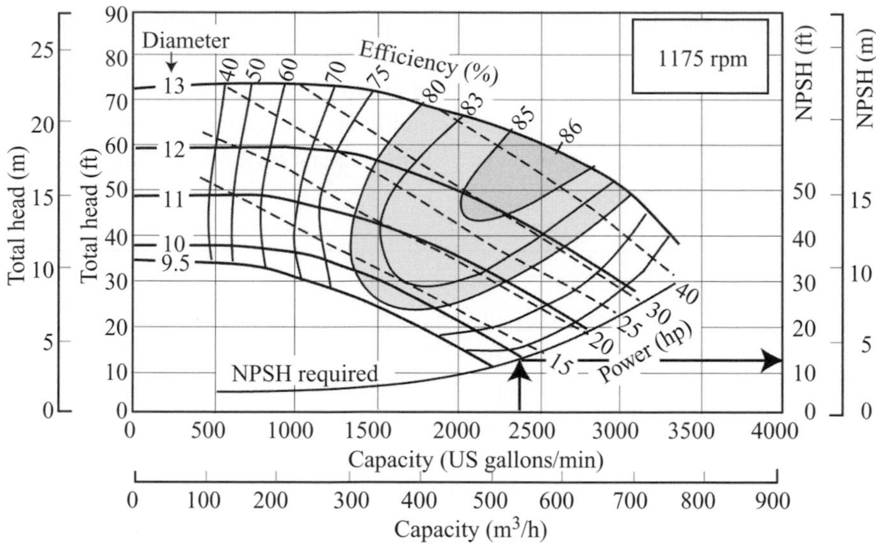

Figure 4.9 Set of pump performance curves for sample pump 2

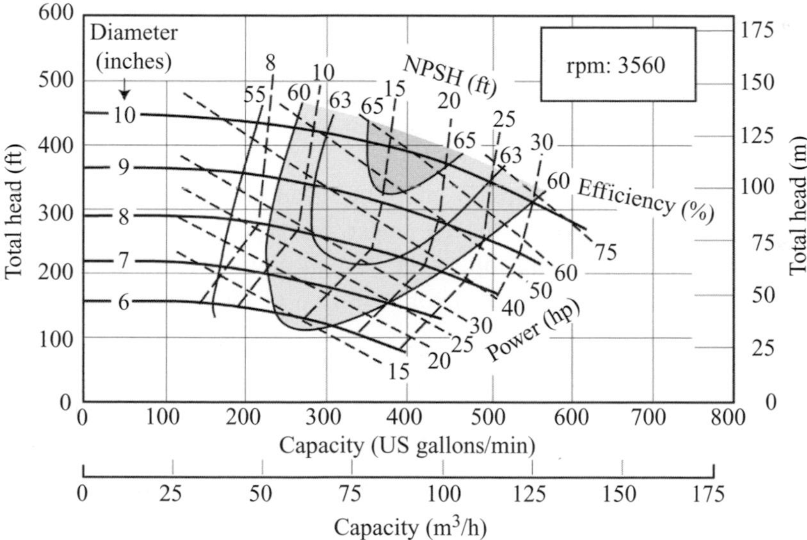

Figure 4.10 Set of pump performance curves for sample pump 3

In most cases, pump curves are displayed for a family of pumps with the same overall casing design but different impeller sizes. Sample pump 2 represents a typical pump type with a flat performance curve (low N_{rpm}) and a high variability in discharges and relatively low heads in Figure 4.9.

Sample pump 3 in Figure 4.10 has a typical steep performance curve (high N_{rpm}) with large heads and a relatively narrow range in discharges. In both curves, the impeller size is indicated as solid lines, the efficiency in shaded areas and the brake horsepower in dashed lines. The highest of these family of curves will usually have the highest efficiency.

When determining the specific speed of a pump, it is preferable to look at the conditions of the head and discharge where the efficiency is maximum. Clearly, pump 2 ($e_p = 86\%$) in Figure 4.9 is more efficient than pump 3 ($e_p = 66\%$). The pump performance curves from the manufacturer usually provide the net positive suction head (NPSH) to be used in Section 4.3. At this point, the NPSH increases with discharge. Example 4.4 shows how to read a pump curve.

♦ **Example 4.4:** Reading a pump curve

Consider pump 3 in Figure 4.10 with an 8-in. line.

Extract the following information from the pump performance curve:

(1) What is the total head at the pump for a discharge of 1 cfs = 450 gpm?
(2) What is the brake horsepower required to run the pump under those conditions?
(3) What is the hydraulic efficiency of this pump under those conditions?
(4) What is the NPSH for those conditions (this will be discussed in Section 4.3)?

Solution: The head is 200 ft, power 40 hp, efficiency 59% and the NPSH is 23 ft.

4.2.3 System Curve and Operating Point

A system curve describes the head at the pump as a function of the discharge. As a starting point, when $Q = 0$, the head at the pump equals the elevation difference ΔE between the two reservoirs. As the discharge increases, the head at the pump becomes $H = \Delta E + \sum kQ^2$ from the sum of friction losses in the pipe as discussed in Chapter 2. The system curve therefore plots the head at the pump as a function of flow discharge from the lower to the upper reservoir. We can then superpose the pump performance curve on top of the system curve, as shown in Figure 4.11. The operation point of a pump is at the intersection of the pump performance curve and the system curve.

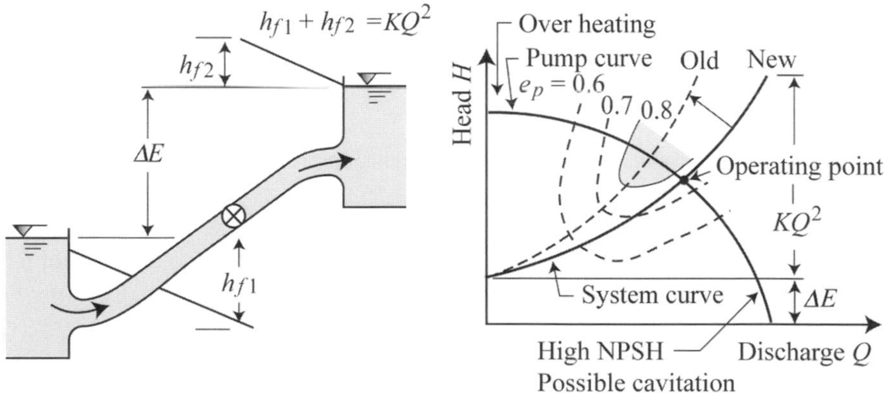

Figure 4.11 Sketch of the operation point

Proper design typically seeks a pump that will have an operating point slightly to the right of the area of maximum efficiency. The reason is that resistance to flow is expected to increase in aging pipes and the system curve should gradually shift upward with time. Accordingly, the operating point will move closer to the point of maximum efficiency of the pump over time. Operating the pump under a high head and a very low discharge will cause overheating of the pump. Conversely, operating the pump at a very high discharge and low head may cause cavitation. Example 4.5 illustrates how to determine the operation point of a hydraulic system.

♦♦ Example 4.5: Pump operating point

Assume $f = 0.015$, and find the operation point in this water system if the pump has the characteristics shown in Fig. E-4.5.

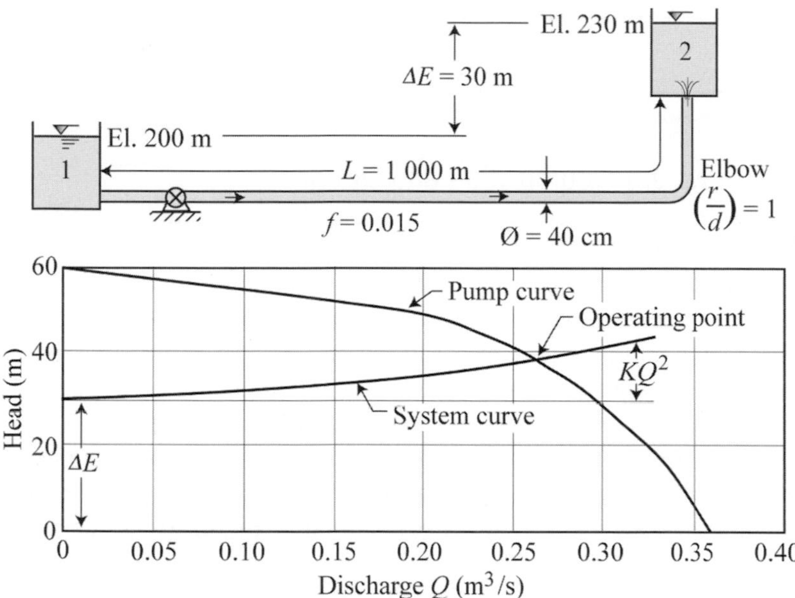

Fig. E-4.5 Pump curve, system curve and operation point

Solution: First, write the energy equation from 1 to 2:

$$\frac{p_1}{\gamma} + z_1 + \frac{V_1^2}{2g} + H = \frac{p_2}{\gamma} + z_2 + \frac{V_2^2}{2g} + \sum h_L.$$

With $z_1 = 200$ m and $z_2 = 230$ m constant in the reservoirs, p_1, p_2, V_1 and V_2 are all zero:

$$0 + 200 + 0 + H = 0 + 230 + 0 + \left(\frac{fL}{D} + K_e + K_b + K_E\right)\frac{Q^2}{2g}\left(\frac{4}{\pi D^2}\right)^2. \quad \text{(E-4.5.1)}$$

The entrance, elbow and exit coefficients are $K_e = 0.5$, $K_b = 0.35$, $K_E = 1.0$, and $H = 30 + \left(\frac{0.015\times1000}{0.4} + 0.5 + 0.35 + 1.0\right)\left[\frac{16Q^2}{2\times9.81\times\left(\pi\times0.4^2\right)^2}\right] = 30 + 127\ Q_{\text{cms}}^2.$

Table E-4.5 shows the discharge Q in m^3/s versus H in m. This defines the system curve plotted in Fig. E-4.5. When the pump performance curve is plotted on the same graph, the intersection point of the two curves defines the operating point at $Q = 0.27$ cms.

Table E-4.5. System curve calculation

Q (m^3/s)	$127Q^2$ (m)	$H = 30 + 127\,Q^2_{cms}$ (m)
0.0	0.0	30.0
0.1	1.3	31.3
0.2	5.1	35.1
0.3	11.4	41.4

4.2.4 Pump System Optimization and Design

We have now reached an important point in the analysis of hydraulic systems. In a pipe and pump system with fixed length, height and discharge, various pipe diameters can be considered. Obviously, the pipe cost increases with the pipe diameter, and the smaller pipe would in theory be desirable. However, a small pipe also increases head losses and would require a larger (and more expensive) pump. Per Figure 4.12, the total material cost (sum of pipe and pump costs) should reach a minimum, which would appear to be most desirable for our design. It is an enjoyable part of hydraulic engineering to find the optimal solution to these systems to maximize performance and/or minimize the costs. Nowadays, spreadsheets can be used given the cost of pipes and pumps (see Problem 4.9) to find an optimal design. The basics of cost analysis are outlined in Appendix A.

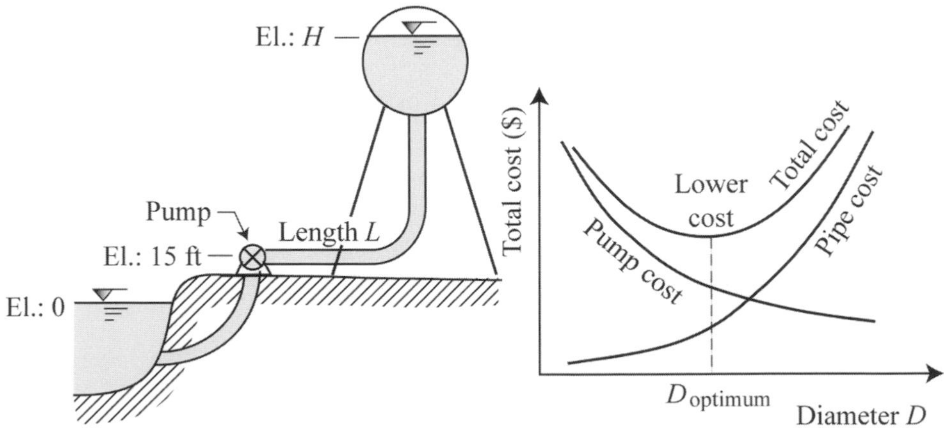

Figure 4.12 Pump and pipe material optimization

Other considerations in the selection of a pump include the ability to start the pump. Some pumps are submersible, which means that they can be placed below the water level. However, most motors must be located above the highest possible water level. Such pumps will require priming such that they can start and function properly. With air in the casing, pumps do not generate enough pressure and simply burn out when running dry for an extended period of time. For the same reason, it is important to make sure that the pressure on the suction sides does not get so low to turn water into vapor and cause cavitation. Motors with low efficiency typically generate a lot of noise and/or heat. Finally, running a pump under very low discharges is also a concern because the water simply turns around in the casing and the energy expenditure simply heats the water, which lowers the boiling point and can also cause cavitation. Finally, environmental engineers should consider: (1) the source of the materials (e.g. metal, concrete, etc.), and (2) how the materials will be disposed of after their service life.

4.3 Pump Cavitation

The cavitation process is described in Section 4.3.1 along with the NPSH in Section 4.3.2. Pumps can also be connected in series for very high heads or in parallel for high discharges (Section 4.3.3).

4.3.1 Cavitation

When the pressure inside a pump locally decreases below the vapor pressure, the water forms vapor bubbles inside the fluid. This typically occurs on the suction side of the blades near the eye of the impeller as sketched in Figure 4.13. The formation of vapor bubbles per se is not a major concern. However, as the pressure increases, the bubbles collapse as the vapor turns back into liquid. It is the presence of collapsing bubbles near the blades that is the major concern. Collapsing vapor bubbles generate local pressures as high as 300,000 psi. Such extreme pressures can pit the steel surface of the rotating blades inside the pump casing and lead to permanent damage to the pump. Cavitation can often be recognized by the crackling and rattling sound that is produced inside the pumps.

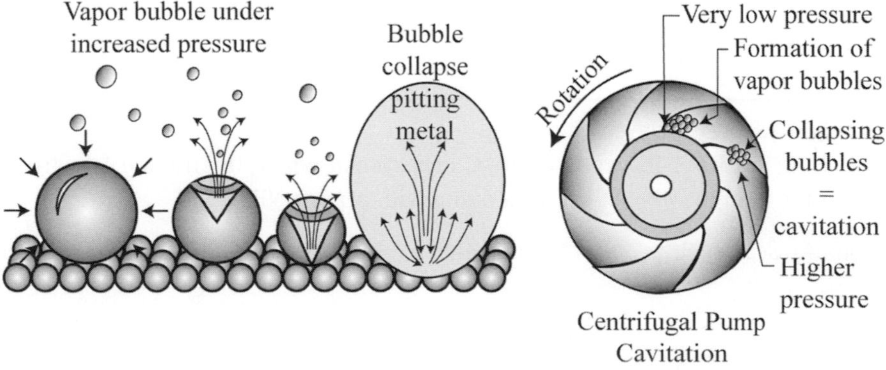

Figure 4.13 Cavitation damage near solid blades of a pump impeller

To prevent cavitation, it is important to make sure that the pressure near the intake of centrifugal pumps is sufficiently high to prevent the vaporization of the fluid. This can be done by considering the NPSH.

4.3.2 Net Positive Suction Head

It is very important to prevent cavitation of a pumping system. The criterion for pump design is called the net positive suction head (NPSH). The NPSH is equivalent to the velocity head at the core of the pump and it increases with the flow discharge in a pump. The NPSH of a pump is determined from the pump performance curve, as discussed in Example 4.4. For any given pump, a low NPSH value is always desirable.

The Bernoulli equation is applied between the intake level at point 1 where $V_1 = 0$ and $p_1 = p_{atm}$, and the location of the pump at point 2, such that $X = z_2 - z_1$:

$$\frac{p_1}{\gamma} + z_1 + \frac{V_1^2}{2g} = \frac{p_2}{\gamma} + z_2 + \frac{V_2^2}{2g} + h_f. \tag{4.3}$$

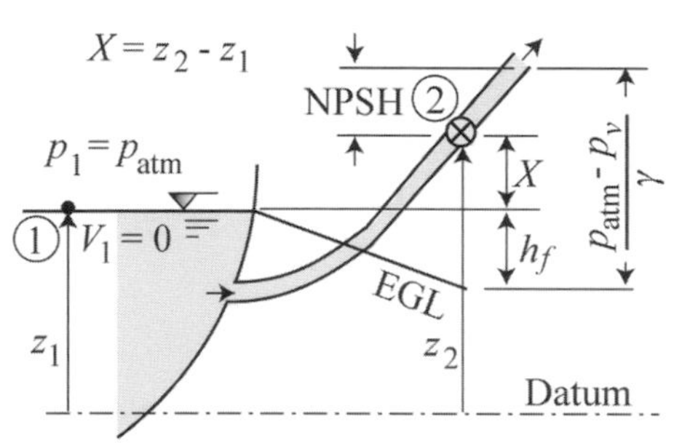

Figure 4.14 NPSH definition sketch

Rearranging gives

$$\frac{p_{atm}}{\gamma} - \frac{p_2}{\gamma} = h_f + X + \frac{V_2^2}{2g}.$$

The NPSH corresponds to the velocity head inside the pump. As discussed in Example 4.6, cavitation will occur when reaching vapor pressure $p_2 = p_v$ inside the pump, i.e. when $X = X_{max}$. Therefore, we obtain the important graphical relationship shown in Figure 4.14:

$$\frac{p_{atm} - p_v}{\gamma} = h_f + X_{max} + NPSH.$$

Note that the suction head is graphically positive. We thus calculate X_{max} as

$$X_{max} = \frac{p_{atm} - p_v}{\gamma} - h_f - NPSH. \tag{4.4}$$

A pump does not cavitate when X is below $X_{max} (X < X_{max})$, see Example 4.6.

♦♦ **Example 4.6:** Net positive suction head

Find X_{max} at sea level for water at 20 °C in a system where the friction loss from the intake to the pump is $h_f = 1.1$ ft for a pump with a 20-ft NPSH.

Solution: At sea level, $p_{atm} = 14.7$ psia and $p_v = 0.34$ psia at this temperature. From Eq. (4.4),

$$X_{max} = \frac{(14.7 - 0.34) \times 144}{62.4} - 1.1 - 20 = 12 \text{ ft}.$$

The pump lowest point of the impeller must remain lower than 12 ft above the intake water level.

4.3.3 Pumps in Series and in Parallel

Pumps in series increase the head at a given discharge as shown in Figure 4.15. Multistage pumps combine the characteristics of a series of pumps on a single axis of rotation, and thus require a single motor for all pumps. The pump requires a single intake but the entire pump must be switched off for maintenance.

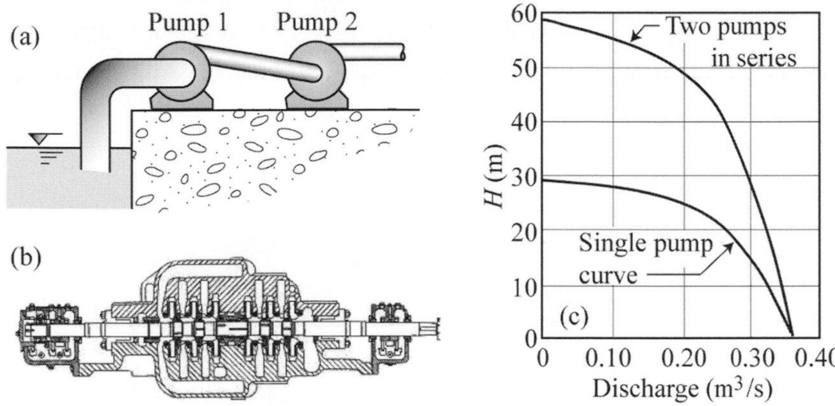

Figure 4.15 Pumps in series and a multi-stage pump

Pumps in parallel increase the discharge at a given head, as shown in Figure 4.16. The discharges from all units are combined into a manifold. The advantage of parallel pumps is that each unit can be turned off separately, which is very convenient for maintenance. The disadvantage is that each unit needs its own motor, intake and screen. Bellmouth intakes are designed to minimize the energy losses at the intake (Padmanabhan 1987).

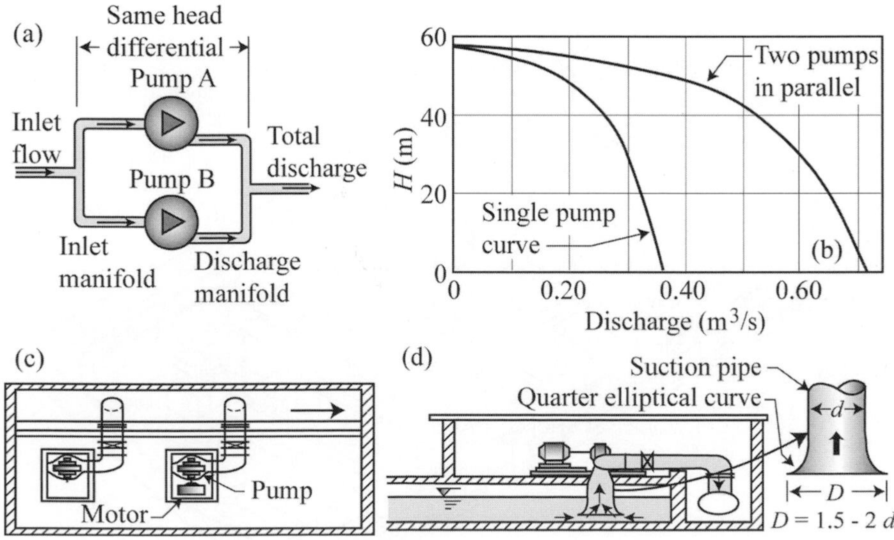

Figure 4.16 Pumps in parallel and a bellmouth intake

Case Study 4.1 provides details of a great example of multi-stage pumps in parallel.

♦♦ Case Study 4.1: The Edmonston Pumping Plant, USA

The A.D. Edmonston Pumping Plant is one of the unique features of the California Department of Water Resources (Deukmejian et al. 1985). Located near the south end of the California Aqueduct, it is a gigantic pumping station. To deliver water to Southern California, the pumps lift 4,410 cfs of water 1,926 ft over the Tehachapi Mountains. The total pumping power is 1,120,000 hp (835 MW) and the cost reached $152 million when constructed from 1967 to 1973.

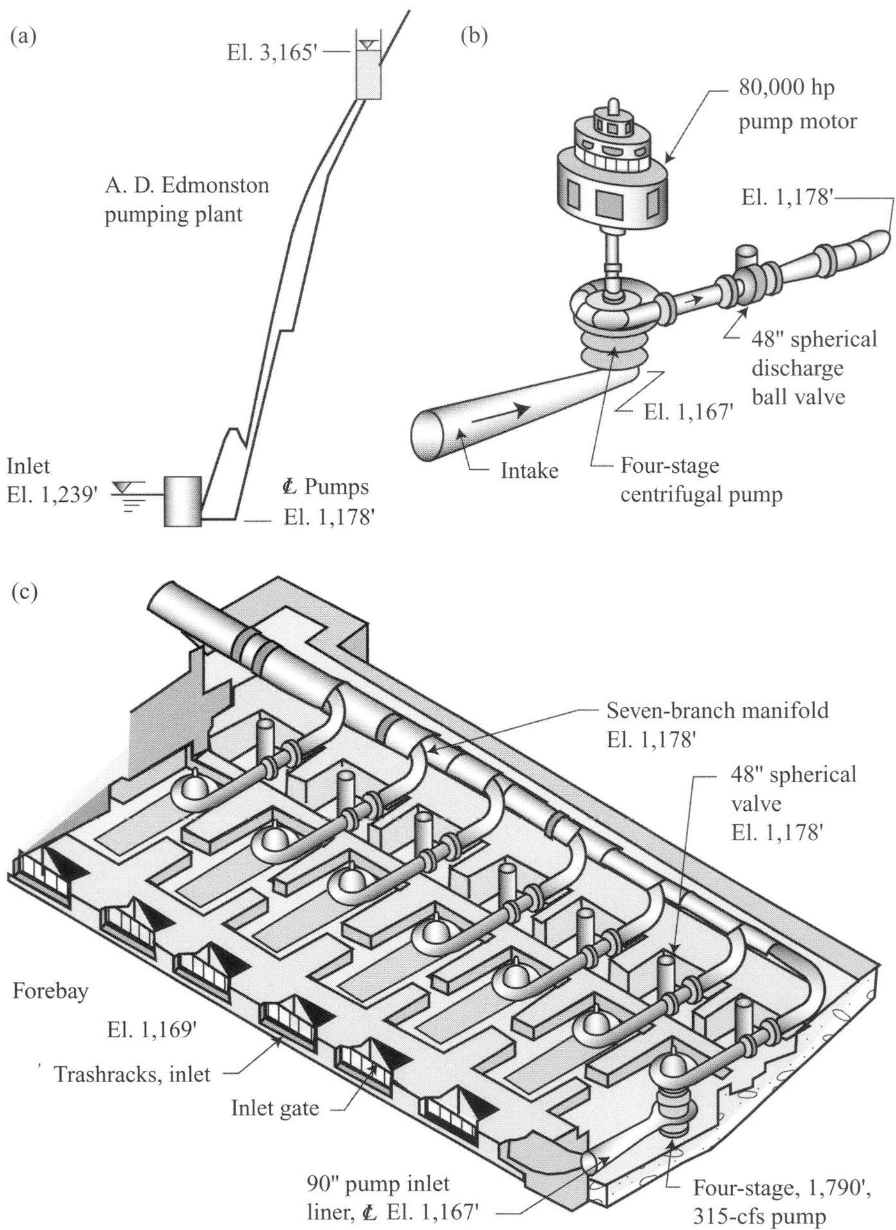

Fig. CS-4.1 Multi-stage pumps in parallel

The pipe system raises water from 1,239 ft at the forebay to a surge tank located at 3,165 ft for a net head of 1,926 ft, Fig. CS-4.1a. The water is pumped into two main pipes starting at the pump outlet at 1,178 ft up to 3,101 ft at the entrance slightly below the surge tank. The total head at the pump is 1,970 ft. At the top, the water enters a cylindrical surge tank 62 ft high and 50 ft in diameter. The normal water level in the surge tank is 3,165 ft with a maximum at 3,180 ft.

There are fourteen 80,000 horsepower (60 MW) centrifugal pumps. Each unit shown in Fig. CS-4.1b rotates at 600 rpm around a vertical axis and discharges 315 cfs of water. Each motor-pump unit stands 65 ft high and weighs 420 tons. Each 16-ft-diameter pump has four stages in series extending 31 ft high. Each pump was designed with an expected efficiency of 92.2% at a cost $1.6 million.

The pumps are housed in two galleries of seven units in parallel and connected into two separate manifolds with separate discharge pipes (Fig. CS-4.1c). The two main pipes are 8,400 ft long. The pipe diameters are 12.5 ft for the lower half and 14 ft for the upper half. Each pipe contains 1.2×10^6 ft^3 of water.

The pumps are started with a motor generator and the starting load is several times the normal running load. Considering the close proximity to the San Andreas fault, the entire structural design included a 0.5g seismic acceleration. At the base, the 65-ton valves next to the pumps control the flow and are designed to withstand a 1,700-psi pressure. Closure of the ball valve under a 500-psi pressure is done in two steps with 80% closure in 10 s and the remainder in 20 s. At a first-stage specific speed of 7,000, the submergence of the impeller had to be 71 ft below the forebay inlet level to prevent cavitation. At the mountain top, the surge tank prevents tunnel damage when the valves to the pumps are open or closed.

EXERCISES

These exercises review the essential concepts from this chapter.

1. What is the purpose of a pump?
2. Explain the power conversion process. What causes energy losses?
3. Why do we need the specific speed of a pump?
4. What are the main attributes of axial pumps?
5. What are the main attributes of radial pumps?
6. Which pump curve best fits the casing?
7. Where does the system curve come from?
8. Why do you calculate the NPSH?
9. What does cavitation do in a pump?
10. What does a multi-stage pump really do?
11. Consider Figure 4.3a; would it be possible to increase N_s and the efficiency of pump 3 (450 ft at 450 gpm) by using a faster motor? Is this a good idea and why?
12. To improve the efficiency of pump 3, would it be better to place pumps in series or in parallel?
13. True or false?
    (a) The centrifugal action increases the pressure in a pump.
    (b) Axial pumps are good at high heads and low discharge.

(c) The pump performance depends on discharge.
(d) The system curve defines the NPSH.
(e) The NPSH of a pump decreases with discharge.
(f) It is the formation of vapor bubbles that causes cavitation.
(g) The first stage of a pump should be placed as low as possible.
(h) Pumps in series increase the discharge.
(i) Bellmouth intakes minimize entrance losses.
(j) A pump should always be placed below X_{max}.

SEARCHING THE WEB

Can you find more information about the following?

1. ♦ Positive displacement pumps.
2. ♦ Axial pumps.
3. Deep-well pumps.
4. ♦ Centrifugal pumps.
5. ♦♦ Axial pump curves.
6. ♦♦ Centrifugal pump curves.
7. ♦♦ Pump cavitations.
8. ♦ Multi-stage pumps.

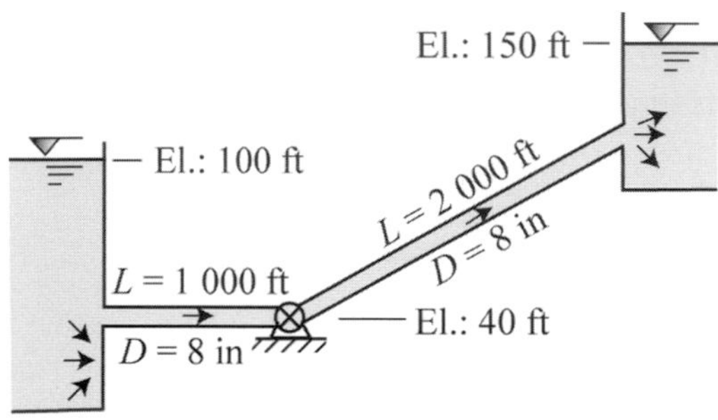

Fig. P-4.1

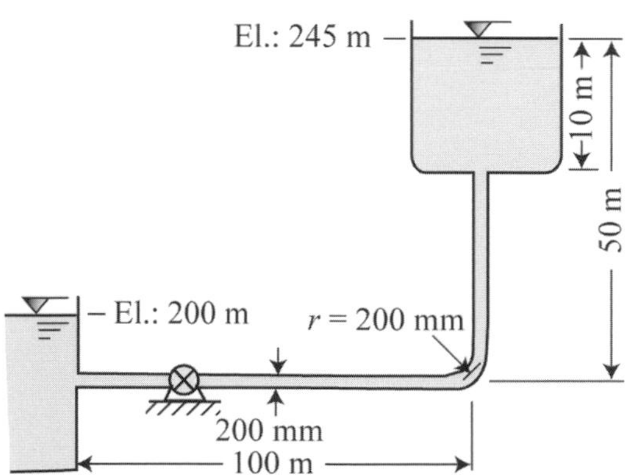

Fig. P-4.2

PROBLEMS

1. ♦ In Fig. P-4.1, determine the power required to pump 2.5 cfs from the lower to the upper reservoir. Assume a friction factor $f = 0.015$ and sketch the EGL.
2. ♦ What power must be supplied by the pump in Fig. P-4.2 if water ($T^\circ = 20°C$) is pumped through a 200-mm-diameter steel pipe from the lower tank to the upper one at a rate of 0.314 cms? Draw the EGL and HGL. [Hint: find f from the Moody diagram.]
3. ♦♦ In Fig. P-4.3 at a flow rate of 0.25 m^3/s and head losses totaling $fL/D = 3.5$, determine the power required at the pump and plot the EGL.
4. ♦♦ For the pump curves 1 to 3, look at conditions near maximum efficiency. Estimate the specific speed and locate each pump on the classification charts.
5. ♦ What type of pump at $N = 1{,}500$ rpm should be used for a discharge of 12 cfs and head of 25 ft.
6. ♦♦ For the pump in Fig. P-4.6, you consider replacement of a burned motor with a faster motor. What is the maximum head that can

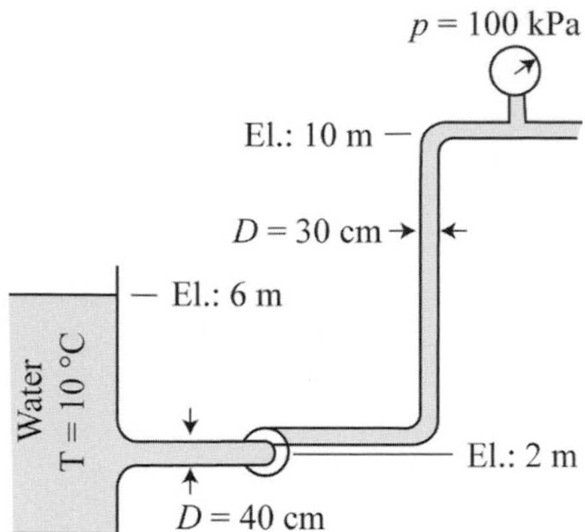

Fig. P-4.3

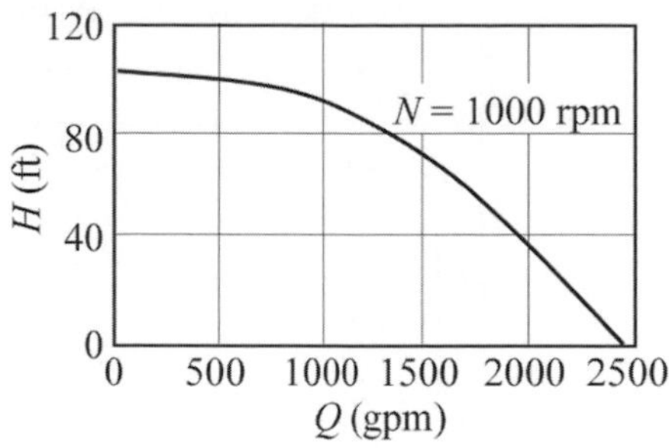

Fig. P-4.6

be generated if the speed is increased to 1,500 rpm? Also, determine the increase in discharge at a given head for the same condition. Finally, would this increase or decrease the potential for cavitation, or NPSH? [Hint: look at the constant values of C_H and C_Q.]

7. ♦♦ For the system in Fig. P-4.7 with the known pump curve, assume $f = 0.02$ and neglect other minor losses. Determine the head losses as a function of discharge and plot the performance on the diagram below. Determine the discharge and head at the pump when the pipe diameter is 16 in. What type of pump would you recommend? If the NPSH is 8 ft for the 16-in. line, what is the maximum elevation X_{max} for this pump. Repeat with a pipe diameter of 24 in.

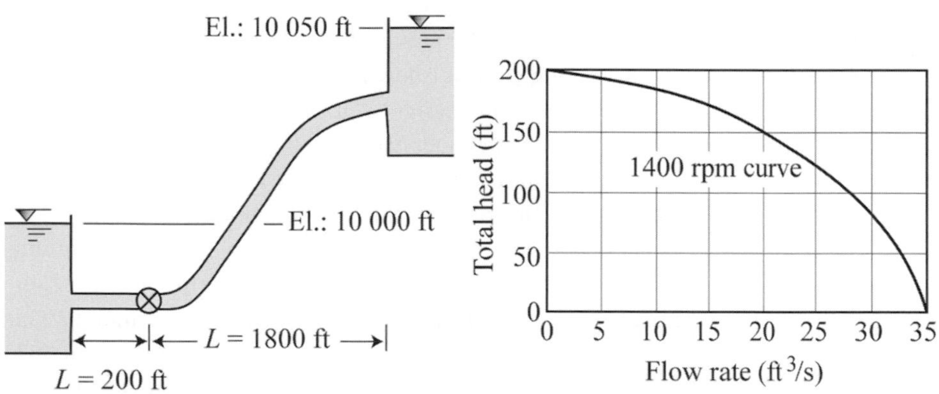

Fig. P-4.7

8. ♦♦ Assume $f = 0.02$ and find the discharge under the conditions shown in Fig. P-4.8 with the given pump characteristics. If the motor is located at an elevation of 23 m, which elevation (propeller or motor) should be considered for the NPSH calculation?

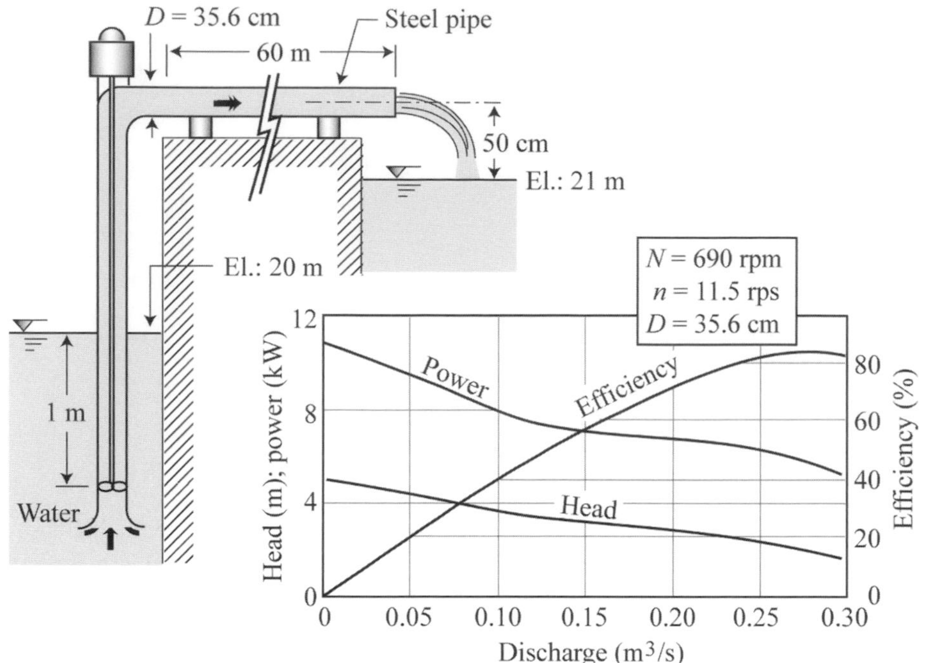

Fig. P-4.8

NB: Problems 9 and 10 are most important!

9. ♦♦♦ Optimization Problem

You live in a district with a water consumption of around 400 liters per person per day. This corresponds to a need for $Q = 0.125$ m^3/s, or 4.42 ft^3/s of water pumped 75% of the time. At a 70% efficiency, the pump brings the water $H = 120$ ft above the source in a $L = 2{,}500$-ft-long pipe as shown in Fig. P-4.9. Your company (JJ Engineering) is asked for a preliminary design of a single pipe and pump system. Your supervisor Jan suggests that you assume $f = 0.02$ and neglect minor losses. Use a spreadsheet to solve for various pipe diameters.

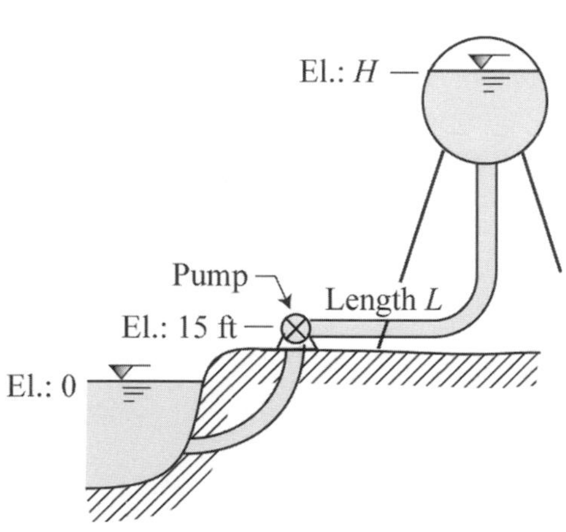

Fig. P-4.9

(a) Jan asks you to estimate the optimum materials cost of each alternative based on a 2014 cost table for galvanized iron pipe: 2 in. ($6/ft), 3 in. ($10/ft), 4 in. ($25/ft), 6 in. ($35/ft), 8 in. ($60/ft), 10 in. ($90/ft), 12 in. ($125/ft), 15 in. ($200/ft), 18 in. ($320/ft), 20 in. ($360/ft), 24 in. ($500/ft), 30 in. ($720/ft), 36 in. ($900/ft), 42 in. ($1,250/ft), and 48 in. ($1,600/ft). The cost of pumps is ~$200 per horsepower, and there is an assumed fixed base cost of $150,000 for the construction. Assume 5% annual inflation rate on all prices.

(b) For the first meeting with your client, Jan has to travel and she asks you to meet your very important client and elected representative Jill. At the meeting, Jill asks you to prepare a two-page summary of your recommendation for presentation at the next public meeting.

10. ♦♦♦ Analytical Solution

The pump fills up the 100-ft-diameter tank through an 800-ft-long, 2-ft-diameter pipe. The pump curve in Fig. P-4.10 can be approximated as $H = 155 - k_p Q^2$. Assume $f = 0.02$ and neglect minor losses to determine the following.

(a) Find the constant k_p of the pump curve.
(b) Find k_s of the system curve $H = h + k_s Q^2$.
(c) Solve both curves to find the operating point.
(d) How long does it take to fill the tank up to an elevation of 255 ft?
(e) What kind of problem emerges between elevations of 250 and 255 ft?

[*Hint: analytically solve the two equations to find the operating point. Also write* Q *as a function of the tank area and the water level* h *in the tank, and integrate* $dt = A_T dh/Q$ *given that* Q *decreases as* h *increases.*]

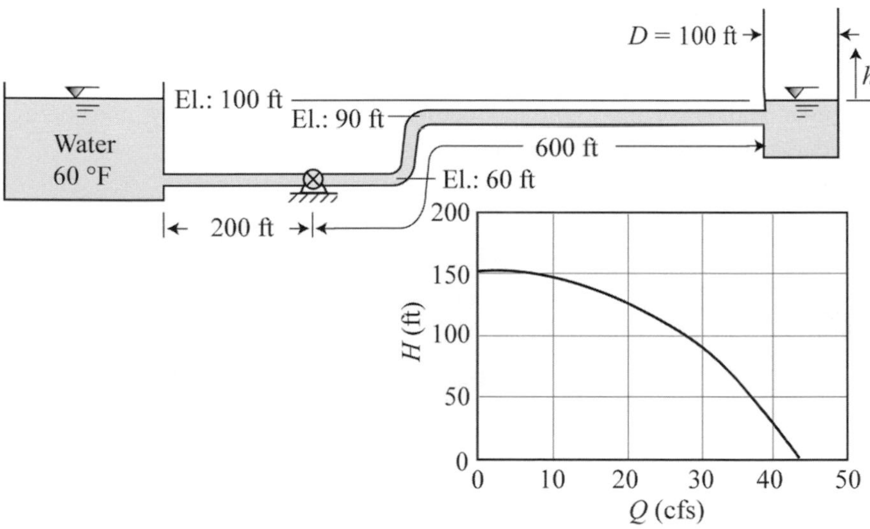

Fig. P-4.10

5 Turbines

Hydraulic turbines generate electricity. This chapter explains hydropower in Section 5.1, turbine types in Section 5.2 and turbine cavitation in Section 5.3.

5.1 Hydropower

The fundamentals of hydropower production are discussed in Section 5.1.1, with the minimization of energy losses explained in Section 5.1.2.

5.1.1 Hydropower Production

In a way, a turbine is the reverse of a pump, and hydropower production consists of converting hydraulic energy into electricity. Figure 5.1 illustrates the various losses from a reservoir to the consumer. Hydraulic losses are found in the intake and penstock. The turbine converts hydraulic energy first into mechanical energy. The generator then converts mechanical energy into electricity. Further electrical losses appear in the transformers and power lines. The hydropower potential is $P = \gamma HQ$, and the hydropower generated $e_t \gamma HQ$ depends on the turbine efficiency e_t.

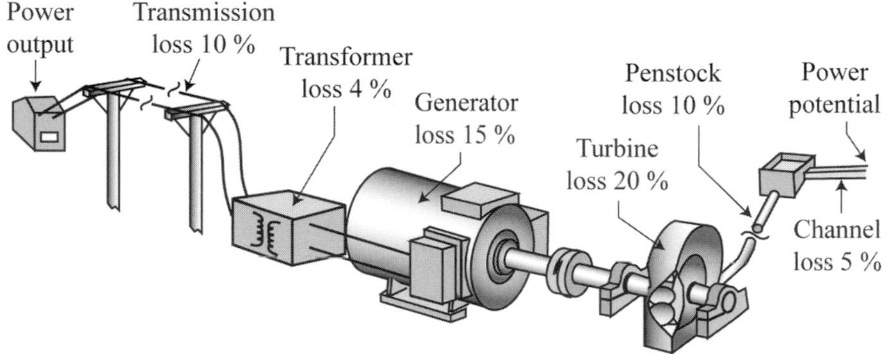

Figure 5.1 Schematic of hydropower production and energy losses

One constraint for hydropower production is the necessity to synchronize the generator to the frequency of the grid. The frequency F of the electric network is measured in cycles per second or hertz, with $F = 50$ Hz in Europe and $F = 60$ Hz in the USA. The electric signal from the generator is linked to the rotational speed N_{rpm} of the turbine. As sketched in Figure 5.2, the two poles correspond to 1 Hz (1 Hz = 1 cycle per second), or $N_{\text{rpm}} =$ 60 rpm. Therefore, $N_{\text{rpm}} = 120\, F/p_0$ where p_0 is the number of poles in the generator.

Note that the poles must be in pairs, e.g. two, four, etc. Increasing the number of poles of a generator is beneficial as it decreases the rotational speed of the turbine.

The angular velocity $\omega_{rad/s}$ of the turbine in radians per second depends on the rotational speed N_{rpm} in rotations per minute (rpm) from

$$\omega_{rad/s} = \frac{N \text{ rotations}}{\text{min}} \cdot \frac{\text{min}}{60 \text{ sec}} \cdot \frac{2\pi \text{ rad}}{\text{rotations}} = \frac{\pi}{30} N_{rpm}.$$

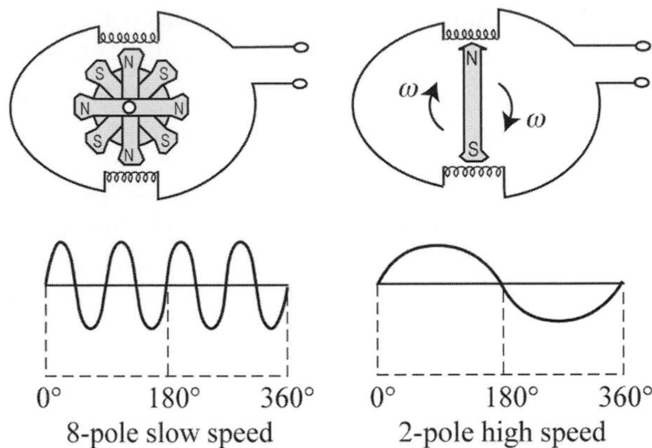

Figure 5.2 Number of poles in a generator

For instance, a generator with 12 pairs of poles ($p_0 = 24$) in the USA ($F = 60$ Hz) rotates at

$$N_{rpm} = \frac{120F}{p_0} = \frac{120 \times 60}{24} = 300 \text{ rpm},$$

and

$$\omega = \frac{\pi \times 300}{30} = 31.4 \text{ rad/s}.$$

Energy losses correspond to revenue losses, and it is usually important to seek ways to maximize the efficiency of the hydropower operations.

5.1.2 Minimizing Hydraulic Losses

Maximizing hydropower production is equivalent to minimizing energy losses. As shown in Figure 5.3, the hydraulic losses from the reservoir to the powerhouse can be

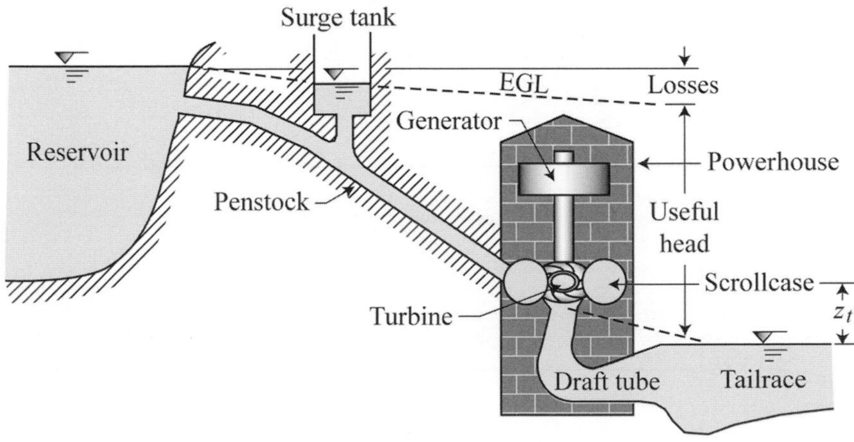

Figure 5.3 Hydraulic losses of a power plant

reduced by enlarging the penstock cross section. Large draft tubes and tailrace areas also reduce the energy losses at the outlet.

Case Study 5.1 illustrates how to optimize the dimension of the penstock by balancing the gain in useful energy against the cost of excavating a larger tunnel.

♦ Case Study 5.1: Optimizing hydropower at Cambambe Dam in Angola

Palu et al. (2018a, b) worked on Cambambe Dam in Angola shown in Fig. CS-5.1. The power plant involved the excavation in bedrock of a penstock between the intake and the powerhouse. The potential head between the reservoir and the Francis turbine outlet is $H_0 = 113.7$ m, the length of the excavated tunnel is $L = 500$ m and the flow discharge is $Q = 100\ \text{m}^3/\text{s}$. To optimize the hydropower revenue, you want to minimize the friction losses. This can be achieved by excavating a larger tunnel to reduce the flow velocity, at the expense of an increase in the cost of bedrock excavation. For the calculations, assume high friction losses in excavated bedrock with $f = 0.03$. Consider the value of electricity in 2012 at \$100 per MWh (megawatt-hour), or \$0.10 per kWh (kilowatt-hour), and the cost of bedrock excavation is \$300 per m^3 (note that the bedrock excavation diameter is 50 cm larger than the circular penstock diameter). It is now possible to determine the optimal penstock diameter that would maximize the revenue of this project over a 30-year period.

For example, a $D = 10$-m tunnel costs $C_\$ = 300AL = 300 \times 0.25\pi D^2 L = \$11.8$ million. The velocity in the 9.5-m penstock is $V = Q/0.25\pi(D - 0.5)^2 = 1.41$ m/s and the corresponding friction loss is $H = [0.03L/(D - 0.5)]\left[V^2/2g\right] = 0.167$ m. The power loss is $P = \gamma HQ = 9.81 \times 0.167 \times 100 = 155$ kW, or 40 GWh (\$4 million loss) in 30 years. In

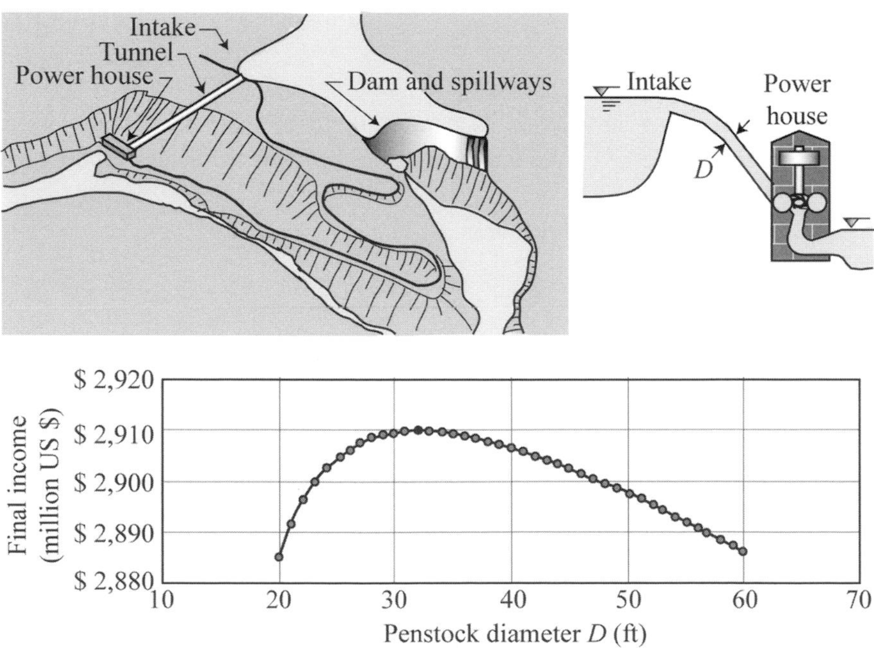

Fig. CS-5.1 Sketch of the tunnel excavation optimal diameter

comparison, a $D = 7$-m tunnel costs \$5.8 million, but the revenue loss from a 1.07-m head loss would be \$27 million. This is a $(5.8+27) - (11.8+4) =$ \$17 million difference over 30 years. Notice the large decrease in revenue from such a small difference in head loss.

Looking back at the Moody diagram, if the roughness height for this excavated penstock ranges between 0.25 ft and 0.5 ft, does it make sense to use $f = 0.03$ for the calculations? [The answer is that the roughness would need to be decreased to less than 0.1 ft to justify using $f = 0.03$ because f is closer to 0.05 when $\varepsilon/D = 0.02$.]

5.2 Turbine Types

Section 5.2.1 presents turbine classification diagrams followed with different turbine types: Pelton wheels (Section 5.2.2), Francis turbines (Section 5.2.3), Kaplan turbines (Section 5.2.4) and bulb turbines (Section 5.2.5).

5.2.1 Turbine Classification

Turbines are classified according to their specific speed N_s.

$$N_s = \frac{N_{\text{rpm}} P_{\text{hp}}{}^{1/2}}{H_{\text{ft}}{}^{5/4}}, \tag{5.1}$$

where N_{rpm} is the rotational speed of the turbine in rotations per minute, P_{hp} is the hydropower of the turbine in horsepower, and H_{ft} is the turbine head in feet. Note the difference between the specific speed of a turbine compared to a pump, and follow the calculations shown in Examples 5.1 and 5.2.

♦♦ **Example 5.1:** Turbine power

Calculate the specific speed for a turbine with $N = 600$ rpm, $Q = 0.5$ cms $= 17.6$ cfs $=$ 7,950 gpm and $H = 1.8$ m $= 6$ ft.

Solution:

$$P = \frac{\gamma HQ}{550} = \frac{62.4 \times 6 \times 17.6}{550} = 12 \text{ hp}$$

and

$$N_s = \frac{N_{\text{rpm}} P_{\text{hp}}{}^{1/2}}{H_{\text{ft}}{}^{5/4}} = \frac{600 \times 12^{1/2}}{6^{5/4}} = 221.$$

Example 5.2: Turbine specific speed

An SI turbine classification with $N_{sq} = \frac{N_{\text{rpm}} Q_{\text{cms}}{}^{1/2}}{H_{\text{m}}{}^{3/4}}$ is based on N_{rpm} in rpm, discharge Q_{cms} in m^3/s and head H_{m} in m. This corresponds to a specific speed $N_{sq} = 1.22 N_s$.

Considering $Q = 3$ cms $= 106$ cfs, $H = 65$ m $= 213$ ft and $N = 1{,}200$ rpm, we find

$$P = \frac{\gamma H_{\text{ft}} Q_{\text{cfs}}}{550} = \frac{62.4 \times 213 \times 106}{550} = 2{,}560 \text{ hp},$$

$$N_s = \frac{N_{\text{rpm}} {P_{\text{hp}}}^{1/2}}{{H_{\text{ft}}}^{5/4}} = \frac{1{,}200 \times 2{,}560^{1/2}}{213^{5/4}} = 75,$$

and

$$N_{sq} = \frac{N_{\text{rpm}} {Q_{\text{cms}}}^{1/2}}{{H_{\text{m}}}^{3/4}} = \frac{1{,}200 \times 3^{1/2}}{65^{3/4}} = 91,$$

and we note that the parameters and exponents for both N_s and N_{sq} are different.
The specific speed in Eq. (5.1) is commonly used in practice.

Most efficient turbine types can be determined as a function of turbine head and specific speed as shown in Figure 5.4.

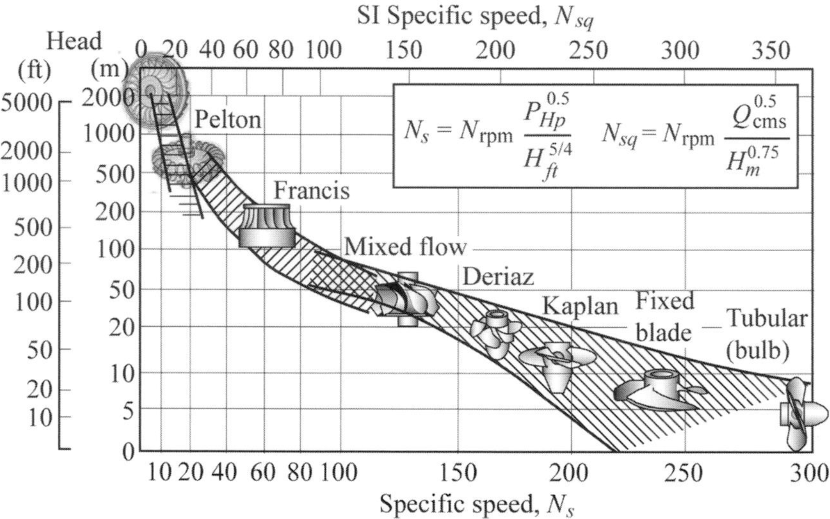

Figure 5.4 Turbine classification from the head and specific speed

The turbine efficiency shown in Figure 5.5 is the rate of conversion of usable hydraulic energy into mechanical energy at the turbine shaft. The efficiency of turbines decreases significantly at discharges lower than the design discharge. For instance, Pelton, Kaplan and Francis turbines maintain a high efficiency as the discharge significantly decreases. However, the efficiency of propeller turbines decreases considerably at discharges less than 75% of the design discharge.

Figure 5.6 shows a turbine classification as a function of the head at the turbine and the hydropower generated. Pelton, Francis, Kaplan and bulb turbines are most effective for small and large hydropower projects (USACE 1985). Deriaz turbines can also be considered in the range between Francis and Kaplan turbines. With the development of mini and micro turbines, Turgo and Banki-Michell can be used for moderate to high heads. Tubular turbines and bulbs are used for a wide range of discharges at low heads. Of the various types, more details and examples are provided for Pelton, Francis, Kaplan and

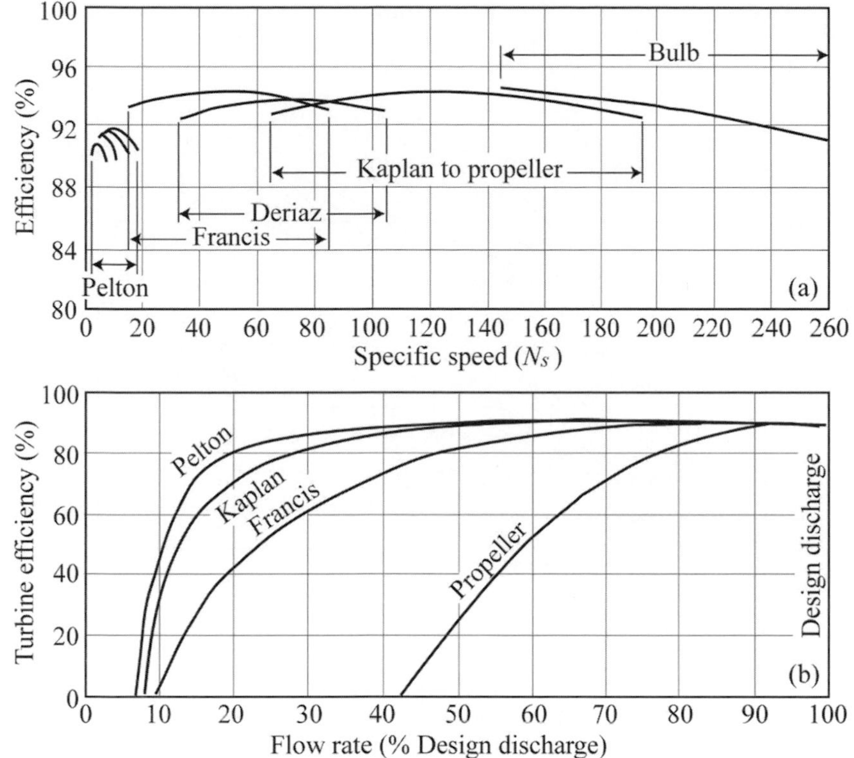

Figure 5.5 Turbine efficiency as a function of turbine type and specific speed

bulb turbines in the following sections of this chapter. These four types have distinct and essential characteristics that are important while others tend to be hybrids of the main types discussed here. The readers can refer to ESHA (1998) for small turbines and/or search the web for additional information on Deriaz, Turgo and Banki-Michell turbines.

5.2.2 Pelton Wheels or Impulse Turbines

In the late nineteenth century, Lester Pelton from Ohio developed the first Pelton wheel. He later participated in the California gold rush. Pelton wheels, also called impulse turbines, are commonly found in mountain areas because they are very efficient for high heads and low discharges. The potential energy is converted into kinetic energy.

As sketched in Figure 5.7, a very fast water jet impinges on a bucket designed to split and deflect the flow. The absolute velocity of the jet is V_j and the absolute velocity of the bucket is V_B. The momentum equation is applied to a control volume moving with the bucket. The force on the bucket is

$$F_B = \rho A_j \left(V_j - V_B\right)^2 (1 + \cos\theta).$$

The force on a single bucket is maximized when $\theta \to 0°$ to give

$$F_B = 2\rho A_j \left(V_j - V_B\right)^2.$$

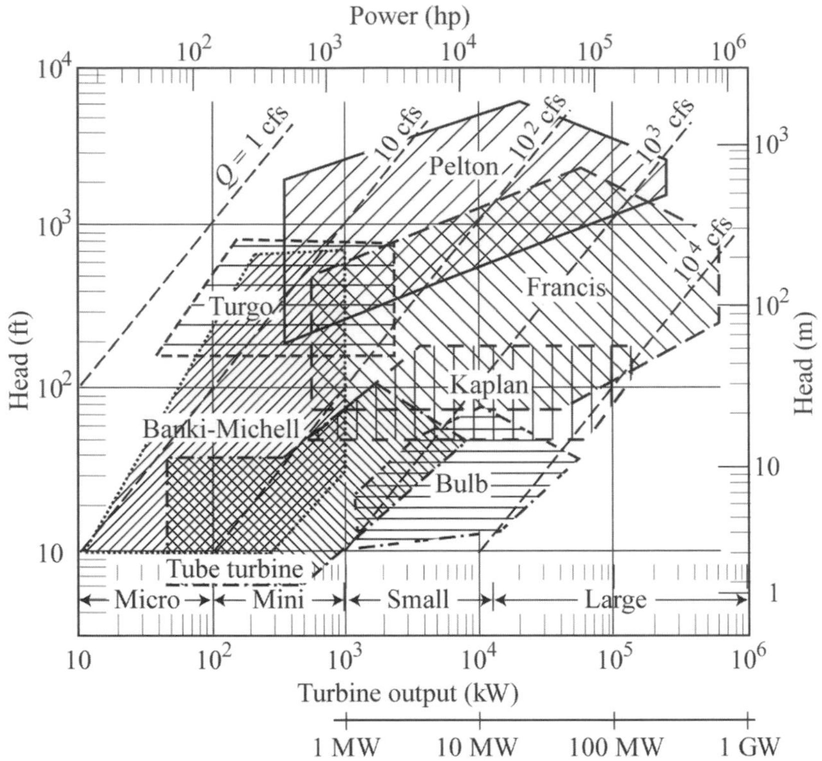

Figure 5.6 Turbine classification chart as a function of head and power (modified after USACE 1985 and ESHA 1998)

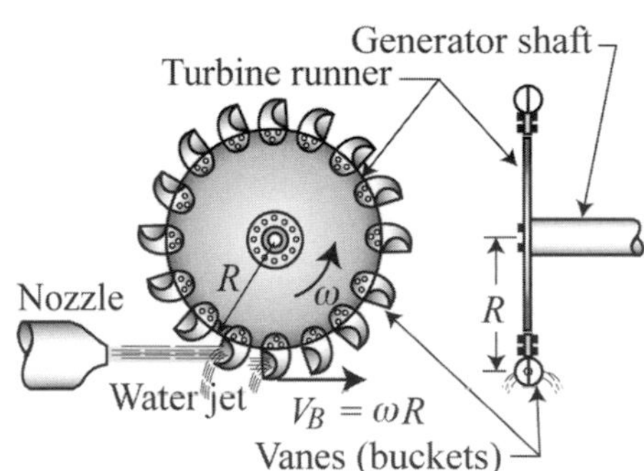

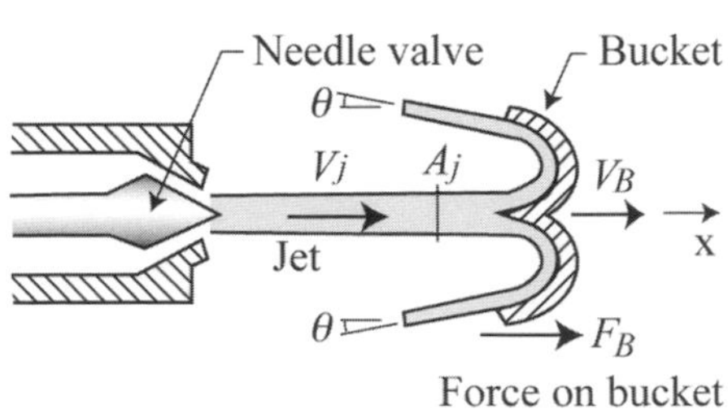

Figure 5.7 Pelton wheel

The quantity $A_j(V_j - V_B)$ represents the discharge of water impacting a single bucket. It is less than the discharge from the nozzle of the turbine. However, if we place the buckets sufficiently close to one another, the water missing the first bucket will impact the next one and the entire flow discharge A_jV_j will transfer momentum to the turbine. Therefore, the total force acting on the buckets is

$$F_B = 2\rho A_j V_j (V_j - V_B). \tag{5.2}$$

The power P is the product of the speed of the bucket and the force acting on it,

$$P = V_B F_B = 2\rho A_j V_j (V_j - V_B) V_B$$
$$= 2\rho A_j (V_j^2 V_B - V_j V_B^2).$$

Note that if $V_B = 0$ or if $V_B = V_j$, the power will be zero. The optimum bucket speed to maximize power is found between these two limits. This speed is determined by differentiating P with respect to V_B and setting the result equal to zero:

$$\frac{dP}{dV_B} = 2\rho A_j\left(V_j^2 - 2V_jV_B\right) = 0.$$

Solving this equation for V_B yields $V_B = 0.5\,V_j$ to generate the maximum power

$$P = V_BF_B = 2\rho\left(A_jV_j\right)\frac{V_j}{2}\left(\frac{V_j}{2}\right) = \rho gQ\left(\frac{V_j^2}{2g}\right) = \gamma QH.$$

This is the potential power γQH because the total head of the jet is $H = V_j^2/2g$. The torque T about the generator shaft is obtained from

$$P = \omega T = \omega\left(\vec{R}\times\vec{F}\right) = V_BF_B. \tag{5.3}$$

From a practical standpoint, the jet is usually turned less than 180° because of interference between the exiting and the incoming jets. Experience suggests: (1) the optimum angle $\theta \simeq 15°$; (2) the speed ratio $V_B \simeq 0.45\,V_j$; and (3) the efficiency (hydraulic power to brake power) of large impulse turbines is near 90%.

Consider Example 5.3 for a detailed analysis of a Pelton wheel.

♦ **Example 5.3:** Pelton turbine design

What power in kilowatts can be developed by the Pelton wheel shown in Fig. E-5.3 for an elevation drop of 670 m if the turbine efficiency is 85%? The 6-km-long and 1-m-diameter penstock has an assumed friction factor $f = 0.015$ and we neglect the nozzle head loss. Find the optimum angular speed $V_{jet} = 2\,V_B$ of the 3-m-diameter wheel for an 18-cm jet diameter. Finally, find the torque on the turbine shaft.

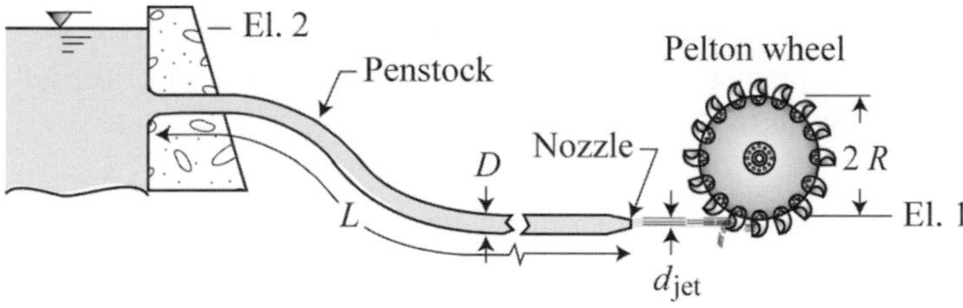

Fig. E-5.3 Sketch for a Pelton wheel

Solution: At the nozzle, the jet area is $A_{jet} = 0.25\pi(0.18)^2 = 0.02545\text{ m}^2$. The penstock area is $A_{pen} = 0.25\pi = 0.785\text{ m}^2$ and the flow velocity in the penstock is $V_{pen} = V_{jet}\left(A_{jet}/A_{pen}\right) = 0.0324V_{jet}$.

The friction losses are

$$h_L = \frac{fL}{D}\left(\frac{V_{pen}^2}{2g}\right) = \frac{0.015\times 6{,}000}{1}\left[\frac{\left(0.0324V_{jet}\right)^2}{2g}\right] = 0.0945\frac{V_{jet}^2}{2g}.$$

The jet velocity 670 m $= 1.0945 \frac{V_{jet}^2}{2g}$ results in

$$V_{jet} = \sqrt{\frac{670 \times 2 \times 9.81}{1.0945}} = 109.6 \text{ m/s}$$

and the discharge $Q = A_{jet} V_{jet} = 0.02545 \times 109.6 = 2.79$ cms.
The net head is $H = 670/1.0945 = 612$ m.
The hydropower potential is $P = \gamma HQ = 9{,}810 \times 670 \times 2.79 = 18.3$ MW.
The gross hydraulic power is

$$P = \gamma Q \left(\frac{V_{jet}^2}{2g} \right) = \frac{9{,}810 \times 2.79 \times 109.6^2}{2 \times 9.81} = 16.75 \text{ MW}.$$

The brake power output of the turbine is $P = 16.75 \times 0.85 = 14.2$ MW.
The tangential bucket speed will be $V_B = 0.5 V_{jet} = 109.6/2 = 54.8$ m/s.
The angular velocity of the $R = 1.5$ m Pelton wheel is

$$\omega = V_B / R = 54.8/1.5 = 36.5 \text{ rad/s},$$

or

$$N = \omega \frac{60}{2\pi} = \frac{36.5 \times 30}{\pi} = 349 \text{ rpm}.$$

We synchronize to the grid with a rotational speed $N = 360$ rpm ($\omega = 37.9$ rad/s) with 10 pairs of poles ($p_0 = 20$) and align the jet at a radius $R = (349/360) \times 1.5 = 1.45$ m.
The torque is obtained from $P = \omega T$ to obtain

$$T = P/\omega = \frac{14{,}250 \text{ kW}}{37.9 \text{ rad/s}} = 376 \text{ kN} \cdot \text{m}.$$

For a more advanced discussion, we can increase the efficiency of a Pelton wheel when the ratio of the diameter of the wheel D_{wheel} to the diameter of the jet d_{jet} follows the empirical relationship

$$\frac{D_{wheel}}{d_{jet}} = \frac{2R}{d_{jet}} = \frac{54}{N_s}$$

from Streeter (1971).

To obtain N, we need the net head $H = 612 \times 3.28 = 2{,}008$ ft, the discharge $Q = 2.79 \times 35.32 = 98.5$ cfs, and the power

$$P = \frac{62.4 \times 2{,}008 \times 98.5}{550} = 22{,}450 \text{ hp},$$

giving

$$N_s = \frac{N_{rpm} {P_{hp}}^{1/2}}{{H_{ft}}^{5/4}} = \frac{360 \times 22{,}450^{1/2}}{2{,}008^{5/4}} = 4.$$

Finally, the optimum nozzle diameter would be

$$d_{jet} = \frac{2RN_s}{54} = \frac{2 \times 1.45 \times 4}{54} = 0.21 \text{ m},$$

which shows that the current design is close to optimum.

At higher discharges (higher N_s), a single jet impacting the low end of a wheel on a horizontal axis may get too large. In such cases, we can design multiple horizontal ports on a single wheel rotating along a vertical axis.

5.2.3 Francis Turbines

In the mid-nineteenth century, James Francis from Massachusetts developed hydraulic turbines that initially fostered the textile industry. As shown in Figure 5.8, a Francis turbine converts the rotational energy (or angular momentum) of the flow entering the turbine into the mechanical energy of rotation of a shaft driving a generator. To some extent, it is a centrifugal pump acting in the reverse direction.

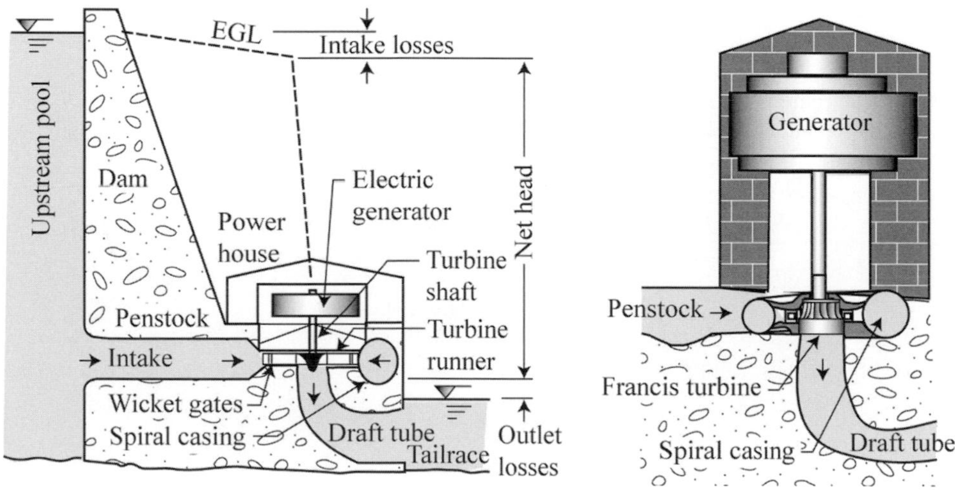

Figure 5.8 Typical Francis turbine

The flow is distributed evenly around the Francis turbine in a spiral casing with the help of adjustable vanes, called wicket gates, placed outside of the runner blade as sketched in Figure 5.9. Adjustable vanes streamline the flow in a direction tangential to the turbine without causing flow separation at the turbine entrance.

The flow field inside a Francis turbine is sketched in Figure 5.10 with subscript 1 at the entrance of the runner and subscript 2 for the outflowing conditions. Outside of the runner at 1, the inflow is controlled by the wicket gates that align the flow velocity V_f in a direction tangential to the runner. Following the streamline along the wicket gate, the radial velocity to the center of the turbine is $V_R = V_f \sin \alpha_1$, and the tangential velocity

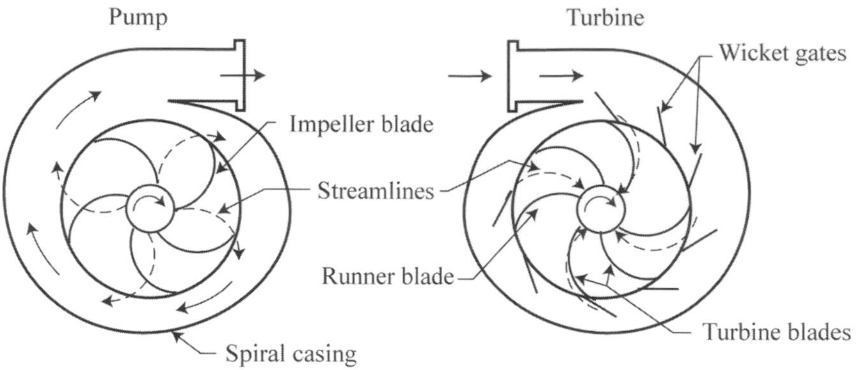

Figure 5.9 Centrifugal pump and Francis turbine similarities

is $V_t = V_f \cos \alpha_1 = V_R \cot \alpha_1$. We must satisfy the condition that the radial velocity $V_R = Q/A_w$ is also equal to the flow discharge Q over the cross-sectional area A_w given by the product of the blade height B and circumference of the runner, or $A_w = 2\pi RB$.

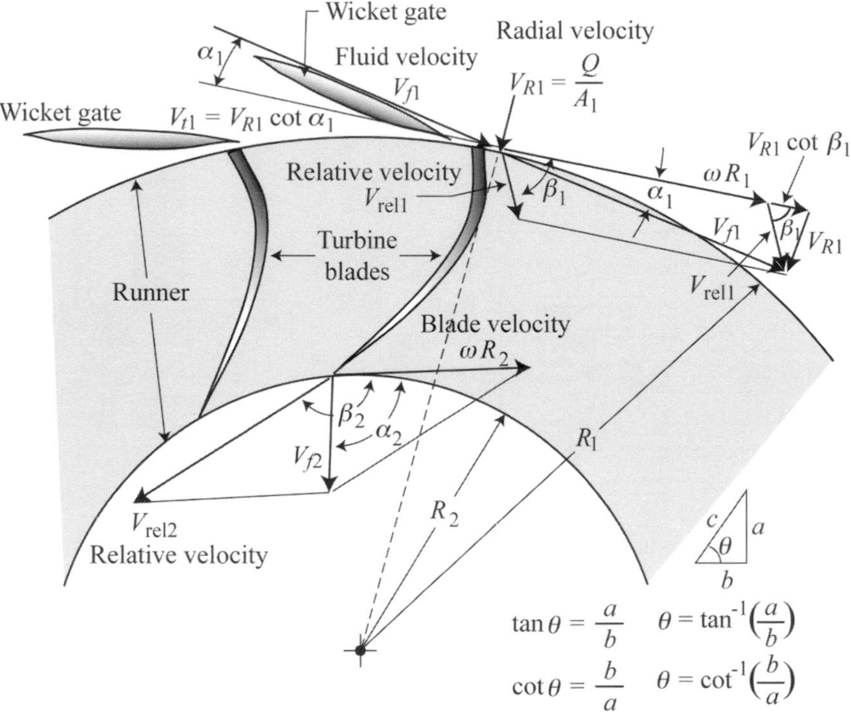

Figure 5.10 Flow field inside a Francis turbine

Inside the runner, the fluid velocity is described by three components: (1) the absolute velocity of the fluid V_f; (2) the absolute velocity of the blade $V_b = \omega R$; and (3) the velocity of the fluid relative to the blade V_{rel} which has to follow the turbine blade.

The direction of the velocity of the fluid relative to the blade V_{rel} is given by the blade orientation angle β. The absolute fluid velocity V_f is obtained from the sum of vectors $\vec{V}_f = \vec{V}_b + \vec{V}_{rel}$.

We can now look at the velocity vector components in the tangential and radial directions. The magnitude of the relative velocity needs to match the radial flow velocity $V_{f1} \sin\beta_1 = V_{R1}$. The tangential velocity inside the runner becomes $V_{t1} = V_{b1} + V_{R1} \cot\beta_1$. To prevent flow separation at the edge of the turbine, the two relationships for V_t are combined to find the wicket gate angle α_1 as a function of the flow condition and geometry of the turbine

$$\frac{V_{t1}}{V_{R1}} = \frac{V_{R1} \cot\alpha_1}{V_{R1}} = \frac{V_{b1}}{V_{R1}} + \frac{V_{R1} \cot\beta_1}{V_{R1}},$$

and thus

$$\cot\alpha_1 = \frac{\omega R_1}{V_{R1}} + \cot\beta_1,$$

or

$$\alpha_1 = \cot^{-1}\left(\frac{\omega R_1}{V_{R1}} + \cot\beta_1\right). \tag{5.4}$$

The blade angle β_1 is fixed by design but $V_R = \frac{Q}{A_w} = \frac{Q}{2\pi RB}$ varies with discharge.

The torque at the shaft is obtained from the difference in angular momentum entering (at section 1) and leaving (at section 2) the turbine, or

$$T_{shaft} = \sum M = \sum\left(\vec{R} \times \vec{V}\right)\rho AV,$$

where $Q = AV$ is the discharge. Therefore,

$$T = T_{in} - T_{out} = \rho Q[(R_1 V_1 \cos\alpha_1) - (R_2 V_2 \cos\alpha_2)].$$

From the torque, the power at the turbine shaft becomes

$$P = T\omega = \rho Q\omega[(R_1 V_1 \cos\alpha_1) - (R_2 V_2 \cos\alpha_2)]. \tag{5.5}$$

We can see from this relationship that the desirable objective in designing a turbine is to maximize power which is obtained when the inflow is as tangential as possible ($\alpha_1 \to 0$) and with outflow in the radial direction ($\alpha_2 \to 90°$). This second condition is equivalent to

$$\beta_2 = 90° + \tan^{-1}\left(\frac{\omega R_2}{V_{R2}}\right) = 90° + \tan^{-1}\left(\frac{2\pi B\omega R_2^2}{Q}\right).$$

In terms of rotational speed, we remember that the Pelton bucket speed for maximum efficiency should be about half the speed of the jet. A similar relationship exists for Francis turbines from the speed ratio

$$\phi = \frac{V_b}{\sqrt{2gH}} = \frac{\omega R}{\sqrt{2gH}} = \frac{\pi D N_{rpm}}{60\sqrt{2gH}},$$

where H is the head drop in the turbine. For maximum efficiency, the optimal rotational speeds for different turbine types are: (1) $0.43 < \phi < 0.47$ for an impulse turbine (Pelton); (2) $0.5 < \phi < 1.0$ for a Francis turbine; and (3) $1.5 < \phi < 3.0$ for a propeller turbine.

This implies that the Francis turbine diameter D depends on N_{rpm} and H from

$$D = \frac{60 \times \phi\sqrt{2gH}}{\pi N_{\mathrm{rpm}}}. \tag{5.6}$$

For installations in the USA ($F = 60$ Hz), the rotational speed $N_{\mathrm{rpm}} = 7{,}200/p_0$ decreases with the number of poles p_0. Consequently, the diameter of a Francis turbine increases with: (1) the number of poles; and (2) the square root of the head. Large turbines are often desirable to accommodate the cavitation requirement. Slowing down the turbine is done by increasing the number of poles in the generator. Useful references include Streeter (1971), Roberson et al. (1997), Moody and Zowski (1984) and Vischer and Sinniger (1999). The calculation in Example 5.4 and Case Study 5.2 are most instructive regarding Francis turbines.

Example 5.4: Wicket-gate angle for a Francis turbine

A Francis turbine is operated at a speed of 600 rpm and a discharge of 4.0 cms. The radius of the runner at the intake is $R_1 = 0.6$ m, the blade angle is designed at $\beta_1 = 110°$ and the blade height $B = 10$ cm. What should be the wicket gate α_1 for a nonseparating flow condition at the runner entrance?

Solution: The velocity $V_{b1} = \omega R_1 = 0.6 \times (2\pi \times 600/60) = 37.7$ m/s, with $\cot(110°) = -0.364$, and the radial velocity is

$$V_{R1} = \frac{Q}{2\pi R_1 B} = \frac{4}{2\pi \times 0.6 \times 0.1} = 10.6 \text{ m/s}.$$

From Eq. (5.4), the angle of the wicket gate becomes

$$\alpha_1 = \cot^{-1}[(37.7/10.6) - 0.364] = 17.4°.$$

The attentive reader will notice that a negative value for $\cot\beta_1$ is not desirable and the blades of this turbine should be redesigned to improve the turbine efficiency.

♦ **Case Study 5.2:** Francis turbine at Grand Coulee

Consider the Francis turbine at Grand Coulee from Streeter (1971) sketched in Fig. CS-5.2. Determine: (1) the torque and flow discharge and (2) the number of poles and turbine diameter.

Solution: The rotational speed is

$$\omega = \frac{\pi N_{\text{rpm}}}{30} = \frac{\pi 120}{30} = 12.6 \text{ rad/s}.$$

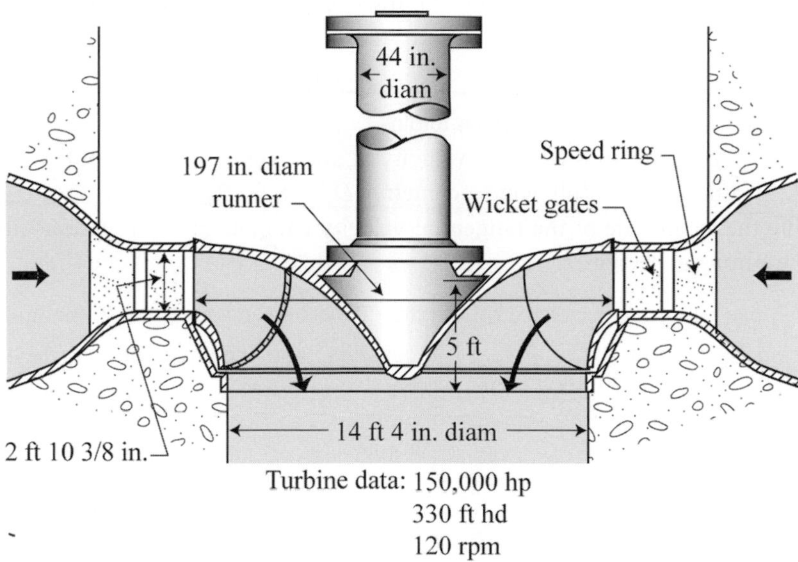

Fig. CS-5.2 Francis turbine at Grand Coulee

The torque is obtained from $T = P/\omega$, or

$$T = \frac{550 \times 1.5 \times 10^5}{12.6} = 6.5 \text{ M lb} \cdot \text{ft}.$$

The discharge is approximately

$$Q = \frac{P}{\gamma H} = \frac{550 \times 150{,}000}{62.4 \times 330} = 4{,}000 \text{ cfs}.$$

The number of poles is $p_0 = 7{,}200/N_{\text{rpm}} = 7{,}200/120 = 60$ poles (30 pairs). Assuming $\phi = 0.75$, the diameter from Eq. (5.6) is

$$D = \frac{60 \times \phi\sqrt{2gH}}{\pi N} = \frac{60 \times 0.75\sqrt{2 \times 32.2 \times 330}}{\pi \times 120} = 17.4 \text{ ft},$$

which compares well with the 16.4-ft (197-in.) turbine diameter. In other words, the speed ratio

$$\phi = \frac{\pi DN}{60\sqrt{2gH}} = \frac{\pi \times 16.4 \times 120}{60\sqrt{2 \times 32.2 \times 330}} = 0.71$$

is within the range $0.5 < \phi < 1.0$ for a Francis turbine.

5.2.4 Kaplan Turbines

In the early twentieth century, the Austrian Viktor Kaplan developed adjustable blades to improve turbine efficiency. Kaplan turbines use the concept of conservation of angular momentum and require large flow volumes to generate a large-scale free vortex (irrotational flow) around the rotating hub. In Figure 5.11, Kaplan turbines perform best under larger discharges and lower heads than Francis turbines. The ratio B/D of the wicket-gate height B to the diameter of the draft tube D is a function of the specific speed as shown in Figure 5.11. The funneling effect of the turbine chamber generates a large irrotational vortex (see Example 5.5), the energy of which is extracted for hydropower production (Example 5.6).

♦ **Example 5.5:** Velocity field of a Kaplan turbine

The wicket gates allow an 8 ft/s flow velocity at an angle of 45° at section 1 shown in Figure 5.11. Consider the following geometry: $D_1 = 8$ ft, $D_{hub} = 1.5$ ft and $D_2 = 4$ ft. Determine the magnitude of the tangential velocity component V_u at various locations along the runner at section 2.

Solution: Since no torque is exerted on the flow between sections 1 and 2, the moment of momentum is constant ($V_{u1}r_1 = V_{u2}r_2$ is constant) and the fluid motion obeys a free vortex. At section 1, the tangential velocity is $V_u = 8\cos 45° = 5.65$ ft/s, and at section 2,

$$V_{u2} = \frac{V_{u1}r_1}{r_2} = \frac{5.65 \times 4}{r_2} = \frac{22.6}{r_2}.$$

Therefore, at the hub ($r_{hub} = 0.75$ ft), $V_{u\,hub} = 22.6/0.75 = 30.1$ ft/s, and at the tip ($r_{tip} = 2$ ft), $V_{u\,tip} = 22.6/2 = 11.3$ ft/s.

Notice that the maximum tangential flow velocity is near the hub.

Example 5.6: Blade angle of a Kaplan turbine

This advanced design example is for a Kaplan turbine rotating at $N = 240$ rpm, or ($\omega = 25.1$ rad/s) with the following geometry: $D_1 = 8$ ft, $D_{hub} = 1.5$ ft, $D_2 = 4$ ft, and $B = 2$ ft. The entrance velocity reaches $V = 8$ ft/s at an angle $\alpha_1 = 45°$. Determine the blade orientation angle to extract the angular velocity of the fluid between sections 2 and 3 shown in Figure 5.11. This example requires the analysis of velocity vectors in three dimensions including: (1) the axial flow velocity V_a; (2) the tangential fluid velocity V_u; and (3) the velocity of the blade V_b. The fluid velocity vector $\vec{V}_f$ has two components V_a and V_u. At section 2, $r_{hub} = 0.75$ ft at the hub, and $r_{tip} = 2$ ft at the blade tip.

Solution:

(1) The discharge through the turbine is $Q = \pi D_1 BV\sin\alpha_1 = 8\pi \times 2 \times 8\sin 45° = 284$ cfs. Hence, the constant axial fluid velocity is $V_a = 284/\pi(2^2 - 0.75^2) = 26.3$ ft/s.

(2) The tangential fluid velocities determined in Example 5.5 are $V_{u\,hub} = 22.6/0.75 = 30.1$ ft/s at the hub, and $V_{u\,tip} = 22.6/2 = 11.3$ ft/s at the blade tip.

(3) The velocity of the blade at $N = 240$ rpm is obtained from $V_b = \omega r = 25.1\,r$ and thus $V_b = 25.1 \times 0.75 = 18.8$ ft/s at the hub, and $V_b = 25.1 \times 2 = 50.2$ ft/s at the blade tip.

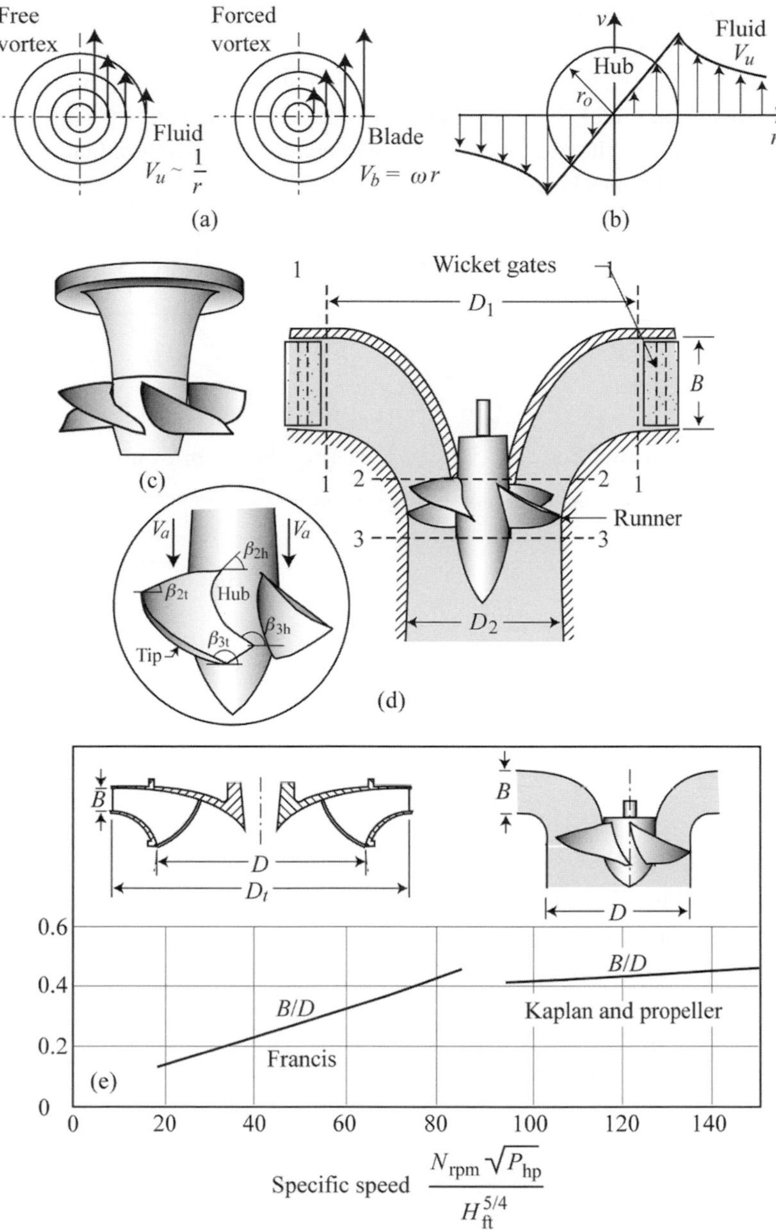

Figure 5.11 Kaplan turbine layout (modified after Moody and Zowski 1984)

The blades are designed to be aligned with the flow to avoid flow separation. From the absolute velocity of the fluid $\vec{V}_f = f(V_a, V_u)$ and the absolute velocity of the blade V_b, the fluid velocity relative to the blade (which determines the blade alignment) is the difference between these two vectors. Graphically, Figure E-5.6 shows how to find the blade angle β at the hub and at the tip. At section 2, the vortex is represented by the tangential fluid velocities V_u and $V_u = 0$ at the exit of the turbine, i.e. section 3.

Notice the large change in blade orientation at the hub ($r = 0.75$ ft on the right) compared to the slight change in angle near the tip of the blade ($r = 2$ ft on the left). It is also clear that changes in discharge will change the optimum blade orientation, which can sometimes be achieved with adjustable blades.

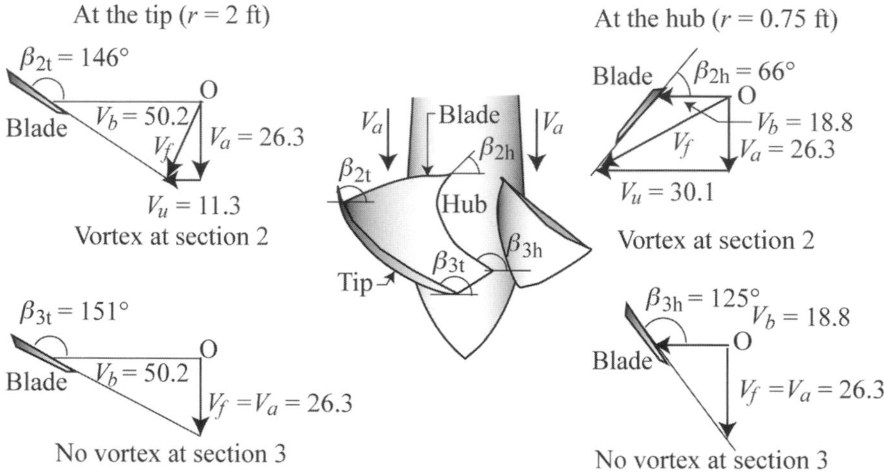

Fig. E-5.6 Blade alignment in a Kaplan turbine

5.2.5 Bulb Turbines

Bulb turbines are used for run-of-the-river operations in large rivers with low heads and large discharges (high specific speed). Bulb turbines have large propellers and a large generator located in a bulb, as sketched in Figure 5.12.

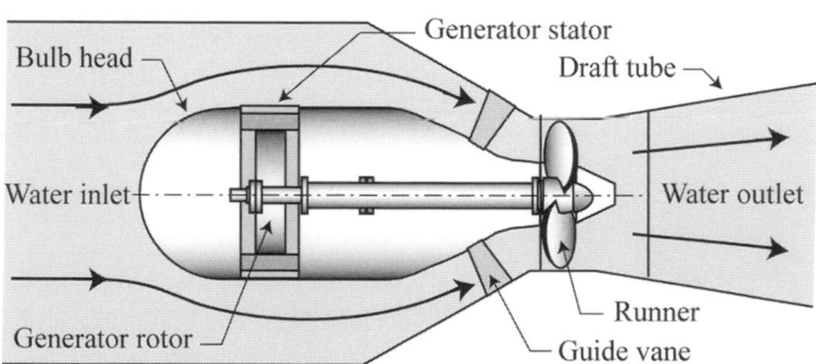

Figure 5.12 Sketch of a tubular (bulb type) turbine

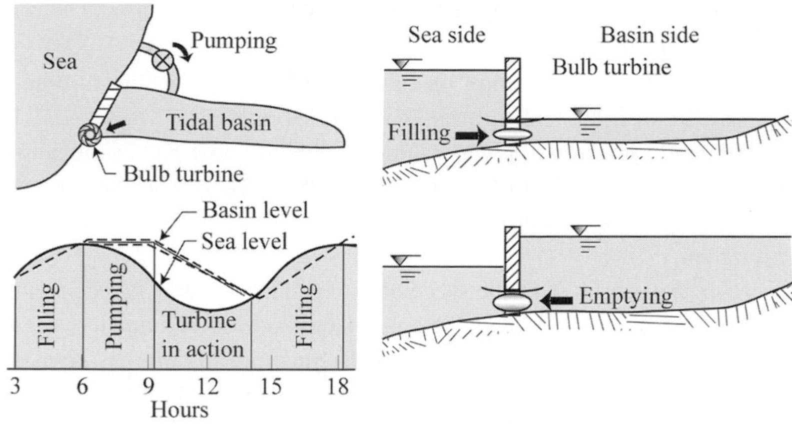

Figure 5.13 Sketch of tidal hydropower production

Tidal hydropower is considered in coastal areas with significant tidal amplitude. As sketched in Figure 5.13, the essential concept consists of filling the tidal basin during the rising tide (flood) and to generate hydropower as the stage falls (ebb). The efficiency of the system is often enhanced by pumping water into the tidal basin at high tide (low head differential) to maximize the volume of water generating hydropower (under a higher head differential). Double action with flow in the opposite direction is sometimes beneficial when water enters the tidal basin. In such cases, the runners of large bulb turbines are equipped with adjustable blades.

Pumped storage operations meet the peak demand in energy. Rather than building large installations used only a few hours daily, pumped-storage systems pump water into storage areas at times of low energetic demand (lower energy price) and turbine the stored water volume during the periods of peak energy demand (higher energy price). Pumped storage systems typically use high lakes or reservoirs perched a short distance above the source of water, see Figure 5.14. These systems are subject to sudden pressure changes resulting from frequent stop–go operations and flow reversals. These pressure fluctuations are handled by surge tanks, discussed in Chapter 6.

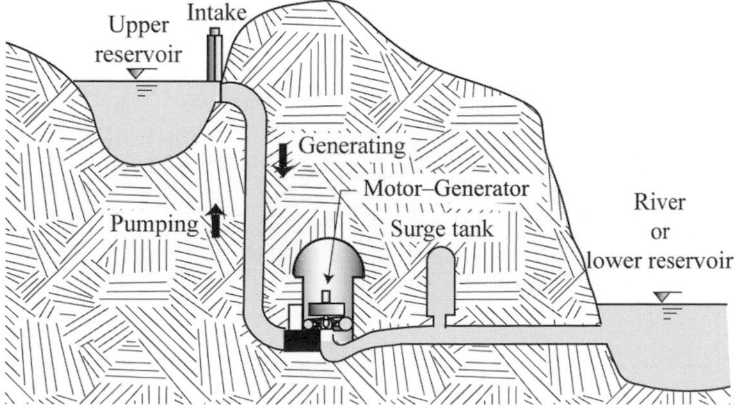

Figure 5.14 Pumped storage hydropower generation

5.3 Turbine Cavitation

Turbines are susceptible to cavitation on the downstream side of the turbine impeller blades. The cavitation index σ for turbines is defined as $\sigma = \frac{1}{H}\left(\frac{p_{atm}-p_v}{\gamma} - z_t\right)$ from the absolute atmospheric pressure p_{atm} and vapor pressure of water p_v, the elevation difference between the turbine blade above the tailrace water level z_t, and the net head across the turbine H (head change at both ends of the turbine). High values of σ are desirable to prevent cavitation.

Because cavitation varies with the speed of the impeller (greater speeds mean greater relative velocities and less pressure on the downstream side of the impeller), the critical σ values depend on the specific speed N_s for different types of turbines. Two empirical equations between the minimum σ and N_s have been determined for Francis and Kaplan turbines (Moody and Zowski 1984):

$$\sigma = 0.006 + 0.55(0.01N_s)^{1.8} \text{ for Francis turbines,} \tag{5.7a}$$

$$\sigma = 0.1 + 0.3(0.01N_s)^{2.5} \text{ for Kaplan turbines.} \tag{5.7b}$$

Both equations can be solved to find the maximum desirable specific speed N_s as a function of H and z_t. The results shown in Figure 5.15 indicate that there are two effective ways to prevent turbine cavitation: (1) low rotational speeds leading to low

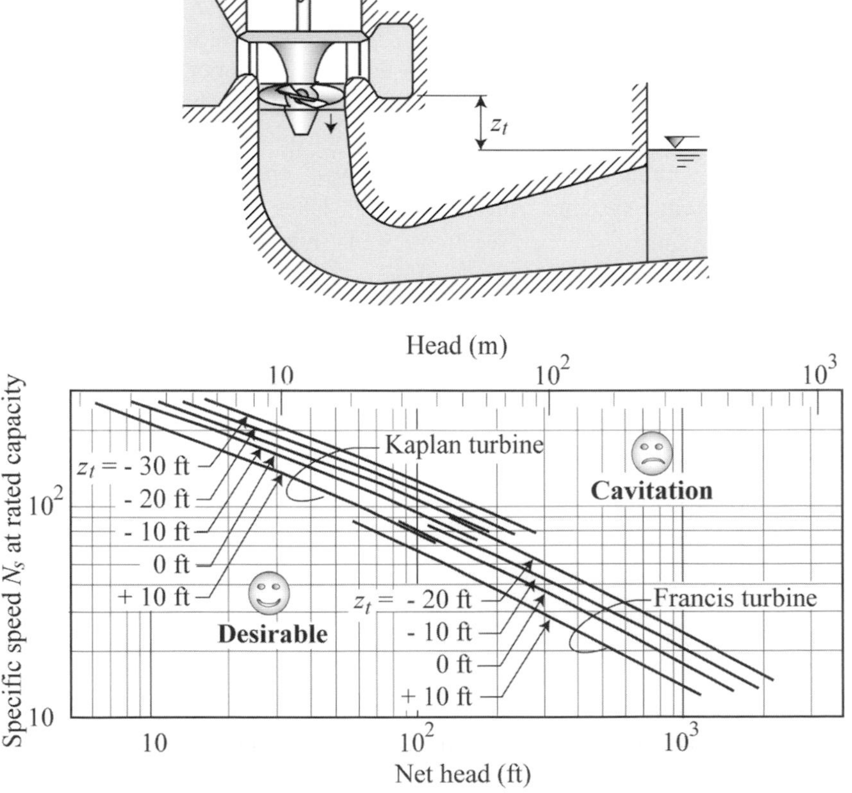

Figure 5.15 Cavitation diagram for Francis and Kaplan turbines at sea level (modified after Moody and Zowski 1984)

values of N_s; and (2) the turbine should be located as low as possible (z below z_t). Note that the chart is valid at sea level and all curves will shift down with altitude (i.e. acceptable N_s values decrease with altitude). More information regarding vorticity around structures can be found in Knauss (1987) and Hite (1992). Case Study 5.3 shows typical turbine calculations to prevent cavitation.

♦ **Case Study 5.3:** Preventing turbine cavitation at Grand Coulee

Select the type, speed and size of a turbine to drive a 60-Hz generator given a net head $H = 330$ ft, $Q = 4{,}300$ cfs and a turbine efficiency of 94%. Find a suitable turbine elevation with reference to the water surface in the tailrace.

Solution: The maximum developed power for the turbine will be

$$P_{\text{hp}} = e\frac{\gamma HQ}{550} = 0.94\frac{62.4 \times 330 \times 4{,}300}{550} = 151{,}330 \text{ hp} = 111 \text{ MW}.$$

From Figures 5.6 and 5.15, a Francis turbine is considered for maximum efficiency and $N_s < 45$ should be considered as a first approximation to prevent cavitation.

Because $N_s = \frac{N_{\text{rpm}} P_{\text{hp}}{}^{1/2}}{H_{\text{ft}}{}^{5/4}}$, we can solve for a maximum value of N:

$$N_{\text{rpm}} = \frac{45 \times 330^{5/4}}{151{,}300^{1/2}} = 163 \text{ rpm}.$$

To operate the turbine at a synchronous speed with the network requires $p_0 = 7{,}200/N = 7{,}200/163 = 44.1$ poles, thus $p_0 = 46$ poles gives $N = 7{,}200/46 = 156$ rpm and $N_s = \frac{N_{\text{rpm}} P_{\text{hp}}{}^{1/2}}{H_{\text{ft}}{}^{5/4}} = \frac{156 \times \sqrt{151{,}300}}{330^{5/4}} = 43.1$.

We also obtain from Figure 5.15 that $-20 \text{ ft} > z_t > -10$ ft would be suitable. More accurate calculations are needed to account for altitude and temperature. From Eq. (5.7a) we obtain $\sigma = 0.006 + 0.55(0.01 \times 43.1)^{1.8} = 0.127$. Considering $p_v = 0.51$ psi at a temperature of 80 °F from Table 1.4, and $p_{atm} = 13.9$ psi at 1,500 ft from Table 1.5, we obtain z_t as

$$z_t = \frac{p_0 - p_v}{\gamma} - \sigma H = \frac{[144(13.9 - 0.51)]}{62.4} - (0.127 \times 330) = -11.1 \text{ ft}$$

meaning that the turbine should be at least 12 ft below the tailrace water level.

Finally, we can also calculate an approximate diameter for the Francis turbine with the information from Section 5.2.3:

$$D = \frac{60 \times \phi\sqrt{2gH}}{\pi N_{\text{rpm}}} = \frac{60 \times 0.75\sqrt{2 \times 32.2 \times 330}}{\pi \times 156} = 13.4 \text{ ft}.$$

This closely matches the Francis turbine at Grand Coulee. In the design, N_{rpm} was appropriately reduced further to prevent cavitation.

As an exercise, the reader can calculate the specific speed N_s, the parameter σ and the elevation z_t when the speed decreases to $N = 120$ rpm. (It is found that $N_s = 33.3$, $\sigma = 0.082$ and $z_t = 6.4$ ft at sea level and $z_t = 3.9$ ft at 80 °F. From comparison with Case Study 5.2, it is noticeable that slowing down the generator results in an increase in the runner diameter.)

EXERCISES

These exercises review the essential concepts from this chapter.

1. Why are poles in pairs?
2. What is the purpose of lining penstocks?
3. Why do we need the specific speed of a turbine?
4. In reservoirs with large water-level fluctuations, which of a Francis or a Kaplan turbine would be preferable?
5. What are the differences between a mini and a micro turbine?
6. Why are the buckets of a Pelton wheel placed so close together?
7. What is the purpose of wicket gates?
8. Do you design the outlet flow of a Francis turbine to be radial or tangential to the runner? Why?
9. What is the difference between power and torque?
10. What is the difference between a free and a forced vortex?
11. When do you use pumped storage?
12. When H and Q are fixed how can you reduce N_s to prevent cavitation?
13. True or false?
    (a) Adding poles to a generator will decrease the rotational speed of a turbine.
    (b) The specific speed formulas for pumps and turbines are identical.
    (c) The specific speed of a turbine increases with the head.
    (d) Pelton wheels are still very efficient when the discharge decreases below the design discharge.
    (e) Large hydropower projects are defined at discharges $Q > 1{,}000$ cfs.
    (f) A Kaplan turbine is also called an impulse turbine.
    (g) A Francis turbine works on the principle of angular momentum.
    (h) The blade of a Kaplan turbine moves like a forced vortex.
    (i) Pumped storage facilities are most effective in coastal areas with high tidal amplitude.
    (j) Adding poles to a turbine generator is beneficial to prevent cavitation.

SEARCHING THE WEB

Find photos of the following features, study them carefully and write down your observation.

1. ♦ (a) A 42-pole electric generator; (b) a Pelton wheel; (c) a Francis turbine; (d) a Kaplan turbine; and (e) a Turgo turbine.
2. ♦ (a) A Deriaz turbine; (b) a bulb turbine; (c) turbine cavitation; (d) a Banki-Michell turbine; and (e) a tidal turbine.

PROBLEMS

Turbines

1. ♦ A turbine discharges 1,200 cfs under a head of 26 ft at an efficiency of 86%. Find the power generated under those conditions.
2. ♦ A shaft produces 200 hp at 600 rpm. Calculate the torque.

3. ♦ The moment of momentum of water is reduced by 20,000 lb · ft in a turbine moving at 400 rpm. Determine the power generated by this turbine.
4. ♦ Determine the power and select the type of turbine for the following conditions:
   (a) the head is 600 ft and the discharge is 10 cfs;
   (b) the head is 200 ft and the discharge is 200 cfs; and
   (c) the head is 50 ft and the discharge is 4,000 cfs.
5. ♦♦ A Pelton wheel is 24 inches in diameter and rotates at 400 rpm. What is the head that is best suited for this wheel? How many pairs of poles are needed? [Hint: can you find the velocity of the wheel, and from it the velocity of the jet?]
6. ♦♦ For the Pelton wheel in Fig. P-5.6, you need to synchronize the wheel speed. Find out what happens with 22 poles. Find N and calculate the N_s value for these conditions. Select the best design if the maximum efficiency is reached when $D_w/d_{jet} = 54/N_s$ where D_w is the diameter of the wheel, d_{jet} is the diameter of the jet and N_s is the specific speed.

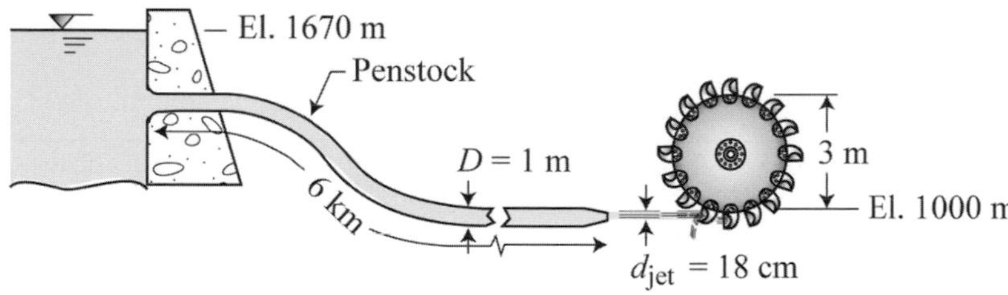

Fig. P-5.6

7. ♦ For the Francis turbine at Hoover Dam ($P = 115{,}000$ hp, $N = 180$ rpm, $H = 480$ ft), find N_s and the discharge if $e = 0.85$.
8. ♦♦ For the Francis turbine at Grand Coulee ($P = 150{,}000$ hp, $H = 330$ ft, $N = 120$ rpm), determine: (a) the angular velocity in rad/s; (b) the specific speed; (c) the discharge if the efficiency is 90%; and search the appropriate information in this chapter to determine (d) the radial velocity in ft/s; and (e) the maximum tangential velocity of the runner.
9. ♦♦ (a) A Francis turbine is designed with the following conditions: $\beta_1 = 60°$, $\beta_2 = 90°$, $r_1 = 5$ m, $r_2 = 3$ m and $B = 1$ m. When the discharge is 126 m^3/s and $N = 60$ rpm, calculate the entrance angle α_1 to prevent separation of the streamlines at the entrance of the runner. Determine the maximum power and torque that can be generated under these conditions.

   ♦♦♦ (b) **Improved design problem!** Under the same discharge and runner size as in Problem 5.9, can you suggest an improvement to the design of the runner blades.
10. ♦♦ A Francis turbine is designed with the following conditions: discharge 113 m^3/s and $N = 120$ rpm, $\beta_1 = 45°$, $r_1 = 2.5$ m and $B = 2.9$ m. Determine the angle α_1 that would avoid separation at the runner inlet.
11. ♦ A Kaplan turbine is designed to generate 24,500 hp at $N = 100$ rpm under 41 ft of head. If the efficiency is 85%, what are the flow discharge, the specific speed and the number of poles of the generator.

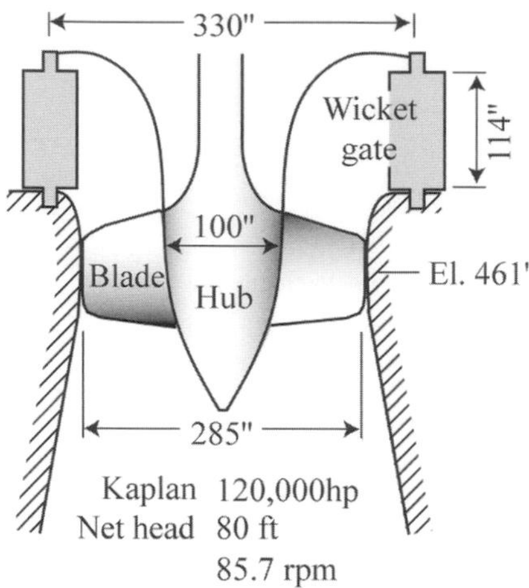

Fig. P-5.13

12. ♦♦♦ For the Kaplan turbine in Fig. E-5.6, draw the velocity diagrams at the trailing edge of the propeller turbine blade such that the tangential velocity $V_{u2} = 0$. Determine the blade angles β_2 for $r = 0.75$ ft, 1 ft, 1.5 ft and 2 ft.

Additional Problems

13. From the Kaplan turbine at Wanapum Dam in Fig. P-5.13, can you determine the blade angles?
14. Consider the following turbines compiled from Moody and Zowski (1984) in Table P-5.14, calculate N_s and compare with the characteristics described in this chapter.

Table P-5.14. Sample of turbine characteristics

Location	Type	Rotation (rpm)	Head (ft)	Power (hp)	Outside diameter (in.)
Paucartambo, Peru	Pelton	450	1,580	28,000	
Smith Mtn Dam	Francis	100	179	204,000	246
Garrison Dam	Francis	90	150	90,000	223
Hoover Dam	Francis	180	440	115,000	
Oxbow power plant	Francis	100	115	73,000	
R. Moses Niagara	Francis	120	300	210,000	
Cullingran, Scotland	Deriaz	300	180	30,500	
Priest Rapids	Kaplan	86	78	114,000	284-in. runner
Wanapum	Kaplan	86	80	120,000	285-in. runner
St-Lawrence Power	Propeller	95	81	79,000	240-in. runner
Pierre Benite, France	Bulb	83	26	27,000	240-in. runner
Ozark L&D	Tube	60	26	33,800	315-in. runner

6 Water Hammer

Water compressibility effects in closed conduits can be devastating and hydraulic structures like surge tanks are specifically designed to minimize the pressure fluctuations in pipe systems. Section 6.1 presents important knowledge on water compressibility. It is followed with the celerity of wave propagation in pipes in Section 6.2. The concept of hydraulic transients and water hammer is detailed in Section 6.3 with prevention measures such as surge tanks in Section 6.4.

6.1 Water Compressibility

The bulk modulus of elasticity E_w is a measure of fluid compressibility. It measures relative changes in volume and fluid density under pressure,

$$E_w = -\frac{\forall dp}{d\forall} = \frac{\rho \, dp}{d\rho}, \tag{6.1}$$

where dp is the increase in pressure required to decrease the volume $d\forall$ from the initial fluid volume $\forall$. From the definition of the mass density $\rho = m/\forall$, we note that the mass change $dm = \rho d\forall + \forall d\rho = 0$ such that $-d\forall/\forall = d\rho/\rho$. A typical elasticity value for water is $E_w = 2.1 \times 10^9$ Pa $\cong 3 \times 10^5$ psi. The bulk modulus of elasticity for water increases slightly with temperature and pressure, as shown in Table 6.1. Example 6.1 illustrates how to calculate the bulk modulus of elasticity of a fluid.

Table 6.1. Water modulus of elasticity E_w in GPa vs temperature and pressure

Pressure	E_w ($T^\circ = 0$ °C)	E_w ($T^\circ = 10$ °C)	E_w ($T^\circ = 20$ °C)
$1 - 25 \times 10^5$ Pa	1.93	2.03	2.07
$25 - 50 \times 10^5$ Pa	1.96	2.06	2.13

♦ **Example 6.1:** Bulk modulus of elasticity of a fluid

A liquid has a volume of 2.0 ft^3 at a pressure of 100 psi. When compressed to 180 psi, the volume decreases to 1.8 ft^3. Find the bulk modulus of elasticity of this fluid.

Solution: $dp = 180 - 100 = 80$ psi $= 11{,}520$ psf and $d\forall = 1.8 - 2 = -0.2$ ft^3, and thus

$$E_{fluid} = -\frac{\forall dp}{d\forall} = -\frac{2 \times 11{,}520}{-0.2} = 115{,}200 \text{ psf}.$$

Note that 1 psi = 144 psf = 6.89 kPa.

6.2 Wave Celerity

We start with the celerity of a compressed water wave in an infinitely rigid pipe in Section 6.2.1, and in elastic pipes in Section 6.2.2. Celerity reduction methods are then discussed in Section 6.2.3.

6.2.1 Celerity in an Infinitely Rigid Pipe

The term celerity describes how fast pressure variations travel in water. It is equivalent to the speed of sound in air. Consider a pressure wave in a large body of water, or an infinitely rigid pipe shown in Figure 6.1. The wave is held stationary by moving with the wave at a constant celerity c, the mass flux entering the compressed area is $\dot{m} = \rho A(c + V)$ and the velocity decreases by $\Delta V = -V$ across the wave front.

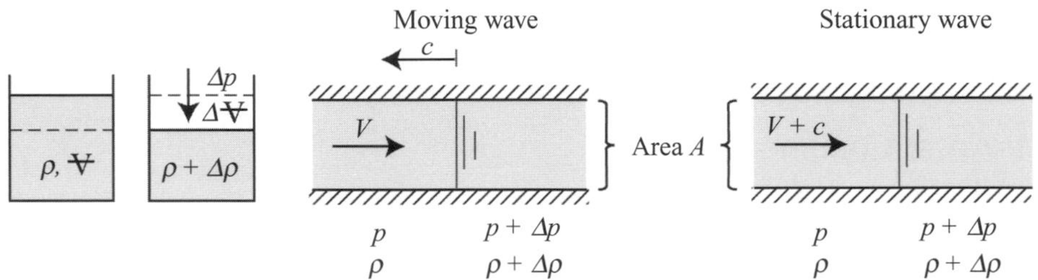

Figure 6.1 Compressible wave celerity in a rigid pipe

The force analysis in the flow direction includes an increase in pressure force from pA to $(p + dp)A$, which corresponds to the rate of change in linear flow momentum of the mass flux $\dot{m} = \rho A(V + c)$, where $c >> V$, or

$$pA - (p + dp)A = \dot{m}\Delta V = \rho A(c + V)(-V) \simeq -\rho AVc.$$

The pressure increase becomes

$$dp = \rho Vc. \tag{6.2}$$

Now, looking at the conservation of mass during a time interval dt, the mass entering the pipe $\rho AVdt$ equals the mass being compressed $d\rho Acdt$, which gives $\rho V = cd\rho$. This is combined with the modulus of elasticity $dp = E_w d\rho/\rho$ to eliminate V as

$$dp = \rho cV = \rho c\left(c\frac{d\rho}{\rho}\right) = \frac{E_w d\rho}{\rho},$$

or

$$c = \sqrt{\frac{E_w}{\rho}} = \sqrt{\frac{dp}{d\rho}}. \tag{6.3}$$

Thus, by application of the momentum and continuity equations, we have derived equations describing the increase in pressure dp (Eq. (6.2)) and the wave celerity c (Eq. (6.3)). From values of E_w in Table 6.1, the compression wave celerity in water is approximately $c = \sqrt{2.1 \times 10^9/1{,}000} \cong 1{,}450$ m/s $= 4{,}750$ ft/s.

6.2.2 Celerity in Elastic Pipes

In elastic pipes, the mass of water is not only stored by compression, but also in the expanded volume of the pipe under increased pressure. Table 6.2 lists values of the modulus of elasticity E_p for various pipe materials. Note that the values change at very high temperature, and the table lists commonly used values to solve hydraulic engineering problems.

Table 6.2. Modulus of elasticity E_p for various pipe materials

Pipe material	E_p(Pa)	E_p(psi)	E_p/E_w
Steel	1.9×10^{11}	28×10^6	90
Reinforced concrete	1.6×10^{11}	25×10^6	83
Cast iron	1.1×10^{11}	16×10^6	52
Copper	9.7×10^{10}	14×10^6	47
Glass	7.0×10^{10}	10×10^6	33
Concrete	3.0×10^{10}	4.3×10^6	14
PVC (polyvinyl chloride)	2.5×10^9	3.6×10^5	1.2
Water	2.1×10^9	3.0×10^5	1.0
HDPE (high-density polyethylene)	1×10^9	1.4×10^5	0.5
Lead	3.1×10^8	4.5×10^4	0.15

Figure 6.2 provides a schematic illustration for the analysis of wave propagation in expanding pipes. The main pipe parameters include the wall thickness e, the tensile force per unit length T_f, the pipe modulus of elasticity E_p, and the radial expansion dr under the pressure increase Δp.

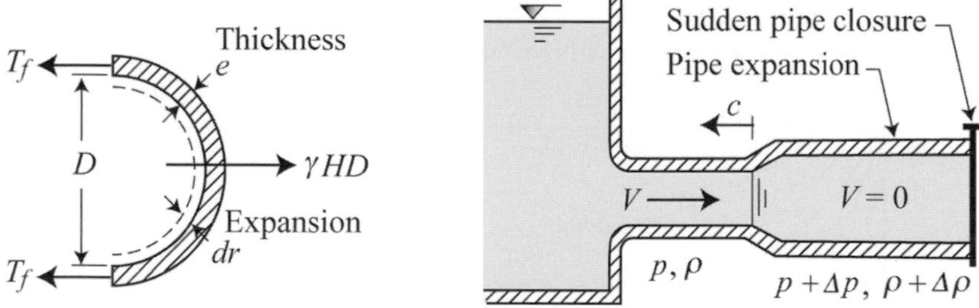

Figure 6.2 Wave propagation in an elastic pipe

Per unit pipe length, the pipe tension is

$$\sigma_p = \frac{T_f}{e} = \frac{pD}{2e} = \frac{\gamma HD}{2e},$$

and its increase is

$$d\sigma_p = \frac{dpD}{2e} = E_p\left(\frac{2dr}{D}\right)$$

or

$$dr = \frac{D^2 dp}{4E_p e}.$$

The expansion volume is given as

$$\frac{d\forall}{\forall} = \frac{4\pi D dr}{\pi D^2} = \frac{D dp}{E_p e} = -\frac{dp}{E_w},$$

which gives

$$\frac{E_w d\forall}{dp \forall} = \frac{E_w D}{E_p e}.$$

The volume change in the pipe therefore has two components: (1) the volume compressed for rigid pipes; and (2) the elastic pipe expansion volume. Therefore, when compared to the rigid-pipe analysis,

$$d\forall_{elastic} = \left[1 + \left(\frac{E_w D}{E_p e}\right)\right] d\forall_{rigid}.$$

Accordingly, the wave celerity in elastic pipes c' will be less than c for rigid pipes:

$$c' = \sqrt{\frac{E_w}{\rho}\left[1 + \left(\frac{E_w D}{E_p e}\right)\right]^{-1}},$$

or

$$\frac{c'}{c} = 1/\sqrt{\left[1 + \left(\frac{E_w D}{E_p e}\right)\right]}. \tag{6.4}$$

Obviously, the celerity in rigid pipes ($E_w/E_p = 0$) reduces to Eq. (6.3). The compressibility wave celerity in elastic pipes is shown in Figure 6.3.

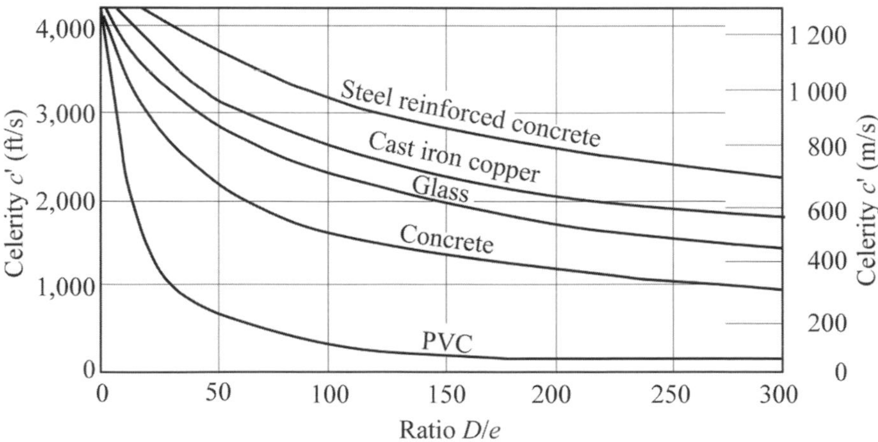

Figure 6.3 Wave celerity in elastic pipes

Example 6.2 indicates how to calculate the wave celerity in an elastic pipe.

♦ **Example 6.2:** Wave celerity in a pipe

Consider a 125-psi cast-iron pipe (Table 2.2) and calculate the wave-propagation celerity for a diameter $D = 42$ in. $= 1{,}068$ mm and thickness $e = 0.5(45.1 - 42.02)$ in. $= 39$ mm.

Solution: Consider $\rho = 1{,}000\ \mathrm{kg/m^3}$, $E_w = 2.1 \times 10^9$ Pa and $E_p = 1.1 \times 10^{11}$ Pa to get

$$c' = \sqrt{\frac{2.1 \times 10^9}{1{,}000}\left[1 + \left(\frac{2.1 \times 10^9 \times 1{,}068}{1.1 \times 10^{11} \times 39}\right)\right]^{-1}} = 1{,}174\ \mathrm{m/s},$$

compared to $c = 1{,}450$ m/s.

Notice that this celerity reduction from the elasticity of the pipe results in significant decreases in pressure surge (Eq. (6.2)) from sudden valve closures.

6.2.3 Wave Celerity Reduction with Air

Let us consider water containing air bubbles. The total volume $\forall = \forall_w + \forall_g$ equals the sum of the water volume $\forall_w$ and the gas volume $\forall_g$, and the concentration of gas $C_g = \forall_g/\forall$. The density of the mixture is $\rho = \rho_w(\forall_w/\forall) + \rho_g(\forall_g/\forall)$. A pressure change brings about a volume change equal to $\Delta\forall = \Delta\forall_w + \Delta\forall_g$. The bulk modulus of elasticity of water $E_w = -\frac{\forall_w\, dp}{d\forall_w} \cong 2.1 \times 10^9$ Pa, compared to the high compressibility of gas $E_g = -\frac{\forall_g\, dp}{d\forall_g} \simeq 9{,}000$ psf (or 430 kPa). Combining these expressions yields the combined air–water bulk modulus $E = E_w/\{1 + C_g[-1 + (E_w/E_g)]\}$, and the celerity with air bubbles is $c' = \sqrt{E/\rho}$. Figure 6.4 illustrates the significant decrease in wave

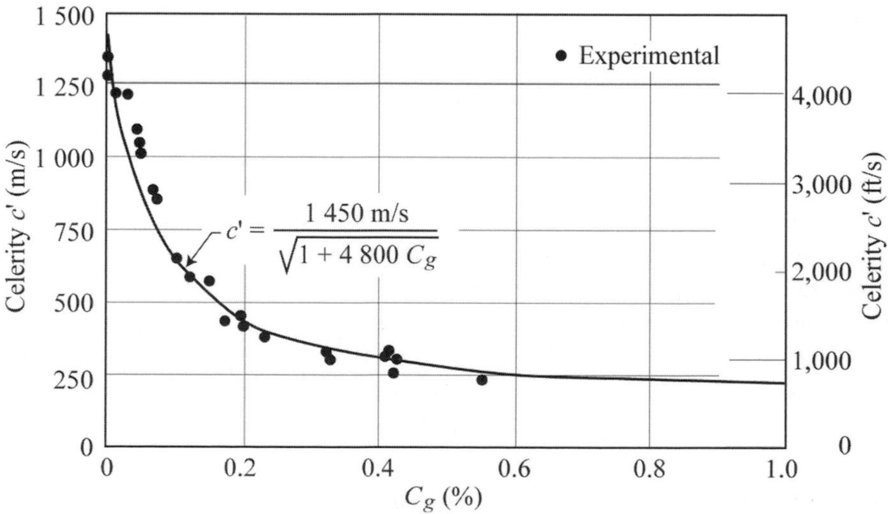

Figure 6.4 Wave celerity propagation in water with air bubbles

celerity as a function of C_g. The presence of air bubbles in water at low concentrations (e.g. $C_g > 0.2\%$) will significantly reduce the celerity of compression waves, as shown in Example 6.3.

Example 6.3: Wave celerity reduction with air
Calculate the wave-propagation speed in water containing 0.5% air ($C_g = 0.005$).

Solution: Consider $E_g = 9{,}000$ psf, $E_w = 3 \times 10^5$ psi $= 4.32 \times 10^7$ psf and the mass densities of air $\rho_g = 0.00238$ slug/ft^3 and water $\rho = 1.94$ slug/ft^3. The modulus is

$$E = \frac{E_w}{1 + C_g\left[-1 + (E_w/E_g)\right]} = \frac{4.32 \times 10^7}{1 + 0.005(4{,}800)} = 1.73 \times 10^6 \text{ psf.}$$

The celerity becomes $c' = \sqrt{E/\rho} = \sqrt{1.73 \times 10^6/1.94} = 943$ ft/s $= 288$ m/s. Air significantly reduces the pressure surge $dp = \rho V c'$ from sudden valve closures in pipes.

6.3 Hydraulic Transients

This section deals with the maximum pressure generated by a sudden valve closure (Section 6.3.1) and a gradual closure (Section 6.3.2), followed by the timescale for valve opening (Section 6.3.3) and emptying large tanks (Section 6.3.4).

6.3.1 Sudden Valve Closure

In a pipe, the time of the valve closure t_C is compared to the travel time $t_T = 2L/c'$ for the wave to propagate back and forth between the valve and the reservoir. A sudden closure occurs when $t_C < 2L/c'$. The pressure increase is $dp = \rho V c'$ and the corresponding pressure-head increase is

$$\Delta H = \frac{\Delta p}{\gamma} = \frac{\rho V c'}{\gamma} = \frac{V c'}{g}.$$

Figure 6.5 sketches the main propagation features of a water compressibility wave in a pipe after a sudden valve closure. At (a), the water enters the pipe and the pressure increase propagates to the reservoir at celerity c'. At (b), the water comes out of the pipe as pressure returns to hydrostatic condition and the wave propagates back to the valve at the same celerity. Upon return (c), and reaching the valve, the water is still pulled out of the pipe and a zone of pipe contraction develops as the change in pressure now becomes negative. This zone of lower pressure propagates from the valve to the reservoir. Upon reaching the reservoir in (d), water reenters the pipe to reestablish hydrostatic pressure conditions and the wave propagates to the valve. At the end of this double sweep, we reach the starting point from which the cycle is repeated. This series of "coup de bélier" or hydraulic ram is called a water hammer, leading to vibrations in short pipes.

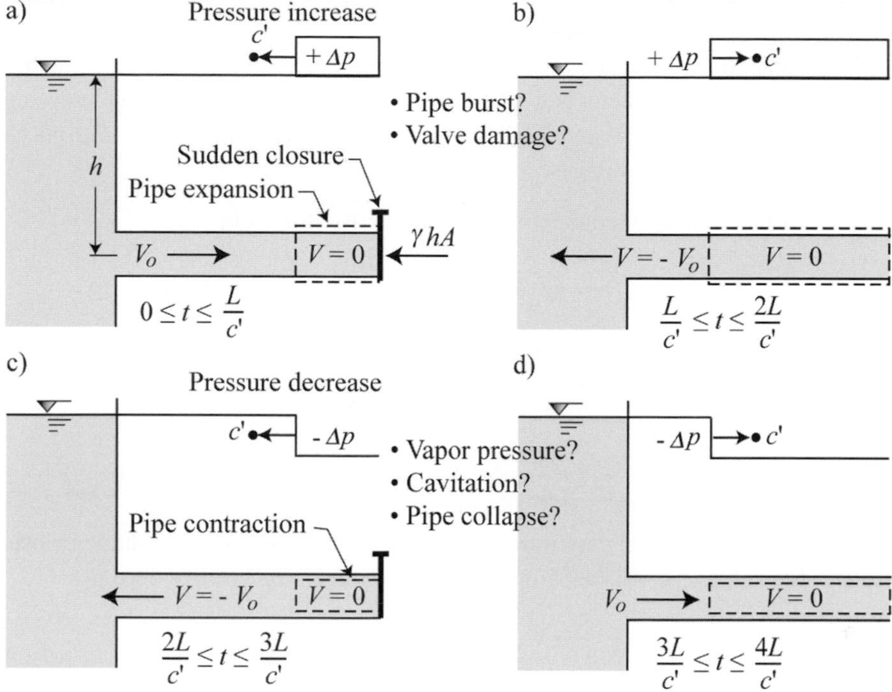

Figure 6.5 Wave propagation in a pipe after a sudden valve closure

Of course, the very high pressures induced by the product of high flow velocities and high wave celerities can lead to pipe bursting. Also, during the phase of pipe contraction, the pressure may become close enough to the vapor pressure to cause cavitation. At all times, the minimum negative pressure head can only be -10 m and cannot be below the absolute zero pressure, as shown from experiments in Figure 6.6. Example 6.4 determines the maximum pressure from a sudden pipe closure.

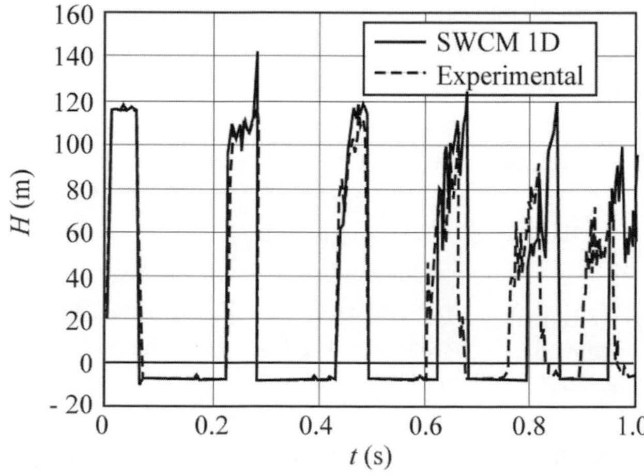

Figure 6.6 Pressure-head measurements from a water hammer in a pipe (Pezzinga and Santoro 2017)

♦♦ **Example 6.4**: Sudden valve closure in a pipe

A 3,000-ft-long, 36-in.-diameter pipe conveys water at a velocity of 4 ft/s. If the initial pressure at the downstream end is $p_0 = 40$ psi, what maximum pressure will develop at the downstream end when a valve is closed in 1 second? Would a 125- or 250-psi cast-iron pipe hold the pressure increase?

Solution: As a first approximation, consider the celerity in a rigid pipe

$$c = \sqrt{E_w/\rho} = \sqrt{(300{,}000 \times 144)/1.94} = 4{,}720 \text{ ft/s}.$$

Next, the travel time for the wave to propagate back in forth in the pipe is

$$t_T = 2L/c = 2 \times 3{,}000/4{,}720 = 1.27 \text{ s}.$$

Since the closure time $t_C < t_T$, the maximum pressure increase $\Delta p = \rho Vc = 1.94 \times 4 \times 4{,}720 = 36{,}630$ psf $= 254$ psi. The maximum pressure $p_{max} = p + \Delta p = 40 + 254 = 294$ psi. (Note that a 125-psi cast-iron pipe would burst open from this sudden closure.)

From Table 2.2, a 250-psi cast-iron pipe is 2 in. thick. The celerity becomes

$$c' = c/\sqrt{\left[1 + \left(\frac{E_w D}{E_p e}\right)\right]} = 4{,}720/\sqrt{\left[1 + \left(\frac{3 \times 36}{160 \times 2}\right)\right]} = 4{,}080 \text{ ft/s}$$

and

$$p_{max} = p_0 + \Delta p = 40 + \rho V c' = 40 + (1.94 \times 4 \times 4{,}080/144) = 260 \text{ psi}.$$

This pipe would burst open from a sudden closure. Options include a slower flow velocity and/or a gradual closure ($t_C > 2L/c' = 2 \times 3{,}000/4{,}080 = 1.5$ s).

6.3.2 Gradual Valve Closure

In the case of a gradual pipe closure without friction losses with ($t_C > 2L/c'$), the attention turns to the force required to slow down a large mass of water moving in the pipe. In general, converting the momentum flux $\rho QV = \rho AV^2$ into pressure $pA = \rho gHA$ gives $V = C\sqrt{H}$. Figure 6.7 shows a valve fully open under the initial pressure head H_0. In this case, the conversion of potential to kinetic energy results in $V_0 = \sqrt{2gH_0}$, or $V_0 = C\sqrt{H_0}$. A gradual valve closure is assumed $V = V_0\left(1 - \frac{t}{t_C}\right)$.

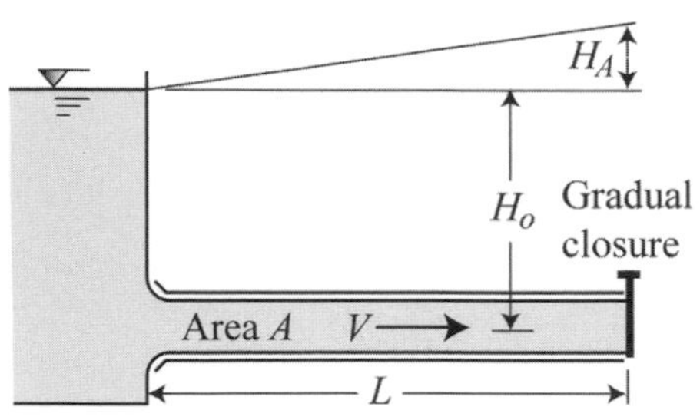

Figure 6.7 Gradual valve closure

Under a slow closure, the head at the valve is expected to increase to $H_0 + H_A$ such that the exit velocity would reach $V = C\sqrt{H_0 + H_A}$. This gives the main approximation for the velocity as a function of time:

$$\frac{V}{V_0} = \left(1 - \frac{t}{t_C}\right)\sqrt{\frac{H_0 + H_A}{H_0}}$$

and its derivative

$$\frac{dV}{dt} = \frac{-V_0}{t_C}\sqrt{\frac{H_0 + H_A}{H_0}}.$$

The maximum pressure-head increase H_A is obtained from the force difference at both ends of the pipe $F = \gamma H_A A$, which decelerates the fluid mass $M = \rho AL$ at an acceleration $a = F/M = -gH_A/L$, or

$$a = \frac{dV}{dt} = \frac{-gH_A}{L} = \frac{-V_0}{t_C}\sqrt{\frac{H_0 + H_A}{H_0}},$$

or

$$\frac{H_A}{H_0} = \frac{LV_0}{gH_0 t_C}\sqrt{1 + \frac{H_A}{H_0}}.$$

The solution to this quadratic equation is known as Allievi's formula, in memory of the Italian Lorenzo Allievi. It defines the maximum head increase H_A (or pressure increase $\Delta p = \gamma H_A$) from

$$\frac{H_A}{H_0} = \frac{\Delta p}{p_0} = \frac{N}{2} + \sqrt{\frac{N^2}{4} + N}, \tag{6.5}$$

where $N = (LV_0/gH_0 t_C)^2 = (\rho LV_0/p_0 t_C)^2$, given the fluid mass density ρ; the initial pressure p_0, velocity V_0 and head H_0; the pipe length L; and the pipe closure time t_C. The maximum head becomes $H_0 + H_A$, and the total pressure is $p = p_0 \pm \Delta p$. Example 6.5 shows how to calculate the pressure increase in a pipe after a gradual valve closure.

♦ **Example 6.5:** Gradual valve closure in a pipe

A 2-km-long cast-iron pipe conveys a discharge of 27 m^3/s in a 5-m-diameter pipe with an exit in air through a 1-m-diameter valve. The initial pressure head is 60 m and the pipe is 5-cm thick. If the valve closes in 5 seconds, what pressure surge would develop?

Solution:

Step (1): The main conditions for this pipe are $D = 5$ m, $A = 19.6\ m^2$. Assume $f = 0.015$, and find the steady friction losses: $V_0 = 4Q/\pi D^2 = 4 \times 2.7/\pi\, 5^2 = 1.37$ m/s,

$$\Delta H = \frac{fL}{D}\frac{V^2}{2g} = \frac{0.015 \times 2{,}000}{5}\,\frac{1.37^2}{2 \times 9.81} = 0.6 \text{ m}$$

and friction is negligible;

or $\Delta p = \gamma H_o = 9{,}810 \times 60 = 589$ kPa (or 85.4 psi).

Step (2): Is it a sudden or gradual closure?

The bulk modulus for water is $E_w = 2.1 \times 10^9$ Pa and, for the cast-iron pipe, $E_p = 1.1 \times 10^{11}$ Pa. The wave celerity in the pipe is

$$c' = \sqrt{\frac{E_w}{\rho}\left[\frac{1}{1+\left(\frac{E_w D}{E_p e}\right)}\right]} = \sqrt{\frac{2.1 \times 10^9}{1{,}000}\left[\frac{1}{1+\left(\frac{2.1 \times 10^9 \times 5}{1.1 \times 10^{11} \times 0.05}\right)}\right]} = 850 \text{ m/s}.$$

The back-and-forth travel time for the wave is $t_T = 2L/c' = 2 \times 2{,}000/850 = 4.7$ s, which is less than the closure time $t_C = 5$ s, resulting in a gradual valve closure.

Step (3): Calculate the pressure increase.

The pressure increase from Allievi's formula is estimated as

$$N = \left(\frac{\rho L V_0}{p_0 t_C}\right)^2 = \left(\frac{L V_0}{g H_0 t_C}\right)^2 = \left(\frac{2{,}000 \times 1.37}{9.81 \times 60 \times 5}\right)^2 = 0.867.$$

Then,

$$\frac{H_A}{H_0} = \frac{N}{2} + \sqrt{\frac{N^2}{4} + N} = \frac{0.867}{2} + \sqrt{\frac{0.867^2}{4} + 0.867} = 1.46.$$

This corresponds to a pressure head $H_A = 1.46 H_0 = 1.46 \times 60 = 88$ m, and a corresponding maximum pressure $p = p_o + \Delta p = p_0 + \gamma H_0 = 589 + (9.81 \times 88) = 589 + 863 = 1{,}452$ kPa (or 210 psi).

For comparison, the sudden closure formula would yield a pressure increase $\Delta p = \rho V c' = 1{,}000 \times 1.37 \times 850 = 1{,}173$ kPa (or 170 psi) and a total pressure $p = p_0 + \Delta p = 589 + 1{,}173 = 1{,}762$ kPa (or 255 psi) which would burst a 250-psi pipe.

With reference to Figure 6.5, each high-pressure pulse for a sudden pipe closure would reoccur every period $T = 4L/c' = 4 \times 2{,}000/850 = 9.4$ s.

The benefits of a gradual closure to prevent pipe bursting can be significant.

6.3.3 Valve Opening

Opening valves in long pipes with friction can require a long time to reach steady flow conditions. After a valve is suddenly opened in Figure 6.8, the total head H accelerates the flow. As the velocity increases, the pressure head is reduced by friction f until steady flow conditions are reached.

The steady velocity V_0 is given from

$$H = \frac{fL}{D}\frac{V_0^2}{2g},$$

or

$$\frac{gH}{L} = \frac{fV_0^2}{2D}$$

and

$$\frac{2D}{f} = \frac{LV_0^2}{gH}.$$

The equation of motion,

$$\gamma A\left(H - \frac{fL}{D}\frac{V^2}{2g}\right) = m\frac{dV}{dt} = \frac{\gamma AL}{g}\frac{dV}{dt},$$

reduces to

$$\frac{dV}{dt} = \frac{gH}{L} - \frac{fV^2}{2D} = \frac{f}{2D}\left(V_0^2 - V^2\right).$$

By solving for dt and integrating gives

$$\int_0^t dt = \frac{2D}{f}\int_0^V \frac{dV}{V_0^2 - V^2} = \frac{LV_0^2}{gH}\int_0^V \frac{dV}{V_0^2 - V^2}$$

or

$$t = \frac{LV_0}{2gH}\ln\left[\frac{1 + (V/V_0)}{1 - (V/V_0)}\right]. \tag{6.6}$$

We can also use the hyperbolic tangent function $\tanh(x) = (e^x - e^{-x})/(e^x + e^{-x})$ since we also know that $\ln[(1 + x)/(1 - x)] = 2\tanh^{-1}(x)$. Therefore, we get

$$t = \frac{LV_0}{gH}\tanh^{-1}\left(\frac{V}{V_0}\right),$$

or

$$\frac{V}{V_0} = \tanh\left(\frac{gHt}{LV_0}\right). \tag{6.7}$$

From Figure 6.8, we notice that $t \approx 2LV_0/gH$, with more details in Example 6.6.

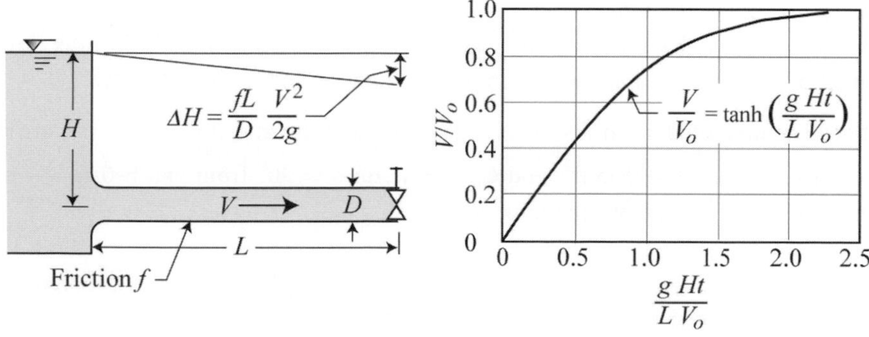

Figure 6.8 Definition sketch for a valve opening

Example 6.6: Valve opening

Consider a pipe $L = 10{,}000$ ft, $D = 8$ ft, $f = 0.03$ and $H = 60$ ft. How long does it take after opening the valve to reach 90% of the steady-state flow velocity $V = 0.9\,V_0$?

Solution: The equilibrium velocity is $V_0 = \sqrt{\frac{2gHD}{fL}} = \sqrt{\frac{2\times 32.2\times 60\times 8}{0.03\times 10{,}000}} = 10.2$ ft/s and after substituting $V = 0.9V_0$ we get

$$t_{0.9} = \frac{LV_0}{2gH}\ln\left(\frac{1.9}{0.1}\right) = \frac{10{,}000\times 10.2}{2\times 32.2\times 60}\ln(19) = 78 \text{ s}.$$

6.3.4 Emptying Large Tanks

To empty a circular tank of diameter D through an orifice of diameter d, as shown in Figure 6.9, we consider the exit flow velocity through the orifice is $V_0 = C_c\sqrt{2gh}$ where C_c is the orifice coefficient. The outflowing jet forms a contracted vein of fluid called "vena contracta," a term coined by Evangelista Torricelli in the seventeenth century. Torricelli was a student of Galileo.

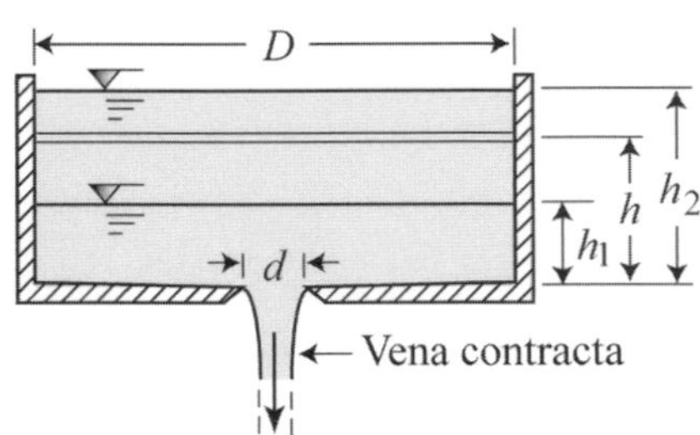

Figure 6.9 Orifice flow

The second condition with $Q = (\pi d^2/4)V_0$ stems from continuity which implies that the volumetric change is $Qdt = A_0V_0dt = (\pi d^2/4)C_c\sqrt{2gh}dt = -(\pi D^2/4)dh$.

Separating t and h gives

$$dt = -\frac{D^2dh}{d^2C_c\sqrt{2gh}},$$

which is integrated from h_2 to h_1 as

$$t_{0-1} = \frac{2D^2}{C_cd^2\sqrt{2g}}\left(\sqrt{h_2} - \sqrt{h_1}\right). \tag{6.8}$$

Orifice flow through a large tank is analyzed in Example 6.7.

Example 6.7: Draining a tank

A 28-ft-diameter tank contains 18 ft of water. What is the time to drain the tank through a 1-inch orifice ($\alpha = 90°$) at the bottom of the tank, as shown in Fig. E-6.7.

Solution: The area of the tank is $A_T = \pi(28)^2/4 = 616$ ft^2 and the area of the orifice is $A_0 = \pi/(4\times 12^2) = 0.00545$ ft^2, and $C_c = 0.61$ for $\alpha = 90°$ from Fig. E-6.7.

The time needed to empty the tank is

$$t_{0-1} = \frac{2D^2}{C_cd^2\sqrt{2g}}\left(\sqrt{h_o} - \sqrt{h_1}\right) = \frac{2\times 28^2\times 12^2}{0.61\sqrt{2\times 32.2}}\left(\sqrt{18} - 0\right) = 195{,}000 \text{ s}$$
$$= 2.26 \text{ days}$$

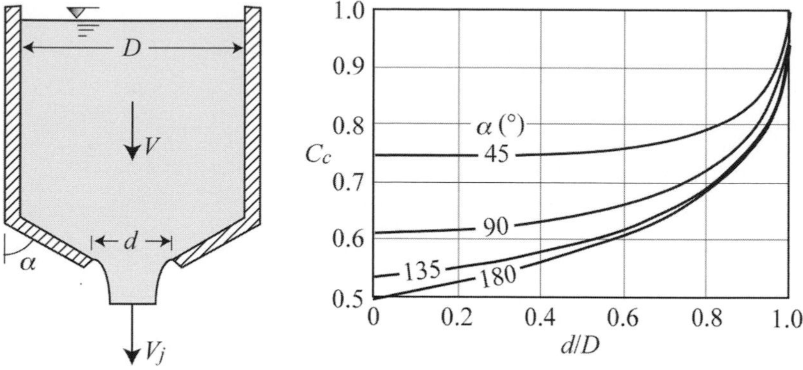

Fig. E-6.7 Vena contracta coefficient

The reader will notice that the tank would empty faster when $\alpha < 90°$.

6.4 Surge Tanks

Surge tanks are covered in Section 6.4.1 and pressure reduction in Section 6.4.2.

6.4.1 Surge Tanks

Surge tanks are a major part of hydropower projects and serve to reduce pressure surges when turbines or pump operations are stopped. A surge tank is typically a large vertical pipe connected to the main pipe. We are concerned with the flow oscillations between the large reservoir and the surge tank. As sketched in Figure 6.10, the water level in the tank rises to a level S above the initial level.

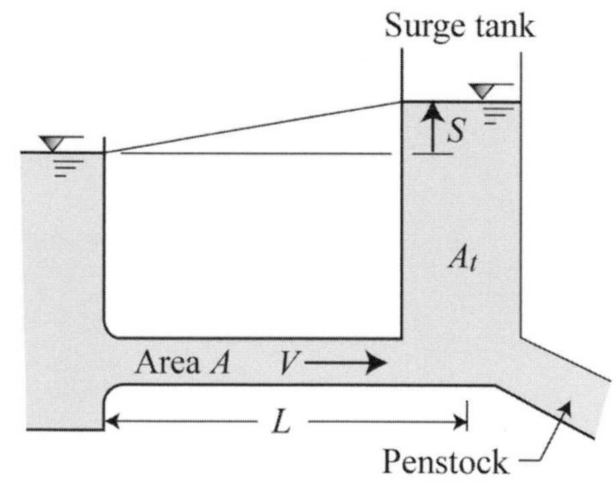

Figure 6.10 Surge tank

The fluid motion is analyzed for the case of negligible friction losses between the reservoir and the surge tank. The flow discharge Q_0 through the pipe is given by the pipe cross section A and steady flow velocity V_0 before the closure. The flow will enter the surge tank of cross-sectional area A_t and the velocity in the tank corresponds to the change in water level S over time dS/dt; therefore, the continuity relationship is written as $\frac{dS}{dt} = \frac{AV}{A_t}$. The momentum equation $F = m\frac{dV}{dt}$ is applied to the fluid mass in the pipe $(\rho AL)\frac{dV}{dt} = -\rho gAS$, which gives

$$\frac{dV}{dt} = \frac{-gS}{L} = \frac{dV}{dS}\frac{dS}{dt} = \frac{AV}{A_t}\frac{dV}{dS}.$$

Integrating $VdV = \frac{-gA_tSdS}{AL}$, with $V = V_0$ and $S = 0$ at $t = t_o$ gives $V^2 = \frac{-gA_tS^2}{AL} + V_0^2$. The maximum surge S_{max} is obtained when $V = 0$,

$$S_{max} = V_0\sqrt{\frac{AL}{A_t g}},$$

and

$$T = 2\pi\sqrt{\frac{A_t L}{Ag}}; \tag{6.9}$$

Chapter 7 will provide more details on the derivation of the oscillation period T. Both S_{max} and T depend on the tank area A_t, the pipe length L, the pipe area A and the initial flow velocity V_0. Typical surge tank calculations are shown in Example 6.8; and Case Study 6.1 provides some details of a surge tank.

♦ **Example 6.8:** Surge tank

A 10-ft-diameter pipe carries 843 cfs over a length of 3,000 ft before reaching a surge tank with a 314-ft^2 cross-sectional area. During a sudden turbine shutdown, what will be the magnitude and time to reach the maximum water level S_{max} in the surge tank?

Solution: The pipe area is $A = 0.25\pi D^2 = 25\pi$ and the velocity $V_0 = Q/A = 843/25\pi = 10.7$ ft/s. The maximum surge height is

$$S_{max} = V_0\sqrt{\frac{AL}{A_t g}} = 10.73\sqrt{\frac{25\pi \times 3{,}000}{314 \times 32.2}} = 51.8 \text{ ft}$$

at a quarter of the flow oscillation period, or

$$t_{max} = \frac{T}{4} = \frac{\pi}{2}\sqrt{\frac{A_t L}{Ag}} = \frac{\pi}{2}\sqrt{\frac{314 \times 3{,}000}{25\pi \times 32.2}} = 30 \text{ s.}$$

♦ **Case Study 6.1:** The Edmonston surge tank, USA

This case study supplements the information presented in Case Study 4.1 from the California Department of Water Resources (Deukmejian et al. 1985). The A.D. Edmonston Pumping Plant has a surge tank located at the top of the two pipelines, as shown in Fig. CS-6.1. At the top, the water enters a cylindrical surge tank which is 62 ft high and 50 ft in diameter. The normal water level in the surge tank is 3,165 ft with a maximum at 3,180 ft. The two main pipes are 8,400-ft long and each conveys 2,205 cfs. The pipe diameters are 12.5 ft for the lower half and 14 ft for the upper half. Each pipe contains 1.2×10^6 ft^3 of water. The closure time for the 4-ft-diameter ball valve at the base of the pipelines is 80% in 10 s and the remainder in the next 20 s. The reader can check that this time of closure is shorter than: (1) the time to fill or drain the surge tank;

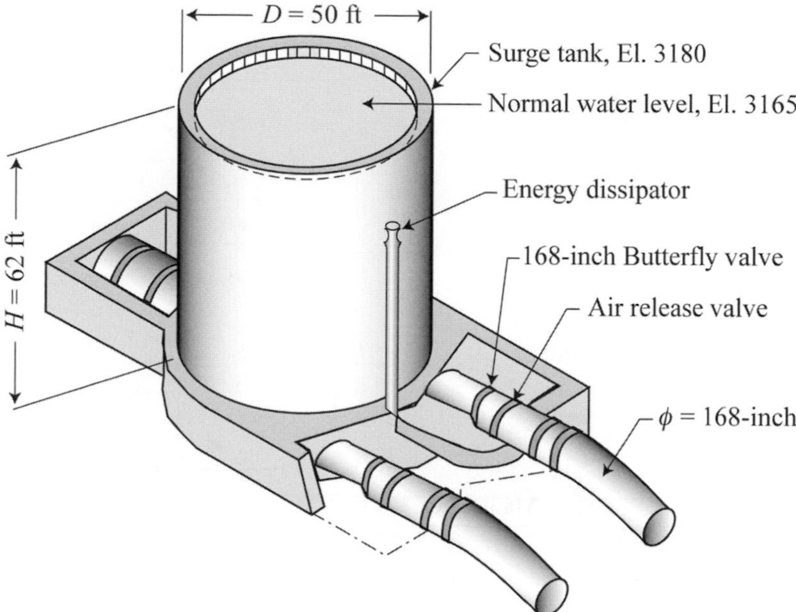

Fig. CS-6.1 Edmonston surge tank

(2) the period of oscillations in the pipeline and surge tank; and the time to drain the pipeline through the ball valve at the base.

6.4.2 Pressure Reduction

Other devices to reduce excessive pressure in pipes include compressed air chambers, and relief valves in Figure 6.11. They are usually placed beside pumps to prevent the pressure surge caused by power outages. These devices are connected to pipe systems to reduce the pressure surge when a pump suddenly stops operating.

Additional Resources

Benjamin Wylie from Michigan advanced the analysis of fluid transients (Wylie and Streeter 1978). Additional resources on water hammer and surge tanks include Parmakian (1963), Rich (1984a and b), Ghidaoui et al. (2005), Chaudhry (2014) and Guo et al. (2017). Recent reviews on hydraulic transients and negative pressure include Chaudhry (2020), Ferras et al. (2020) and Karney (2020).

EXERCISES

These exercises review the essential concepts from this chapter.

1. What does the modulus of elasticity measure?
2. What is a wave celerity?
3. Is the wave celerity of water greater in the ocean or in a pipe? Why?
4. Why is the wave celerity in pipes important?
5. Why should we care about air bubbles in pipes?

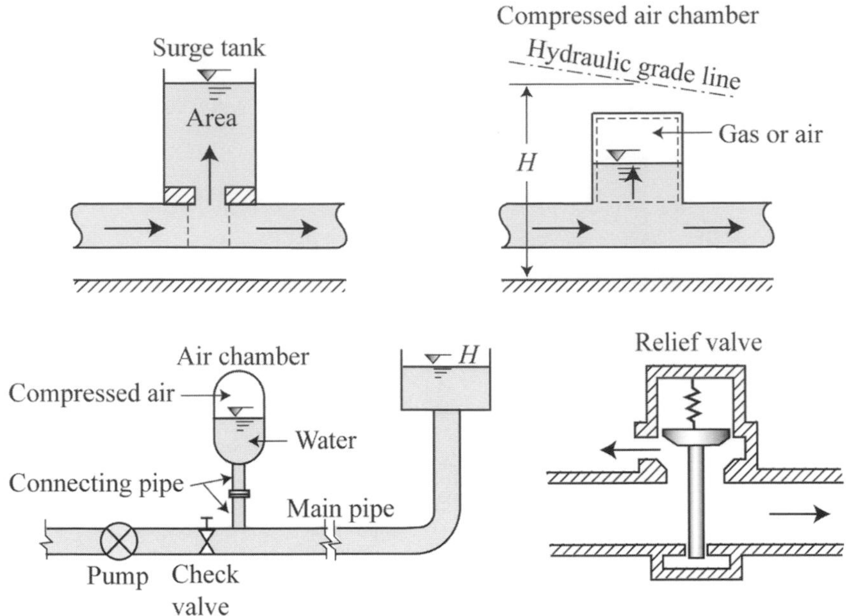

Figure 6.11 Pressure reduction in pipes

6. Why do we use twice the pipe length in the analysis of water hammer?
7. What is the lowest pressure generated from a sudden valve closure?
8. Why is the time of closure of valves important?
9. Are the water compressibility effects included in Allievi's formula?
10. Is the pipe friction loss included in Allievi's formula?
11. What is a vena contracta? Does it increase or decrease the flow rate of a given opening?
12. What does a relief valve do?
13. What is the purpose of a surge tank?
14. True or false?
    (a) The modulus of elasticity of water does not change significantly with temperature or pressure.
    (b) The wave celerity in a pipe is approximately one mile per hour.
    (c) Water is more elastic than steel.
    (d) The pipe expansion under pressure increases with pipe thickness.
    (e) Thicker pipes contribute to higher pressures from sudden closures.
    (f) Lower wave celerities reduce the pressure increase from a sudden valve closure.
    (g) Compared with steel, PVC pipes reduce the pressure from sudden valve closures.
    (h) The time required to establish the flow in a valve opening is approximately $2LV/gH$.
    (i) Compressed air is an effective way to reduce the pipe pressure fluctuations.
    (j) The present worth analysis distributes the cost of infrastructure over long periods of time.

SEARCHING THE WEB

Find photos of the following features, study them carefully and write down your observations.

1. ♦ Burst pipes.
2. ♦ Surge tanks.

PROBLEMS

Hydraulic Transients and Surge Tanks

1. ♦♦ A cast-iron 18-in.-diameter pipeline carries water at 70 °F over 1,000 ft from a reservoir to a powerhouse. The flow velocity is 5 ft/s and the initial pressure is 46 psi. If the cast-iron pipe is 1-in. thick, consider the following questions.
   (a) The wave celerity in this pipe.
   (b) The added pressure generated by a sudden closure.
   (c) Would the sudden closure cause cavitation or burst the 175-psi pipe open?
   (d) How long would the closure time have to be to reduce the maximum pressure?
   (e) What is the increased pressure if the time of closure is 1 second?
   (f) A 10-ft-diameter surge tank is built halfway in the line. Find the maximum surge height?
   (g) What is the period of oscillations in the surge tank?
2. ♦♦ The City of Thornton considers a water pipeline. From a 2015 newspaper article, you extract the main statements: (a) current population 138,000; (b) demographic expansion up to 242,000 in 10 years; (c) pipeline delivery of 14,000 ac. · ft of water per year; (d) 60-mile length; (e) 48-inch pipe diameter; (f) preliminary cost estimate $400 million; and (g) on line in 2025. A second newspaper article in 2017 indicates that 1 ac. · ft of water meets the demand of three–four urban households and the value of water recently increased from $6,500 to $16,700 per ac. · ft. Consider the following questions.
   (a) What is the cost of the pipeline per linear foot?
   (b) If the population starts at 138,000 residents, what is the annual population increase?
   (c) What is the continuous equivalent discharge: (i) in cfs for 14,000 ac. · ft per year; and (ii) in gallons per household per day?
   (d) Assuming $p_0 = 40$ psi, what is the pressure head in the pipe?
   (e) How long would it take to establish 95% of the equilibrium flow velocity after a sudden opening?
   (f) Assuming that friction is negligible, what is the period of oscillations in this pipe?
   (g) Assuming a rigid pipe with $f = 0.02$ and $p_0 = 40$ psi, what would be the pressure generated from a sudden closure?
   (h) What if a valve closure requires 10 seconds?
   (i) Based on the value of water from the second article, do you think this project has a high or low benefit to cost ratio?

Cost Analysis (see also Appendix A)

3. ♦♦ The cost to build a hydropower plant is $50,000,000. The annual energy generation is equal to 52,000 MWh (megawatt-hours) and the value of electricity is constant at $70 per MWh. The period of contract for energy supply is 20 years and the annual interest rate is 5%.
   (a) Is this contract lucrative? [Hint: use present worth analysis.]
   (b) What should be the minimum value of electricity to break even in this contract?
   (c) Suppose that value of electricity is fixed, but you can sell the hydropower plant at the end of the contract. What should be the minimum sale value?
4. ♦♦ Consider the 7.5-m-diameter penstock in Case Study 5.1. You could increase your revenue by $250,000 per year over a period of 30 years by enlarging the penstock diameter to 9.5 m. Consider interest rates of 0% and 5% and compare the results.
   (a) What is the incremental construction cost?
   (b) Is this contract lucrative over 30 years if the interest rate is 0%?
   (c) Is this expansion project valuable if the interest rate is 5%?
   (d) Is this contract profitable if the interest rate is 4% but you lose the first year of revenue because of the project construction?

7 Pipe Flow Oscillations

This brief chapter supplements Chapter 6 and is normally not covered in undergraduate courses. The material helps graduate and honors students bridge the gap between spring–mass systems covered in engineering mechanics and flow oscillations in pipes. This more advanced treatment focuses on fluid oscillations in pipes without friction in Section 7.1, with laminar friction covered in Section 7.2, turbulent friction in Section 7.3, and oscillations between reservoirs considered in Section 7.4.

7.1 Oscillations without Friction

We first review spring–mass systems in Section 7.1.1, with applications to water in Section 7.1.2.

7.1.1 Spring–Mass Oscillations

Let's consider the free vibrations of a spring–mass system in Figure 7.1.

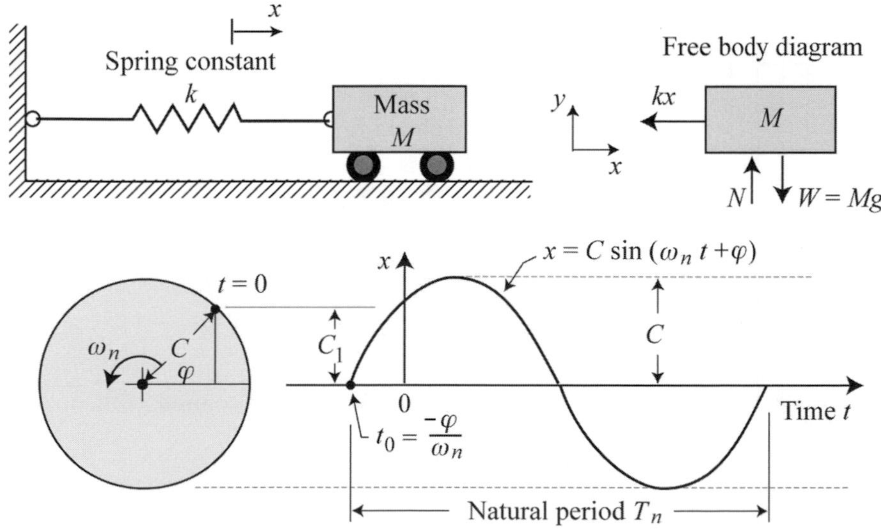

Figure 7.1 Oscillations of a spring–mass system

The short-hand notation uses dots for time derivatives, $\dot{x} = dx/dt$ and $\ddot{x} = d^2x/dt^2$, and the force balance from the free-body diagram is $\sum F_x = -kx = M\ddot{x}$, or

$$\ddot{x} + \frac{k}{M}x = 0. \tag{7.1}$$

The solution for the displacement x as a function of time t is $x = C\sin(\omega_n t + \varphi)$ where C is the amplitude of the motion, ω_n is the natural circular frequency and φ is the initial angle at $t = 0$. The velocity and acceleration are $V = \dot{x} = \omega_n C\cos(\omega_n t + \varphi)$ and $a = \ddot{x} = -\omega_n{}^2 C\sin(\omega_n t + \varphi)$. Substitution into Eq. (7.1) gives $-\omega_n{}^2 C\sin(\omega_n t + \varphi) + \frac{k}{M}C\sin(\omega_n t + \varphi) = 0$ from which we obtain the natural circular frequency ω_n in radians per second,

$$\omega_n = \sqrt{\frac{k}{M}}. \tag{7.2}$$

The corresponding natural frequency is $f_n = \frac{\omega_n}{2\pi}$ in cycles per second or hertz (1 Hz = 60 rpm), and the natural period of oscillations without friction is $T_n = \frac{1}{f_n} = \frac{2\pi}{\omega_n}$. Instead of using the initial angle φ, the equation of motion can be written as $x = C_1\cos(\omega_n t) + C_2\sin(\omega_n t)$. The initial displacement and velocity ($t = 0$) are $x_0 = C_1 = C\sin\varphi$ and $V_0 = \dot{x}_0 = \omega_n C_2 = \omega_n C\cos\varphi$, or $\varphi = \tan^{-1}(\omega_n x_0/V_0)$.

7.1.2 Flow Oscillations without Friction

The oscillations of a liquid without friction in the U-tube sketched in Figure 7.2 are now considered. Given the cross-sectional area A of the tube and the length L of the water column, the mass to be accelerated is ρAL. Euler's equation of motion considers that the pressure force $\rho g(z_2 - z_1)A$ equals the fluid mass times its acceleration:

$$pA = \rho g(z_2 - z_1)A = -\rho AL\frac{dV}{dt}.$$

Dividing by ρAL, and considering both $(z_2 - z_1) = 2z$ and $a = \ddot{z} = \frac{dV}{dt}$, we obtain the main equation $\ddot{z} + \frac{2g}{L}z = 0$.

This equation is simply Eq. (7.1) where $\omega_n^2 = \frac{k}{M} = \frac{2g}{L}$, with the natural period of oscillations,

$$T_n = \frac{2\pi}{\omega_n} = 2\pi\sqrt{\frac{L}{2g}}. \tag{7.3}$$

The fluid elevation z in a water column without friction varies with time t as

$$z = C_1\cos\sqrt{\frac{2g}{L}}t + C_2\sin\sqrt{\frac{2g}{L}}t. \tag{7.4}$$

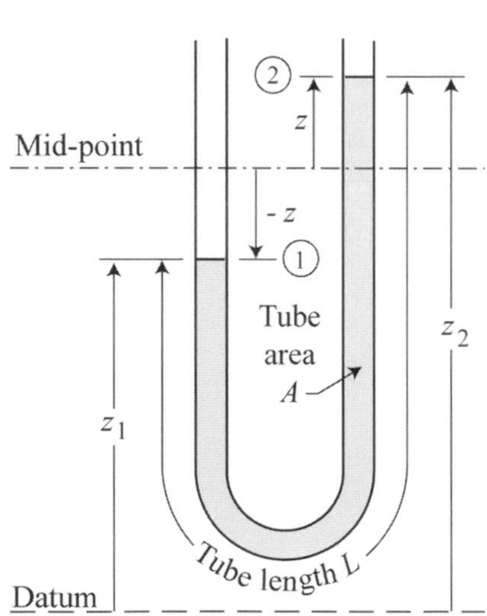

Figure 7.2 Oscillations without friction

The position z_0 at $t = 0$ defines $C_1 = z_0$ and the initial velocity V_0 gives

$$C_2 = V_0/\omega_n = V_0\sqrt{L/2g}.$$

Numerical solutions to the equation of motion are also possible (and often preferable). The acceleration $\ddot{z}$ is first solved and used to calculate the velocity from $V = V_0 + \ddot{z}\Delta t$. The displacement over this short time interval Δt is then simply obtained from $z = z_0 + V\Delta t$. The method becomes increasingly accurate as $\Delta t \to 0$. Example 7.1 shows detailed calculations for pipe flow oscillations without friction.

♦♦ Example 7.1: Pipe flow oscillations without friction

A frictionless fluid column 4.025-ft long has an initial upward velocity of $V_0 = 4$ ft/s at $z_0 = 1$ ft. Find: (1) the period; (2) the equation of motion; and (3) the maximum elevation z_{max} and velocity V_{max}.

Solution:

(1) The natural circular frequency is

$$\omega_n = \sqrt{\frac{2g}{L}} = \sqrt{\frac{2 \times 32.2}{4.025}} = 4 \text{ rad/s}$$

and the natural period of oscillations is

$$T_n = \frac{1}{f_n} = \frac{2\pi}{\omega_n} = 1.57 \text{ s.}$$

(2) From the initial conditions $C_1 = z_0 = 1$ and $C_2 = V_0/\omega_n = 4/4 = 1$, we obtain the equation of motion $z = \cos(4t) + \sin(4t)$.

Alternatively, $C = \sqrt{C_1^2 + C_2^2} = \sqrt{2}$ and $\varphi = \tan^{-1} C_1/C_2 = \pi/4 = 45°$, and $z = \sqrt{2}\sin(4t + \pi/4)$.

(3) The maximum elevation z_{max} when $\sin(4t_m + \pi/4) = 1$ gives $z_{max} = C = \sqrt{2} =$ 1.41 ft.

The time for z_{max} occurs when $\dot{z} = -4\sin(4t) + 4\cos(4t) = 0$ or $\tan(4\,t_{max}) = 1$ and $t_{max} = \frac{\pi}{16} = 0.196$ s.

The maximum velocity occurs when the acceleration $\ddot{z} = -16\cos(4t) - 16\sin(4t) = 0$, which is $\tan(4t_{Vmax}) = -1$, or the time where the vecocity is maximum is

$$t_{Vmax} = -\frac{\pi}{16}, \text{ or } = \frac{7\pi}{16} = 1.374 \text{ s,}$$

and the maximum velocity becomes

$$V_{max} = \dot{z}_{max} = -4\sin\left(\frac{-4\pi}{16}\right) + 4\cos\left(\frac{-4\pi}{16}\right) = 5.66 \text{ ft/s.}$$

Alternatively, we simply obtain the maximum velocity from $V_{max} = \omega_n C = 4\sqrt{2} =$ 5.66 ft/s as shown in Fig. E-7.1.

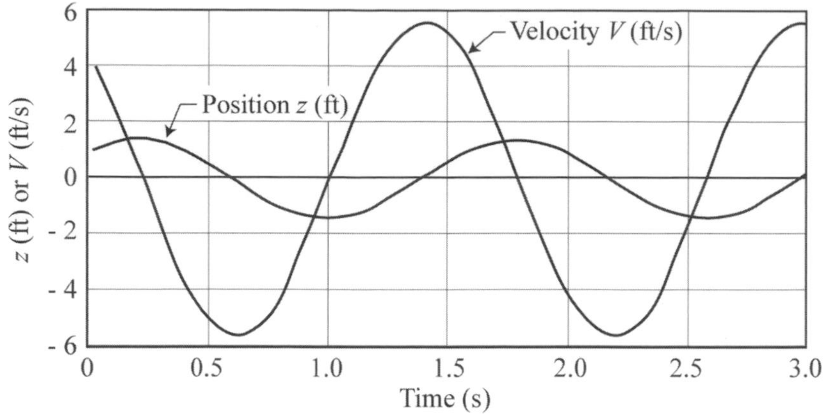

Fig. E-7.1 Pipe flow oscillations without friction

7.2 Oscillations for Laminar Flow

We first review damped spring–mass systems in Section 7.2.1 with applications to laminar flow in capillary tubes in Section 7.2.2.

7.2.1 Damped Spring–Mass Oscillations

Let us first consider the damped vibrations of the spring–mass system shown in Figure 7.3. The damping coefficient represents resistance to the motion. The damping

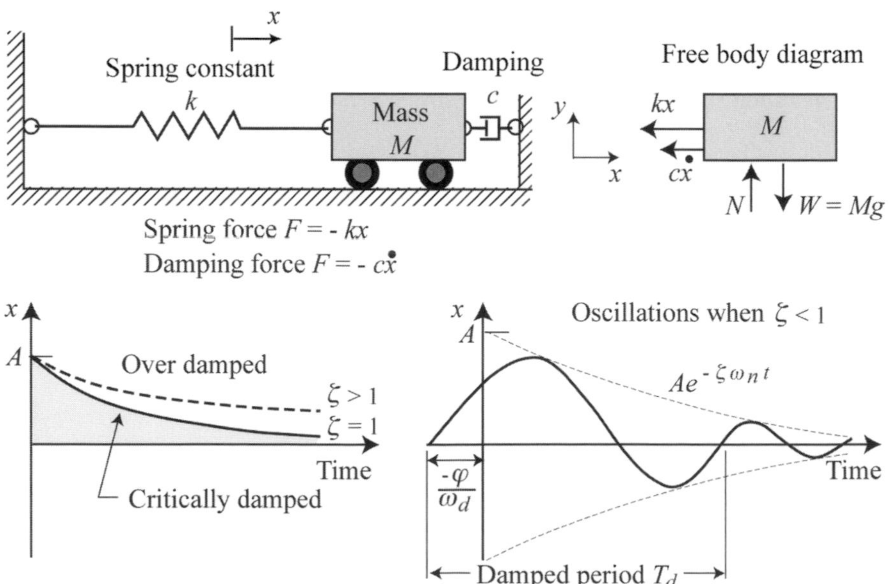

Figure 7.3 Oscillations of a damped spring–mass system

force is opposite to the velocity such that when the mass moves to the right ($\dot{x} > 0$), the force is applied to the left ($F = -c\dot{x}$).

The sum of forces in the x direction includes the damping force component of magnitude $-c\dot{x}$ in the direction opposite to the velocity $\dot{x}$, or $\sum F_x = -kx - c\dot{x} = M\ddot{x}$, or $\ddot{x} + \frac{c}{M}\dot{x} + \frac{k}{M}x = 0$.

The general form of this differential equation is

$$\ddot{x} + 2\zeta\omega_n\dot{x} + \omega_n^2 x = 0, \tag{7.5}$$

with the dimensionless damping coefficient $\zeta = \frac{c}{2M\omega_n}$ and the natural circular frequency $\omega_n = \sqrt{\frac{k}{M}}$.

The general equations for the position and velocity are, respectively,

$$x = Ae^{-\zeta\omega_n t}\sin(\omega_d t + \varphi), \tag{7.6}$$

$$\dot{x} = Ae^{-\zeta\omega_n t}\{[\omega_d\cos(\omega_d t + \varphi)] - [\zeta\omega_n\sin(\omega_d t + \varphi)]\}, \tag{7.7}$$

where $\omega_d = \omega_n\sqrt{1-\zeta^2}$ is the damped circular frequency (the system only vibrates when $\zeta < 1$), with the phase angle φ, damped frequency $f_d = \frac{\omega_d}{2\pi}$ and damped period $T_d = \frac{1}{f_d} = \frac{2\pi}{\omega_d}$.

The solution to the problem of damped oscillations with laminar flow is illustrated for two practical cases: (1) case A with initial displacement; and (2) case B with initial velocity.

Case A: initial displacement at $t = 0$, $x_0 = C$ and $V_0 = \dot{x}_0 = 0$.
From the displacement in Eq. (7.6) at $t = 0$, we obtain $A = C/\sin\varphi$.
From the velocity in Eq. (7.7) at $t = 0$, $\tan\varphi = \frac{\omega_d}{\zeta\omega_n}$, or $\varphi = \tan^{-1}\left(\frac{\omega_d}{\zeta\omega_n}\right)$.
The successive maximum/minimum positions are at times $t = 0, \frac{T_d}{2} = \frac{\pi}{\omega_d}, T_d = \frac{2\pi}{\omega_d} \ldots$
The successive maximim/minimum velocities are at times $t = \frac{T_d}{4} = \frac{\pi}{2\omega_d}, \frac{3T_d}{4} = \frac{3\pi}{2\omega_d} \ldots$
The graph of the position as a function of time is sketched in Figure 7.4.

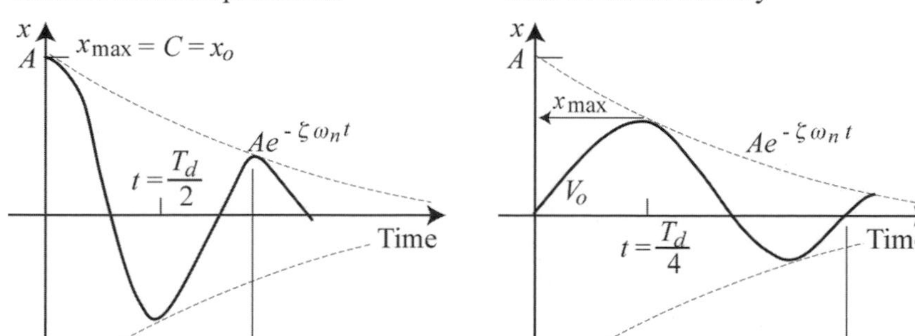

Figure 7.4 Position of a damped spring–mass for initial displacement and velocity

Case B: initial velocity at $t = 0$, $x_0 = 0$ and $V_0 = A\omega_d$.
From the position and velocity in Eqs. (7.6) and (7.7), we obtain $\varphi = 0$ and $A = V_0/\omega_d$.
The successive maximum/minimum positions are at times $t = \frac{T_d}{4} = \frac{\pi}{2\omega_d}, \frac{3T_d}{4} = \frac{3\pi}{2\omega_d} \ldots$
The successive maximum/minimum velocities are at times $t = 0, \frac{T_d}{2} = \frac{\pi}{\omega_d}, T_d = \frac{2\pi}{\omega_d} \ldots$
We are now ready to study the unsteady fluid motion in capillary tubes with viscous damping.

7.2.2 Damped Flow Oscillations in Capillary Tubes

This section is applicable to damped oscillations for laminar flow (Re < 2,000) in capillary tubes of constant area A. With reference to Figure 7.5, the equation of motion includes the friction losses h_f:

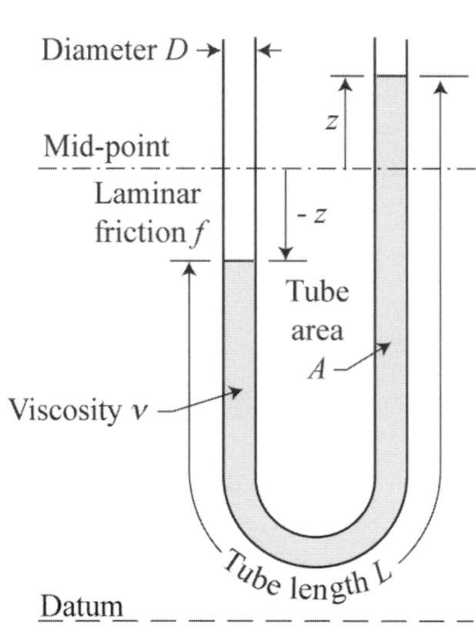

Figure 7.5 Oscillations with viscous friction

$$\rho g(z_2 - z_1)A + \rho g h_f A = -\rho AL\frac{dV}{dt},$$

where the friction loss are

$$h_f = \frac{fL}{D}\frac{V|V|}{2g}.$$

For laminar flow, the Darcy–Weisbach friction factor $f = \frac{64}{\text{Re}} = \frac{64\upsilon}{VD}$ depends on the kinematic viscosity υ of the fluid. Therefore, the friction loss is given by

$$h_f = \frac{64\upsilon}{VD}\left(\frac{L}{D}\right)\frac{V|V|}{2g}.$$

Dividing the equation of motion by ρA with $z_2 - z_1 = 2z$ gives

$$L\frac{dV}{dt} + \frac{32\upsilon LV}{D^2} + 2gz = 0.$$

This is simplified further with $V = dz/dt = \dot{z}$ and $dV/dt = \ddot{z}$; thus,

$$\ddot{z} + \frac{32\upsilon}{D^2}\dot{z} + \frac{2g}{L}z = 0,$$

which is equivalent to Eq. (7.5) where $\omega_n = \sqrt{2g/L}$, and $2\zeta\omega_n = 32\upsilon/D^2$ gives $\zeta = 16\upsilon/\omega_n D^2$, and $\omega_d = \omega_n\sqrt{1-\zeta^2}$.

In summary, the displacement can be analytically defined from Eq. (7.6)

$$z = Ae^{-\zeta\omega_n t}\sin(\omega_d t + \varphi) \tag{7.8}$$

and the velocity is given as

$$V = \dot{z} = Ae^{-\zeta\omega_n t}\{[\omega_d\cos(\omega_d t + \varphi)] - [\zeta\omega_n\sin(\omega_d t + \varphi)]\}, \tag{7.9}$$

where $\omega_n = \sqrt{2g/L}$, $\zeta = 16\upsilon/\omega_n D^2$ and $\omega_d = \omega_n\sqrt{1-\zeta^2}$.

Finally, A and φ depend on the initial displacement and velocity conditions. For instance, the boundary conditions for an initial displacement C without velocity is $A = C/\sin\varphi$, and $\varphi = \tan^{-1}(\omega_d/\zeta\omega_n)$. This may sound complicated without the application detailed in Example 7.2.

♦ Example 7.2: Laminar flow oscillations in a capillary tube

Consider oscillations in a 10-ft-long and 1.0-in.-diameter U-tube containing a fluid more viscous than water, $\upsilon = 1 \times 10^{-4}$ ft^2/s. The initial head difference between both ends of the tube is 16 in. Can you find the equation for the position z as a function of time? Also find the maximum velocity and Reynolds number to double check that the flow is laminar.

Solution:

Step (1): The natural circular frequency is

$$\omega_n = \sqrt{\frac{2g}{L}} = \sqrt{\frac{2 \times 32.2}{10}} = 2.54 \text{ radians per second.}$$

The damping coefficient is

$$\zeta = \frac{16\,\upsilon}{\omega_n D^2} = \frac{16 \times 10^{-4} \times 12^2}{2.54} = 0.091,$$

with oscillations, because $\zeta < 1$.

The damped circular frequency is

$\omega_d = \omega_n\sqrt{1-\zeta^2} = 2.54\sqrt{1-0.091^2} = 2.53$ rad/s and $\zeta\omega_n = 0.091 \times 2.54 = 0.231$.

Step (2): As in case A, the initial conditions give the phase angle:

$$\varphi = \tan^{-1}\left(\frac{\omega_d}{\zeta\omega_n}\right) = \tan^{-1}\left(\frac{2.53}{0.231}\right) = 1.48 \text{ rad} = 85°.$$

The initial conditions are $C = 0.5 \times 16/12 = 0.667$ ft, and

$$A = \frac{C}{\sin\varphi} = \frac{0.667}{\sin 85°} = 0.669 \text{ ft.}$$

Step (3): From Eqs. (7.8) and (7.9), the height in ft and velocity in ft/s at time t in seconds are, respectively,

$$z = 0.669e^{-0.231t}\sin(2.53\,t + 1.48 \text{ rad})$$

and

$$V = 0.669e^{-0.231t}\{[2.53\cos(2.53\,t + 1.48 \text{ rad})] - [0.231\sin(2.53t + 1.48)]\}.$$

Step (4): The minimum velocity V_{min} at $t = T_d/4 = \pi/2\omega_d = \pi/(2 \times 2.53) = 0.62$ s gives $V_{min} = -1.46$ ft/s and

$$\mathrm{Re}_{min} = \frac{|V|D}{\upsilon} = \frac{1.46 \times 1}{12 \times 10^{-4}} = 1{,}220$$

and the flow is laminar because Re < 2,000.

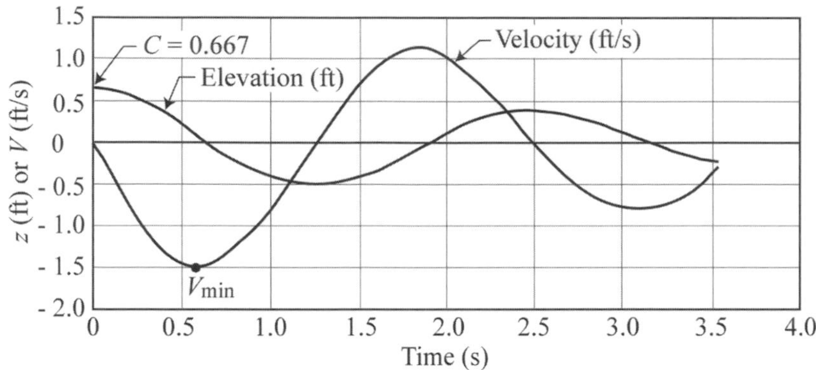

Fig. E-7.2 Pipe flow with viscous oscillations

7.3 Oscillations for Turbulent Flow

The case of oscillations for turbulent flows in large pipes is far more complicated because the equation of motion becomes nonlinear. The governing equation was derived in Section 7.2.2, except that the Darcy–Weisbach friction coefficient is now constant:

$$\rho g(z_2 - z_1)A + \rho g\left(\frac{fL}{D}\frac{V|V|}{2g}\right)A = -\rho AL\frac{dV}{dt}.$$

Dividing by ρAL and given $(z_2 - z_1) = 2z$, with $V = \dot{z}$ and $a = dV/dt = \ddot{z}$, we obtain

$$\ddot{z} + \frac{f}{2D}\dot{z}|\dot{z}| + \frac{2gz}{L} = 0. \tag{7.10}$$

This is a nonlinear differential equation because of the squared velocity term. The absolute value of the velocity term is needed to ensure that the resistance opposes the velocity in both flow directions. The equation can be integrated once with respect to t and the first integration is given here without derivation (see Rainville 1964, Streeter 1971):

$$\left(\frac{dz}{dt}\right)^2 = \frac{4gD^2}{f^2L}\left(1 + \frac{fz}{D}\right) + Ce^{\left(\frac{fz}{D}\right)}$$

The integration constant C is evaluated for minimum/maximum values, i.e. $z = z_m$ at $dz/dt = 0$ (where the subscript m represents the maximum or minimum),

$$C = -\frac{4gD^2}{f^2L}\left(1 + \frac{fz_m}{D}\right)e^{\left(-\frac{fz_m}{D}\right)},$$

and the velocity relationship becomes

$$V^2 = \left(\frac{dz}{dt}\right)^2 = \frac{4gD^2}{f^2L}\left\{\left(1+\frac{fz}{D}\right) - \left(1+\frac{fz_m}{D}\right)\exp\left[\frac{f(z-z_m)}{D}\right]\right\}. \tag{7.11}$$

This equation is useful to determine the successive peaks (high z_m^+ and low z_{m+1}^-) obtained when $V = \dot{z} = 0$, and the equation simplifies to

$$\left(1+\frac{fz_m}{D}\right)\exp\left(-\frac{fz_m}{D}\right) = \left(1+\frac{fz_{m+1}}{D}\right)\exp\left(-\frac{fz_{m+1}}{D}\right).$$

The main equation to be solved with $\phi = fz/D$ is

$$F(\phi) = (1+\phi)e^{-\phi}. \tag{7.12}$$

With changes in flow direction, the successive values of ϕ are obtained by alternating the signs of ϕ and solving $F(\phi) = (1+\phi)e^{-\phi} = (1-\phi)e^{+\phi}$. This is plotted on Figure 7.6 with an illustrated example.

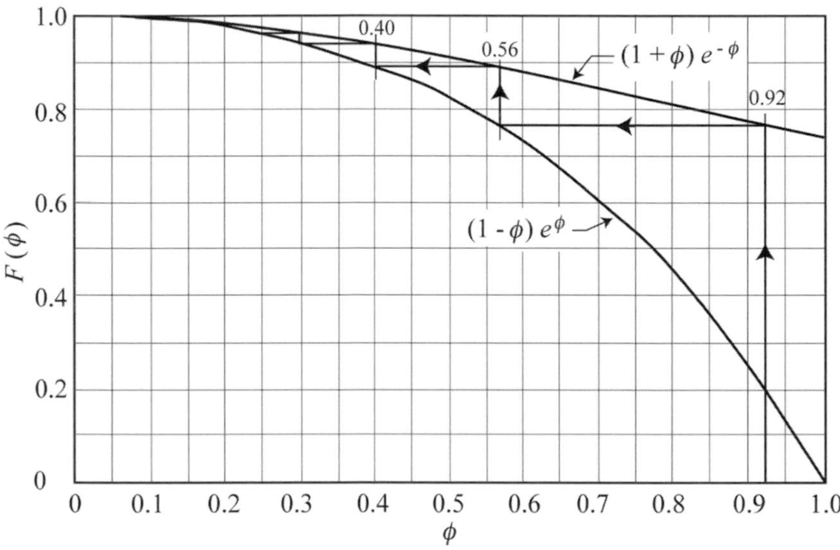

Figure 7.6 Plot of $F(\phi) = (1+\phi)e^{-\phi}$ for turbulent pipe flow oscillations

The graphical procedure to determine successive peaks is illustrated in Figure 7.6. With an example starting at $\phi_1 = 0.92$, we obtain $F(\phi) = (1+\phi)e^{-\phi} = 1.92\,e^{-0.92} = 0.77$, which corresponds to $\phi_2 = -0.56$ because $F(\phi) = (1+\phi)e^{-\phi} = (1-0.56)\,e^{+0.56} = 0.77$.

To find the next minimum, we reverse the sign of the last value $\phi_2 = +0.56$ to find $F(0.56) = (1+\phi)e^{-\phi} = 1.56\,e^{-0.56} = 0.89$, and the subsequent value is $\phi_3 = -0.40$ because $F(-0.40) = (1+\phi)e^{-\phi} = (1-0.40)\,e^{+0.40} = 0.89$, and this results in $\phi_3 = 0.40$, and so on.

The successive peaks are therefore $\phi_1 = 0.92$, $\phi_2 = -0.56$ and $\phi_3 = 0.40$, etc. and the corresponding successive maximum/minimum elevations using f and D are $z_1 = \phi_1 D/f$, $z_2 = \phi_2 D/f$, etc.

The maximum value of V is found by equating $dV^2/dt = 0$ from Eq. (7.11) at the position z':

$$\frac{dV^2}{dt} = 0 = \frac{f}{D} - \left(1 + \frac{f z_m}{D}\right)\left\{\exp\left[\frac{f(z' - z_{max})}{D}\right]\right\}\frac{f}{D},$$

which gives

$$z' = z_{max} - \frac{D}{f}\ln\left(1 + \frac{f z_{max}}{D}\right).$$

The result is substituted back into Eq. (7.11) for the maximum velocity:

$$V_{max}^2 = \frac{4gD^2}{f^2 L}\left[\left(\frac{f z_{max}}{D}\right) - \ln\left(1 + \frac{f z_{max}}{D}\right)\right]. \quad (7.13)$$

This procedure looks awfully complicated, and Example 7.3 should be very helpful.

Example 7.3: Turbulent flow oscillations in a large pipe

A 1,000-ft-long U-tube consists of 2.0-ft-diameter pipe with $f = 0.03$. The initial water-level difference between both ends is $z_1 = 20$ ft. Find the successive minimum and maximum elevations.

Solution:

Step (1): The initial value is $\phi_1 = \frac{f z_1}{D} = \frac{0.03 \times 20}{2} = 0.3$ and $F(0.3) = (1 + \phi)e^{-\phi} = 1.3\,e^{-0.3} = 0.963$.

This corresponds to $\phi_2 = -0.25$ because $F(\phi) = (1 + \phi)e^{-\phi} = (1 - 0.25)\,e^{+0.25} = 0.963$.

Step (2): The sign of $\phi_2 = 0.25$ is reversed to give $F(0.25) = (1 + \phi)e^{-\phi} = 1.25\,e^{-0.25} = 0.973$.

The third peak is $\phi_3 = -0.215$ because $F(-0.215) = (1 - 0.215)\,e^{+0.215} = 0.973$, and $\phi_3 = 0.215$.

Step (3): The corresponding sequence of maximum/minimum elevations for $\phi_1 = 0.3$, $\phi_2 = -0.25$ and $\phi_3 = 0.215$ is $z_1 = 20$ ft, $z_2 = \phi_2 D/f = -0.25 \times 2/0.03 = -16.7$ ft, and $z_3 = \phi_3 D/f = +0.215 \times 2/0.03 = +14.3$ ft.

Step (4): The maximum velocity from Eq. (7.13) corresponds to the first maximum $\phi_1 = 0.3$, and

$$V_{max}^2 = \frac{4gD^2}{f^2 L}\left[\left(\frac{f z_{max}}{D}\right) - \ln\left(1 + \frac{f z_{max}}{D}\right)\right] = \frac{4 \times 32.2 \times 2^2}{0.03^2 \times 1{,}000}[(0.3) - \ln(1 + 0.3)]$$

$$= 21.5\ \text{ft}^2/\text{s}^2,$$

or

$V_{max} = -4.64$ ft/s; this occurs when

$$z' = z_{max} - \frac{D}{f} \ln\left(1 + \frac{f z_{max}}{D}\right) = 20 - \frac{2}{0.03} \ln(1 + 0.3) = 2.5 \text{ ft.}$$

The basic equation of motion (Eq. (7.10)) can also be solved numerically at very short time increments to show the displacement and the velocity as a function of time. Figure E-7.3 shows a comparison of the numerical method and the successive maximum/minimum elevation and velocity values.

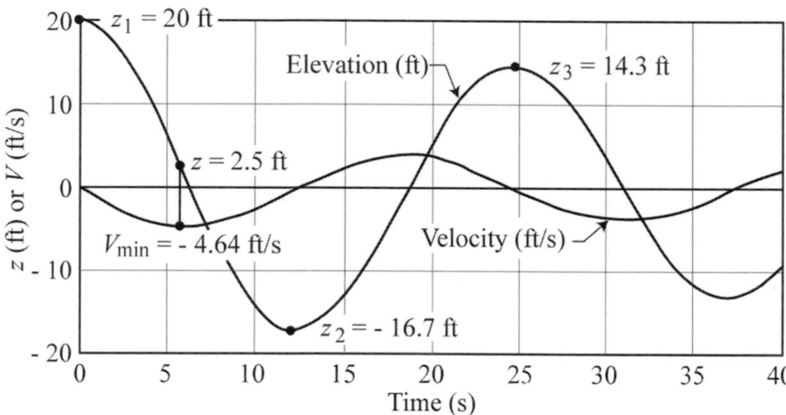

Fig. E-7.3 Turbulent pipe flow oscillations

7.4 Oscillations between Reservoirs

Two reservoirs are connected by a pipeline of length L and diameter D in Figure 7.7. We study the flow oscillations given the reservoir surface areas A_1 and A_2 with water elevations z_1 and z_2, respectively. For the analysis, the equilibrium position from the volumetric relationship $zA = z_1A_1 = z_2A_2$, gives $z_1 = zA/A_1$ and $z_2 = zA/A_2$. Also, given $H = z_1 + z_2$, we find $z_1 = HA_2/(A_1 + A_2)$ and $z_2 = HA_1/(A_1 + A_2)$. Minor losses are considered by using the equivalent friction factor $f_e = f + KD/L$ with the pipe friction factor f and minor-loss coefficient K.

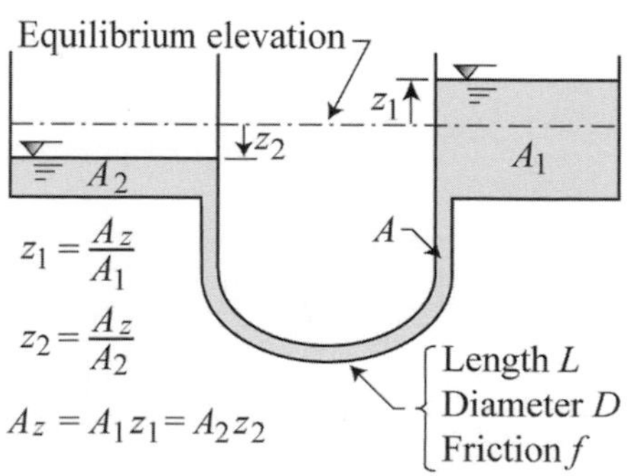

Figure 7.7 Oscillations between reservoirs

The equation of motion when neglecting the momentum flux ($\rho QV << pA$) is

$$-\gamma(z_2 + z_1)A - \frac{\gamma A f_e L}{2gD} \frac{dz}{dt}\left|\frac{dz}{dt}\right| = \frac{\gamma A L}{g} \frac{d^2z}{dt^2},$$

which gives

$$\frac{d^2z}{dt^2} + \frac{f_e}{2D}\frac{dz}{dt}\left|\frac{dz}{dt}\right| + \frac{gA}{L}\left(\frac{1}{A_2} + \frac{1}{A_1}\right)z = 0.$$

We are already very familiar with this formulation and the solution is the same as in Section 7.3 after replacing f with f_e, using $\phi = f_e\frac{z}{D}$, and replacing $2g/L$ by $gA[(1/A_1) + (1/A_2)]/L$. Notice that oscillations can be independent of the original elevation z_{10}, because $F(\phi) = (1+\phi)e^{-\phi} \to 0$ when $\phi > 5$. The application in Example 7.4 considers the oscillations between two large tanks.

♦ Example 7.4: Oscillations between large tanks

A valve is opened in a pipe connecting two water tanks of surface areas $A_1 = 200$ ft^2 and $A_2 = 300$ ft^2. The initial head difference between the two tanks is 66.7 ft. The 2,000-ft pipe has a 3-ft diameter with a friction factor $f = 0.024$ and minor losses $3.5\ V^2/2g$. Find the high and low water levels in tank 1.

Solution:

Step (1): The initial head of tank 1 is $z_{10} = HA_2/(A_1 + A_2) = 66.7 \times 300/(200 + 300) =$ 40 ft and the initial level in tank 2 is $z_{20} = \frac{z_{10}A_1}{A_2} = 40 \times \frac{200}{300} = 26.7$ ft below the reference elevation.

Step (2): The equivalent friction factor is $f_e = f + KD/L = 0.024 + (3.5 \times 3/2{,}000) =$ 0.02925.

The initial high level in the first tank $z_{10} = 40$ ft gives

$$z_{m0} = \frac{z_{10}A_1}{A} = \frac{40 \times 200 \times 4}{\pi \times 3^2} = 1{,}132 \text{ ft}.$$

The corresponding ϕ is

$$\phi_0 = \frac{f_e z_{m_0}}{D} = \frac{0.029 \times 1{,}132}{3} = 11.0;$$

note that $\phi_0 > 5$ here, and $F(\phi_0) = (1+\phi_0)\,e^{-\phi_0} = (1+11)\,e^{-11} = 0.0002.$

The first oscillation starts when $\phi_0 < 5$, or after

$$z_{10} = \frac{z_{m0}A}{A_1} = \frac{A\phi_0 D}{A_1 f_e} = \frac{\pi \times 3^2 \times 5 \times 3}{4 \times 200 \times 0.029} = 18.1 \text{ ft}.$$

Step (3): The first minimum will happen when $\phi_1 = -1$, which corresponds to

$$z_{m1} = \frac{\phi_1 D}{f_e} = \frac{-1 \times 3}{0.02925} = -102.6 \text{ ft},$$

and thus

$$z_{11} = \frac{z_{m1}A}{A_1} = \frac{-102.6 \times \pi \times 3^2}{200 \times 4} = -3.62 \text{ ft}.$$

Step (4): For the next maximum, the absolute value of $\phi_1 = +1$ gives $F(\phi_1) = (1+\phi_1)e^{-\phi_1} = (1+1)e^{-1} = 0.736$, which corresponds to $\phi_2 = 0.593$ given that $F(\phi_2) = (1-\phi_2)e^{\phi_2} = (1-0.593)e^{+0.593} = 0.736$. This second peak is at

$$z_{m2} = \frac{\phi_2 D}{f_e} = \frac{0.593 \times 3}{0.02925} = 60.9 \text{ ft},$$

which is at elevation

$$z_{12} = \frac{z_{m2} A}{A_1} = \frac{60.9 \times \pi \times 3^2}{200 \times 4} = 2.15 \text{ ft}.$$

Step (5): The maximin/minimum sequence in the first tank is: $z_{10} = 40$ ft, $z_{11} = -3.62$ ft and $z_{12} = 2.15$ ft, etc.

Step (6): The numerical methods become increasingly accurate as $\Delta t \to 0$. The numerical scheme consists of three parts:

(a) Eq. 7.10 is solved for the acceleration $\ddot{z}$ given the initial values of elevation and velocity.

(b) The velocity is then calculated from $V = V_0 + \ddot{z}\Delta t$.

(c) The displacement is then simply obtained from $z = z_0 + V\Delta t$.

This maximum/minimum calculation sequence is compared with numerical calculations in Fig. E-7.4.

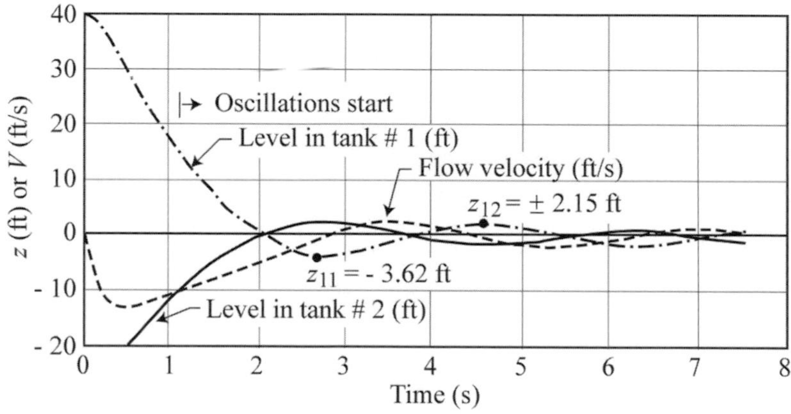

Fig. E-7.4 Turbulent flow in a pipe connecting two reservoirs

Additional Resources

Additional information on transients and flow oscillations in pipes can be found in Streeter (1971), Wylie and Streeter (1978), Naudascher and Rockwell (1994), Ghidaoui et al. (2005) and Chaudhry (2014). Numerical methods which successively solve Eq. (7.10) for acceleration, velocity and position at very short time intervals are quite popular and effective nowadays on fast computers. Alternatively, analytical solutions for flow oscillations with turbulent friction losses are now becoming possible with the use of Lambert functions and elliptic integrals readily available in Matlab (Guo et al. 2017).

EXERCISES

These exercises review the essential concepts from this chapter.

1. What is the source of elasticity in pipe flow oscillations?
2. What is the effect of friction on the frequency of oscillations?
3. When do oscillations start in laminar flow?
4. When do oscillations start in turbulent flow?
5. True or false?
   (a) The pressure at the reference elevation (midpoint) in the pipe remains constant during the oscillation.
   (b) The period of oscillations in pipes without friction only depends on the pipe length.
   (c) Friction increases the frequency of oscillations.
   (d) Friction decreases the period of oscillations.

PROBLEMS

1. ♦♦ A 10-m-long, 5-mm-diameter U-shaped plastic tube is holding water at 20 °C. There is a 1-m head difference between both ends when the pressure is suddenly released at $t = 0$. Determine the following: (a) the natural circular frequency of the oscillations; (b) the damping factor; (c) the circular frequency of the damped oscillations; (d) the period of the damped oscillations; (e) the lowest water level; (f) the maximum flow velocity; and (g) whether or not the flow is laminar.
2. ♦♦ A 25-m-long, 25-cm-diameter U-shaped pipe has an 8-m difference between both ends as the system is released from rest. If $f = 0.04$, calculate the successive maxima during the oscillations and find the maximum velocity in the pipe. Compare the calculations with $f = 0.08$.

8 Steady Uniform Flow

As opposed to pressurized flow in closed conduits, open-channel flows convey water by gravity in man-made channels and natural waterways. The cross-sectional area of open channels varies with discharge as described in Section 8.1. Section 8.2 examines resistance to flow, the normal depth is considered in Section 8.3 and shear stress in Section 8.4.

8.1 Open-Channel Geometry

The cross section of a channel is measured perpendicular to the main flow direction. Figure 8.1 depicts the geometric elements of a typical cross section. The main parameters are: flow depth y, surface width W, wetted perimeter P, cross-sectional area A, averaged depth $h = A/W$ and hydraulic radius $R_h = A/P$.

The geometry of open-channel cross sections is summarized in Figure 8.2.

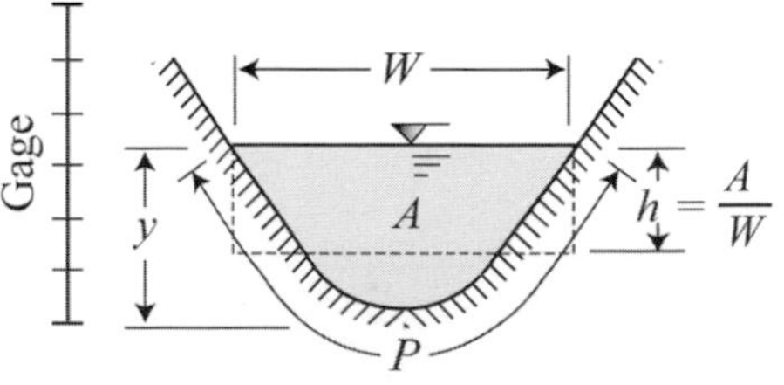

Figure 8.1 Cross-sectional geometry

For the compound sections sketched in Figure 8.3a, the hydraulic radius is calculated from the sums of partial areas and partial wetted perimeters. Typical river cross-section profiles in Figure 8.3b indicate the substrate material and floodplain vegetation types in terms of deciduous and coniferous trees, shrubs and grasses. The bankfull elevation is usually important because floodplains can extend laterally over long distances. The vegetation on the floodplain also increases roughness while the main channel will tend to limit vegetation growth.

Three typical calculation examples are presented for: (1) a circular section for sewers and culverts in Example 8.1; (2) a trapezoidal canal in Example 8.2; and a compound section in Example 8.3.

Example 8.1: Circular cross section

Define the hydraulic geometry of a circular open-channel cross section shown in Fig. E-8.1.

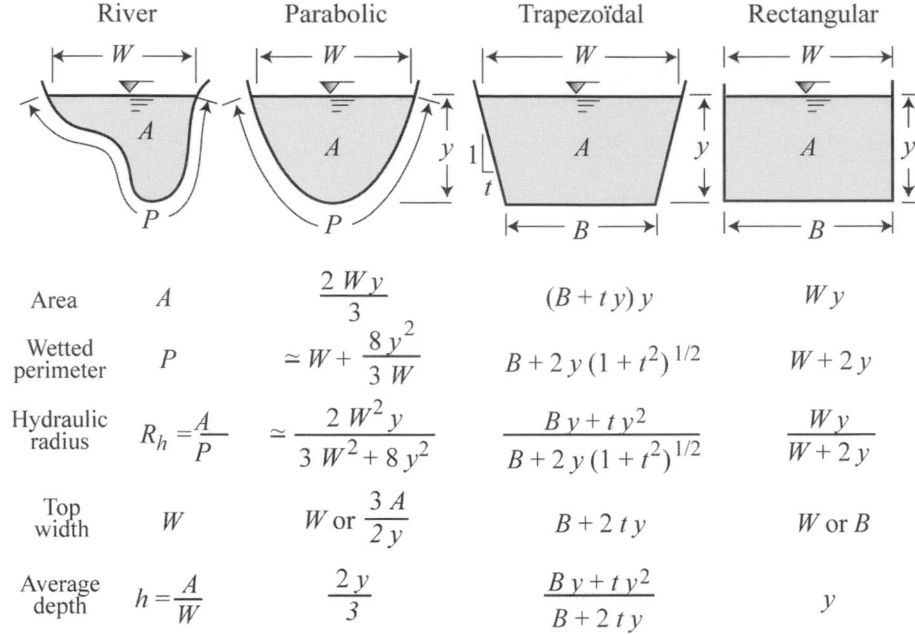

River
Parabolic
Trapezoïdal
Rectangular
W
A
P
y
1
t
B
Area
Wetted perimeter
Hydraulic radius
Top width
Average depth
A
P
$R_h = \frac{A}{P}$
W
$h = \frac{A}{W}$
$\frac{2\,W\,y}{3}$
$\simeq W + \frac{8\,y^2}{3\,W}$
$\simeq \frac{2\,W^2\,y}{3\,W^2 + 8\,y^2}$
W or $\frac{3\,A}{2\,y}$
$\frac{2\,y}{3}$
$(B + t\,y)\,y$
$B + 2\,y\,(1 + t^2)^{1/2}$
$\frac{B\,y + t\,y^2}{B + 2\,y\,(1 + t^2)^{1/2}}$
$B + 2\,t\,y$
$\frac{B\,y + t\,y^2}{B + 2\,t\,y}$
$W\,y$
$W + 2\,y$
$\frac{W\,y}{W + 2\,y}$
W or B
y

Figure 8.2 Geometric parameters of typical cross sections

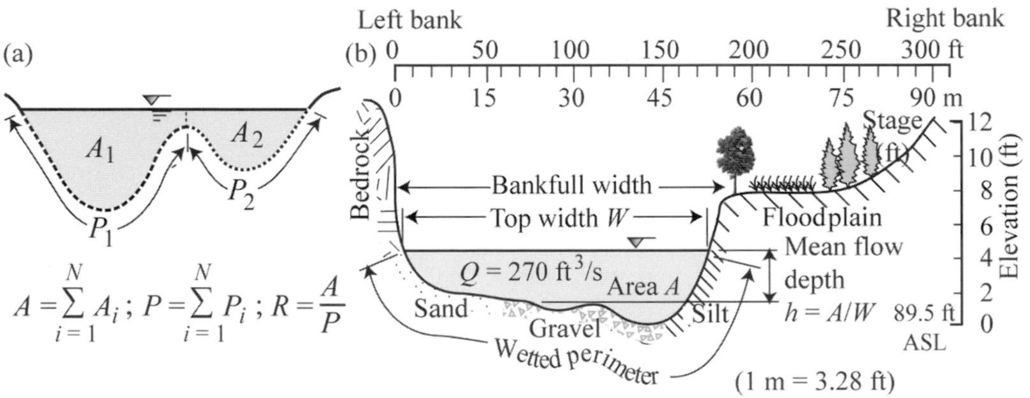

Figure 8.3 Matamek River cross section (Frenette and Julien 1980)

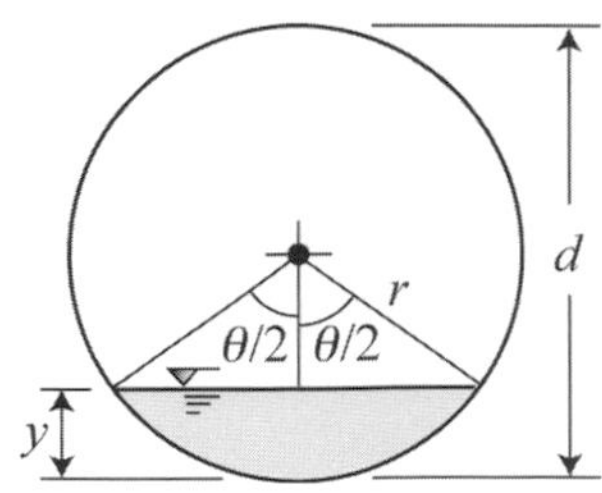

Fig. E-8.1 Circular section

Solution:

The angle $\theta = 2\cos^{-1}\left(1 - \frac{2y}{d}\right)$ is in radians.

The flow area $A = 0.5\theta r^2 - r^2 \sin(\theta/2)\cos(\theta/2)$ or $A = 0.5(\theta - \sin\theta)r^2$.

The wetted perimeter is $P = \theta r$.

The hydraulic radius is $R_h = \frac{A}{P} = 0.5\left(1 - \frac{\sin\theta}{\theta}\right)r$.

The top width $W = 2r\sin(\theta/2)$.

The average flow depth is $h = \frac{A}{W} = \frac{0.25\,(\theta - \sin\theta)\,r}{\sin(\theta/2)}$.

Example 8.2: Trapezoidal cross section

Compute the hydraulic radius and average depth for a trapezoidal channel with a 9.14-m bottom width, a 3.0-m flow depth and a side slope $t = 2\text{H}:1\text{V}$ in Fig. E-8.2.

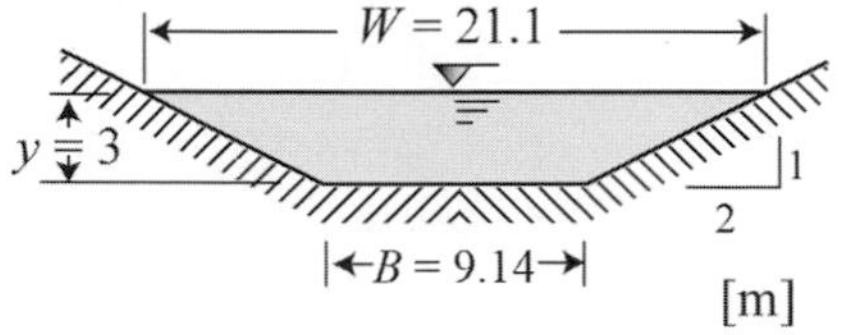

Fig. E-8.2 Trapezoidal channel

Solution:

The flow area $A = (9.14 \times 3) + \left(2 \times 3^2\right) = 45.4 \text{ m}^2$.
The wetted perimeter is $P = 9.14 + 2 \times 3\sqrt{5} = 22.5$ m.
The hydraulic radius is $R_h = A/P = 45.4/22.5 = 2.01$ m.
The top width is $W = 9.14 + 2 \times 2 \times 3 = 21.1$ m.
The average depth is $h = A/W = 45.4/21.1 = 2.15$ m.

Example 8.3: Composite cross section

Compute the hydraulic geometry parameters for the following main channel and a floodplain in Fig. E-8.3.

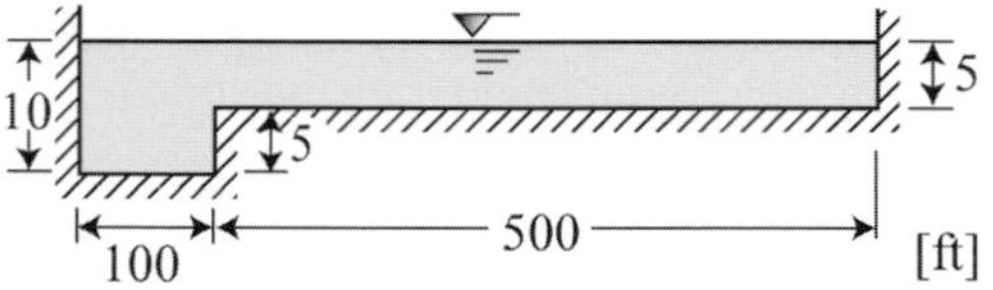

Fig. E-8.3 Channel with floodplain

Solution:

The flow area $A = 1{,}000 + 2{,}500 = 3{,}500 \text{ ft}^2$.
The wetted perimeter is $P = 600 + 20 = 620$ ft.
The hydraulic radius is

$$R_h = \frac{A}{P} = \frac{3{,}500}{620} = 5.65 \text{ ft}.$$

The top width $W = 600$ ft, and the average flow depth is

$$h = \frac{A}{W} = \frac{3{,}500}{600} = 5.83 \text{ ft}.$$

8.2 Resistance to Flow

In open channels, the mean flow velocity $V = Q/A$ is a power of the hydraulic radius R_h and the friction slope S_f, which is the slope of the energy grade line. In the eighteenth century, the French engineer Antoine de Chézy developed the formula $V = CR_h^{1/2}S_f^{1/2}$, where C is the Chézy coefficient. For comparison with the Darcy–Weisbach formula for flow in pipes of diameter D, the hydraulic radius is $R_h = A/P = \pi D^2/4\pi D = D/4$ and the formula from Chapter 2 is rewritten as $V = \sqrt{8g/f}R_h^{1/2}S_f^{1/2}$, such that $C = \sqrt{8g/f}$. In 1867, the French engineer Philippe Gauckler found that the Chézy coefficient increased with the 1/6th power of the hydraulic radius. His contribution became obscured by a more complex and popular formulation by Ganguillet and Kutter. By 1890, in Ireland, Robert Manning made similar observations and, in his formula, velocity became inversely proportional to a coefficient n. In the twentieth century, the Swiss engineer Albert Strickler found that Manning's coefficient n increases with the grain diameter. For gravel and cobble-bed streams, $n \cong 0.064 d_{sm}^{1/6}$ given the bed material size d_{sm} in meters.

The Gauckler–Manning–Strickler equation is simply called Manning's equation. A unit conversion coefficient m is used for different systems of units ($m = 1$ for SI and $m = 1.49$ for customary units):

$$V = \frac{m}{n} R_h^{2/3} S_f^{1/2} \quad m = 1 \text{ when } R_h \text{ in m and } V \text{ in m/s,} \tag{8.1}$$

$$V = \frac{1.49}{n} R_h^{2/3} S_f^{1/2} \quad m = 3.28^{1/3} = 1.49 \text{ when } R_h \text{ in ft and } V \text{ in ft/s.}$$

When comparing the three formulas in dimensionless form, it is noticed that

$$\frac{C}{\sqrt{g}} \equiv \sqrt{\frac{8}{f}} \equiv \frac{m\, R_h^{1/6}}{n\, \sqrt{g}} \cong 5 \left(\frac{R_h}{d_s}\right)^{1/6}. \tag{8.2}$$

Note that the Manning n coefficient has units in s/m$^{1/3}$ and Darcy–Weisbach f is dimensionless. Both increase with resistance to flow while the Chézy C coefficient has fundamental dimensions $L^{1/2}/T$. It describes flow conveyance and decreases with flow resistance. The advantage of Manning's formulation is to keep the same parameter value for all systems of units but at the price of introducing the unit conversion parameter m.

Figure 8.4 shows a graph of resistance to flow parameters as a function of the ratio of flow depth h to median grain diameter d_{50} of the bed material of a stream. The measurements are for plane surfaces. Manning's approximation is very good for most fluvial conditions $10 < h/d_{50} < 10^4$. Deviations are observed for steep mountain channels with low discharge and very coarse bed material. Other factors increasing Manning's n coefficient include vegetation, channel obstructions and sinuosity.

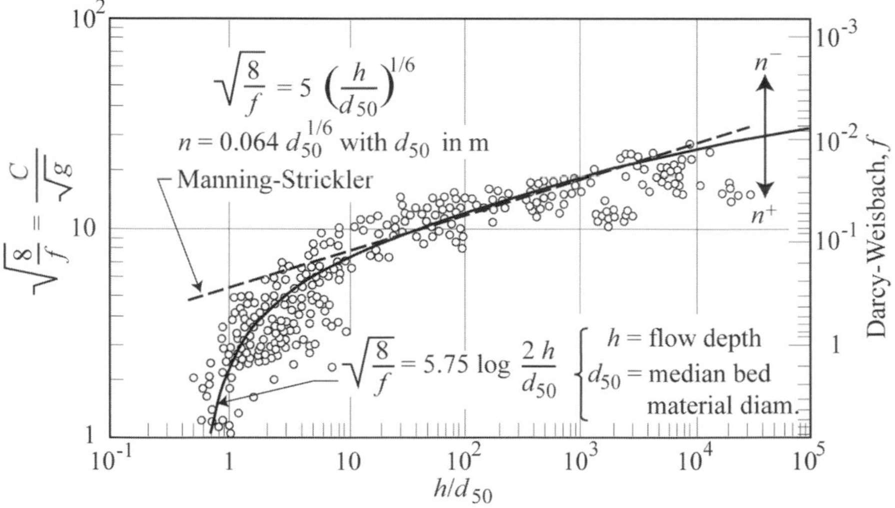

Figure 8.4 Resistance to flow for plane surfaces (modified after Julien 2018)

Typical n values for man-made and natural channels are listed in Table 8.1 and Figure 8.5 under steady and uniform flows. Example 8.4 extracts Manning n values from a velocity profile. Example 8.5 calculates the flow parameters for a compound river section with a wide floodplain. Case Study 8.1 also summarizes an analysis of resistance to flow for 2,604 different rivers.

Table 8.1. Manning n values for open channels

Conduits and lined channels	Manning n
Plexiglass	0.010–0.014
Metal, wood and smooth concrete	0.010–0.015
Rough concrete	0.015–0.020
Riveted metal	0.015–0.020
Corrugated metal	0.020–0.030
Brickwork	0.020–0.030
Excavated channels	**Manning n**
Straight and clean earth channel	0.013–0.025
Sand dunes and ripples	0.018–0.040
Gravel bottom	0.018–0.030
Bedrock	0.025–0.050
Not maintained (weeds and brush)	0.040–0.140
Natural streams	**Manning n**
Straight minor streams	0.025–0.040
Overbank short grass	0.025–0.035
Tall grasses and reeds	0.030–0.050
Floodplain crops	0.020–0.050
Sparse vegetation	0.035–0.050
Medium to dense brush	0.070–0.160

		Manning n		
		Min.	Mean	Max.
Natural channels	Sand	0.014	**0.036**	0.151
	Gravel	0.011	**0.045**	0.250
	Cobble	0.015	**0.051**	0.327
	Boulder	0.023	**0.080**	0.444
Vegetated channels	Grass	0.015	**0.045**	0.250
	Shrub	0.016	**0.057**	0.250
	Tree	0.018	**0.047**	0.310

Figure 8.5 Manning n estimates

Example 8.4: Resistance parameter from a velocity profile

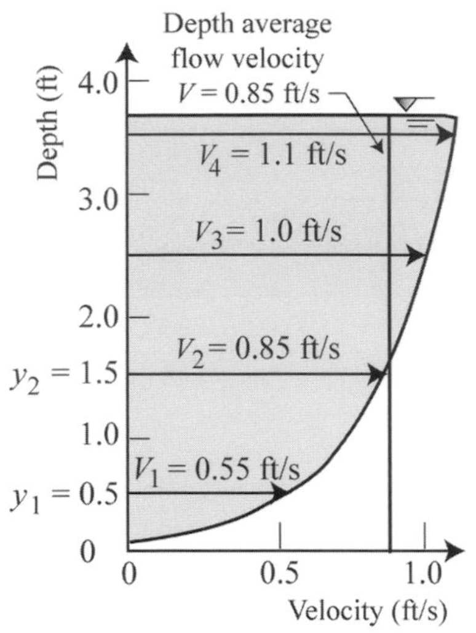

Fig. E-8.4 Velocity profile

In the case of steady and uniform flow conditions, resistance to flow parameters can be determined from a velocity profile. Consider the measured velocity profile in Fig. E-8.4 for a 200-ft-wide rectangular channel at a slope $S_f = 1 \times 10^{-4} =$ 10 cm/km.

Calculate: (1) the flow depth, (2) the hydraulic radius, (3) the mean flow velocity, (4) the unit and total discharge, (5) the Manning n coefficient and (6) the Chézy C coefficient.

Solution:

(1) Note that $dy_i = 1$ ft, except for the uppermost velocity measurement ($dz_4 = 0.7$ ft) and the flow depth is $h = \sum_i dy_i = 3.7 \text{ ft} = 1.13 \text{ m ft}.$

(2) The hydraulic radius:

$$R_h = \frac{A}{P} = \frac{Wh}{W + 2h} = \frac{200 \times 3.7 \text{ ft}^2}{200 + (2 \times 3.7) \text{ ft}} = 3.57 \text{ ft} = 1.09 \text{ m}.$$

The hydraulic radius is very close to the flow depth $h =$ 3.7 ft, because the width–depth ratio is large (i.e. $W/h =$ $200/3.7 = 54$).

(3) The depth-averaged flow velocity:

$$V = \frac{1}{A}\int_A v_i \, \text{d}A \cong \frac{1}{h}\sum_i v_i \, \text{dy}_i = \frac{1}{3.7 \text{ ft}}\left[0.55 + 0.85 + 1.0 + (1.1 \times 0.7)\right]\frac{\text{ft}^2}{\text{s}} = 0.85\frac{\text{ft}}{\text{s}}$$
$$= 0.26\frac{\text{m}}{\text{s}}.$$

(4) The unit and total discharge are respectively:

$$q = Vh = 0.85 \text{ ft/s} \times 3.7 \text{ ft} = 3.1 \text{ ft}^2/\text{s} = 0.29 \text{ m}^2/\text{s},$$
$$Q = Wq = 200 \text{ ft} \times 3.1 \text{ ft}^2/\text{s} = 630 \text{ ft}^3/\text{s} = 17.8 \text{ m}^3/\text{s}.$$

(5) The Manning n coefficient is

$$n = \frac{m}{V}R_h^{2/3}\,S_f^{1/2} = \frac{1.49 \text{ ft}^{1/3}\text{s}}{0.85 \text{ ft m}^{1/3}}(3.57 \text{ ft})^{2/3}\left(1 \times 10^{-4}\right)^{1/2} = 0.041.$$

Note that the units of n are in s/m$^{1/3}$. Because the conversion factor also has dimensions ($m = 1.49 \text{ ft}^{1/3}/\text{m}^{1/3}$), the numerical value of n is the same in both systems of units. Therefore, the units of both m and n are not displayed in practice.

(6) The Chézy C coefficient in both systems of units is

$$\frac{C}{\sqrt{g}} \equiv \sqrt{\frac{8}{f}} \equiv \frac{m}{n}\frac{R_h^{1/6}}{\sqrt{g}} = \frac{1.49}{0.041}\frac{3.57^{1/6}}{\sqrt{32.2}} = 7.9,$$

which gives $C = 7.9\sqrt{g} \equiv 7.9\sqrt{32.2} = 45\ \text{ft}^{1/2}/\text{s}$ in US customary units, and a different value $C = 7.9\sqrt{g} \equiv 7.9\sqrt{9.81} = 25\ \text{m}^{1/2}/\text{s}$ in SI units.

Example 8.5: Resistance-to-flow calculations in a compound channel

Consider the compound channel geometry from Example 8.3. Assume a channel slope $S = 0.005$, with the Manning coefficient $n = 0.02$ for the main channel and $n = 0.1$ for the floodplain, as shown in Fig. E-8.5. Separate the flow into two separate areas for the main channel and floodplain. Calculate every 3 inches until a maximum stage elevation of 10 ft. Calculate (1) the hydraulic radius, mean flow depth and velocity; (2) the discharge for each component; and (3) plot the flow velocity and the floodwave celerity $\Delta Q/\Delta A$ as a function of depth.

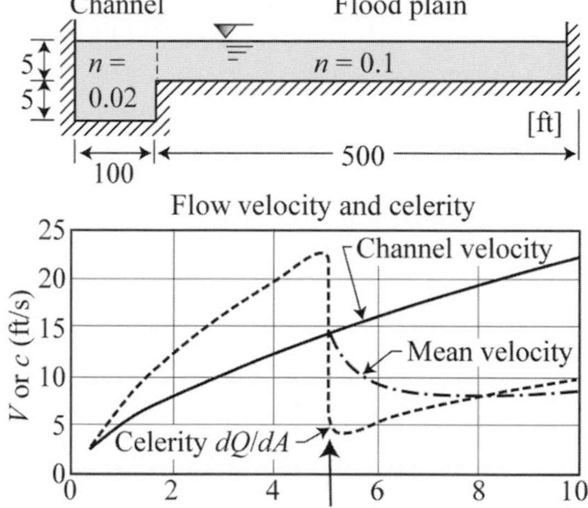

Fig. E-8.5 Velocity and celerity in a compound channel

Solution: Calculations are shown for two depths: (1) flow below bankfull depth; and (2) flow above bankfull depth.

(1) The flow below the bankfull depth ($y < 5$ ft) remains entirely in the main channel, and for instance $y = 2$ ft, $A = 200\ \text{ft}^2$, $P = 104$ ft, $R_h = A/P = 1.92$ ft, $V = (m/n)R_h^{2/3}S_f^{1/2} = (1.49/0.02)(1.92)^{2/3}(0.005)^{0.5} = 8.13$ ft/s and $Q = AV = 1{,}627$ cfs.

(2) When the flow is above the bankfull depth ($y > 5$ft), we consider the channel and floodplain separately. For instance, when $y = 8$ ft, $A_c = 800\ \text{ft}^2$, $P_c = 113$ ft and $V_c = (m/n)(A_c/P_c)R_{hc}^{2/3}S_f^{1/2} = (1.49/0.02)(800/113)^{2/3}(0.005)^{0.5} = 19.4$ ft/s with a channel discharge $Q_c = A_c V_c = 15{,}538$ cfs.

On the floodplain, the properties are $A_f = 1{,}500\ \text{ft}^2$, $P_f = 503$ ft, the flow velocity $V_f = (m/n)(A_f/P_f)R_{hf}^{2/3}S_f^{1/2} = (1.49/0.1)(1{,}500/503)^{2/3}(0.005)^{0.5} = 2.18$ ft/s and $Q_f = A_f V_f = 3{,}274$ cfs. The total discharge $Q_t = Q_c + Q_f$ and $A_t = A_c + A_f$.

The mean flow properties are

$\bar{V} = Q_t/A_t = (15{,}538 + 3{,}274)/(800 + 1{,}500) = 8.18$ ft/s.

(3) To determine the celerity, consider the flow at $y = 8.25$ ft, $A_c = 825\ \text{ft}^2$, $P_c = 113.5$ ft, $V_c = 19.76$ ft/s and $Q_c = 16{,}300$ cfs in the channel and $A_f = 1{,}625\ \text{ft}^2$, $P_f = 503.3$ ft, $V_f = 2.302$ ft/s and $Q_f = 3{,}740$ cfs on the floodplain. In comparison with the conditions at $y = 8$ ft, $\Delta Q = (16{,}300 + 3{,}740) - (15{,}540 + 3{,}275) = 1{,}230$ cfs and $\Delta A = (825 + 1{,}625) - (800 + 1{,}500) = 150\ \text{ft}^2$, and the floodwave celerity is $\Delta Q/\Delta A = 1{,}230/150 = 8.2$ ft/s, which is now far less than the velocity in the main channel ($V_c = 19.8$ ft/s). Chapter 11 will present a more detailed discussion of the importance of the floodwave celerity.

Case Study 8.1: Resistance to flow for bankfull conditions in rivers

The analysis of resistance to flow of 2,604 rivers by Lee et al. (2012) and summarized in Julien (2018). The database from Lee and Julien (2017) includes only bankfull measurements. The range of parameters for bankfull conditions is summarized in Table CS-8.1. Natural channels (1,865 different rivers) are described in terms of their substrate or bed material classified into four types: sands, gravels, cobbles and boulders. Also, 739 vegetated channels are sorted into three types: grasses, shrubs and trees.

Table CS-8.1. River database for resistance to flow (after Lee and Julien 2017)

Type	Number	Discharge Q (m^3/s)	Friction slope S_f (cm/km)	Bed material d_{50} (mm)	Velocity V (m/s)	Flow depth h (m)
Natural channels						
Sand	172	0.14–26,560	10–2,860	0.01–1.64	0.02–3.64	0.10–15.7
Gravel	989	0.01–14,998	9–8,100	2–63.6	0.04–4.70	0.04–11.2
Cobble	651	0.02–3,820	1–5,080	64–253	0.07–4.29	0.10–6.94
Boulder	53	2–1,700	2,060–3,730	263–945	0.32–5.11	0.28–4.09
Vegetated channels						
Grass	281	0.01–750	7–1,790	0.33–305	0.06–3.66	0.16–3.96
Shrub	150	0.38–542	1–3,400	16–893	0.1–3.64	0.04–3.08
Tree	308	0.02–3,220	10–4,050	0.17–397	0.07–5.11	0.10–9.17

The results in terms of the Manning coefficient n vs slope and bankfull discharge are shown in Fig. CS-8.1. It is observed that the Manning n for different rivers decreases slightly with bankfull discharge Q and increases slightly with the channel slope S_f. When both parameters are combined, the approximation for natural rivers is

$$n \cong 0.25 S_f^{1/6} / Q^{1/8}. \tag{8.3}$$

Considering the 25th and 75th percentile of the distribution in Fig. CS-8.1, the range of coefficient values remains between 0.15 and 0.3 for all types of bed material and vegetation. Therefore, a value 0.25 provides a good first approximation of resistance to flow for a wide variety of river types. For instance, it is possible to estimate the Manning n coefficient for the Matamek River from Example 8.4. At a bankfull discharge of 18 m^3/s and 10^{-4} slope, Eq. (8.3) gives the following estimate: $n \cong 0.25 \times 0.0001^{1/6}/18^{1/8} = 0.037$, which is relatively close to the measured value $n = 0.041$.

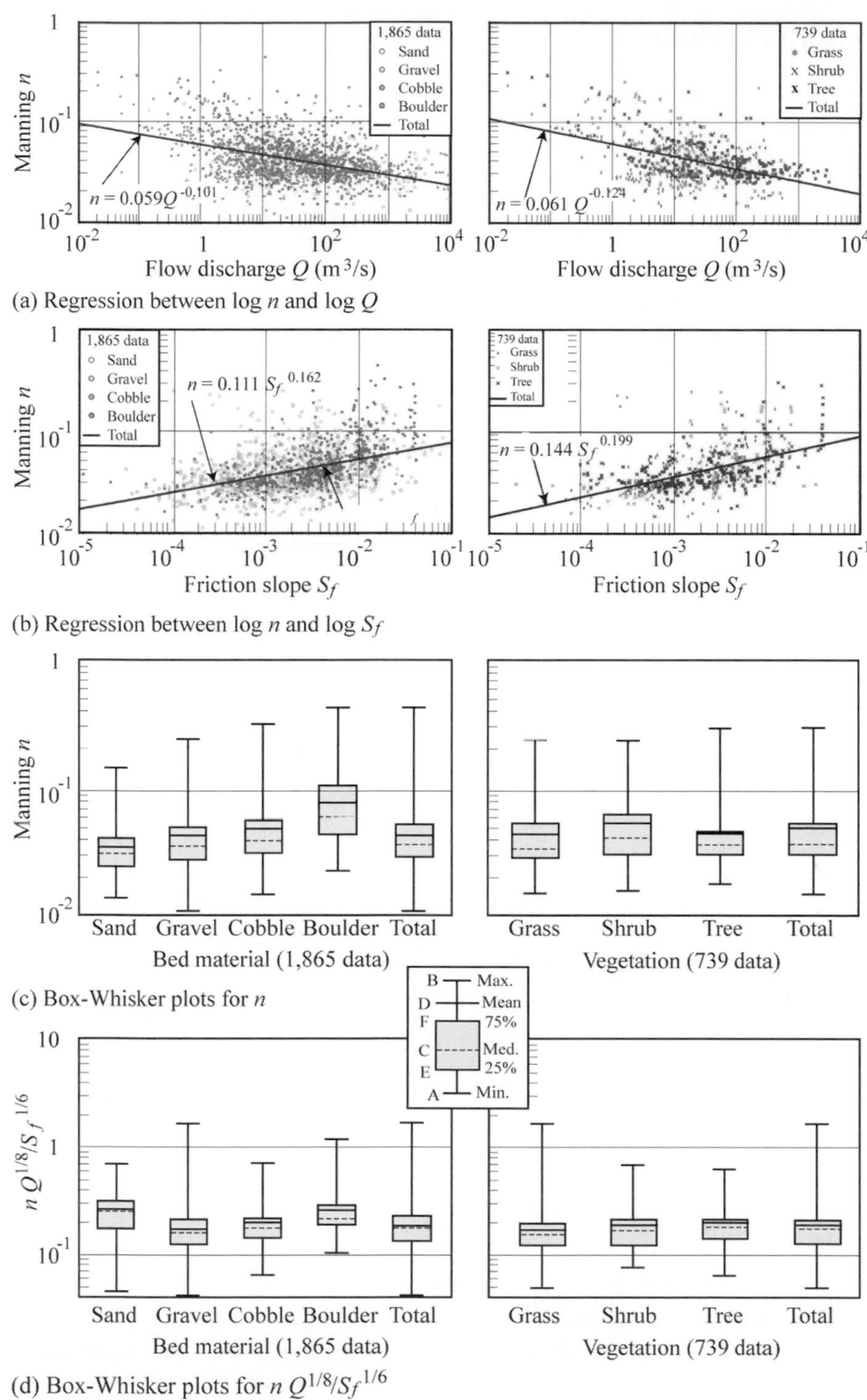

Fig. CS-8.1 Manning n roughness coefficients for rivers (after Julien 2018)

8.3 Normal Depth

This section describes channel conveyance and the normal depth of a channel. The normal depth physically represents the flow depth that open-channel flows strive to attain. Using Manning's equation, the flow discharge Q becomes

$$Q = AV = \frac{1}{n} A R_h^{2/3} S_f^{1/2} (\text{SI units}) = \frac{1.49}{n} A R_h^{2/3} S_f^{1/2} (\text{customary units}).$$

Flow conveyance describes the ability of a given cross section to deliver a certain flow rate. It is obtained from the discharge Q as a function of friction slope S_f:

$$Q = K S_f^{1/2}, \tag{8.4}$$

where K is the conveyance coefficient, which depends on the cross-sectional hydraulic geometry and roughness coefficient:

$$K = \frac{m}{n} A R_h^{2/3} = \frac{1}{n} A R_h^{2/3} (\text{SI units}) = \frac{1.49}{n} A R_h^{2/3} (\text{customary units}).$$

Similarly, the friction slope S_f depends on the discharge Q and the conveyance coefficient K:

$$S_f = \frac{Q^2}{K^2} = \left(\frac{nQ}{mAR_h^{2/3}}\right)^2 = \left(\frac{nV}{R_h^{2/3}}\right)^2 (\text{SI units}) = \left(\frac{nV}{1.49 R_h^{2/3}}\right)^2 (\text{customary units}).$$

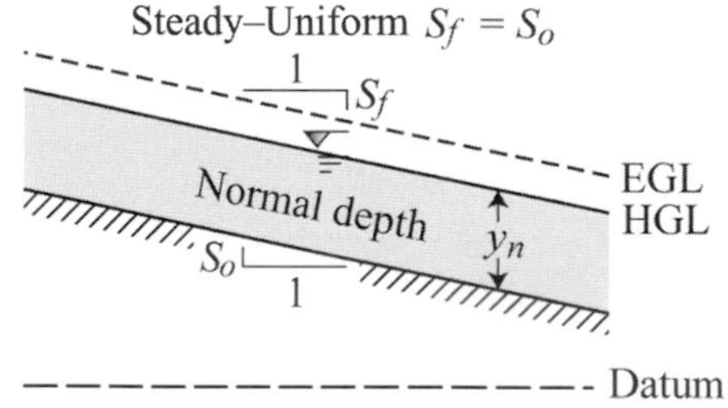

Figure 8.6 Steady–uniform flow

The normal depth is the flow depth corresponding to steady–uniform flow conditions. Steady flow means that the flow rate does not change with time. Uniform flows do not change in the downstream direction. Figure 8.6 shows that for steady–uniform flow, the energy grade line (EGL) is parallel to the channel bed and the friction slope S_f becomes identical to the bed slope S_0, i.e. $S_f = S_0$. The corresponding flow depth in the channel is called the normal depth h_n.

For any cross section, the normal depth is calculated from Eq. (8.4) when $S_f - S_0$

$$A_n R_{hn}^{2/3} = \frac{A_n^{5/3}}{P_n^{2/3}} = \frac{nQ}{m S_0^{1/2}}. \tag{8.5}$$

Depending on the cross-sectional geometry, the area A_n and wetted perimeter P_n at normal depth correspond to the normal depth. In general, the normal depth is obtained with equation solvers or by trial-and-error after changing the flow depth until the conveyance equals the known term on the right-hand side of Eq. (8.5).

Analytical solutions to the normal depth are only possible for simple geometries such as wide rectangular channels for which $R_h = h_n$ because $W >> h$, and $AR_h^{2/3} = Wh_n^{5/3}$. Also, the unit discharge $q = Q/W$ and thus $Q/AR_h^{2.3} = q/h_n^{5/3}$. In wide rectangular channels, the normal depth $h_n = y_n$ is simply

$$h_n = \left(\frac{nq}{S_0^{1/2}}\right)^{3/5} = \left(\frac{nQ}{WS_0^{1/2}}\right)^{3/5} \text{ in SI or}$$
$$h_n = \left(\frac{nq}{1.49S_0^{1/2}}\right)^{3/5} = \left(\frac{nQ}{1.49WS_0^{1/2}}\right)^{3/5} \text{ in customary units.} \quad (8.6)$$

The normal depth increases with discharge and friction but decreases with slope.

For instance, consider a channel slope $S_0 = 0.01$, roughness $n = 0.025$ and width $W = 50$ m. The normal depth when $Q = 200$ m^3/s is simply $h_n = \left(nQ/WS_0^{1/2}\right)^{3/5} = \left[(0.025 \times 200)/(50\sqrt{0.01})\right]^{3/5} = 1$ m, and $W/h = 50$.

8.4 Shear Stress

The concepts of resistance to flow and normal depth allow the determination of the shear stress applied on the wetted perimeter of open channels. Figure 8.7 illustrates the force balance in an open channel with a uniform cross-sectional area.

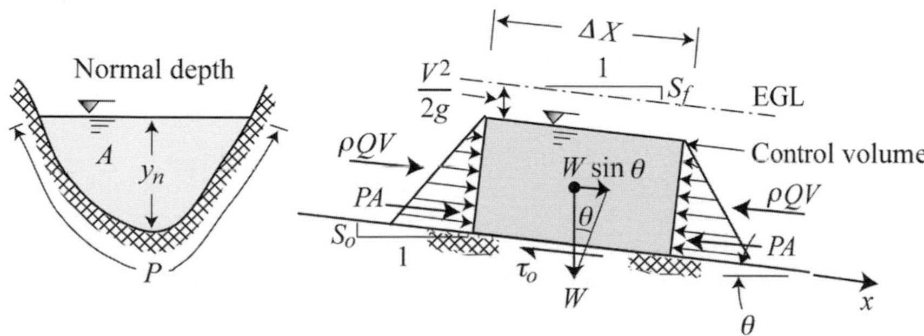

Figure 8.7 Shear stress in steady–uniform open channels

Under steady–uniform flow conditions, the flow depth is the normal depth and the velocity is constant along the reach of length ΔX. The velocity head is constant such that the bed slope and the friction slope are equal ($S_0 = S_f$) and the hydrostatic and momentum forces at both ends of the control volume are equal. In the control volume, the volume of fluid is $\forall = A\Delta X$ and the weight is $W = \gamma\forall$. The force component in the downstream x direction is $W \sin\theta = \gamma A \Delta X \sin\theta$. This force has to be balanced by the shear force $\tau_0 P \Delta X$ exerted on the wetted surface $P\Delta X$. The average bed shear stress τ_0 is applied by the boundary onto the fluid in the upstream direction. This average shear stress relationship for small angles ($S_f = \sin\theta$) is

$$\tau_0 = \gamma \frac{A}{P} S_f = \gamma R_h S_f. \quad (8.7)$$

It is now interesting to describe the shear stress in terms of the friction factors. From the Darcy-Weisbach friction factor in pipes, knowing that the hydraulic radius in pipes is $R_{h\,pipe} = A/P = \pi D^2/4\pi D = D/4$, the friction slope is

$$S_f = \frac{h_f}{L} = \frac{f}{D}\frac{V^2}{2g} = \frac{f}{8}\frac{V^2}{gR_h}.$$

This relationship is linked to Eq. (8.2) to describe the friction slope and shear stress in terms of the Darcy–Weisbach f, Manning n or Chézy C coefficients. In the particular case of wide rectangular channels, $R_h = h_n$, the friction slope becomes a simple function of the Darcy–Weisbach f and Froude number Fr as $S_f = (f/8)\mathrm{Fr}^2$. Shear-stress calculations become particularly important in relation to sediment transport in rivers with ample more details in Julien (2010).

We conclude with two more complex applications of normal depth in a trapezoidal channel in Example 8.6 and a circular cross section in Case Study 8.2.

Example 8.6: Normal depth in a trapezoidal channel

The US Department of Agriculture's Agricultural Research Service (SCS 1986) uses steep trapezoidal channels to measure the flow discharge at Goodwin Creek, Mississippi. For a 21.4-km^2 watershed, the trapezoidal channel is 9.14 m wide at the base with 2H:1V side slopes (see Example 8.2), the downstream slope is $S_0 = 0.20$ and the Manning coefficient $n = 0.015$. Find the normal depth at a 100-year discharge $Q = 176\ \mathrm{m^3/s}$.

Solution: To find the normal depth, we need the flow area $A = By + ty^2 = 9.14y + 2y^2$ from Figure 8.2 and the wetted perimeter $P = B + 2y\sqrt{1+t^2} = 9.14 + 2\sqrt{5}y$.

The normal depth ($y = y_n$) requires solving Eq. (8.5) when $S_f = S_0$, or

$$\frac{A_n^{5/3}}{P_n^{2/3}} = \frac{nQ}{mS_0^{1/2}} = \frac{\left(9.14y_n + 2y_n^2\right)^{5/3}}{\left(9.14 + 2\sqrt{5}y_n\right)^{2/3}} = \frac{0.015 \times 176}{1\sqrt{0.2}} = 5.9.$$

Using a solver (e.g. goal seek in Excel), we obtain the normal depth $y_n = 0.75$ m, and from the formulas in Figure 8.2, we obtain $A_n = 8.0\ \mathrm{m^2}$, $P_n = 12.5$ m, $R_{hn} = 0.64$ m, $W_n = 12.1$ m, $h_n = 0.66$ m and $V_n = Q/A_n = 22.1$ m/s.

♦Case Study 8.2: Normal flow depth in the Tehachapi Tunnel, CA

The Edmonston plant from Case Study 4.1 pumps water into a surge tank feeding a series of tunnels through the Tehachapi Mountains in California. Tunnel 1 conveys 5,360 cfs in a lined circular conduit 23.5 ft in diameter as sketched in Fig. CS-8.2. Find the normal depth assuming $n = 0.012$.

Solution: The bed slope of the tunnel is $S_0 = (3{,}090 - 3{,}080.5)/7{,}933 = 0.0012$.

To find the normal depth, we need the relationships for the area and wetted perimeter from Example 8.1. The area is $A = 0.5r^2(\theta - \sin\theta)$, and the perimeter is $P = \theta r$.

The normal depth is obtained from Eq. (8.5) for $\theta = \theta_n$ when $S_f = S_0 = 0.0012$, $n = 0.012$, $m = 1.49$ in customary units, $Q = 5{,}360$ cfs and $r = 23.5/2 = 11.75$ ft:

$$\frac{A^{5/3}}{P^{2/3}} = \frac{\left[0.5 \times 11.75^2(\theta_n - \sin\theta_n)\right]^{5/3}}{(11.75\theta_n)^{2/3}} = \frac{nQ}{mS_0^{1/2}} = \frac{0.012 \times 5{,}360}{1.49 \times \sqrt{0.0012}} = 1{,}246.$$

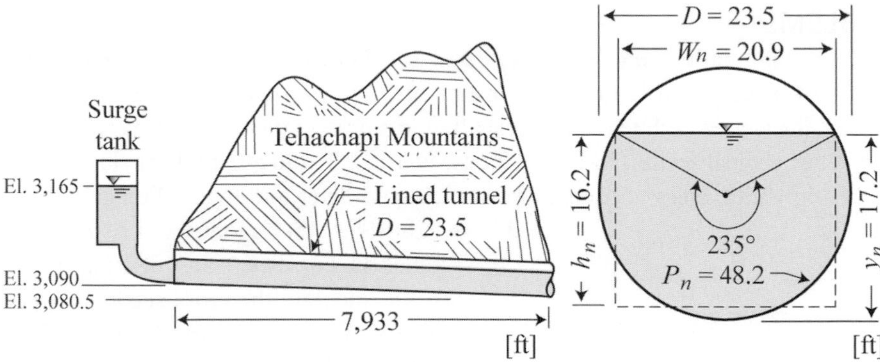

Fig. CS-8.2 Tehachapi tunnel

Using a solver (or goal seek in Excel), we obtain the angle $\theta_n = 4.1$ rad $= 235°$ and the corresponding normal depth is $y_n = r[1 - \cos(\theta_n/2)] = 11.75[1 - \cos(4.1/2)] = 17.2$ ft.

Under the normal flow conditions, the relationships from Example 8.1 give a cross-sectional area $A_n = 340$ ft^2, a wetted perimeter $P_n = 48.2$ ft, a hydraulic radius $R_{hn} = 7.05$ ft, a top width $W_n = 20.9$ ft and a mean flow depth $h_n = 16.2$ ft.

Additional Resources

Additional information on open-channel flow can be found in Chow (1959), Henderson (1966), French (1985), Sturm (2001) and Cruise et al. (2007). Useful references on resistance to flow in open channels include Chow (1959), Bray (1979), Arcement and Schneider (1984), Barnes (1987), FHWA (1984, 2001), Jarrett (1985), Ferguson (2007), Cheng (2015), and Lee and Julien (2012a, 2012b, 2017).

EXERCISES

These exercises review the essential concepts from this chapter.

1. Why should the cross section be perpendicular to the flow direction?
2. What is the difference between the average depth and the hydraulic radius?
3. Is the wetted perimeter smaller or larger than the surface width?
4. Which of *C, n, K* and *f* are resistance coefficients?
5. What is the difference between the friction and bed slopes in steady-uniform flows?
6. True or false?
   (a) The average depth of a parabolic distribution is 2/3 of the maximum depth.
   (b) The hydraulic radius of a pipe flowing full is *D*/4.
   (c) Manning coefficients vary for different systems of units.
   (d) Chézy coefficients vary for different systems of units.
   (e) Darcy–Weisbach coefficients vary for different systems of units.
   (f) The Manning *n* coefficient can be less than 0.01.
   (g) The coefficients *C* and *f* are dimensionless parameters.
   (h) Increasing slope increases the normal depth.

PROBLEMS

1. ♦ A rectangular channel is 4 m wide and 2.5 m deep. The water level in the channel is 1.75 m deep and is flowing at a rate of 21 m^3/s. Determine the cross-sectional area, the wetted perimeter and the hydraulic radius. Is the flow laminar or turbulent?
2. ♦ A trapezoidal section has a 5.0-ft base width, 2.5-ft depth, and 1:1 side slope. Find the top width, cross-sectional area, wetted perimeter and hydraulic radius.
3. ♦♦ Consider the geometry of the channel sketched in Fig. P-8.3 and develop a spreadsheet to calculate the cross-sectional area and wetted perimeter as a function of flow depth. Assume $n = 0.02$ for both the channel and the floodplain. Calculate every 3 inches until a maximum stage elevation of 10 ft. Determine the channel width, average flow depth and hydraulic radius for these conditions. Plot the results for each parameter.

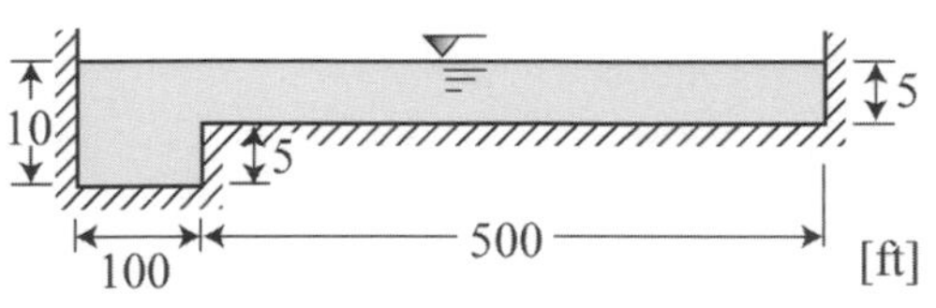

Fig. P-8.3

4. ♦♦ The flow depth in a river cross section starting from the left bank to the right bank is measured every 40 m (see Table P-8.4). Use a spreadsheet and plot the river section and determine the hydraulic radius and hydraulic depth of the river.

Table P-8.4.

Distance	Depth	Elevation	dP	dA
0	0	24		
40	1.2	22.8	40.018	24
80	4.9	19.1	40.17076	122
120	6.9	17.1	40.04997	236
160	9.2	14.8	40.06607	322
200	11.7	12.3	40.07805	418
240	14	10	40.06607	514
280	16.5	7.5	40.07805	610
320	18.8	5.2	40.06607	706
360	21	3	40.06045	796
400	17	7	40.1995	760
440	15.8	8.2	40.018	656
480	14	10	40.04048	596
520	12.5	11.5	40.02812	530
560	10.2	13.8	40.06607	454
600	7.3	16.7	40.10499	350
640	2.3	21.7	40.31129	192
680	1	23	40.02112	66
705	0	24	25.01999	12.5

5. ♦ A trapezoidal channel has a bed width of 10 ft and side slope 2H:1V. The channel has a smooth cement surface with $n = 0.011$. If the channel is laid on a slope of 0.0001 and carries a uniform flow of depth 2 ft, determine the discharge.
6. ♦ A trapezoidal channel has a bed width of 10 ft and side slope 2H:1V. The channel is paved with a smooth cement surface at $n = 0.011$. If the channel has a slope of 0.0001 and carries a uniform flow of depth 4 ft, determine the discharge.
7. ♦♦ A trapezoidal earthen channel with a bottom width of 5 ft and side slope 2H:1V is carrying a discharge of 200 cfs at normal depth. If the channel is running on a slope of 0.0001 ft/ft and has a Manning coefficient $n = 0.025$, find the normal depth.
8. ♦♦ A trapezoidal channel is designed to carry 25 m^3/s on a slope of 0.0015 m/m. The channel is unlined, and the maximum allowable velocity to prevent erosion is 1.5 m/s. The side slope must be no steeper than 2H:1V and the Manning n value is 0.03. What flow depth and bottom width would meet these requirements? [Hint: write V and Q as a function of b and y and solve the two equations and two unknowns.]

Problems 9 and 10 are most important!

9. ♦♦♦ **Pump and pipe system design!** Let's return to Problem 4.9 in Chapter 4. This is part (c). Your first meeting with Jill went very well. Upon return from travel, Jan asks you whether you considered the energy cost in your calculations. The energy cost adjusted for inflation is assumed to remain constant at ~10 cents per kilowatt-hour ($100/MWh). Recalculate the cost for each pipe diameter, and plot the total cost of your project vs time over the next 50 years,

   Your elected representative Jill finds the project reasonably priced and affordable for her district. She mentions that her very good friend Jacques "Jack" (a local entrepreneur selling pipes in your city) can supply 5,000 ft of 6-in. diameter pipe at $15/ft with a 50% discount on the purchase of any suitable pump. Jack also told Jill that it would be best to design a system with two parallel pipelines. In cutting the discharge in half, he said, the energy losses would drop a lot. Conduct your own analysis of Jack's idea.

   Prepare a revised two-page summary of your analysis. Your recommendation should be a maximum two-page summary letter addressed to Jill. It will be presented to the public at the next council meeting (i.e. it will likely be requested at the next mid-term exam). [Hint: look at Appendix B.]
10. ♦♦♦ **Design problem.** The concrete-lined canal in Fig. P-8.10 has one side vertical and one side sloping at 2H:1V. The canal carries a steady discharge of 10 m^3/s with $n = 0.014$ and the velocity should not exceed 0.715 m/s. Design the dimensions and the bed slope of the canal to minimize the construction per unit length of 100 m. The excavation cost is 2.0 $/$m^3$ and the cost of lining the wetted perimeter is 6.0 $/$m^2$.

[Hint: first define the cost as a function of width b and depth y, then replace b as a function of y. Plot the cost as a function of y and find the minimum. Your answer should delineate your problem formulation, assumptions, methods, accuracy and presentation.]

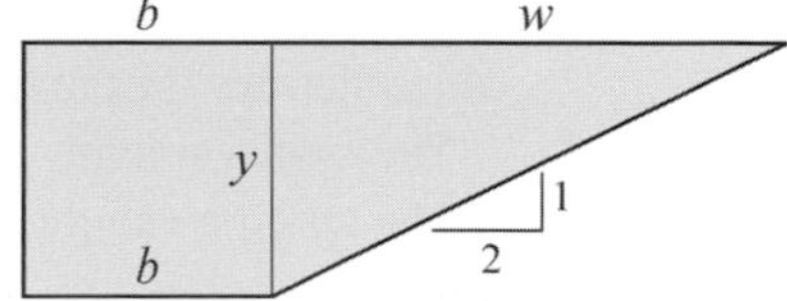

Fig. P-8.10

9 Rapidly Varied Flow

Rapidly varied flow refers to nonuniform flow conditions changing suddenly over short distances. As opposed to gradually varied flow with gradual changes in hydraulic conditions in Figure 9.1, the analysis of rapidly varied flow requires the application of the principles of conservation of energy in Section 9.1 and conservation of momentum in Section 9.2. This leads us to the definition of hydraulic controls in Section 9.3.

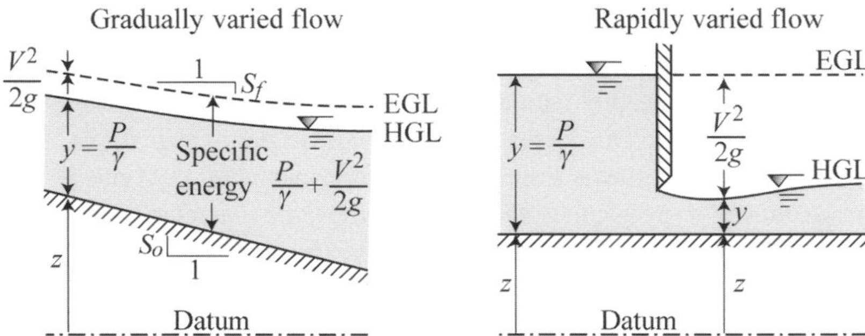

Figure 9.1 Comparison of gradually and rapidly varied flow conditions

9.1 Energy in Open Channels

Section 9.1.1 defines specific energy in open channels. Two important new depths termed the critical depth (Section 9.1.2) and alternate depths (Section 9.1.3) are defined, and choking conditions are discussed in Section 9.1.4.

9.1.1 Specific Energy

Under small vertical accelerations, the hydrostatic pressure distribution implies $y = p/\gamma$ in Figure 9.1. The total energy level corresponds to the Bernoulli sum H:

$$H = z + \frac{p}{\gamma} + \frac{V^2}{2g} = z + y + \frac{V^2}{2g},$$

which defines the energy grade line (EGL).

The first two terms $(y + z)$ define the free surface or hydraulic grade line (HGL).

The specific energy is defined as the energy level above the channel floor:

$$E = \frac{p}{\gamma} + \frac{V^2}{2g} = y + \frac{V^2}{2g} = y + \frac{Q^2}{2gA^2}. \tag{9.1}$$

The flow area A increases with flow depth y such that, at a given discharge, we can define a relationship between the specific energy E and the flow depth y.

The specific energy diagram in Figure 9.2 describes the relationship between specific energy E and flow depth y, considering that A increases with y for constant Q and g.

The minimum of the specific energy diagram defines the critical flow depth y_c. When the energy level is below the minimum energy level, there is insufficient energy to pass the desired flow discharge. We refer to this condition as choking the flow. This means that the water will accumulate upstream and raise the water level until the minimum energy level is reached.

Beyond the minimum specific energy level, it is possible to have two different flow depths with the same specific energy level. These two depths with the same specific energy level are called alternate depths.

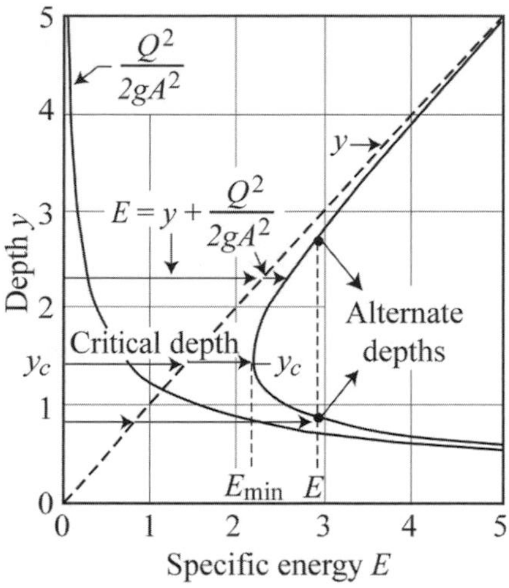

Figure 9.2 Specific energy diagram

9.1.2 Critical Depth

A general relationship can be derived for the critical depth y_c for any cross-sectional geometry with $dA = Wdy$, as sketched in Figure 9.3. The specific energy at a constant discharge Q is $E = y + \frac{Q^2}{2gA^2}$ and the critical depth $y = y_c$ corresponds to $dE/dy = 0$, or

$$\frac{dE}{dy} = 1 + \frac{Q^2}{2g}\left(\frac{-2\,dA}{A^3\,dy}\right) = 1 - \frac{Q^2 W}{gA^3} = 0.$$

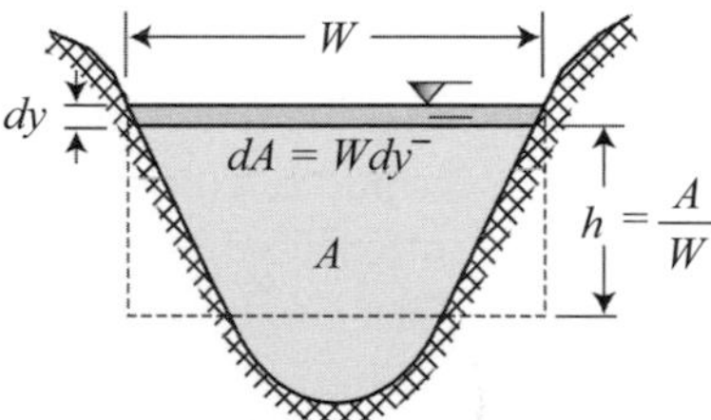

Figure 9.3 Channel cross section

The general condition is now based on the critical average depth h_c since

$$\frac{Q^2 W}{gA_c^3} = \frac{V_c^2 W}{gA_c} = \frac{V_c^2}{gh_c} = \mathrm{Fr}^2 = 1. \quad (9.2)$$

Note here that the average flow depth h is used to define the Froude number Fr. Also, the critical velocity head is half of h_c since $\frac{V_c^2}{gh_c} = 1 \Rightarrow \frac{V_c^2}{2g} = \frac{h_c}{2}$ (when $\mathrm{Fr} = 1$). The flow is subcritical when $\mathrm{Fr} < 1$ and supercritical when $\mathrm{Fr} > 1$.

For rectangular channels shown in Figure 9.4, the geometry is simply $W = B$ and $y = h$ with area $A = By$, the discharge $Q = AV = ByV$ and the unit discharge $q = Q/B = Vy$. Therefore, the specific energy for a rectangular channel is

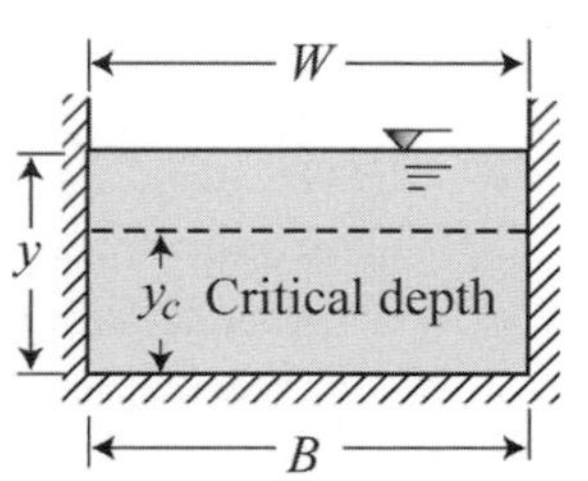

Figure 9.4 Rectangular channel

$$E = y + \frac{Q^2}{2gA^2} = y + \frac{Q^2}{2gB^2y^2} = y + \frac{q^2}{2gy^2}.$$

We find the critical depth $y = y_c$ for a constant unit discharge q from

$$\frac{dE}{dy} = 1 + \frac{q^2}{2g}\left(\frac{-2}{y^3}\right) = 0,$$

which gives $y = y_c$ when $\frac{q^2}{gy_c^3} = 1$ and therefore, the critical depth in a rectangular channel is

$$y_c = \left(\frac{q^2}{g}\right)^{1/3}.$$

Note that the critical depth is independent of slope and resistance to flow. For rectangular channels, the average flow depth and Froude number are linked

$$\mathrm{Fr}^2 = \frac{V^2}{gy} = \frac{q^2}{gy^3} = \left(\frac{y_c}{y}\right)^3. \tag{9.3}$$

The Froude number is the criterion that separates subcritical and supercritical flow. It is now clear that the flow is subcritical when $y > y_c$, or $\mathrm{Fr} < 1$, and it goes without saying that the flow is supercritical when $y < y_c$ or $\mathrm{Fr} > 1$. Example 9.1 illustrates basic critical depth calculations for a rectangular channel. The concept of critical depth is also often used for measuring the flow discharge at stream gauging stations as shown in Case Study 9.1.

Example 9.1: Critical depth in a rectangular channel

A rectangular channel is 50 m wide and 2 m deep with the Manning coefficient $n = 0.025$. Calculate the critical depth and the Froude number at a flow rate of 200 m^3/s.

Solution: The unit discharge $q = Q/B = 200/50 = 4 \mathrm{m}^2/\mathrm{s}$ and the critical depth is

$$y_c = \left(\frac{q^2}{g}\right)^{1/3} = \left(\frac{16}{9.81}\right)^{1/3} = 1.18 \text{ m}.$$

The Froude number is

$$\mathrm{Fr} = \left(\frac{y_c}{y}\right)^{3/2} = \left(\frac{1.18}{2}\right)^{3/2} = 0.45,$$

the flow is subcritical because $\mathrm{Fr} < 1$, or $y > y_c$, and neither roughness nor slope factor in the critical-depth calculations.

Case Study 9.1: Gauging station at Goodwin Creek, Mississippi

Goodwin Creek is an experimental watershed of the US Department of Agriculture's Agricultural Research Service (SCS 1986). Two types of gauging stations were constructed: trapezoidal and triangular sections.

A trapezoidal gauging section was constructed at the watershed outlet. The drainage area covers 21.4 km^2 and the flow discharge is estimated at 176 m^3/s for a 100-year flood. The base is 9.14 m wide with 2H:1V side slopes. The critical depth can be determined given the main geometric properties of trapezoidal channels from Examples 8.2 and 8.6. Accordingly, we have the top width $W = B + 2ty = 9.14 + 4y$, and the cross-sectional area $A = By + ty^2 = 9.14y + 2y^2$. The critical depth $y = y_c$ corresponds to Fr = 1 and is obtained by solving Eq. (9.2) written as

$$\frac{Q^2 W}{gA^3} = \frac{176^2(9.14 + 4y_c)}{9.81(9.14y_c + 2y_c{}^2)^3} = 1,$$

which, using a solver, gives $y_c = 2.75$ m.

For smaller drainage areas, V-notch weirs were also constructed. For instance, a 100-year discharge of 57 m^3/s is expected on a 4.3-km^2 drainage area. A triangular gauging station with a 2H:1V side slope has been constructed. In the case of V-notch weirs, the critical-depth calculations are using the equation for trapezoidal channels with a base width equal to zero. The critical depth $y = y_c$ corresponds to Fr = 1 from

$$\frac{Q^2 W}{gA^3} = \frac{Q^2(2ty)}{g(ty^2)^3} = \frac{57^2(4y_c)}{9.81(2y_c{}^2)^3} = 1,$$

or

$$y_c = \left[\frac{Q^2(2t)}{gt^3}\right]^{1/5} = \left(\frac{57^2 \times 4}{9.81 \times 8}\right)^{0.2} = 2.78 \text{ m}.$$

Both types of stream gauges are illustrated in Fig. CS-9.1.

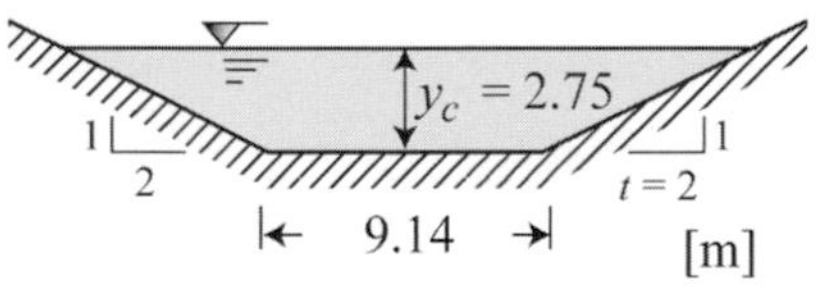

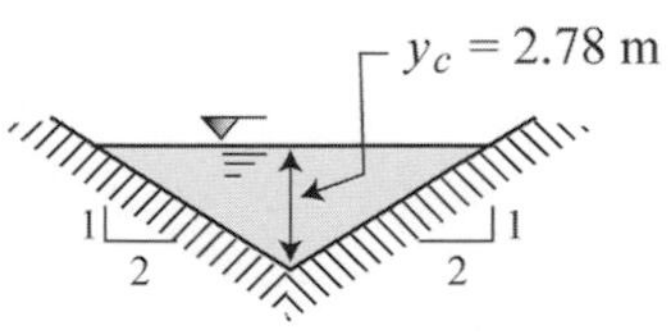

Fig. CS-9.1 Goodwin Creek gauging stations

9.1.3 Alternate Depths

In rapidly varied flow, as for the sluice gate shown in Figure 9.5, the conversion of potential to kinetic energy occurs without major friction losses because of the short distance. As a first approximation, the specific energy is the same on both sides of the sluice gate.

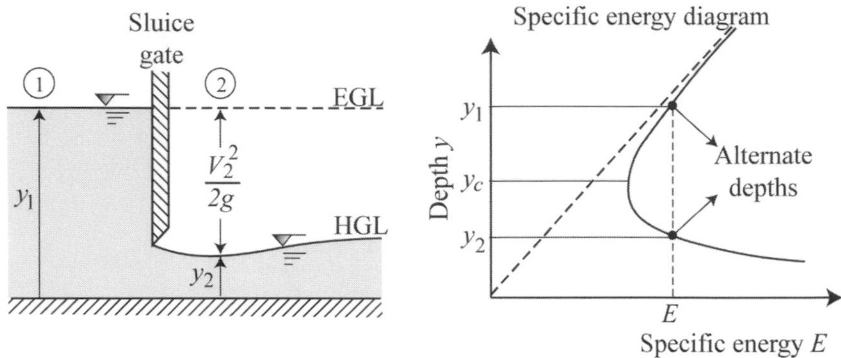

Figure 9.5 Flow depths near a sluice gate

We can analytically define these two alternate depths in a rectangular channel. The specific energy on both sides of the sluice gate remains constant; thus,

$$E = y_1 + \frac{q^2}{2gy_1^2} = y_2 + \frac{q^2}{2gy_2^2},$$

which is solved for y_2 as a function of y_1. After defining $X = \frac{y_1}{y_2}$, the equation is multiplied by $c = \frac{2gy_2^2}{q^2}$ and rearranged to give $X - X^3 = X(1+X)(1-X) = cy_1(1-X)$. After dividing both sides by $(1-X)$ we obtain the quadratic equation $X^2 + X - cy_1 = 0$ from which, the + root is

$$X = \frac{y_1}{y_2} = 0.5\left(-1 + \sqrt{1 + 4cy_1}\right)$$

or

$$y_2 = \frac{2y_1}{-1 + \sqrt{1 + \frac{8gy_1^3}{q^2}}} = \frac{2y_1}{-1 + \sqrt{1 + \frac{8}{Fr_1{}^2}}}.$$

Also,

$$y_1 = \frac{2y_2}{-1 + \sqrt{1 + \frac{8}{Fr_2{}^2}}}. \tag{9.4}$$

This relationship is only applicable for rectangular open channels (Moglen 2015). For other cross-sectional geometries, the use of solvers (or Excel spreadsheet with the goal seek function) is recommended – see Example 9.2 for a circular conduit.

Example 9.2: Circular open channel

A 10-m-diameter circular concrete tunnel ($n = 0.012$) conveys 100 m^3/s at a 0.01% slope. Find: (1) the normal depth; (2) the critical depth; and (3) the alternate depth.

Solution: With reference to Example 8.1, the main hydraulic parameters sketched in Fig. E-9.2a. are:
the flow depth $y = r[1 - \cos(\theta/2)]$ with θ in radians,
the free surface width $W = 2r \sin(\theta/2)$,
the wetted perimeter $P = \theta r$,
cross-sectional area $A = 0.5(\theta - \sin\theta)r^2$ and
the hydraulic radius $R_h = A/P$.

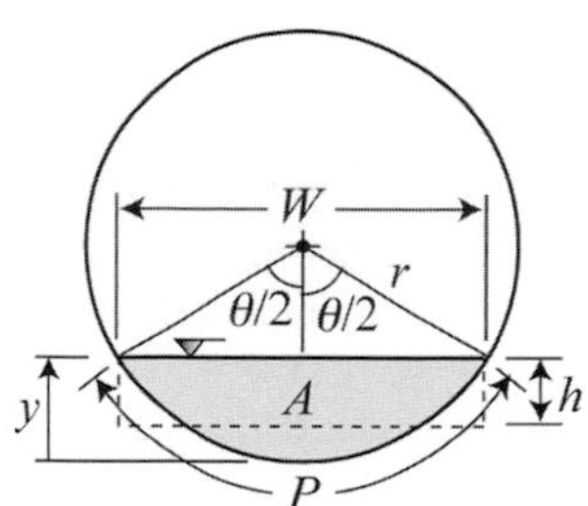

Fig. E-9.2a Circular cross-sectional area

(1) The normal depth is obtained from Manning's equation ($m = 1$ in SI) with $r = 5$ m:

$$Q = AV = \frac{mAR_h^{2/3}S_0^{1/2}}{n} = \frac{[0.5r^2(\theta - \sin\theta)]^{5/3}}{(r\theta)^{2/3}}\frac{m\sqrt{S_0}}{n} = \frac{[12.5(\theta_n - \sin\theta_n)]^{5/3}}{(5\theta_n)^{2/3}}\frac{\sqrt{0.0001}}{0.012} = 100.$$

Using a solver (or goal seek in Excel), we obtain the angle $\theta_n = 3.943$ rad and the normal depth $y_n = r[1 - \cos(\theta_n/2)] = 5[1 - \cos(3.943/2)] = 6.95$ m, as sketched in Fig. E-9.2b.

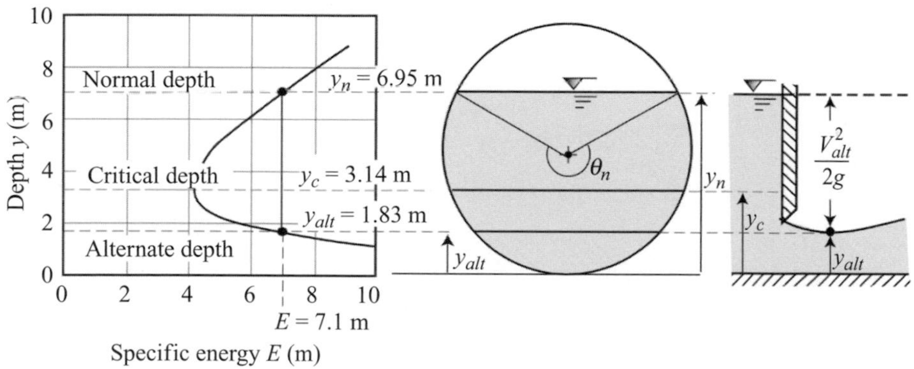

Fig. E-9.2b Critical, normal and alternate depths in a circular conduit

(2) The critical depth corresponds to Fr = 1 or

$$Fr_c^2 = \frac{Q^2 W_c}{gA_c^3} = \frac{100^2[2r\sin(\theta_c/2)]}{9.81[0.5r^2(\theta_c - \sin\theta_c)]^3} = 1,$$

which, from a solver, gives $\theta_c = 2.38$ rad and the corresponding critical depth $y_c = 5[1 - \cos(1.19)] = 3.14$ m.

From Fig. E-9.2b, the critical depth is at the minimum of the specific energy diagram.

(3) The alternate depth y_{alt} has the same specific energy $E_n = E_{alt}$ as the normal depth, or

$$E_n = y_{alt} + \frac{Q^2}{2gA_{alt}^2} = 5\left[1 - \cos\left(\frac{\theta_{alt}}{2}\right)\right] + \frac{100^2}{2 \times 9.81\left[0.5 \times 5^2(\theta_{alt} - \sin\theta_{alt})\right]^2}$$
$$= 7.1 \text{ m}.$$

The solver gives $\theta_{alt} = 1.77$ rad and alternate depth $y_{alt} = 5[1 - \cos(1.77/2)] = 1.83$ m.

Referring to the specific energy diagram, the normal depth is above the critical depth (subcritical flow $\mathrm{Fr} < 1$), the alternate depth is below the critical depth (supercritical flow, $\mathrm{Fr} > 1$).

9.1.4 Choking

Choking occurs when the energy level is below the minimum specific energy. It means that the flow discharge cannot pass through this critical cross section. The water will accumulate upstream of the choking section until the minimum energy level is reached. Sill-raising and channel contraction can impact the water level in wide rectangular open channels, as sketched in Figure 9.6.

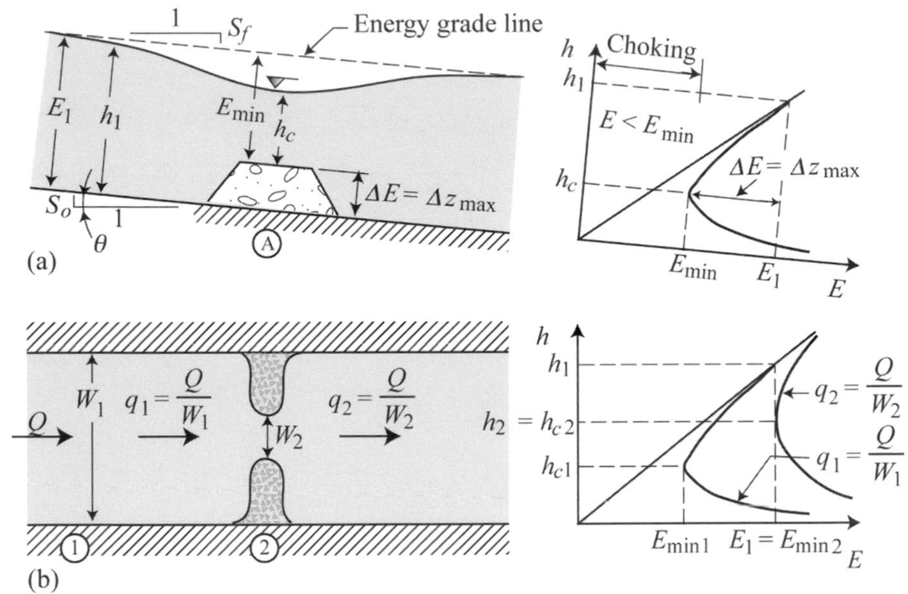

Figure 9.6 Sill-raising and channel contraction

In open channels, two main questions regarding choking are: (1) what is the maximum sill height Δz_{max} that will not raise the upstream water level, i.e. without choking the flow?; and (2) what is the minimum width of a channel contraction that will not raise the upstream water level?

Both questions can be answered with the use of the specific energy diagram. Noting that with $h = y$ in rectangular channels, the specific energy is

$$E_1 = h_1 + \frac{q^2}{2gh_1^2},$$

with the upstream flow depth h_1 and unit discharge $q = Q/W$.

(1) To prevent choking in the case of sill-raising, the specific energy above the step height Δz needs to be at least equal to the minimum specific energy level. Therefore, the maximum sill height $\Delta z_{\text{max}} = E - E_{\text{min}}$.

(2) In the case of a channel contraction, the minimum channel width is obtained when the specific energy at section 1 equals the specific energy of the contracted section 2, or $E_1 = E_{2\,min}$. For rectangular sections, given that $h_c = 2E_{2\,\text{min}}/3$,

$$1 = \frac{q_2^2}{gh_{c2}^3} = \left(\frac{Q}{W_{2\,min}}\right)^2 \frac{1}{gh_{2c}^3} = \left(\frac{Q}{W_{2\,min}}\right)^2 \frac{1}{g}\left(\frac{3}{2E_1}\right)^3,$$

and the minimum channel width without choking the flow is

$$W_{2\,min} = \frac{Q}{\sqrt{g(0.667E_1)^3}}.$$

9.2 Momentum

In this section, the principle of conservation of momentum is applied to open channels. As discussed in Section 3.2, the concept of momentum corresponds to hydrodynamic forces. The magnitude of the hydrostatic force on a surface is given by the pressure at the centroid of the surface times the surface area. In open channels, our first challenge in Section 9.2.1 is to determine the position of the centroid for any cross-sectional geometry. Then, we cover specific momentum in Section 9.2.2 and conjugate depths in Section 9.2.3.

9.2.1 Centroid Position for Open-Channel Flow

For symmetrical cross sections, like a circular pipe flowing full or a rectangular cross section, the centroid is simply located at the mid depth. However, for open channels with different geometries, the centroid can be calculated using the trapezoidal approximation $\bar{h}A = \bar{h}_1A_1 + \bar{h}_2A_2 + \bar{h}_3A_3\ldots$, where the total area is the sum of partial areas $A = A_1 + A_2 + A_3\ldots$ and $\bar{h}_1, \bar{h}_2, \bar{h}_3\ldots$ are the respective centroids of each sub-area. Notice that the distances are measured from the free surface and not from the datum, because the hydrostatic pressure at the centroid is given by $p = \gamma\bar{h}$, and $F = \gamma\bar{h}A$. Typical centroid positions for asymmetrical geometries are listed in Figure 9.7. Applications are shown for trapezoidal channels in Example 9.3 and circular conduits in Example 9.4.

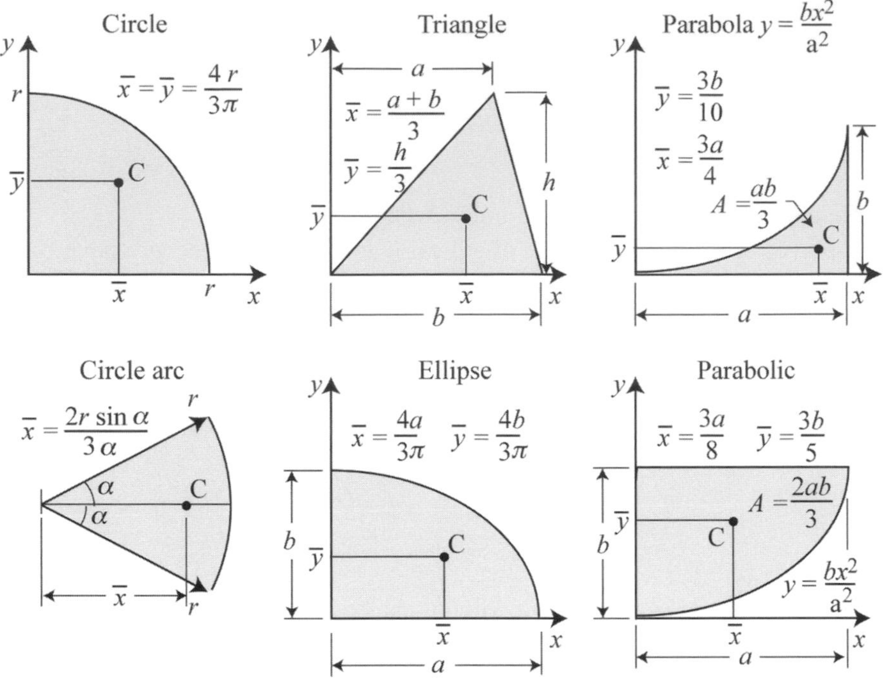

Figure 9.7 Centroid for circular, triangular, elliptic and parabolic geometries

---

Example 9.3: Centroid of a trapezoidal section

Determine the position of the centroid below the free surface for a trapezoidal channel of base width B and side slope t, as sketched in Fig. E-9.3. Then calculate the hydrostatic force on this surface area.

Solution: For $\bar{h}A = \bar{h}_A A_A + \bar{h}_B A_B + \bar{h}_C A_C$ with $\bar{h}_A = \bar{h}_C = h/3$ and $\bar{h}_B = h/2$,

$$\bar{h}(B + th)h = \frac{h}{3}\frac{th^2}{2} + \frac{h}{2}Bh + \frac{h}{3}\frac{th^2}{2},$$

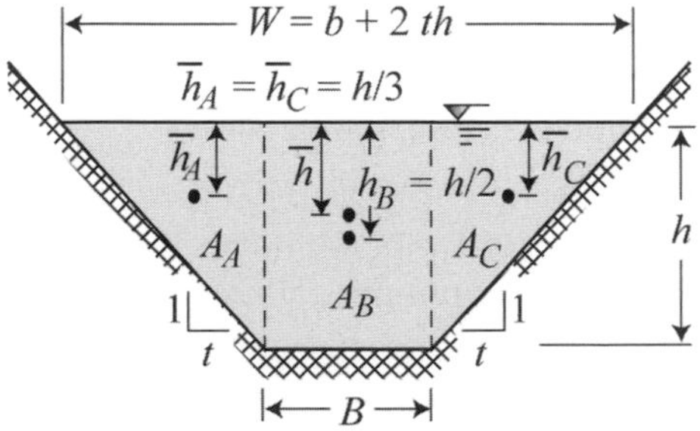

$$\bar{h}\left(Bh + th^2\right) = \frac{Bh^2}{2} + \frac{2th^3}{6}$$

or

$$\bar{h} = \frac{3Bh^2 + 2th^3}{6\left(Bh + th^2\right)} = \frac{h}{6}\left(\frac{3B + 2th}{B + th}\right).$$

Note that this centroid relationship reduces to $\bar{h} = h/2$ for a rectangular section $(t = 0)$, and $\bar{h} = h/3$ for a triangular section $(B = 0)$.

Fig. E-9.3 Centroid of a trapezoidal channel

The hydrostatic force is

$$F_h = \gamma \bar{h}(Bh + th^2) = \gamma\left(\frac{Bh^2}{2} + \frac{th^3}{3}\right),$$

which reduces to $F_h = \gamma\left(\frac{Bh^2}{2}\right)$ for a rectangle ($t = 0$), and $F_h = \gamma\left(\frac{th^3}{3}\right)$ for a triangle ($B = 0$).

Example 9.4: Centroid of a circular conduit
Determine the position of the centroid in a circular conduit sketched in Fig. E-9.4.

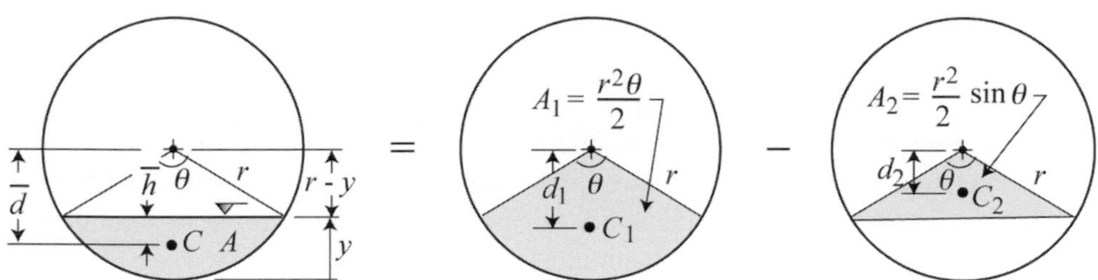

Fig. E-9.4 Centroid of a circular conduit

Solution: The areas are

$$A_1 = \frac{r^2\theta}{2}, \; A_2 = \frac{r^2}{2}\sin\theta, \; A = \frac{r^2}{2}(\theta - \sin\theta),$$

and

$$d_1 = \frac{2r\sin(\theta/2)}{3(\theta/2)}, \; d_2 = \frac{2r\cos(\theta/2)}{3};$$

from $\bar{d}A = d_1A_1 - d_2A_2$, we obtain

$$\bar{d} = \frac{d_1A_1 - d_2A_2}{A} = \frac{4r}{3}\left[\frac{\sin^3(\theta/2)}{(\theta - \sin\theta)}\right],$$

and the position of the centroid below the free surface becomes $\bar{h} = \bar{d} - r\cos(\theta/2)$.

We can check that $\bar{h} = \bar{d} = \frac{4r}{3\pi}$ when the pipe is flowing half full ($\theta = \pi$).

Finally, we get $\bar{d} = 0$ and $\bar{h} = r$ when the pipe is flowing full, i.e. $\theta = 2\pi$.

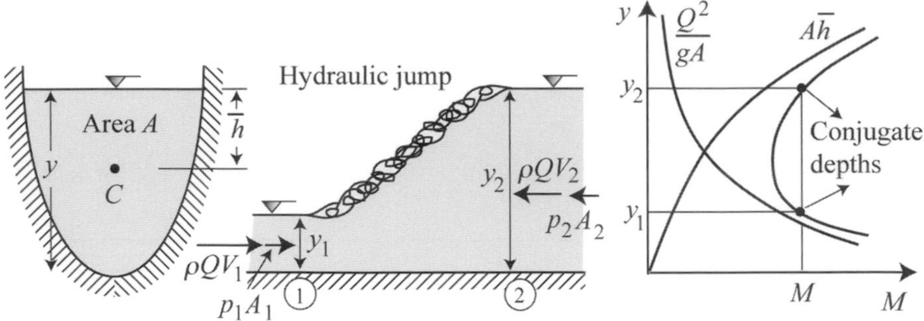

Figure 9.8 Momentum force and specific momentum diagram in open channels

9.2.2 Specific Momentum

As discussed in Chapter 3, the momentum force in open channels is the sum of two force components: (1) the hydrostatic force component $F = pA = \gamma\bar{h}A$; and (2) the hydrodynamic force or momentum flux component $F = \rho QV$. These two components are sketched in Figure 9.8 for a hydraulic jump.

The concept of momentum balances the pressure and momentum forces,

$$F = pA + \rho QV = \gamma\bar{h}A + \rho QV = \gamma\bar{h}A + \frac{\gamma Q^2}{gA}. \tag{9.5}$$

The specific momentum M is defined as the momentum force divided by the specific weight of the fluid γ:

$$M = \frac{F}{\gamma} = A\bar{h} + \frac{Q^2}{gA}. \tag{9.6}$$

When the flow discharge is constant, the specific momentum diagram resembles the specific energy diagram, because the area increases with flow depth. The two depths with the same specific momentum value are called conjugate depths or sequent depths, as shown in Figure 9.8. In rapidly decelerating flows such as in a hydraulic jump, we assume that the friction force on a flat surface is negligible. We can thus consider that the momentum forces on both sides of the hydraulic jump are equal. Therefore, we have the same value of specific momentum on both sides of the jump and these two flow depths correspond to the conjugate depths on the specific momentum diagram. We can analytically determine the conjugate depths in rectangular channels, as detailed in Section 9.2.3.

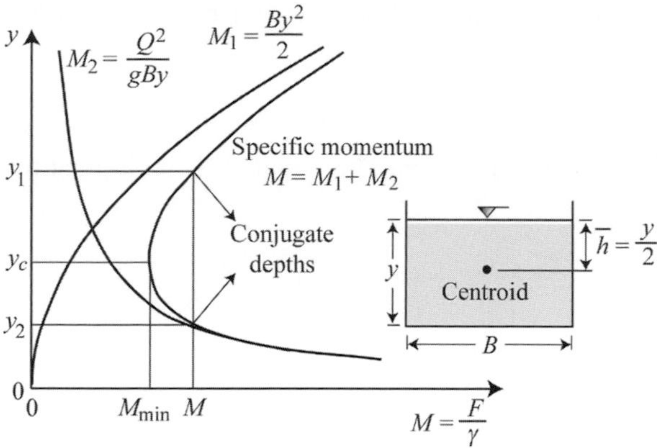

Figure 9.9 Specific momentum diagram

9.2.3 Conjugate Depths

We can analytically define the two conjugate depths in the case of simple geometries such as rectangular channels. Conjugate (sometimes called sequent) depths are two flow depths with the same specific momentum. As shown in Figure 9.9, for a rectangular channel, $A = By$, with B, Q and g constant, the specific momentum equation becomes

$$M = \frac{By^2}{2} + \frac{Q^2}{gBy}.$$

After dividing the specific momentum by the channel width, we can use the constant unit discharge $q = Q/B$ and seek flow-depth values upstream y_1 and downstream y_2 of a hydraulic jump:

$$\frac{M}{B} = \frac{{y_1}^2}{2} + \frac{q^2}{gy_1} = \frac{{y_2}^2}{2} + \frac{q^2}{gy_2}. \tag{9.7}$$

This equation is similar to the specific energy relationship from Section 9.1.3 and it can be solved in a similar fashion. In rapidly varied flow, like the hydraulic jump sketched in Figure 9.10, the momentum balance between both sides is assumed because we can neglect the friction losses on the bed surface over such a short distance. In a rectangular channel, we derive a relationship for the two conjugate depths.

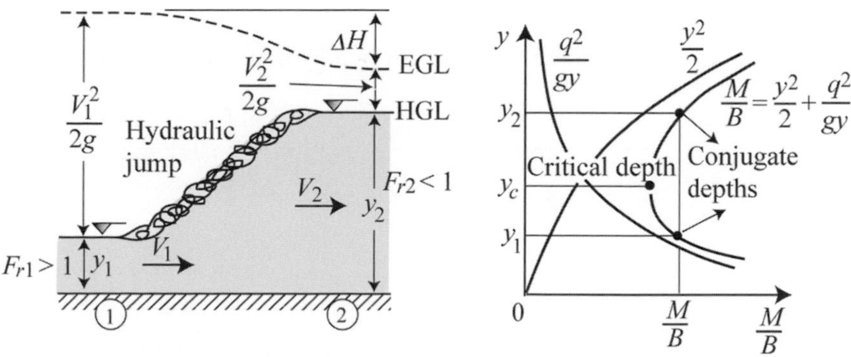

Figure 9.10 Conjugate (sequent) depths of a hydraulic jump

The specific momentum on both sides of the sluice gate remain constant; thus,

$$\frac{M}{B}=\frac{y_1^2}{2}+\frac{q^2}{gy_1}=\frac{y_2^2}{2}+\frac{q^2}{gy_2}$$

and we solve for y_2 as a function of y_1. This equation is multiplied by $\frac{2}{y_2^2}$ and rearranged with $X=\frac{y_1}{y_2}$ as

$$1-X^2=\frac{2q^2}{gy_2^3}\left(\frac{1-X}{X}\right).$$

Dividing both sides by $(1-X)$, and defining

$$c=\frac{2q^2}{gy_2^3}=2\mathrm{Fr}_2^2$$

(as in Section 9.1.3), we obtain the quadratic equation, $X^2+X-c=0$, from which the positive root is

$$X=y_1/y_2=0.5\left(-1+\sqrt{1+4c}\right),$$

giving

$$y_1=\frac{y_2}{2}\left(-1+\sqrt{1+8\mathrm{Fr}_2^2}\right),\ y_2=\frac{y_1}{2}\left(-1+\sqrt{1+8\mathrm{Fr}_1^2}\right). \tag{9.8}$$

Note that this equation is only applicable for rectangular open channels. It has been known in France since the early nineteenth century as the Bélanger equation. Furthermore, the energy loss in the jump is

$$\Delta E=\left(y_1+\frac{q^2}{2gy_1^2}\right)-\left(y_2+\frac{q^2}{2gy_2^2}\right)=(y_1-y_2)+\frac{q^2}{2g}\left(\frac{1}{y_1^2}-\frac{1}{y_2^2}\right),$$

which is combined with the specific momentum Eq. (9.7) rewritten as

$$\frac{q^2}{g}=\frac{y_1y_2(y_1+y_2)}{2}$$

in order to eliminate q and g. The head loss ΔE in a hydraulic jump reduces to

$$\Delta E=(y_1-y_2)+\frac{y_1y_2(y_1+y_2)}{4}\left(\frac{y_2^2-y_1^2}{y_1^2y_2^2}\right)=(y_2-y_1)\left[-1+\frac{(y_1+y_2)^2}{4y_1y_2}\right],$$

$$\Delta E=\frac{(y_2-y_1)^3}{4y_1y_2}. \tag{9.9}$$

The length of the hydraulic jump depends on the upstream Froude number and is scaled to the downstream flow depth as shown in Figure 9.11.

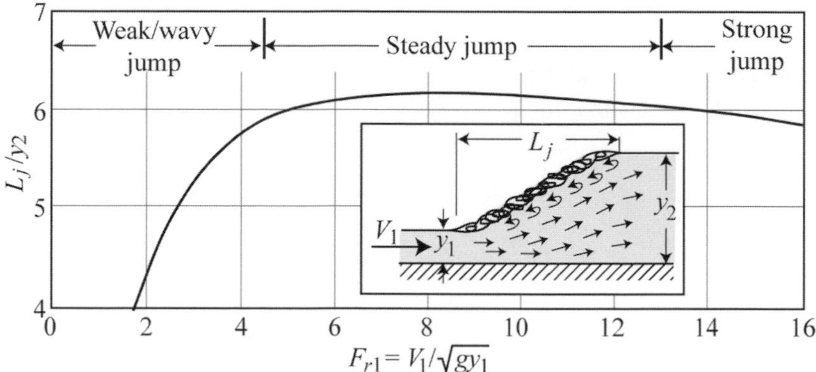

Figure 9.11 Length of a hydraulic jump (USBR 1977)

For most cross-sectional geometries, the use of solvers (e.g. Excel spreadsheet with the goal seek function) is recommended – see Example 9.5 for a trapezoidal channel.

Example 9.5: Conjugate depths in a trapezoidal channel

Determine the equation of specific energy and specific momentum for a trapezoidal channel. Plot the specific momentum and specific energy diagrams when the flow discharge is 6 m^3/s in a trapezoidal channel with a 2-m base width and side slopes at 1H:1V.

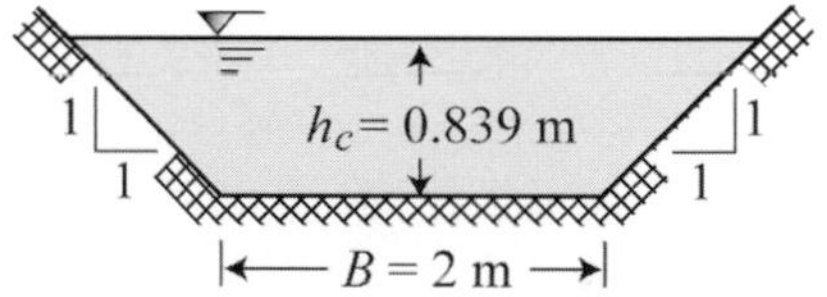

Fig. E-9.5a Centroid of a trapezoidal channel

Solution: With reference to Example 9.3, the position of the centroid for the trapezoidal section is

$$\bar{h} = \frac{h}{6}\left(\frac{3B + 2th}{B + th}\right),$$

the cross-section area is $A = Bh + th^2$, and the specific energy and momentum functions are, respectively,

$$E = h + \frac{V^2}{2g} = h + \frac{1}{2g}\left(\frac{Q}{Bh + th^2}\right)^2$$

and

$$M = \frac{F}{\gamma} = A\bar{h} + \frac{Q^2}{gA} = \left(\frac{Bh^2}{2} + \frac{th^3}{3}\right) + \frac{Q^2}{gh(B + th)}.$$

To define the critical depth, it is helpful to calculate

$$\frac{Q^2 W}{gA^3} = \frac{Q^2(B+2th)}{9.81\left(Bh+th^2\right)^3},$$

which equals 1 at the critical depth h_c.

For the conditions at this site, $Q = 6$ m^3/s, $B = 2$ m and $t = 1$, we obtain

$$E = h + \frac{1}{2g}\left(\frac{Q}{Bh+th^2}\right)^2 = h + \frac{6^2}{2\times 9.81\left(2h+h^2\right)^2},$$

$$M = \left(\frac{Bh^2}{2}+\frac{th^3}{3}\right) + \frac{Q^2}{gh(B+th)}$$
$$= \left(h^2+\frac{h^3}{3}\right) + \frac{36}{9.81h(2+h)},$$

and, using a solver, the critical depth from

$$\frac{Q^2 W}{gA^3} = \frac{6^2(2+2h_c)}{9.81\left(2h_c+h_c{}^2\right)^3} = 1$$

is $h_c = 0.839$ m (see Fig. E-9.5a).

We can also use spreadsheets and plot the specific energy and specific momentum curves as a function of flow depth, as shown in Fig. E-9.5b. Notice that the critical flow depth also corresponds to the minimum value of both the specific energy and the specific momentum.

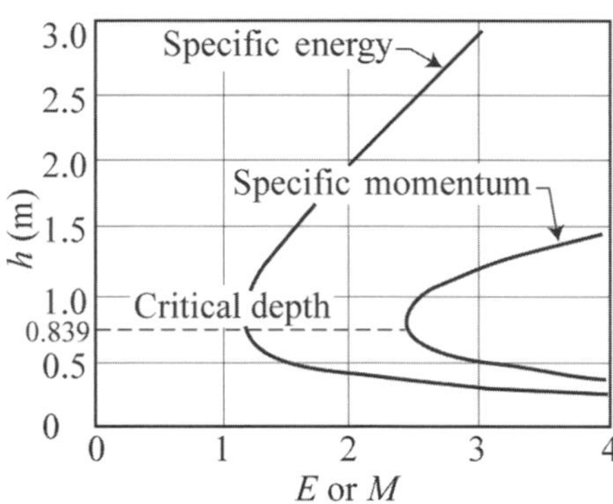

Fig. E-9.5b Specific energy E and momentum M for a trapezoidal channel

9.3 Hydraulic Controls

Hydraulic controls describe whether the flow conditions are predetermined from upstream or downstream conditions. The concept of wave celerity is introduced in Section 9.3.1, followed by upstream and downstream control in Section 9.3.2, surge celerity in Section 9.3.3, a brief discussion on momentum and energy in Section 9.3.4, and nonhydrostatic flow conditions in Section 9.3.5.

9.3.1 Small-Wave Propagation

Consider a solitary wave produced by a sudden horizontal displacement of a vertical gate in a rectangular laboratory flume. Assume a frictionless channel and a wave traveling without changing shape or velocity. As sketched in Figure 9.12, the solitary wave travels to the right with celerity c in a stationary fluid. An observer moving along

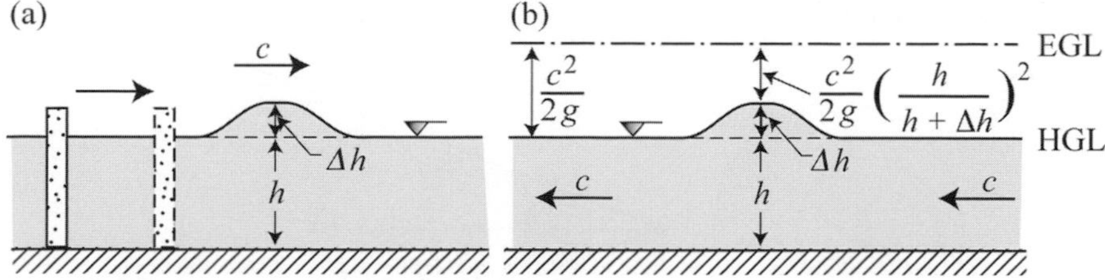

Figure 9.12 Definition sketch for a solitary wave without friction

the wave crest at a velocity equal to the celerity will perceive steady flow, where the wave appears to stand still with the flow moving to the left at a velocity equal to c.

When the continuity relationship is applied to the steady-flow case, a constant discharge implies that the relative velocity under the crest is $ch/(h + \Delta h)$. The equation of motion is then applied to the steady relative motion. When friction is neglected, the specific energy equation between the normal section and the wave crest gives

$$h + \frac{c^2}{2g} = h + \Delta h + \frac{c^2}{2g}\left(\frac{h}{h + \Delta h}\right)^2,$$

which is solved as

$$c = \sqrt{\frac{2g(h + \Delta h)^2}{2h + \Delta h}}.$$

The propagation of small waves in still water can be simplified further for waves of small amplitude, $\Delta h \ll h$, and the wave celerity simply becomes

$$c = \sqrt{gh}. \tag{9.10}$$

This has been known since the eighteenth century as the Lagrange celerity relationship (in honor of Joseph-Louis Lagrange) to describe the celerity of small-amplitude waves in a shallow channel without friction. The Froude number is often described as $\mathrm{Fr} = V/\sqrt{gh} = V/c$, given by the ratio of the mean flow velocity to the celerity of small perturbations. Thus, the Froude number bears practical significance in the ability to allow perturbations to propagate downstream.

When considering a surface perturbation from a point source (a bridge pier for instance), the perturbation will develop circular patterns expanding at celerity c. When the perturbation moves with the main flow velocity V, two general flow patterns develop, depending on the Froude number, as sketched in Figure 9.13.

For subcritical flow, $\mathrm{Fr} < 1$, the perturbation will migrate upstream of the point source. However, for supercritical flow, $\mathrm{Fr} > 1$, the perturbation will form a wave front at an angle $\sin\alpha = c/V$. These typical patterns can be recognized in the field to find out whether the flow is supercritical or not.

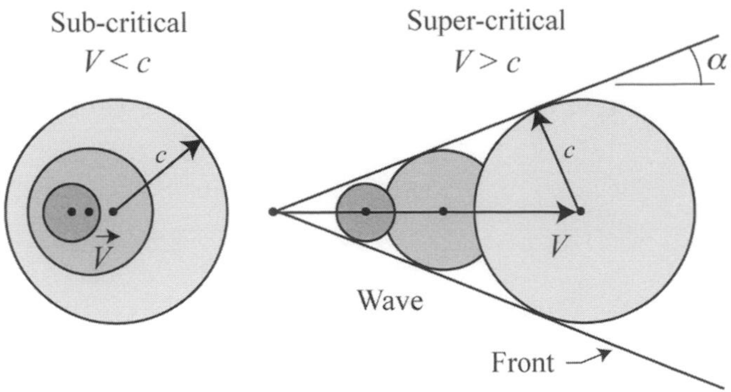

Figure 9.13 Wave propagation from point sources

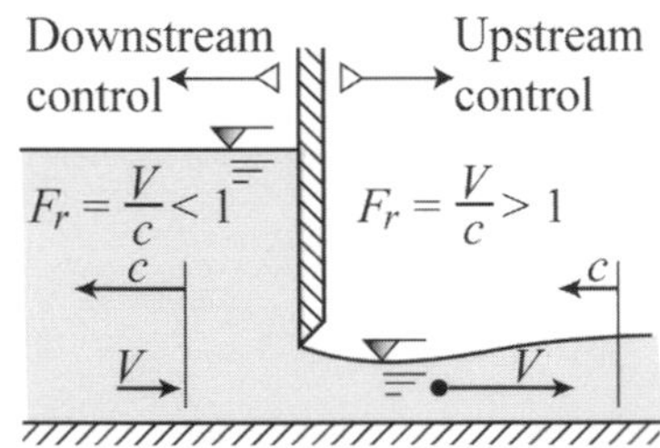

Figure 9.14 Upstream and downstream control

9.3.2 Upstream and Downstream Control

From the previous discussion, the comparison of the flow velocity V and the propagation speed or celerity c of flow perturbations depends on the Froude number. As sketched in Figure 9.14, the flow perturbation can propagate upstream when the celerity $c = \sqrt{gh}$ exceeds the flow velocity V. Therefore, we have downstream control in subcritical flow, i.e. when $\text{Fr} = \frac{V}{\sqrt{gh}} < 1$. Conversely, the effects of flow perturbations cannot travel upstream when the flow velocity exceeds the celerity. The flow conditions will be controlled upstream in supercritical flow $\text{Fr} = \frac{V}{\sqrt{gh}} > 1$.

9.3.3 Surge Propagation

Surges occur from sudden changes in flow conditions in open channels. Practical cases occur in irrigation canals, shutdown of power plants, tidal bores, etc. Consider a gate closure in the rectangular canal sketched in Figure 9.15.

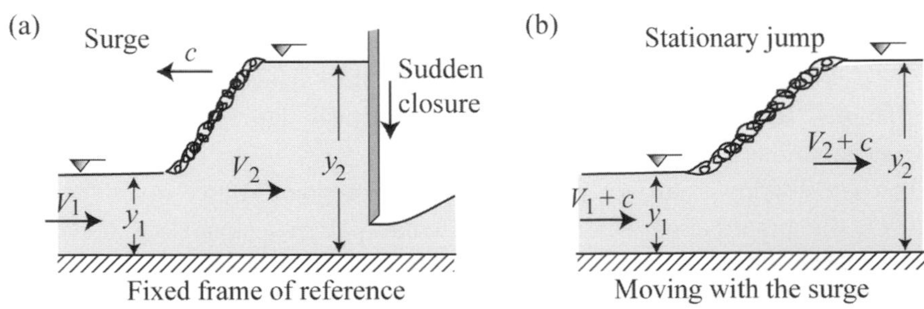

Figure 9.15 Surge propagation from a gate closure

To an observer on the ground, the flow is unsteady as the surge front propagates in the upstream direction at a celerity c. The flow is made steady by moving with the surge front. The relative motion of the upstream flow increases to $V_1 + c$, and the conservation of mass gives $(V_1 + c)y_1 = (V_2 + c)y_2$, or

$$c = \frac{V_1 y_1 - V_2 y_2}{y_2 - y_1} = \frac{\Delta q}{\Delta y}. \tag{9.11}$$

It is noted that the celerity equals the change in unit discharge $q = Vy$ over the change in flow depth. The case of a sudden gate closure corresponds to $V_2 = 0$.

For the momentum equation, we reached the condition previously analyzed in Figure 9.10 except that the upstream velocity is now $V_1^* = V_1 + c$, and the conjugate depth y_2 of a flow surge from Eq. (9.8) becomes

$$y_2 = \frac{y_1}{2}\left(-1 + \sqrt{1 + 8Fr_1^{*2}}\right)$$

or

$$\mathrm{Fr}_1^{*2} = \frac{(V_1 + c)^2}{g y_1} = \frac{y_2}{2y_1}\left(1 + \frac{y_2}{y_1}\right), \tag{9.12}$$

where the upstream surge Froude number is $\mathrm{Fr}_1^* = (V_1 + c)/\sqrt{g y_1}$ and the downstream surge Froude number is $\mathrm{Fr}_2^* = (V_2 + c)/\sqrt{g y_2}$. Of course, $\mathrm{Fr}_1^* > 1$ and $\mathrm{Fr}_2^* < 1$ just like the Froude numbers upstream and downstream of a hydraulic jump. Example 9.6 illustrates detailed calculations in a rectangular canal.

Example 9.6: Surge from a gate closure in a straight canal

A gate is suddenly closed in a canal with a steady uniform flow depth of 2 m and flow velocity of 1 m/s. Determine the celerity of the surge and the water-level rise at the gate.

Solution: From Eq. (9.11) with $V_2 = 0$ at the gate, we obtain $c = 2/(y_2 - 2)$ and from Eq. (9.12), the momentum equation reduces to

$$(V_1 + c)^2 = \left[1 + \frac{2}{(y_2 - 2)}\right]^2 = \frac{9.81 y_2}{2}\left(1 + \frac{y_2}{2}\right).$$

We find $y_2 = 2.475$ m from a solver and the water-level rise at the gate is $\Delta y = y_2 - y_1 = 2.475 - 2 = 0.475$ m.

The wave celerity is $c = 4.21$ m/s. The surge Froude numbers are, respectively, $\mathrm{Fr}_1^* = (V_1 + c)/\sqrt{g y_1} = (1 + 4.21)/\sqrt{9.81 \times 2} = 1.17$ upstream and $\mathrm{Fr}_2^* = (V_2 + c)/\sqrt{g y_2} = (4.21)/\sqrt{9.81 \times 2.475} = 0.85$ downstream of the surge front.

9.3.4 Standing Waves in Supercritical Flows

The design of supercritical flow channels is quite complicated because obstructions and channel alignment changes can cause the formation of standing waves, also called oblique waves. As illustrated in Figure 9.16, a change in flow alignment in supercritical flow will cause the formation of a standing wave front. For the analysis, we consider a rectangular channel with uniform supercritical upstream flow velocity V_1 and flow depth y_1 subjected to a sudden change in the flow orientation angle θ. This perturbation causes the formation of a standing wave front at an angle β from the initial flow direction.

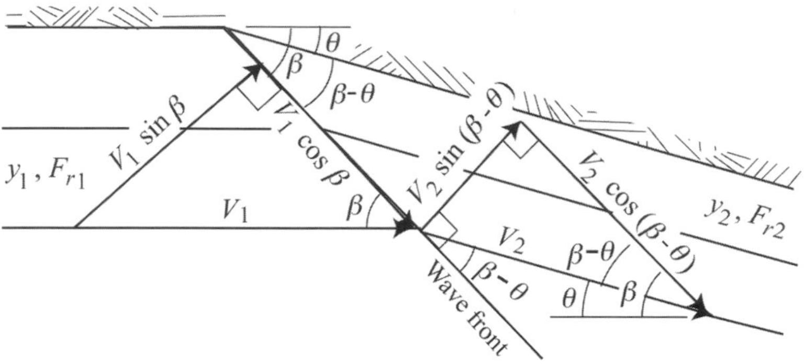

Figure 9.16 Standing supercritical wave front

The velocity component normal to the wave front is $V_1 \sin\beta$ and the component parallel to the wave is $V_1 \cos\beta$. Likewise, the velocity components on the downstream side of the front are $V_2 \sin(\beta - \theta)$ normal and $V_2 \cos(\beta - \theta)$ parallel.

From continuity, the unit discharge normal to the wave front must be equal on both sides of the wave since no water is accumulated, and $y_2 V_2 \sin(\beta - \theta) = y_1 V_1 \sin\beta$. Also, the two parallel velocity components must be equal since no force is exerted and $V_2 \cos(\beta - \theta) = V_1 \cos\beta$. Combining these two equations yields

$$\frac{y_2}{y_1} = \frac{\tan\beta}{\tan(\beta - \theta)}. \tag{9.13}$$

The momentum relationship in the normal direction is

$$\frac{y_2^2}{2} + \frac{y_2 V_2^2 [\sin(\beta - \theta)]^2}{g} = \frac{y_1^2}{2} + \frac{y_1 V_1^2 (\sin\beta)^2}{g}, \tag{9.14}$$

which, after substitution of $V_2 \sin(\beta - \theta) = y_1 V_1 \sin\beta / y_2$, reduces to a modified version of Eq. (9.8) written as

$$\frac{y_2}{y_1} = \frac{1}{2}\left(-1 + \sqrt{1 + 8\mathrm{Fr}_1^2 \sin^2\beta}\right)$$

or

$$\sin\beta = \frac{1}{\mathrm{Fr}_1}\left[\frac{y_2}{2y_1}\left(\frac{y_2}{y_1} + 1\right)\right]^{1/2},$$

which with the use of Eq. (9.13) finally gives

$$\sin\beta = \frac{1}{\mathrm{Fr}_1}\left[\frac{1}{2}\left(\frac{\tan\beta}{\tan(\beta-\theta)}\right)\left(\frac{\tan\beta}{\tan(\beta-\theta)}+1\right)\right]^{1/2}, \tag{9.15}$$

as illustrated in Figure 9.17. Once the wave-front angle β is known, the downstream Froude number Fr_2 is obtained from

$$\mathrm{Fr}_2 = \mathrm{Fr}_1 \frac{\sin\beta}{\sin(\beta-\theta)}\left(\frac{y_1}{y_2}\right)^{3/2},$$

and the downstream flow velocity is $V_2 = \mathrm{Fr}_2\sqrt{gy_2}$.

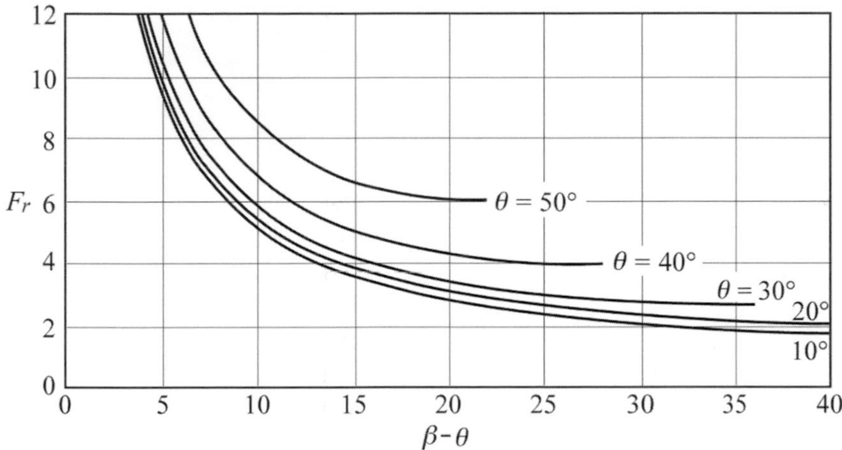

Figure 9.17 Oblique waves in supercritical flows

9.3.5 Nonhydrostatic Flows in Curved Channels

Curved channels pose unique problems in the analysis of pressure and forces on curved surfaces because of the vertical and/or lateral accelerations. Let us consider a few examples below.

First, a curvilinear channel with radius R is sketched in Figure 9.18a. The centripetal acceleration accelerates the fluid to the center of rotation. For instance, in a channel of width W, a lateral pressure gradient develops in a curved channel towards the center of rotation. In a curved channel of radius R and mean flow velocity V, the centripetal acceleration is equal to $a_c = V^2/R$ and the pressure gradient in the radial direction is $(1/\rho)dp/dr = 2g\Delta h/W = a_c = V^2/R$. The superelevation Δh over the channel width is

$$\Delta h = W\frac{V^2}{gR}. \tag{9.16}$$

The vertical pressure is hydrostatic but varies laterally along the channel width.

In the case of vertical accelerations, the vertical pressure distribution is no longer hydrostatic. For instance, Figure 9.18b illustrates the pressure distribution at the base of a flip-bucket spillway, where the centripetal acceleration is added to the gravitational acceleration along the vertical to give $p = \rho h\left[g + \left(V^2/R\right)\right]$.

A third example, illustrated in Figure 9.18c and d, occurs for spillways designed at a fixed discharge and water level in the reservoir. When the reservoir stage is above the design condition, the pressure on the spillway surface becomes negative and the suction increases the discharge coefficient. Case Study 9.2 illustrates calculations for supercritical flow in a steep channel.

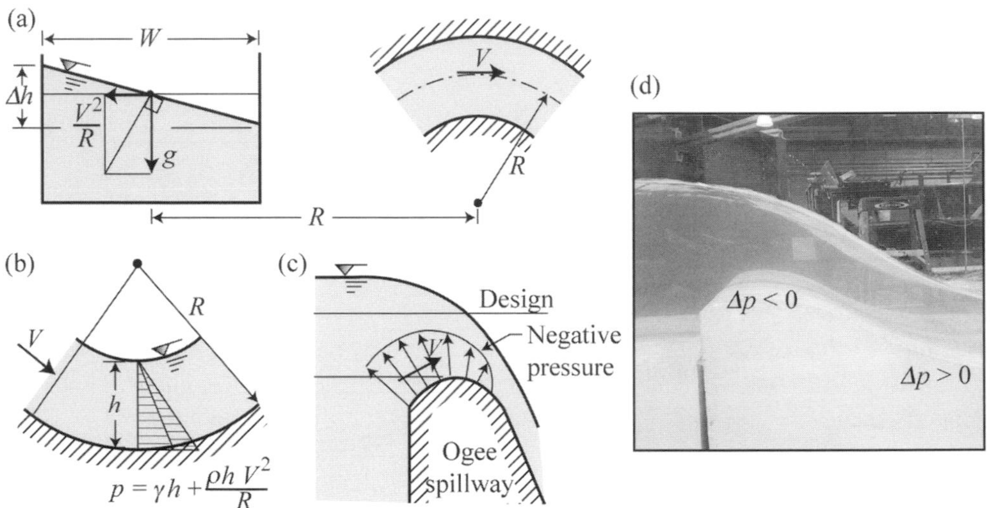

Figure 9.18 Effects of curvature on pressure distribution

Case Study 9.2: Supercritical waves in a steep channel in Montana

A steep lined channel is designed to pass a six-hour probable maximum flood reaching $Q = 7{,}590$ cfs in a $W = 31$ wide rectangular channel. The bed slope is $S_0 = 0.0442$ and the Manning coefficient $n = 0.013$ is assumed. The unit discharge is $q = Q/W = 7{,}590/31 = 245$ ft^2/s. The normal depth from $q = m(By_n)^{5/3}\sqrt{S_0}/\left[n(B + 2y_n)^{2/3}\right]$ is $y_n = 4.45$ ft. The flow velocity reaches $V_n = 55$ ft/s and the flow is supercritical with a Froude number $\mathrm{Fr} = 4.6$. The channel is curved with a radius of curvature $R \simeq 100$ ft as shown in Fig. CS-9.2.

The outside superelevation above the normal depth in a continuous channel bend is estimated from $\Delta h/2 = WV^2/2gR = 31 \times 55^2/32.2 \times 100 = 14.5$ ft.

When assuming an abrupt alignment change $\theta = 36°$ and $\mathrm{Fr}_1 = 4.6$ in Figure 9.17, we obtain $\beta - \theta = 15°$ and the standing-wave angle $\beta = 51°$. When compared to the results of a computational fluid dynamics model in Fig. CS-9.2b and c, the wave-front angle is relatively well simulated. The flow velocity of the computer model (53 ft/s in b) compares well with the flow velocity of 55 ft/s calculated above using Manning's equation. The superelevation (16.7 ft in c) compares well with the 14.5 ft estimated in the above formula.

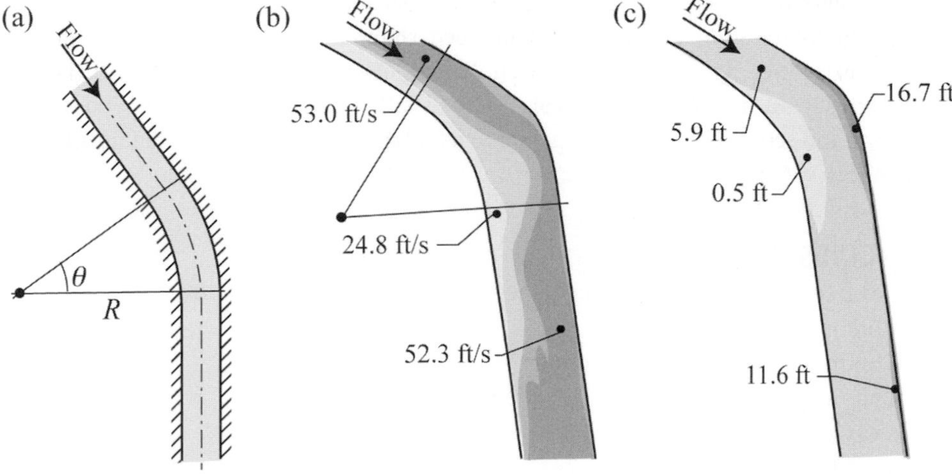

Fig. CS-9.2 Supercritical flow in a steep channel

9.3.6 Conservation of Momentum or Energy?

The question often arises in rapidly varied flows as to whether we should use conservation of momentum or conservation of energy. In general, potential energy can be converted into kinetic energy in accelerating flows, and hence conservation of energy can be used. However, the conversion of kinetic energy into potential energy usually involves energy dissipation through turbulence while momentum is conserved.

Figure 9.19 illustrates the difference between the conservation of energy and the conservation of momentum in rapidly varied flow for the flow around a sluice gate. The momentum force on the gate is not zero because there is no conservation of momentum, and thus the force on the sluice gate can be determined by the difference in specific momentum between both sides of the gate. The presence of a hydraulic jump downstream of the gate requires the conservation of momentum while energy is dissipated.

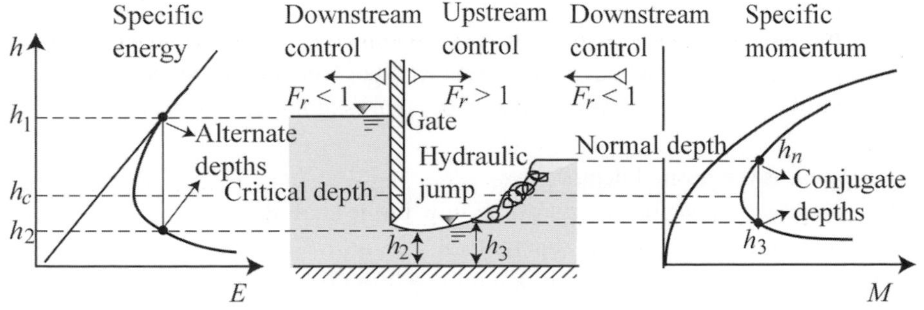

Figure 9.19 Conservation of energy/momentum and hydraulic controls

Additional Resources

Additional information can be found in open-channel textbooks like Chow (1959), Henderson (1966), Jain (2001), Chanson (2004), Akan (2006), Cruise et al. (2007), Chaudhry (2008) and Moglen (2015). There is also ample information about river crossings and scour prevention in the manuals of the Federal Highway Administration (FHWA) (FHWA 2001, 2009, 2012).

EXERCISES

These exercises review the essential concepts from this chapter.

1. What is specific energy?
2. What is a specific energy diagram?
3. What defines critical flow?
4. What are the two depths with the same specific energy called?
5. What happens when $E < E_{min}$?
6. What is the centroid used for?
7. What does specific momentum describe?
8. What are two depths with the same specific momentum called?
9. What are the two depths of a hydraulic jump called?
10. True or false?
    (a) The specific energy is the same on both sides of a sluice gate.
    (b) The critical depth depends on the channel slope.
    (c) The critical depth does not depend on resistance to flow.
    (d) The celerity of a small wave is $c = \sqrt{gh}$.
    (e) Definite wave fronts form when Fr < 1.
    (f) A surge is similar to a moving hydraulic jump.
    (g) Oblique waves only satisfy conservation of mass and momentum.
    (h) Resistance to flow can cause nonhydrostatic conditions.
    (i) Sequent and conjugate depths are synonymous.
    (j) Channel curvature can cause flow superelevation.

PROBLEMS

1. ♦ For a trapezoidal channel given the bed width $b = 6.0$ ft and the side slope $t = 2$, find the critical depth if $Q = 75$ ft^3/s.
2. ♦ For a trapezoidal channel given the bed width $b = 6.0$ ft and the side slope $t = 2$, find the critical depth if $Q = 300$ ft^3/s.
3. ♦♦ A trapezoidal channel having a bottom width of 5 m and side slopes $t = 2$ runs on a slope of 0.0005 and carries a discharge of 50 m^3/s at normal depth. Take Manning's n as 0.021.
    (a) What is the normal depth?
    (b) What is the specific energy of the flow in the channel?
    (c) Is the flow subcritical or supercritical?
4. ♦♦ A trapezoidal channel having a bottom width of 5 m and side slopes $t = 2$ runs on a slope of 0.0005 and carries a discharge of 50 m^3/s at normal depth. Take Manning's n as 0.021.

(a) What is the normal depth?
(b) What is the alternate depth to the normal depth?
(c) What is the critical depth in the channel?

5. ♦ Water flows in a rectangular channel at a rate of 250 ft^3/s. The depth of flow is 5 ft and the bed width is 10 ft. If the channel bed elevation is locally increased by 0.5 ft using a smooth step, what is the depth and Froude number at the raised section?

6. ♦♦ Consider a flow of 10 m^3/s in a 10-m-wide rectangular channel at a normal flow depth of 0.788 m. Determine the following:
(a) the channel slope if Manning's coefficient n is 0.02;
(b) the maximum elevation of a weir Δz_{max} that would not cause backwater; and
(c) the maximum channel contraction without an elevation change that would be possible without choking the flow.

7. ♦♦ Draw the specific momentum diagram for a triangular channel with an opening angle of 90° and discharge of 10 m^3/s. [Hint: find the centroid of a triangular section.]

8. ♦♦ A trapezoidal channel has a 3-m bottom width and a 1:1 side slope. It carries a discharge of 10 m^3/s at a slope of 12 cm/km. A hydraulic jump forms at an upstream depth of 0.6 m. Calculate the specific momentum function, the critical flow depth and the conjugate depth. If you place baffle blocks to induce the hydraulic jump where the upstream depth is 0.4 m and the flow depth downstream of the baffles is 1.2 m, what is the force on the baffle blocks? [Hint: did you properly locate the center of pressure?]

9. ♦♦ A rectangular channel 5.0 m wide carries a specific discharge of 4.0 m^2/s. A hydraulic jump is fixed in a certain position by placing concrete blocks at the base of the spillway. The initial depth of the jump is 0.5 m. Determine the force acting on the blocks if the depth of water after the jump is 2.0 m. If the concrete blocks are removed, what will be the new water depth?

10. ♦♦ Water flows at a rate of 45 ft^3/s along a horizontal rectangular channel 4.5 ft in width. A hydraulic jump forms where the depth is 0.25 ft. Find the rise in water level and the power lost in the jump.

10 Gradually Varied Flow

Gradually varied flows in open channels change slowly in the downstream direction. The main equation in Section 10.1 leads to the classification of water-surface profiles in Section 10.2, with calculations in Section 10.3. Energy losses at bridge crossings are covered in Section 10.4 and numerical models are introduced in Section 10.5.

10.1 Gradually Varied Flow Equation

Based on the conservation of momentum, the general gradually varied flow equation in Section 10.1.1 is applied to rectangular channels in Section 10.1.2.

10.1.1 Gradually Varied Flow

The hydraulic conditions (velocity and depth) of gradually varied flows in open channels change slowly in the downstream direction as shown in Figure 10.1.

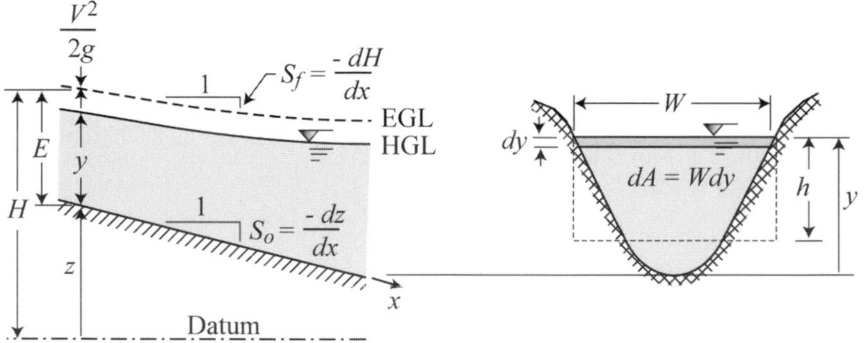

Figure 10.1 Gradually varied flow in open channels

The governing equation describing gradually varied flow stems from the energy grade line (EGL) in the downstream x direction $H = z + E$ and we remember $\mathrm{Fr}^2 = Q^2W/gA^3$ with $dA = Wdy$, $S_0 = -dz/dx$ and $S_f = -dH/dx$. Thus,

$$\frac{dH}{dx} = \frac{dz}{dx} + \frac{dE}{dx} = \frac{d}{dx}\left(z + y + \frac{V^2}{2g}\right) = -S_f = -S_0 + \frac{d}{dx}\left(y + \frac{Q^2}{2gA^2}\right).$$

This can be rewritten as

$$S_0 - S_f = \frac{dy}{dx} + \frac{d}{dy}\left(\frac{Q^2}{2gA^2}\right)\frac{dy}{dx},$$

with Q and g constant;

$$S_0 - S_f = \frac{dy}{dx} + \left[\frac{Q^2}{2g}\left(\frac{-2}{A^3}\right)\frac{dA}{dy}\right]\frac{dy}{dx} = \left[1 - \frac{Q^2 W}{gA^3}\right]\frac{dy}{dx} = \left(1 - \mathrm{Fr}^2\right)\frac{dy}{dx};$$

and, finally,

$$\frac{dy}{dx} = \frac{S_0 - S_f}{1 - \mathrm{Fr}^2}. \tag{10.1}$$

This is the general form of the gradually varied flow equation for open channels. We owe seminal contributions on gradually varied flow to Jacques Bresse and Henri Bazin from France and Boris Bakhmeteff from Russia.

10.1.2 Wide Rectangular Channels

Consider the particular case of gradually varied flow in a wide rectangular channel ($B \gg y$). In Figure 10.2, the geometry of wide rectangular channels is quite simple because the flow depth y, the mean flow depth h, and the hydraulic radius R_h are identical. The free surface width W and base width B are equal. From the definition of critical depth $h_c = \left(q^2/g\right)^{1/3}$, the Froude number is

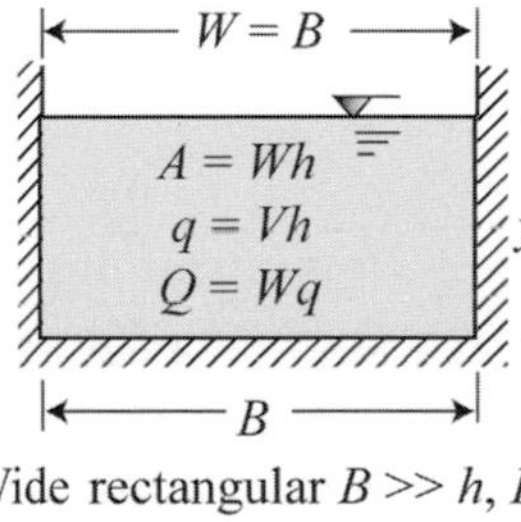

$$\mathrm{Fr}^2 = \frac{Q^2 W}{gA^3} = \frac{V^2}{gh} = \frac{q^2}{gh^3} = \left(\frac{h_c}{h}\right)^{M=3}. \tag{10.2}$$

$W = B$
$A = Wh$
$q = Vh$
$Q = Wq$
$y = h$
B
Wide rectangular $B >> h$, $R_h = h$

Figure 10.2 Wide rectangular channel

Resistance to flow (the Manning equation) defines the normal flow depth h_n in a wide rectangular channel:

$$q = Vh = \frac{m}{n}h^{2/3}S_f^{1/2}h = \frac{m}{n}h^{5/3}S_f^{1/2},$$

with $m = 1$ in SI and $m = 1.49$ in customary units.

We can define the friction slope S_f at any depth h from Manning's equation as

$$h = \left(\frac{nq}{mS_f^{1/2}}\right)^{3/5} \text{ or } S_f = \left(\frac{nq}{mh^{5/3}}\right)^2. \tag{10.3}$$

In steady–uniform flow, the normal depth $h = h_n$ corresponds to $S_f = S_0$ or

$$h_n = \left(\frac{nq}{mS_0^{1/2}}\right)^{3/5}$$

or

$$S_0 = \left(\frac{nq}{m{h_n}^{5/3}}\right)^2,$$

and this gives

$$\frac{S_f}{S_0} = \left(\frac{h_n}{h}\right)^{N=10/3}.$$

Note that $N = 10/3$ when using Manning's equation and decreases to $N = 3$ when using Chézy's equation. The exponent $M = 3$ is independent of resistance to flow. With the help of Eqs. (10.2) and (10.3), the gradually varied flow equation (Eq. (10.1)) for wide rectangular channels becomes

$$\frac{dh}{dx} = \frac{S_0\left[1 - \left(\frac{S_f}{S_0}\right)\right]}{\left[1 - Fr^2\right]} = \frac{S_0\left[1 - \left(\frac{h_n}{h}\right)^{N=10/3}\right]}{\left[1 - \left(\frac{h_c}{h}\right)^{M=3}\right]}. \tag{10.4}$$

It is interesting that, at a given bed slope, the gradually varied flow equation becomes a function of the critical and normal depths.

10.2 Gradually Varied Flow Profiles

We classify various profiles in Section 10.2.1, locate control points and hydraulic jumps in Section 10.2.2, and sketch water-surface profiles in Section 10.2.3.

10.2.1 Classification of Gradually Varied Flows

There are three depths involved in gradually varied flow calculations: (1) the flow depth y; (2) the normal depth y_n; and (3) the critical depth y_c. We can then examine whether the change in flow depth dy/dx is positive or negative from

$$\frac{dy}{dx} = S_0\left[1 - \left(\frac{y_n}{y}\right)^N\right] \Big/ \left[1 - \left(\frac{y_c}{y}\right)^M\right],$$

with $N = 10/3$ and $M = 3$ in rivers:

the numerator is + when $y > y_n$, and $dy/dx = 0$ when $y = y_n$;
the denominator is + when $y > y_c$, and $dy/dx = \infty$ when $y = y_c$; and
open-channel flows naturally tend towards the normal depth.

The two most common water-surface profiles in open channels are the mild and steep slopes sketched in Figure 10.3. Three zones are recognized depending on whether the profile is above (zone 1), between (zone 2) or below (zone 3) the normal and critical depths. All profiles are asymptotic to the normal depth.

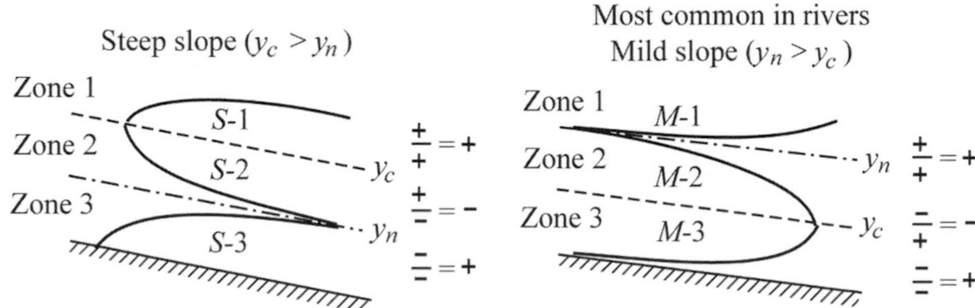

Figure 10.3 The two most common backwater types

Broadly speaking, however, there are five general gradually varied flow profiles: (1) mild slopes when $y_n > y_c$; (2) steep slopes when $y_n < y_c$; (3) critical slopes when $y_n = y_c$; (4) horizontal slopes when $S_0 = 0$; and (5) adverse slopes when $S_0 < 0$. Any given flow depth will either be above both, below both or between the critical and normal depths. Keeping in mind the signs of the numerator and denominator as well as the slope of the gradually varied flow profile $dy/dx = 0$ at normal depth and $dy/dx = \infty$ at critical depth, we obtain the main profiles sketched in Figure 10.4. Notice that in zones 1 and 3, the flow depth always increases in the downstream direction, while it always decreases in zone 2. Less-common water-surface profiles are also sketched in Figure 10.5.

10.2.2 Control Points and Hydraulic Jumps

Control points are defined when the flow depth crosses the critical depth. As discussed in Section 9.3.2, downstream control (DC) occurs in subcritical flow $\mathrm{Fr} < 1$, or $y > y_c$, which means that the flow depth depends on a downstream boundary condition (at a dam, a waterfall, etc.). Likewise, supercritical flow $\mathrm{Fr} > 1$, or $y < y_c$, requires upstream control (UC), as sketched in Figure 10.6. Natural control points refer to converging (accelerating) flows passing through the critical depth. In the case of human-made control points like gates, the alternate depths upstream and downstream of the gate are determined by the conservation of specific energy. The flow condition upstream of the gate is subcritical. The gate therefore serves as the control point for the backwater profile upstream of the gate. Conversely, the flow depth downstream of the gate is supercritical and it is the gate that controls the flow conditions and is the starting point for the backwater profile downstream of the gate.

Decelerating flows forming hydraulic jumps are identified as J in Figure 10.7. The two depths upstream and downstream of the jump are conjugate depths, i.e. with the same

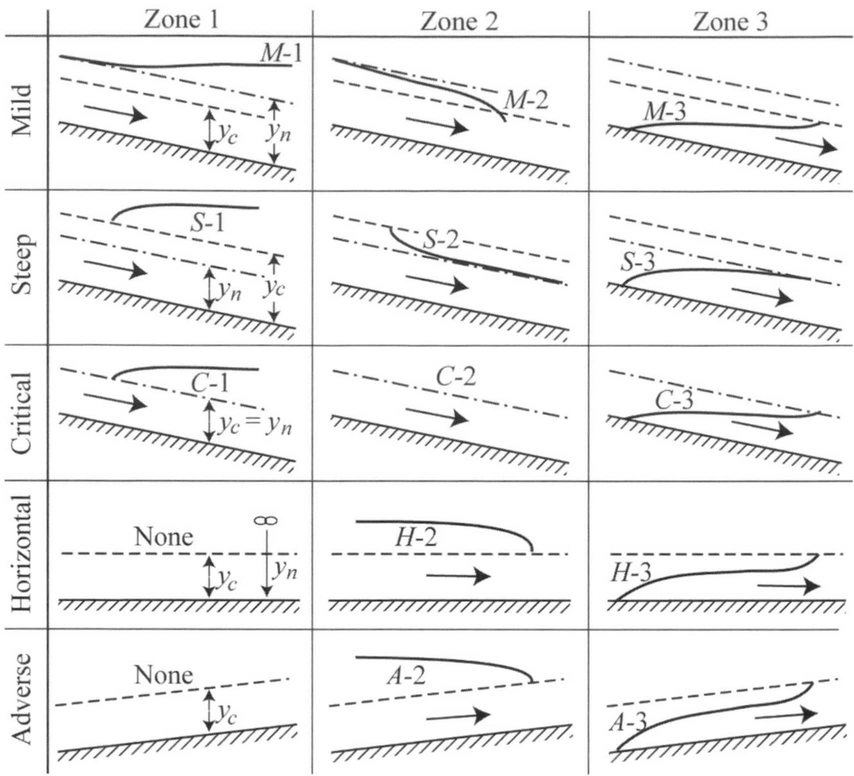

Figure 10.4 Backwater classification for open channels

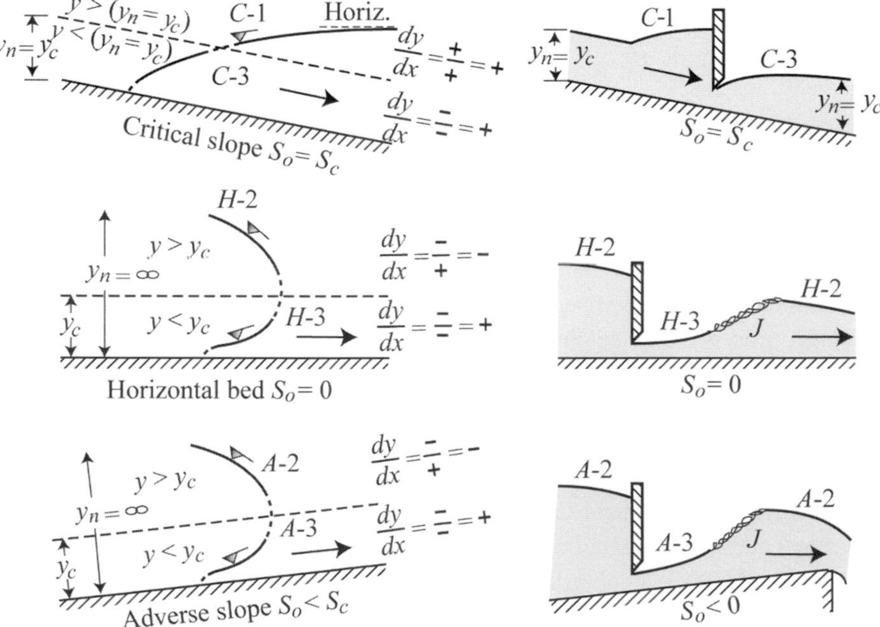

Figure 10.5 Less-common water-surface profiles

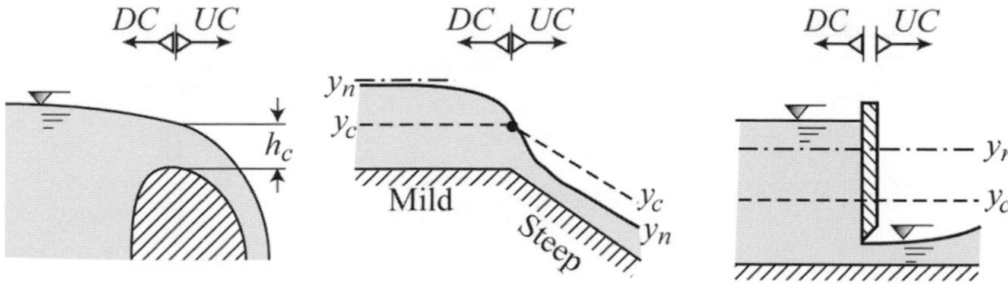

Figure 10.6 Examples of upstream and downstream controls

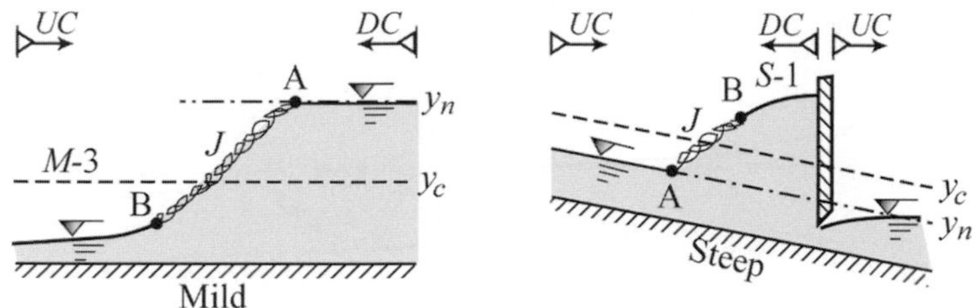

Figure 10.7 Calculation of conjugate depths in hydraulic jumps from A to B

value of specific momentum. The upstream side of the hydraulic jump has supercritical flow and is controlled farther upstream. The downstream side of the hydraulic jump is subcritical and is controlled farther downstream. The jump is located where the two conjugate depths have the same specific momentum. In both cases, we use the flow depth at A to calculate the flow depth at B from the conjugate-depth relationship (e.g. the Bélanger equation in wide rectangular channels). For instance, on a mild slope, we use the normal depth at A downstream to calculate the conjugate depth at B. Conversely, on a steep slope, we calculate the downstream flow depth at B from the normal flow depth at A.

10.2.3 Sketching Water-Surface Profiles

The typical procedure to sketch water-surface profiles consists of the following steps, also illustrated with numerous examples in Figure 10.8.

(1) Define the geometry, slope and roughness of each subreach.
(2) Calculate and sketch the critical depth.
(3) Calculate the normal depth.
(4) Identify the profile type (mild, steep, etc.).
(5) Identify the control points.
(6) Identify the zones (M-1, S-2, etc.).
(7) Locate hydraulic jumps.
(8) Sketch the profile starting at the control points.

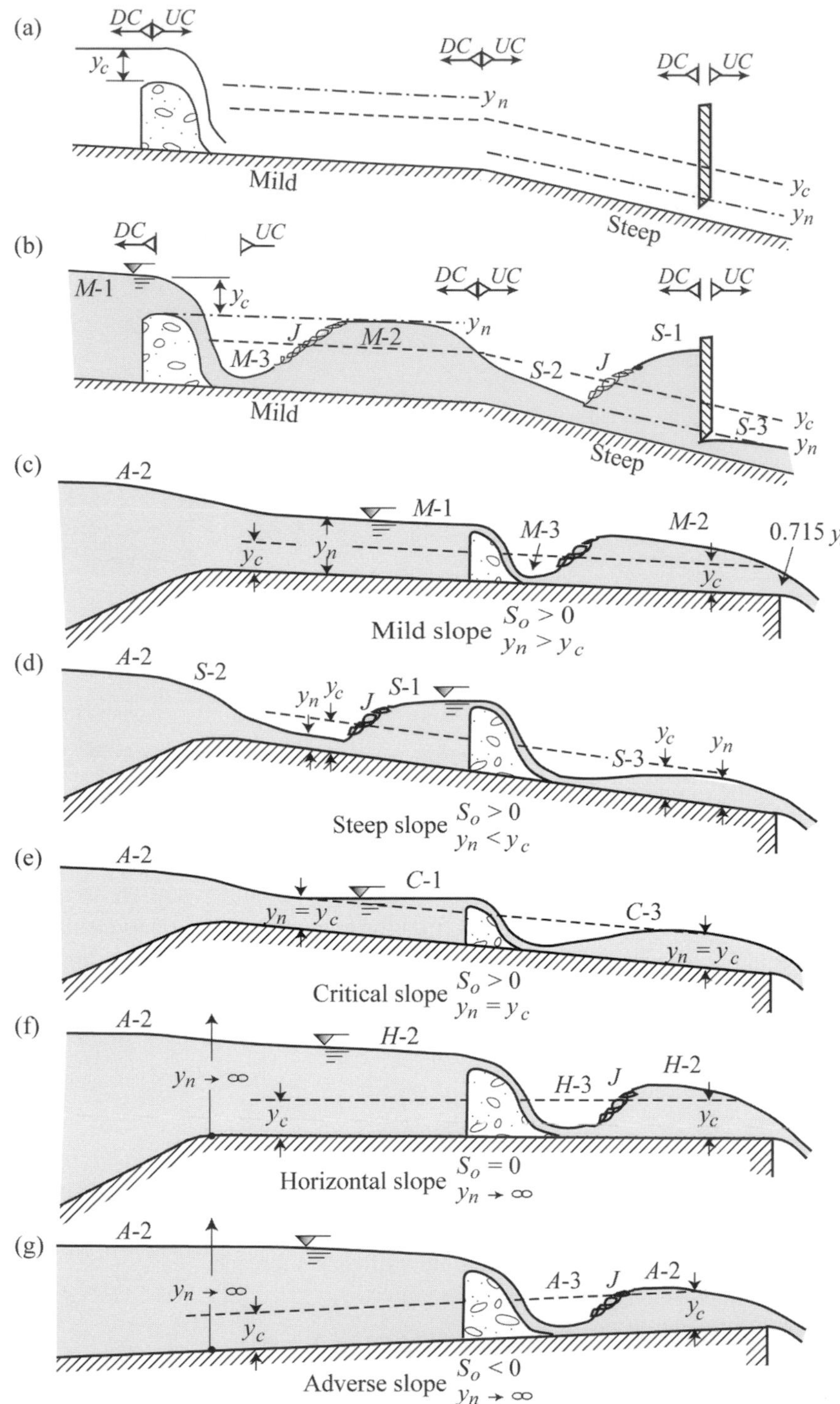

Figure 10.8 Sketches of gradually varied flow profiles

10.3 Gradually Varied Flow Calculations

In practice, there are two basic numerical methods to calculate water-surface profiles in open channels: (1) the standard-step method in Section 10.3.1; and (2) the direct-step method in Section 10.3.2. The numerical scheme for gradually varied flow calculations is sketched in Figure 10.9. The calculations start at or near a control point. Note that the boundary conditions need to be consistent. For instance, the calculations on both sides of a gate are called alternate depths.

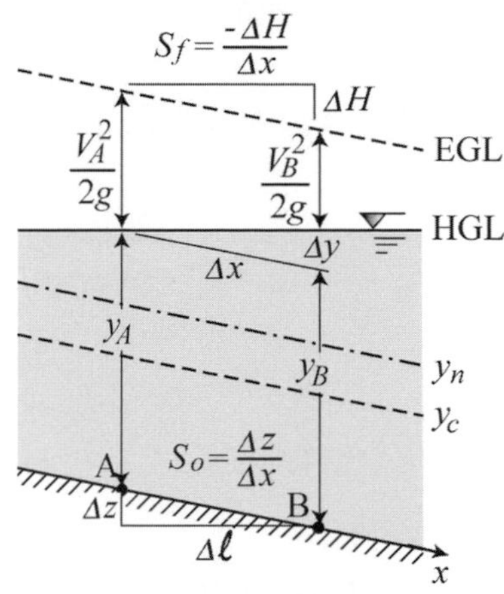

Figure 10.9 Backwater algorithm

10.3.1 Standard-Step Method

The standard-step method calculates the change in flow depth Δy over a constant interval Δx. The general formulation is

$$\Delta y = \frac{\Delta x\left[S_0 - S_f\right]}{\left[1 - \mathrm{Fr}^2\right]}, \tag{10.5}$$

where S_0 is the bed slope, S_f is the friction slope, Fr is the Froude number and Δy is the change in flow depth over a distance Δx.

For the particular case of a wide rectangular channel ($h = y$), we can use the formulation from Section 10.1.2 (Eq. (10.4)):

$$\Delta y = \frac{S_0 \Delta x\left[1 - \left(\frac{y_n}{y}\right)^N\right]}{\left[1 - \left(\frac{y_c}{y}\right)^M\right]} \tag{10.6}$$

where $M = 3$ and $N = 10/3$ for Manning's equation, y_n is the normal depth, y_c is the critical depth, S_0 is the bed slope and Δx is the grid size.

When the flow is supercritical, $\mathrm{Fr} > 1$, the calculations start at the control point (upstream) and proceed in the downstream direction with $y_B = y_A + \Delta y$. Notice that Δy can be negative (e.g. M-2 backwater curve). The procedure marches onward in the downstream direction. Note that calculations should not start at the critical depth because of the indetermination in dividing by zero.

When the flow is subcritical, $\mathrm{Fr} < 1$, the calculations start at the control point (now downstream) and proceed in the upstream direction ($\Delta x < 0$) using the algorithm $y_A = y_B - \Delta y$. The procedure marches in the upstream direction.

10.3.2 Direct-Step Method

The direct-step method calculates variable values of Δx from fixed values of Δy. The difference between direct-/standard-step methods is sketched in Figure 10.10.

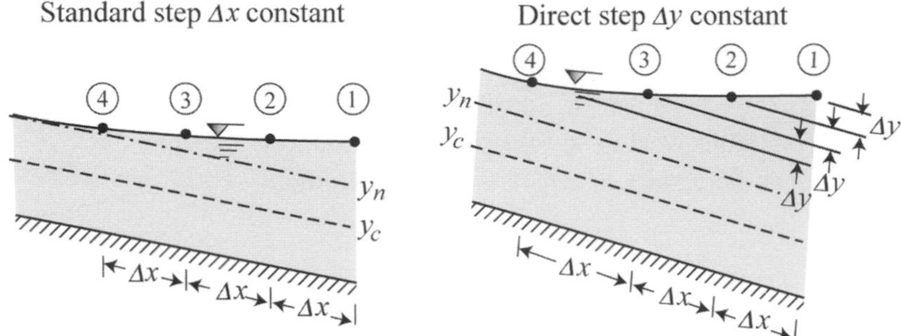

Figure 10.10 Standard-step and direct-step methods

The direct-step method uses a constant change in flow depth Δy and calculates the corresponding interval Δx. The general formulation is

$$\Delta x = \frac{\Delta y\left[1 - \mathrm{Fr}^2\right]}{\left[S_0 - S_f\right]}, \tag{10.7}$$

where S_0 is the bed slope, S_f is the friction slope and Fr is the Froude number.

In the case of a wide rectangular channel, $y = h$, we can use the formulation

$$\Delta x = \frac{\Delta h\left[1 - \left(\frac{h_c}{h}\right)^{M=3}\right]}{S_0\left[1 - \left(\frac{h_n}{h}\right)^{N=10/3}\right]}, \tag{10.8}$$

where h_n is the normal depth and h_c is the critical depth. The calculations proceed downstream from the upstream control point for supercritical flow, $\mathrm{Fr} > 1$, and proceed upstream from the downstream control point in subcritical flow, $\mathrm{Fr} < 1$. With the direct-step method, caution is required near the normal depth as the denominator approaches 0 and Δx becomes infinitely large. Near the critical depth, calculations should start slightly above or below the critical depth to avoid $\Delta x = 0$.

Two simpler examples are presented for subcritical and supercritical flows.

Example 10.1 illustrates M-1-profile calculations with the standard-step method. Example 10.2 uses the direct-step method for an S-2 backwater profile.

More complex examples are presented to locate hydraulic jumps and with different cross-sectional geometries. Example 10.3 calculates an M-3 backwater curve in a rectangular channel. Example 10.4 calculates an S-1 backwater curve in a steep trapezoidal channel.

Example 10.1: Standard-step calculation in a wide rectangular channel

Consider a smooth and wide rectangular channel with a unit discharge of 22 ft^2/s, a normal depth $y_n = 3.5$ ft and bed slope 1:2,000. If the water level is raised to a depth $y_0 = 10$ ft, use the standard-step method to calculate the backwater profile.

Solution: Assuming wide rectangular conditions, we calculate the critical depth $y_c = \left(\frac{q^2}{g}\right)^{1/3} = \left(\frac{22^2}{32.2}\right)^{1/3} = 2.47$ ft and we have a mild slope because $y_n > y_c$. We also have an M-1 backwater curve because $y_0 > y_n$.

Let us use a coarse mesh for this first calculation with $\Delta x = 1{,}000$ ft. At each step, the bed elevation will be raised by $S_0\Delta x = 1{,}000/2{,}000 = 0.5$ ft. Using the standard-step method starting at $x = 0$, $z = 0$ and $y = 10$ ft, we obtain

$$\Delta y = \frac{S_0\Delta x\left[1 - \left(\frac{y_n}{y_0}\right)^{10/3}\right]}{\left[1 - \left(\frac{y_c}{y_0}\right)^3\right]} = \frac{0.5\left[1 - \left(\frac{3.5}{10}\right)^{10/3}\right]}{\left[1 - \left(\frac{2.47}{10}\right)^3\right]} = 0.492 \text{ ft.}$$

The flow is subcritical and we move upstream from $y_0 = 10$ ft. At our first calculation node, we get a bed elevation $z_1 = z_0 + S_0\Delta x = 0 + 0.5 = 0.5$ ft. The flow depth is $y_1 = y_0 - \Delta y = 10 - 0.492 = 9.508$ ft. The corresponding water-surface elevation (or hydraulic grade line [HGL]) is $HGL_1 = z_1 + y_1 = 0.5 + 9.508 = 10.008$ ft. The flow velocity is $V_1 = q/y_1 = 22/9.508 = 2.314$ ft/s and $V^2/2g = 2.314^2/2 \times 32.2 = 0.0831$ ft. The EGL is $EGL_1 = HGL_1 + V_1^2/2g = 10.008 + 0.0831 = 10.091$ ft. And the procedure is repeated successively with the last calculated flow depth (i.e. $y_1 = 9.508$ ft) to plot the backwater profile shown in Fig. E-10.1.

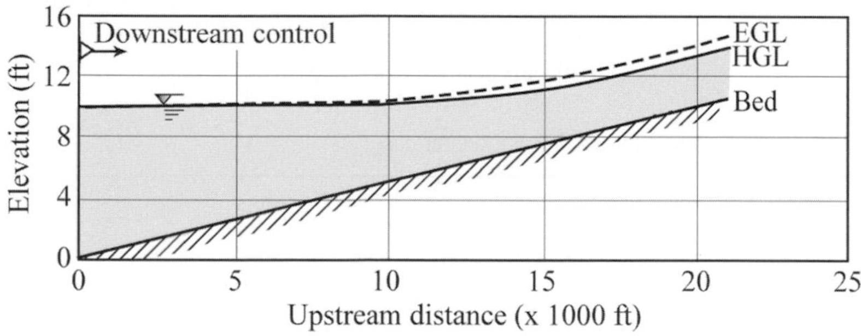

Fig. E-10.1 Standard-step calculation for an M-1 backwater curve

Example 10.2: Direct-step calculation for a wide rectangular channel

Consider a smooth and near-horizontal wide rectangular channel with a unit discharge of 3 m^2/s, a normal depth $h_n = 2$ m and Manning's coefficient $n = 0.012$ and slope $S_0 = 0.00013$. The direct-step method is used for the water-surface profile calculation near a vertical drop.

Solution: Assuming wide rectangular conditions $h = y$, we calculate the critical depth $h_c = \left(\frac{q^2}{g}\right)^{1/3} = \left(\frac{3^2}{9.81}\right)^{1/3} = 0.97$ m and we have a mild slope because $h_n > h_c$. We also have an M-2 backwater curve approaching the critical depth at the downstream end.

The flow is subcritical and we move upstream from a depth $h_1 = 1$ m at $x_1 = 0$, i.e. slightly above the critical depth to avoid the discontinuity caused by $\Delta x = 0$ at the critical depth ($h = h_c$). The flow depth will range from 0.97 m to 2 m taken in a small flow-depth increment $\Delta h = 0.05$ cm with the direct-step method:

$$\Delta x_1 = \frac{\Delta h\left[1 - \left(\frac{h_c}{h_1}\right)^3\right]}{S_0\left[1 - \left(\frac{h_n}{h_1}\right)^{10/3}\right]} = \frac{0.05\left[1 - \left(\frac{0.97}{1}\right)^3\right]}{0.00013\left[1 - \left(\frac{2}{1}\right)^{10/3}\right]} = -3.7 \text{ m}.$$

At a distance $x_1 = -3.7$ m (negative as we move upstream), the bed elevation is $z_1 = z_0 - S_0\Delta x_1 = 0 + 0.00013 \times 3.7 = 0.00048$ m, the water surface is at elevation $HGL_1 = z_1 + h_1 = 0.00048 + 1 = 1.00048$ m and the EGL is at $EGL_1 = HGL_1 + q^2/2gh_1^2 = 1.00048 + 3^2/(2 \times 9.81 \times 1^2) = 1.459$ m. The distances are short near the critical depth, but increase very rapidly as larger flow depths are selected. Obviously, the normal depth is reached asymptotically upstream as shown in Fig. E-10.2.

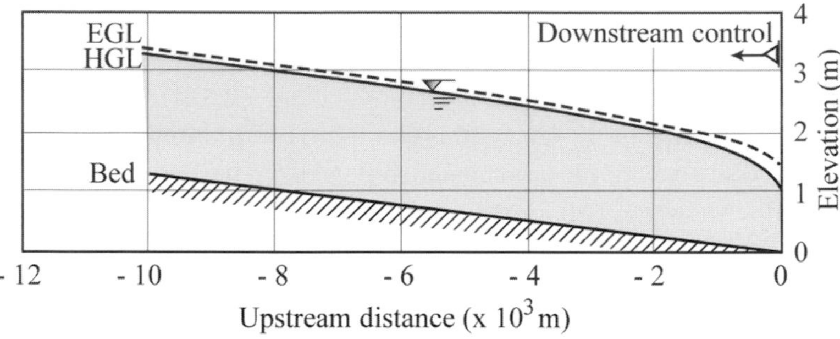

Fig. E-10.2 Direct-step method for an M-2 backwater curve

Example 10.3: Hydraulic-jump location in a mild rectangular channel

Locate the hydraulic jump on a smooth ($n = 0.012$) surface below a spillway. The rectangular channel is 60-ft wide and the flow depth is 1 ft at a flow velocity of 50 ft/s. The normal depth farther downstream is 6 ft. To force the jump 75 ft below the control point, what would be the force on the baffle blocks?

Solution: We now have a more demanding problem. The main parameters at the upstream end are $W = 60$ ft, $V_{up} = 50$ ft/s, $h_{up} = 1$ ft, $\mathrm{Fr}_{up} = V_{up}/\sqrt{gh_{up}} = 50/\sqrt{32.2 \times 1} = 8.81$. The unit discharge is $q = Vh = 50$ ft^2/s and the total discharge $Q = Wq = 3{,}000$ ft^3/s. The critical depth is $h_c = (q^2/g)^{1/3} = (50^2/32.2)^{1/3} = 4.27$ ft and the flow is supercritical ($h < h_c$) at the upstream end with upstream control.

At the downstream end, we have subcritical flow at the normal depth ($h_n = 6$ ft $> h_c$), and this corresponds to a mild slope. The corresponding parameters at normal depth are the wetted perimeter $P_n = W + 2h_n = 60 + 2 \times 6 = 72$ ft, cross-sectional area $A_n = Wh_n = 60 \times 6 = 360$ ft^2, hydraulic radius $R_{hn} = A_n/P_n = 360/72 = 5$ ft and the normal flow velocity is $V_n = q/h_n = 50/6 = 8.33$ ft/s. This gives a downstream Froude number $\mathrm{Fr}_n = V_n/\sqrt{gh_n} = 8.33/\sqrt{32.2 \times 6} = 0.6$. The bed slope is calculated from the Manning resistance formula for a smooth ($n = 0.012$) surface:

$$S_0 = \left(\frac{nV_n}{1.49{R_{hn}}^{2/3}}\right)^2 = \left(\frac{0.012 \times 8.33}{1.49 \times 5^{2/3}}\right)^2 = 0.00053$$

and

$$S_f = \left(\frac{nQP^{2/3}}{1.49A^{5/3}}\right)^2 = \left(\frac{0.012 \times 3{,}000 \times (60 + 2h)^{2/3}}{1.49(60h)^{5/3}}\right)^2.$$

From the Bélanger equation (Eq. (9.8)), the conjugate depth to the downstream normal depth is

$$h_{con} = \frac{h_n}{2}\left(-1 + \sqrt{1 + 8\mathrm{Fr}_n^2}\right) = \frac{6}{2}\left(-1 + \sqrt{1 + 8 \times 0.6^2}\right) = 2.91\,\mathrm{ft}.$$

We have an M-3 backwater curve because h_{con} is below both the normal and critical depths. The control is upstream and we start at $h_{up} = 1$ ft until $h_{con} = 2.91$ ft.

Backwater calculations proceed downstream using a direct-step approach and, with a vertical increment $\Delta h = 0.1$ m, the first value of Δx is

$$\Delta x = \frac{\Delta h\left[1 - \left(\frac{h_c}{h}\right)^3\right]}{S_0\left[1 - \left(\frac{h_n}{h}\right)^{10/3}\right]} = \frac{0.1\left[1 - \left(\frac{4.27}{1}\right)^3\right]}{0.00053\left[1 - \left(\frac{6}{1}\right)^{10/3}\right]} = 37 \text{ ft downstream.}$$

We calculate and plot the M-3 profile in Fig. E-10.3. The jump is located where the calculated flow depth matches the conjugate depth $h_{con} = 2.91$ ft. This occurs approximately 800 ft downstream of the control point, which would require a very long apron. The water-surface profile without baffles is shown in Fig. E-10.3.

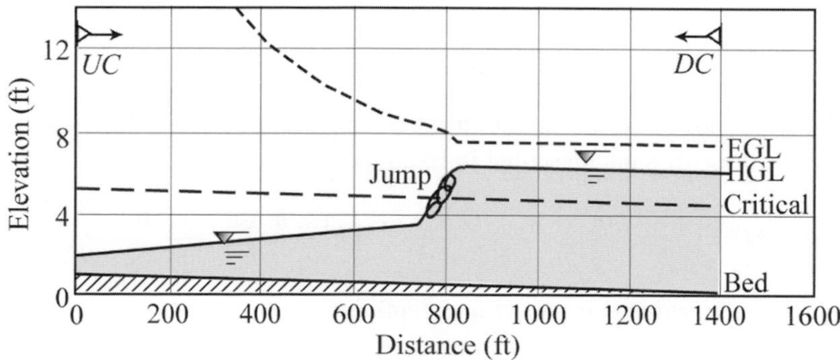

Fig. E-10.3 Hydraulic-jump location in an M-3 backwater curve

Baffles would force a hydraulic jump closer to the base of the spillway. The question becomes what force F would be exerted on the baffles. This can be done by using the specific momentum previously discussed. At a distance of 75 ft downstream of the spillway, the flow depth is $h_{75} = 1.2$ ft, and the area is $A_{75} = 72$ ft^2, and the specific momentum M_{75} at a distance of 75 ft and M_n for the normal depth are, respectively,

$$M_{75} = \frac{F_{75}}{\gamma} = A_{75}\bar{h} + \frac{Q^2}{gA_{75}} = \left(72 \times \frac{1.2}{2}\right) + \frac{3{,}000^2}{32.2 \times 72} = 3{,}925 \text{ ft}^3$$

and

$$M_n = \frac{F_n}{\gamma} = A_n\bar{h} + \frac{Q^2}{gA_n} = \left(6 \times 60 \times \frac{6}{2}\right) + \frac{3{,}000^2}{32.2 \times 6 \times 60} = 1{,}856 \text{ ft}^3.$$

The force on the baffles in a 60-ft-wide channel would be

$$F = \gamma(M_{75} - M_n) = 62.4 \times (63{,}925 - 1{,}856) = 129{,}100 \text{ lb}$$

or

$$F/W = 129{,}100/60 = 2{,}150 \text{ lb per foot of width.}$$

Example 10.4: Hydraulic-jump location in a steep trapezoidal channel
Locate the hydraulic jump in a smooth ($n = 0.012$) trapezoidal channel 2 m wide at the base with a 1V:1H side slope and a 0.01 downstream slope. The flow discharge is 6 m^3/s with normal depth upstream and the downstream flow depth is 2 m.

Solution: The trapezoidal geometry complicates this problem and a numerical solution is indicated.

With reference to Example 9.5, the critical depth is $h_c = 0.839$ m and the normal depth is calculated from Manning's equation for a trapezoidal channel with $B = 2$ m and $t = 1$:

$$Q = AV = \left(2h_n + h_n^{\,2}\right)\frac{1}{0.012}\left(\frac{2h_n + h_n^{\,2}}{2 + 2\sqrt{2}h_n}\right)^{2/3} 0.01^{1/2} = 6,$$

and, with a solver, $h_n = 0.535$ m. It is a steep slope ($h_n < h_c$) with normal depth upstream.

Starting at the downstream end, the specific momentum in the S-1 backwater curve is calculated from the formula presented in Example 9.5:

$$M = \left(\frac{Bh^2}{2} + \frac{th^3}{3}\right) + \frac{Q^2}{gh(B + th)} = \left(h^2 + \frac{h^3}{3}\right) + \frac{6^2}{9.81h(2 + h)}.$$

The specific momentum for the normal depth is

$$M = \left(h_n^{\,2} + \frac{h_n^{\,3}}{3}\right) + \frac{36}{9.81h_n(2+h_n)} = \left(0.535^2 + \frac{0.535^3}{3}\right) + \frac{36}{9.81 \times 0.535(2+0.535)} = 3.04.$$

The hydraulic jump is located where the specific momentum in the S-1 curve equals 3.04.

As a second method, the jump is located where the flow depth downstream of the jump is equal to the conjugate depth of the normal depth:

$$M = \left(h_{con}{}^2 + \frac{h_{con}{}^3}{3}\right) + \frac{36}{9.81 h_{con}(2 + h_{con})} = 3.04$$

and, from a solver, $h_{con} = 1.225$ m.

At the downstream end, the backwater curve starts at $h = 2$ m which corresponds to a top width $W = 2 + 2h = 6$ m, a cross-sectional area $A = 2h + h^2 = 8\,\text{m}^2$ and a mean flow velocity $V = Q/A = 6/8 = 0.75$ m/s. The bed elevation is zero, the HGL is 2 m and the EGL is $z + y + V^2/2g = 0 + 2 + 0.75^2/2 \times 9.81 = 2.029$. The average flow depth is $h = A/W = 8/6 = 1.33$ m and the Froude number is $\text{Fr} = V/\sqrt{gh} = 0.75/\sqrt{9.81 \times 1.33} = 0.207$. The friction slope can be calculated from differences in the EGL over successive calculation points. We can also calculate the friction slope at each node from Manning's equation, rewritten as

$$S_f = n^2 Q^2 \left(\frac{P^{4/3}}{A^{10/3}}\right) = (nQ)^2 \left[\frac{\left(B + 2\sqrt{1 + t^2}h\right)^{4/3}}{\left(Bh + th^2\right)^{10/3}}\right] = (0.072)^2 \left[\frac{\left(2 + 2\sqrt{2}h\right)^{4/3}}{\left(2h + h^2\right)^{10/3}}\right].$$

The friction slope equals the bed slope ($S_f = S_0$) when $h = h_n$.

Backwater calculations proceed upstream from the downstream end using the general backwater formula and a direct-step approach. Therefore, with a vertical increment $\Delta y = -0.03$ m, the first value of Δx is calculated from $\Delta x = \Delta y \frac{[1 - Fr^2]}{[S_0 - S_f]} = -0.03 \frac{[1 - 0.207^2]}{[0.01 - 0.000076]} = -2.9$ m upstream.

The S-1 backwater is calculated and plotted in Fig. E-10.4. The jump is located 68 m upstream of the control point, where the flow depth on the S-1 backwater curve matches the conjugate depth $h_{con} = 1.225$ m.

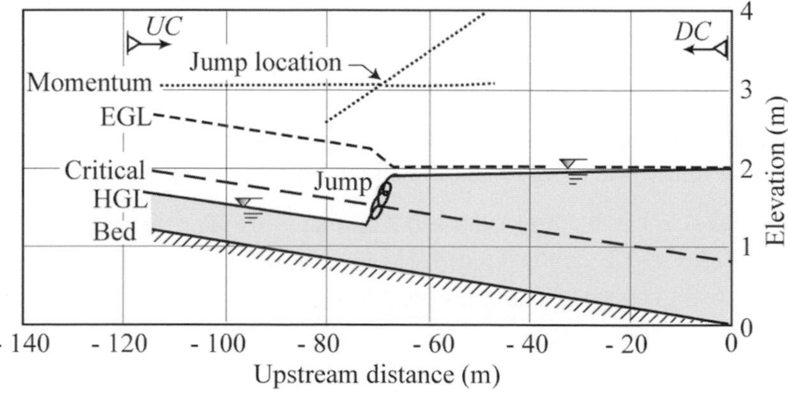

Fig. E-10.4 Locating the hydraulic jump in an S-1 backwater curve

10.4 Bridge Hydraulics

Bridge crossings are usually located in straight and uniform river reaches to avoid the potential lateral migration of alluvial rivers. The first condition at bridge contractions is to make sure that the abutments do not choke the flow as discussed in Section 9.1.4. Choking would raise the water level due to the backwater and cause potential flooding upstream of the bridge during large floods. Bridge contractions also induce head losses that can be written as a function of the velocity V_b in the contracted section. When considering floods with significant flow on the floodplains, the head loss at bridge openings is primarily caused by the channel contraction and by the presence of bridge piers. The head loss at a bridge opening can be approximated by

$$h_f \simeq \left(K_b + \Delta K_p\right) \frac{V_b^2}{2g},$$

where K_b is the contraction coefficient and ΔK_p is the additional energy loss coefficient due to the presence of bridge piers. It is considered here that the reach is uniform and additional losses from the bridge-crossing eccentricity and skewness are negligible.

As sketched in Figure 10.11, the discharge parameter $M_0 = Q_b/Q_t$ represents the ratio of the discharge Q_b in the opening area without abutments to the total flow discharge Q_t of the river. The contraction coefficient K_b depends primarily on the discharge parameter M_0 and the type of abutment. Spill-through abutment minimizes abutment head losses. The pier coefficient $\Delta K_p = \sigma K_p$ depends on two factors: (1) σ is a function of M_0, which reduces to unity when $M_0 = 1$; and (2) K_p is a function of the obstruction caused by the bridge piers. This pier obstruction is $J = A_p/A_t$ where A_p is the cross-sectional area obstructed by the piers in the direction normal to the flow, and A_t is the total cross-sectional area normal to the flow direction. Example 10.5 shows basic head-loss calculations.

Example 10.5: Head losses at a bridge crossing

Consider a 200-ft-wide spill-through bridge opening with a flow of 7,000 cfs through the opening and 3,000 cfs outside of the bridge-crossing area. The flow is aligned with three long tight rows of 5-ft-diameter bridge piers obstructing the 10-ft-deep rectangular section area. Estimate the energy loss through the bridge opening.

Solution: $M_0 = Q_b/Q_t = 7{,}000/10{,}000 = 0.7$ and $K_b = 0.5$. For the piers, $\sigma \simeq 0.75$ and $K_p \simeq 0.28$ when $J = (3 \times 5)/200 = 0.075$. The mean flow velocity is $V_b = Q/\left(A_t - A_p\right) = 10{,}000/10(200 - 15) = 5.4$ ft/s, and the head loss is

$$h_f \simeq \left(K_b + \sigma K_p\right) \frac{V_b^2}{2g} = [0.5 + (0.75 \times 0.28)] \frac{5.4^2}{2 \times 32.2} = 0.3\,\text{ft}.$$

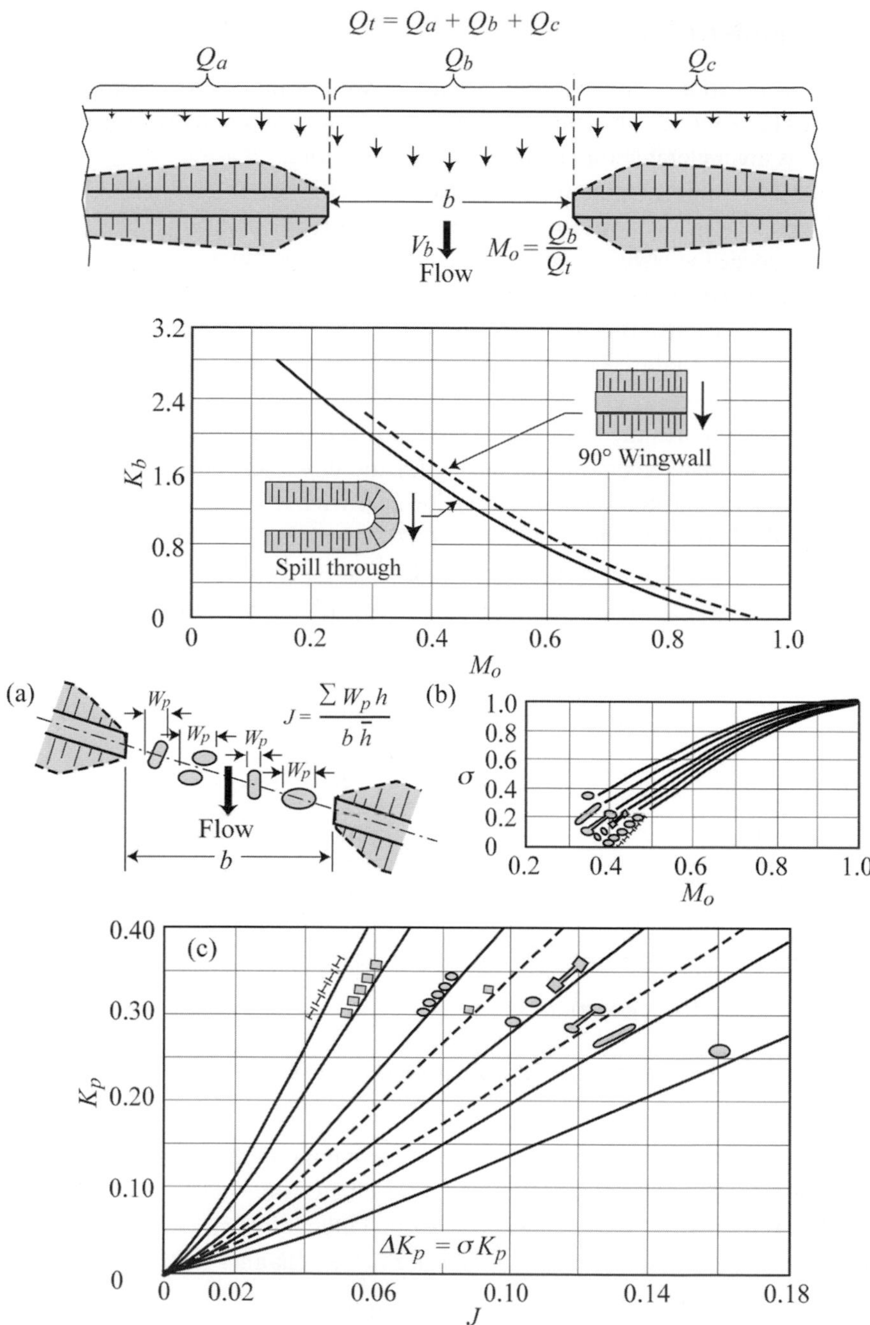

Figure 10.11 Energy losses for bridge openings (modified after Bradley 1978)

10.5 Computer Models

Backwater calculations can be performed by a variety of computer programs. One-dimensional (1-D) models are generally used to solve practical problems. Two-dimensional (2-D) models are preferable for complex fluvial geometries.

Arguably the most commonly used model is HEC–RAS (Hydraulic Engineering Center–River Analysis System) from the US Army Corps of Engineers (USACE). The freeware is downloaded at www.hec.usace.army.mil/software/hec-ras/.

HEC-RAS allows users to perform 1-D steady and unsteady flow with sediment transport computations. The software package features: (1) a graphical user interface (GUI); (2) hydraulic analysis components; (3) data storage and management; (4) graphics generation; and (5) report generation.

The program requires the following information: (1) a schematic of the river system; (2) the cross-sectional hydraulic geometry; (3) the cross-sectional spacing and reach length; (4) hydraulic structures and obstruction coefficients; (5) the Manning coefficient; (6) the flow discharge (steady/unsteady); and (7) boundary conditions.

Figure 10.12 from Bender and Julien (2011) shows a typical HEC-RAS cross section and longitudinal profile of the Middle Rio Grande with main channel, floodplains and points of banks above which the water can reach the floodplain.

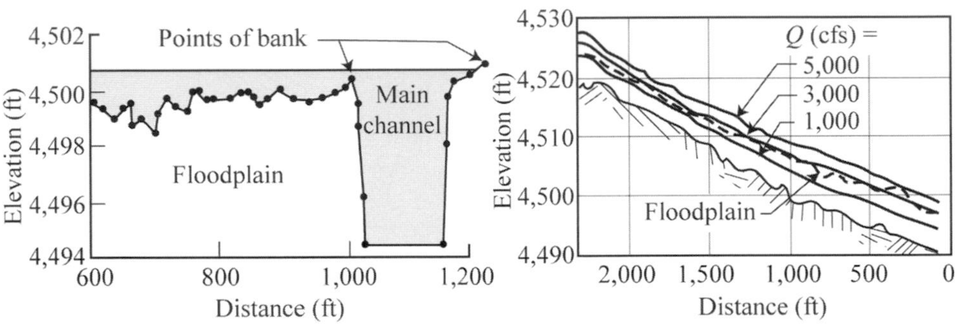

Figure 10.12 Typical HEC-RAS cross-sectional and longitudinal profile

HEC-RAS uses the standard-step method to compute the water-surface profile. The model: (1) uses an iterative technique to calculate water-surface elevation at two adjacent cross sections; (2) assumes 1-D steady gradually varied flow; (3) assumes constant velocity and horizontal water surface across channel sections; (4) solves the energy equation section by section.

The standard-step general procedure: (1) assumes an initial water-surface elevation; (2) starts calculations upstream if the flow is supercritical flow; (3) starts calculations downstream if the flow is subcritical; (4) uses the resulting cross-section geometry to calculate conveyance and velocity head; (5) computes representative friction slope and energy losses; (6) solves 1-D energy equation for water surface at cross sections 2; and (7) compares the assumed and calculated values and iterates until agreement within a user-defined tolerance (often 0.01 ft).

The program is constrained by a maximum number of iterations for balancing the water elevation. Common sources of error include: (1) insufficient cross sections; (2) inadequate cross-section data; and (3) incorrect boundary conditions.

Figure 10.13 shows a sample of successive cross sections (aggradation/degradation lines) of the Rio Grande, NM, with flood mapping using lidar surveys and RAS Mapper.

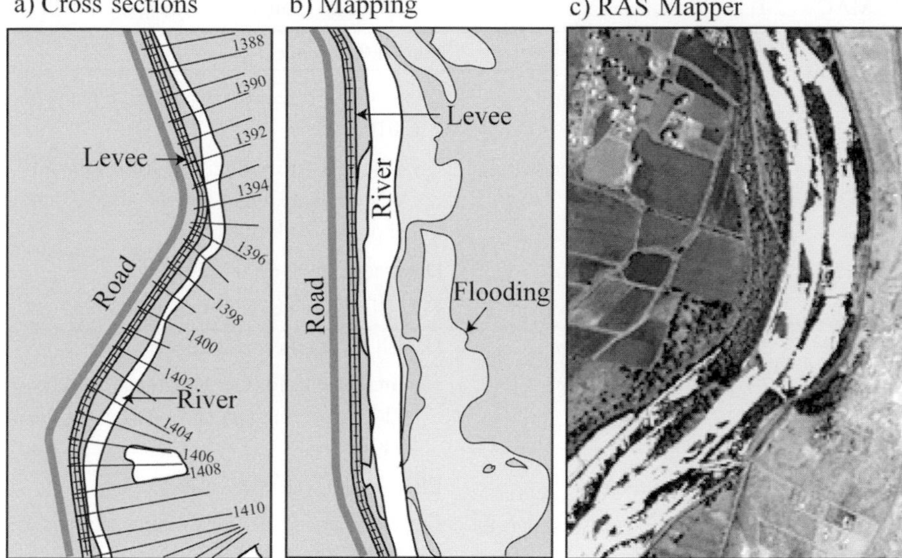

Fig. 10.13 HEC-RAS examples for the Middle Rio Grande, NM

Other commonly used 1-D models include GSTARS, FLUVIAL12 and commercial codes like MIKE11. Two-dimensional (2-D) river models solve continuity and momentum equations with a resistance to flow relationship (e.g. the Manning equation). The calculations can be performed in a raster-based or vector-based format depending on the Geographic Information Systems (GIS) database available for the calculations.

A fair number of river models have been developed and some are readily available in the public domain. Table 10.1 provides a list of river-engineering models subdivided into: freeware, where executables can be easily downloaded; and commercial codes. The list is far from exhaustive but is representative of the current state of the art. Web users can readily access a wealth of information regarding the details of each model.

Table 10.1. List of river models

Freeware	Type	Features	Source/References
HEC–RAS,	1-D	River analysis	USACE–Hydrologic Engineering Center
HEC–HMS	1-D	Watershed	USACE–Hydrologic Engineering Center
WSPRO	1-D	Bridges	Federal Highway Administration (FWHA) (Arneson and Shearman 1998)
SRH	2-D	Rivers and sediment	US Bureau of Reclamation (USBR) (Lai 2008)
GSTARS	2-D	Sediment transport in rivers and reservoirs	USBR- Colorado State University (Yang et al. 2008)

Table 10.1. *(cont.)*

Freeware	Type	Features	Source/References
BRI-STARS	2-D	Bridges and rivers	FHWA (Molinas 2000)
iRIC	2-D, 3-D	River flow and riverbed variation	USGS–University of Hokkaido (Nelson et al. 2006, Shimizu et al. 2000)
TELEMAC	2-D	Mascaret	Electricité de France
Commercial	**Type**	**Features**	**Source/Reference**
FLUVIAL12	1-D	Rivers, sediment	Chang (2006); see, for example, Julien et al. (2010)
GSSHA, WMS, TABS	2-D	Watershed, rivers, coastal	USACE, US Army Engineer Research and Development Center Coastal and Hydraulics Laboratory (ERDC–CHL)
HydroSed	2-D	Rivers, sediment	Duan and Julien (2005, 2010)
FLO-2D	2-D	Mudflows, floodplain	FLO-2D software (O'Brien et al. 1993)
MIKE11, 21	2-D	Rivers, sediment	Danish Hydraulic Institute
CCHE	2-D	Rivers, sediment	National Center for Computational Hydroscience and Engineering (Jia and Wang 1999, Jia et al. 2009)
CH3D-SED	3-D	Large rivers	ERDC–CHL (Gessler et al. 1999)
FLOW-3D	3-D	Fluid mechanics, density currents	FLOW Science; see, for example, An and Julien (2014), An et al. (2015)

Additional Resources

This chapter expands upon many textbooks, including Chow (1959), Henderson (1966), Fennema and Chaudhry (1990), Jain (2001) and Sturm (2001). Additional references include: Shearman (1990) and Arneson and Shearman (1998) on bridge hydraulics; ASCE (2013) for navigable waterways; Duan and Julien (2005, 2010) for meandering rivers; ASCE (2008) and Julien (2010) for sedimentation engineering; Battjes and Labeur (2017) for unsteady flow; and Palu and Julien (2019, 2020) for dam-break problems.

Extensive recent HEC-RAS modeling applications on the Middle Rio Grande in New Mexico include Doidge (2019), Beckwith and Julien (2020), Fogarty (2020), LaForge et al. (2020) and Yang et al. (2020) for the aquatic habitat, Holste and Baird (2020) for the modeling of perched channels and Mortensen et al. (2020) for the integration of hydro-biological habitat for endangered species.

EXERCISES

These exercises review the essential concepts from this chapter.

1. What is gradually varied flow?
2. When does $dy/dx = 0$?
3. When does $dy/dx \to \infty$?
4. What are the two most common gradually varied flow profiles?
5. Why is there no H-1 profile?
6. When do you have upstream control?

7. What are the two common methods to calculate gradually varied flow in open channels?
8. What are the two main types of energy losses at bridge openings?
9. Which computer program is arguably the most commonly used for water-surface calculations?
10. True or false?
   (a) The critical depth depends on resistance to flow.
   (b) The slope $dy/dx \rightarrow \infty$ when $\mathrm{Fr} = 1$.
   (c) Profile C-2 does not exist because $y_n = y_c$.
   (d) Profiles H-1 and A-1 do not exist because $y_n \rightarrow \infty$.
   (e) Hydraulic jumps always occur when $y = y_c$.
   (f) In zone 2, the flow depth always decreases in the downstream direction.
   (g) Downstream control corresponds to supercritical flow.
   (h) The direct-step method does not work when $y = y_c$.
   (i) The HGL always decreases in the downstream direction.
   (j) Head losses at bridge crossings are minimal because the flow is converging.

PROBLEMS

1. ♦ A spillway has a specific discharge of 70 ft^2/s. A hydraulic jump is formed over the apron of the spillway, as shown in Fig. P-10.1. If the water depth just upstream of the spillway is 15.0 ft, determine the sequent depths of the jump. Neglect friction losses over the spillway. Where are the controls?

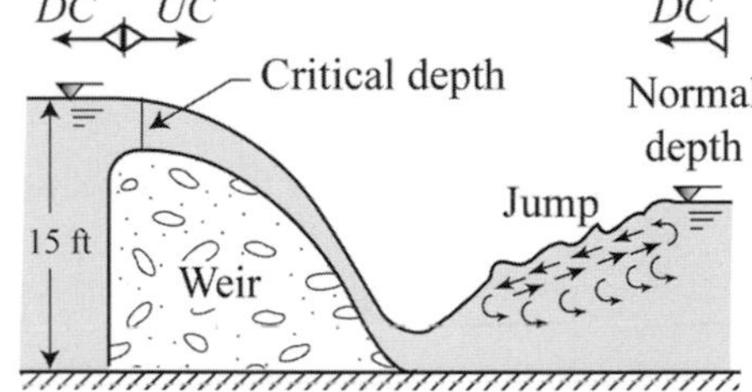

Fig. P-10.1

2. ♦ Water is flowing in a 5-m-wide rectangular channel on a bed slope 0.0065 m/m with a Manning coefficient $n = 0.035$. The discharge is 21 m^3/s. At a certain point the depth is observed to be 3 m.
   (a) Classify the water profile in this vicinity.
   (b) Where is the control located (upstream or downstream)?
3. ♦ Water flows underneath a sluice gate to a mild reach in a wide rectangular open channel. At the end of this reach, a weir maintains the required upstream flow depth. Downstream of the weir, the channel has a steep slope. Sketch the water-surface profile on Fig. P-10.3.

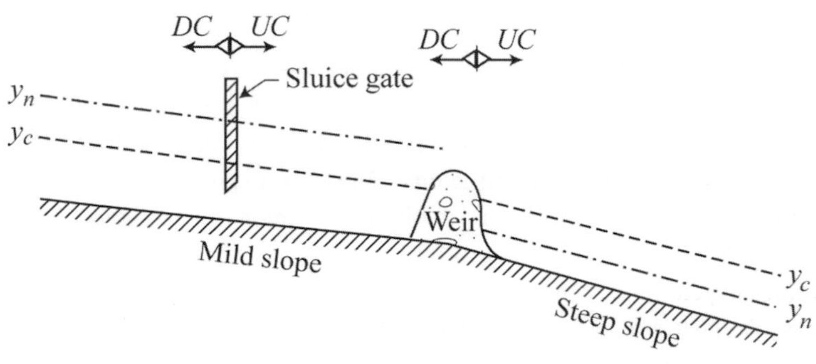

Fig. P-10.3

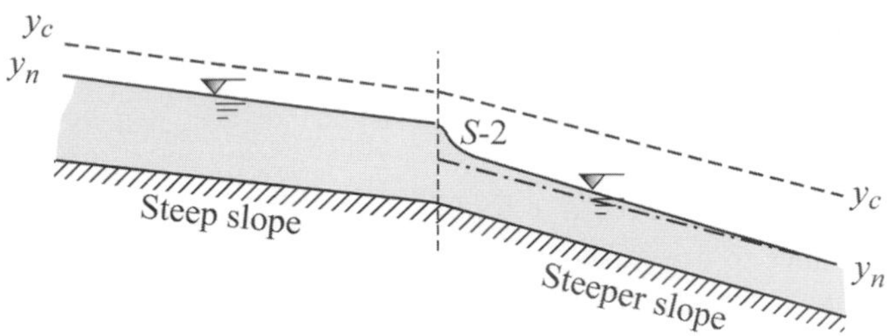

Fig. P-10.4

4. ♦ An open channel consists of two long reaches with steep slopes as shown in Fig. P-10.4. The second reach is steeper than the first and terminates into a free overfall. Draw the water-surface profile for the following cases:
   (a) no gate exists;
   (b) a gate is placed at the break point in the slope between the two reaches and the outflow depth is half the normal depth; and
   (c) two gates are placed midway in each reach with outflow depths at half their respective normal depths.

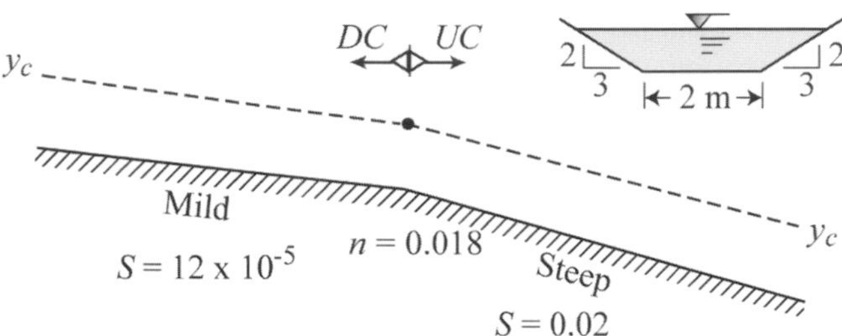

Fig. P-10.5

5. ♦♦ A trapezoidal channel with a bed width of 2.0 m and a side slope of 3H:2V carries a discharge of 300,000 m^3/d. The channel shown in Fig. P-10.5 has a bed slope of 12×10^{-5} and the Manning coefficient is 0.018. At a certain point the bed slope increases to 20×10^{-3}. Calculate and plot the water-surface profile and determine the length of the water profile in the two reaches.

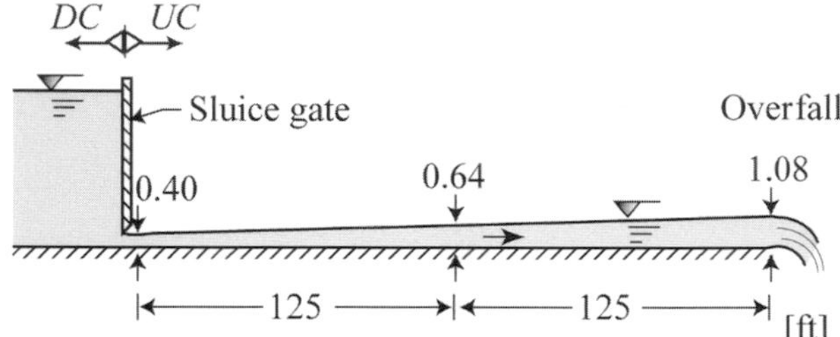

Fig. P-10.6

6. ♦♦ A wide and smooth horizontal channel below a sluice gate carries a 10.8 ft^2/s unit discharge over a distance of 250 ft before a free overfall in Fig. P-10.6. What is the water level behind the gate for this flow condition? Calculate the water-surface profile over the entire apron starting from a depth of 0.4 ft at the gate.

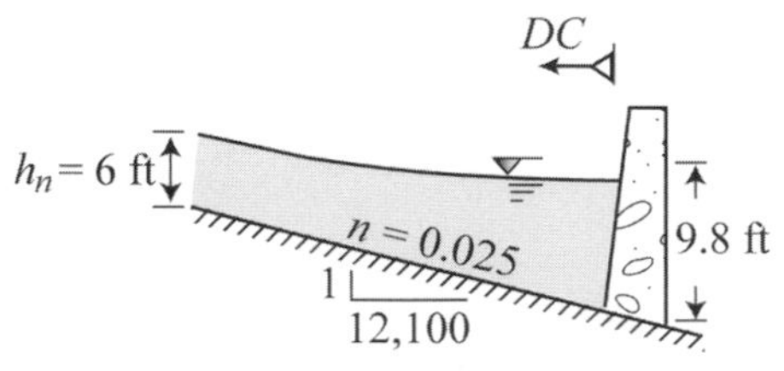

Fig. P-10.7

7. ♦ On Fig. P-10.7, a rectangular channel has a width of 30.0 ft and a bed slope of 1:12,100. The normal depth is 6 ft. A dam across the river elevates the water surface and produces a depth of 9.8 ft just upstream of the dam. Develop a spreadsheet with the standard-step method and plot the backwater profile with $n = 0.025$.
8. ♦ Solve Problem 10.7 by using the direct-step method.

9. ♦ Solve Example 10.1 with $y_0 = 12$ ft and $S_0 = 1:1{,}500$.
10. ♦ Solve Example 10.2 with $n = 0.035$.
11. ♦♦ Solve Example 10.3 with a hydraulic jump 50 ft from the spillway toe.
12. ♦♦ Solve Example 10.4 for $n = 0.035$.
13. ♦♦♦ In Problem 10.6, what would happen if you built a 1-foot-high weir at the downstream end. Calculate the new water-surface profile and plot the results.
14. ♦♦♦ In Problem 10.6, what would happen if you built a 2-foot-high weir at the downstream end. Calculate the new water-surface profile and plot the results.

Refer to the Professional Engineering Obligations in Appendix C for the following questions

15. ♦♦ **Engineering obligations – is there an issue?**
    When considering the following professional situations, can you identify which (there could be several) of the nine professional engineering obligations listed in Appendix C would cause a conflict.
    (a) You use hydraulic engineering methods set 25 years ago because this is how things have always been done.
    (b) You engage in public denouncing of engineers from a competing firm.
    (c) You would like to promote a solution that has the lowest cost in the first 2 years but becomes twice as expensive as other solutions after 10 years.
16. ♦♦ **Engineering obligations – is there an issue?**
    When considering the following professional situations, can you identify which (there could be several) of the nine professional engineering obligations listed in Appendix C would be a cause of conflict.
    (a) Your project requires the use of contaminants that may impact children's health in your community.
    (b) Upon changing employer, you would like to keep all records and data of the project you previously worked on.
    (c) You consider an association with nonengineers to free yourself from professional obligations?
17. ♦♦ **Engineering obligations – what would you do?**
    What would you do in the following situation?
    (a) A competing firm offers you a 20% pay raise to join them and compete with your current employer.
    (b) Your company proposes a solution that is not in the best interest of the public.
    (c) You learn that your supervisor promotes a solution that would be far more expensive than it should be, but would yield a significant profit to his family.
18. ♦♦ **Engineering obligations – what would you do?**
    What would you do in the following situation?
    (a) You are asked to alter your calculations to promote an alternative solution that is objectively not the best.
    (b) You developed software for a client who claims at the project's end that the source code is his own property.
    (c) Who should you notify when you are about to accept a position from a different employer?

19. ♦ **Comprehensive review problem.** Consider the 200-ft-wide rectangular spillway sketched in Fig. P-10.19 with a 100-ft drop and a discharge capacity of 16,000 cfs. Assume very low friction on the spillway, $n = 0.03$ and a slope 70 cm/km downstream of the spillway base.

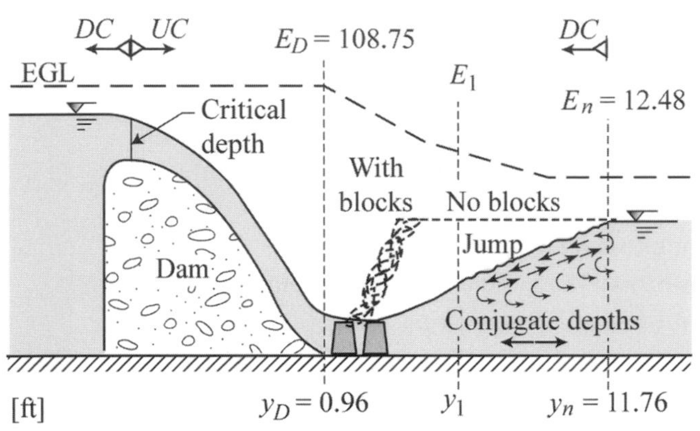

Fig. P-10.19

Determine the following:

(a) the critical depth;
(b) the flow depth in the reservoir;
(c) the flow depth at the base of the spillway;
(d) the flow velocity at the base of the spillway;
(e) plot the specific energy diagram and the momentum diagram as a function of flow depth from spreadsheet calculations;
(f) the flow momentum at the base of the spillway;
(g) the Froude number at the base of the spillway;
(h) whether the flow depths at the base of the spillway and in the reservoir are alternate or conjugate depth.
(i) the normal depth in the channel downstream of the spillway;
(j) the Froude number at normal depth;
(k) whether the slope of the downstream channel should be considered steep or mild;
(l) the momentum force in the downstream channel at normal depth;
(m) the conjugate depth upstream of the hydraulic jump;
(n) the head loss in the jump;
(o) the power lost in the hydraulic jump;
(p) what type of hydraulic jump would occur at that location;
(q) if the jump is induced at the base of the spillway with the help of baffle blocks, what the force on the blocks would be;
(r) sketch the water-surface profile and energy grade line for the case without baffles; and
(s) sketch the hydraulic and energy lines in the case with baffles.

11 Unsteady Flow

In this chapter, the governing equation for floodwave propagation is derived in Section 11.1, with solutions to the advection–diffusion equation outlined in Section 11.2.

11.1 Floodwave Equation

Three relationships describe unsteady flow in open channels: (1) the conservation of mass in Section 11.1.1; (2) flow resistance in Section 11.1.2; and (3) momentum in Section 11.1.3. They combine into a diffusion equation in Section 11.1.4.

11.1.1 Continuity for Unsteady Flow

The principle of conservation of mass indicates that the mass of water remains constant. In Figure 11.1, we identify the top width W and the wetted perimeter P, and the flow discharge Q is the product of the mean flow velocity V and cross-sectional area A. We can add complexity with rainfall intensity i_r, infiltration i_b through the wetted perimeter and the unit discharge from lateral inflow q_l.

The volume of water in the control volume is Adx. Over a reach length dx, the inflow discharge Q differs from the outflow discharge $Q + (dQ/dx)dx$. When including rain i_r and lateral inflow q_l while losing water through infiltration i_b, the volumetric balance is

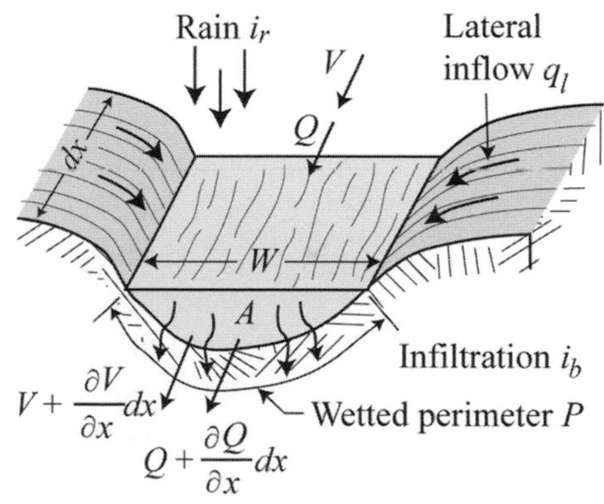

Figure 11.1 Continuity for open channels

$$Q + q_l dx + i_r W dx - i_b P dx - [Q + (\partial Q/\partial x)dx] = [\partial(Adx)/\partial t].$$

After division by dx, it reduces to

$$\frac{\partial A}{\partial t} + \frac{\partial Q}{\partial x} = i_r W + q_l - i_b P.$$

Of course, when rainfall precipitation, infiltration and lateral flow are negligible, we obtain the main relationship describing continuity, or conservation of mass in rivers:

$$\frac{\partial A}{\partial t} + \frac{\partial Q}{\partial x} = 0. \tag{11.1}$$

For rectangular channels of constant width W, it reduces further to

$$\frac{\partial h}{\partial t} + \frac{1}{W}\frac{\partial Q}{\partial x} = 0 \tag{11.1a}$$

or

$$\frac{\partial^2 h}{\partial t \partial x} = -\frac{1}{W}\frac{\partial^2 Q}{\partial x^2}. \tag{11.1b}$$

Notice that the conservation of mass links space and time derivatives.

11.1.2 Flow Resistance

Resistance to flow in open channels is described by Manning's equation

$$Q = A\frac{m}{n}R_h^{2/3}S_f^{1/2} = k\,S_f^{1/2},$$

where

$$k = A\frac{m}{n}R_h^{2/3}.$$

In SI units, $m = 1$ and k is the conveyance coefficient. For wide rectangular channels, we can simply write the discharge per unit width as a power function of flow depth

$$q = \frac{Q}{W} = Vh = \frac{m}{n}h^{5/3}S_f^{1/2} = \alpha h^{\beta},$$

where $\alpha = \frac{m}{n}S_f^{1/2}$ and $\beta = 5/3$.

The advantage of this formulation for wide rectangular channels is that α and β remain constant while k varies with flow depth; hence,

$$k = \frac{Q}{\sqrt{S_f}} = \frac{W}{\sqrt{S_f}}\alpha h^{\beta}$$

or

$$S_f = \frac{Q^2}{k^2}. \tag{11.2}$$

Values of W, α and S_f do not change with h, and we obtain

$$\frac{\partial k}{\partial h} = \frac{W}{\sqrt{S_f}}\alpha\beta h^{\beta-1} = \frac{\beta k}{h}.$$

Since k is only a function of h, we can combine with Eq. (11.1a) to get

$$\frac{\partial k}{\partial t} = \frac{\partial k}{\partial h}\frac{\partial h}{\partial t} = \frac{\beta k}{h}\left(-\frac{1}{W}\frac{\partial Q}{\partial x}\right). \tag{11.2a}$$

We can now examine the time derivative of $S_f = Q^2/k^2$ when both Q and k vary with time. We have the derivative of a ratio:

$$\frac{d}{dt}\left(\frac{u}{v}\right) = \frac{vu' - uv'}{v^2},$$

which gives

$$\frac{\partial S_f}{\partial t} = \frac{\partial}{\partial t}\left(\frac{Q^2}{k^2}\right) = \frac{2Q}{k^2}\frac{\partial Q}{\partial t} - \frac{2Q^2}{k^3}\frac{\partial k}{\partial t}, \tag{11.2b}$$

and this is combined with Eq. (11.2a) to give

$$\frac{\partial S_f}{\partial t} = \frac{2Q}{k^2}\frac{\partial Q}{\partial t} - \frac{2Q^2}{k^3}\left[\frac{\beta k}{h}\left(-\frac{1}{W}\frac{\partial Q}{\partial x}\right)\right]. \tag{11.2c}$$

The conveyance relationship only includes terms in $\frac{\partial Q}{\partial t}$ and $\frac{\partial Q}{\partial x}$. At this point we have only combined the conservation of mass and resistance to flow. As we are about to find out, the resistance to flow terms only contribute to advective terms of the advection–diffusion equation. This means that they only contribute to pure floodwave translation without diffusion.

11.1.3 Momentum

The momentum equation is derived in Julien (2018) and is presented here without derivation. The derivation by Barre de Saint-Venant in France focused on the main terms of the Navier–Stokes equation for one-dimensional flow. The five terms include local and convective acceleration terms, as well as gravity, pressure gradient and shear-stress terms. The resulting Saint-Venant equation is written in dimensionless form as

$$\begin{array}{llllll} S_f & \cong S_0 & - \dfrac{\partial h}{\partial x} & - \dfrac{V\partial V}{g\partial x} & - \dfrac{1}{g}\dfrac{\partial V}{\partial t} \\ (1) & (2) & (3) & (4) & (5) \end{array} \tag{11.3}$$

| kinematic |

| diffusive |

| quasi-steady |

| full dynamic |,

where S_f is the friction slope, S_0 is the bed slope, h is the flow depth, V is the mean flow velocity and g is the gravitational acceleration. From this relationship, we want to describe the flow depth h as a function of distance x and time t. The attentive reader will also recognize that this equation reduces to Eq. (10.1) for steady gradually varied flow in open channels since $\mathrm{Fr}^2 = V^2/gh$.

Different floodwave propagation types can be identified depending on the various approximations of Eq. (11.3). The full dynamic-wave approximation of the Saint-Venant equation describes flows where the unsteady term is significant. For all steady flows, the quasi-steady dynamic-wave approximation includes the first four terms of Eq. (11.3); this also corresponds to the gradually varied flow equation in Chapter 10. The diffusive-wave approximation, $S_f = S_0 - \partial h/\partial x$, is most commonly used for floodwave propagation in rivers. Finally, the kinematic-wave approximation is obtained when all but the first two terms vanish. This is typically the case when the channel slope is very large, e.g. in mountain streams.

When considering the quasi-steady formulation from Eq. (10.1), Eq. (11.3) can also be written as

$$S_f = S_0 - \left(1 - \mathrm{Fr}^2\right)\frac{\partial h}{\partial x} = S_0 - \Omega\frac{\partial h}{\partial x},$$

where $\Omega = 1 - \mathrm{Fr}^2$. It is also considered that the Froude number Fr and thus Ω remain constant at different flow depths. Taking the time derivative gives

$$\frac{\partial S_f}{\partial t} = -\Omega\frac{\partial^2 h}{\partial x \partial t}. \tag{11.3a}$$

We are now ready to derive the unsteady flow equation for open channels.

11.1.4 Flood Routing

Equations (11.1) to (11.3) are now combined to describe unsteady flow and floodwave propagation in wide rectangular channels.

$$\text{Continuity}\ \frac{\partial h}{\partial t} = -\frac{1}{W}\frac{\partial Q}{\partial x} \tag{11.1a}$$

$$\text{Conveyance}\ S_f = \frac{Q^2}{k^2} \tag{11.2}$$

$$\text{Momentum}\ S_f = S_0 - \Omega\frac{\partial h}{\partial x} \tag{11.3}$$

The last equation, Eq. (11.3), is also called the diffusive-wave approximation, for a reason we are about to discover. The strategy adopted to solve these differential equations is to eliminate h from Eqs. (11.1a) and (11.3). This is done through differentiating Eq. (11.1) in space x and differentiating Eq. (11.3) in time t. Thus, combining Eqs. (11.1b) and (11.3a) and comparing with Eq. (11.2c) gives

$$\frac{\partial S_f}{\partial t} = \frac{2Q}{k^2}\frac{\partial Q}{\partial t} + \frac{2Q^2}{k^3}\left(\frac{\beta k}{hW}\frac{\partial Q}{\partial x}\right) = \frac{\Omega}{W}\frac{\partial^2 Q}{\partial x^2}.$$

Multiplying by $k^2/2Q$ (same as $Q/2S_f$) with algebraic simplifications yields

$$\frac{\partial Q}{\partial t} + \beta V \frac{\partial Q}{\partial x} = \frac{\Omega Q}{2WS_f} \frac{\partial^2 Q}{\partial x^2}. \tag{11.4}$$

This basic relationship describes unsteady flow propagation in a wide rectangular channel. The attentive reader will notice here that the diffusion term $\partial^2 Q/\partial x^2$ stems from the momentum equation via Eq. (11.3a). Physically, it is the pressure gradient that causes the diffusion of floodwaves. Clearly, the diffusion term vanishes when $\Omega = 1 - \text{Fr}^2 = 0$. In turn, this means that floodwave attenuation will be significant in large rivers with a low Froude number. Conversely, the attenuation of floodwaves will be minimal in mountain streams where $\text{Fr} \approx 1$, and floodwaves may even amplify in supercritical flows ($\text{Fr} > 1$).

This advection–diffusion (or advection–dispersion) equation is

$$\frac{\partial Q}{\partial t} + c \frac{\partial Q}{\partial x} = K \frac{\partial^2 Q}{\partial x^2}, \tag{11.5}$$

where $c = \beta V$ is the flood celerity, and $K = \frac{\Omega Q}{2WS_f}$ is the flood diffusion coefficient.

We learn that the celerity of the floodwave in open channels is faster than the flow velocity because $c = \beta V$ and $\beta = 5/3$ in wide rectangular channels, from Manning's equation. The second important characteristic of the floodwave propagation equation is that the diffusion coefficient K describes the attenuation of the floodwave, as shown in Figure 11.2.

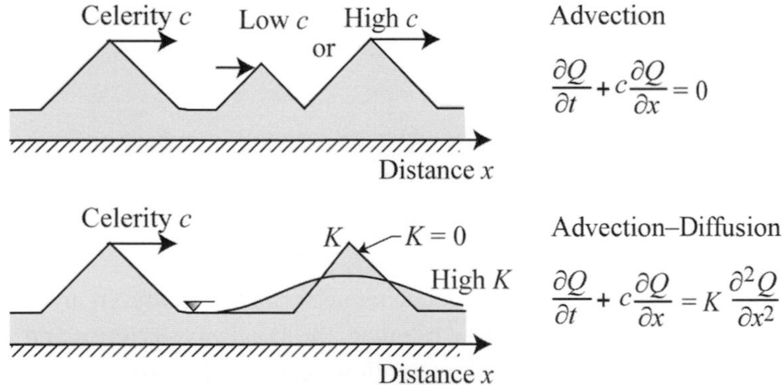

Figure 11.2 Floodwave propagation in wide open channels

Diffusivity K increases when the discharge Q increases and the slope S_f decreases. The floodwaves of large flat rivers attenuate greatly in comparison with smaller floods in steep mountain channels, as discussed in Julien (2018). The term $\Omega = 1 - \text{Fr}^2$ in the parameter K also indicates that floodwave attenuation increases in rivers with a low Froude number. Example 11.1 illustrates how to get a first estimate for the parameters of the advection–diffusion equation.

Example 11.1: Parameter estimation for the advection–diffusion equation
The main river parameters of the Doce River in Brazil include the channel slope $S_0 = 0.0005$, channel width $W = 150$ m and Manning's coefficient $n = 0.05$. Estimate the parameters of the advection–diffusion equation for calculations of floodwave propagation with a flow discharge $Q \approx 300$ m^3/s.

Solution: We can first estimate the mean flow velocity, depth and Froude number for steady flow:

$$Q = AV = Wh_n \frac{1}{n}\left[\frac{Wh_n}{(W + 2h_n)}\right]^{2/3} S_0^{1/2} = 300 = 150h_n \frac{1}{0.05}\left[\frac{150h_n}{(150 + 2h_n)}\right]^{2/3} \sqrt{0.0005},$$

which, with a solver, gives $h_n = 2.5$ m, $V = 0.81$ m/s and Fr $= 0.16$. In this case, $\beta = 5/3$ and the celerity is $c = \beta V = (5 \times 0.8)/3 = 1.34$ m/s, $\Omega = 1 - \mathrm{Fr}^2 = 1 - 0.17^2 = 0.97$ and $K \simeq \frac{\Omega Q}{2WS_f} = \frac{0.97\times300}{2\times150\times0.0005} = 2{,}000$ m^2/s, and a first approximation for the advection–diffusion equation becomes

$$\frac{\partial Q}{\partial t} + 1.34\frac{\partial Q}{\partial x} \simeq 2{,}000\frac{\partial^2 Q}{\partial x^2},$$

where the discharge Q is in m^3/s, x in m and t in s.

It is noticed that the value of the diffusion coefficient will increase primarily with discharge. It is not clear at this point whether low or high discharge values should be used for calculations, a subject that will be discussed further in this chapter.

11.2 Floodwave Propagation

We explore an analytical solution for floodwave propagation in Section 11.2.1 followed with a numerical solution in Section 11.2.2.

11.2.1 Analytical Solution for Floodwave Propagation

The propagation of floodwaves in open channels can be analyzed by solving the advection–diffusion Eq. (11.5), where c in m/s is the floodwave celerity and K in m^2/s is the dispersion coefficient. Given the mean flow celerity c in a river, a constant pulse of water at a discharge Q_0 over a duration T starting at $t = 0$, the discharge $Q(x, t)$ is calculated at a distance x as a function of time t as

$$Q(x,t) = \frac{Q_0}{2}\left\{erfc\left[\frac{x - ct}{2\sqrt{Kt}}\right] - erfc\left[\frac{x - c(t-T)}{2\sqrt{K(t-T)}}\right]\right\} + \frac{Q_0}{2}e^{\frac{cx}{K}}\left\{erfc\left[\frac{x + ct}{2\sqrt{Kt}}\right] - erfc\left[\frac{x + c(t-T)}{2\sqrt{K(t-T)}}\right]\right\} \tag{11.6}$$

where $erfc(x) = 1 - erf(x)$ is the complementary error function from the error function $erf(x) = \frac{2}{\sqrt{\pi}} \int_0^x e^{-\alpha^2} d\alpha$. Figure 11.3 plots the normal distribution, the error function and the complementary error function. Error functions are calculated with any mathematical package (e.g. erf.precise and erfc.precise in Excel). As time increases, the second line in Eq. (11.6) becomes very small because $erfc(x > 3) \to 0$. Example 11.2 calculates the propagation of a single flow pulse.

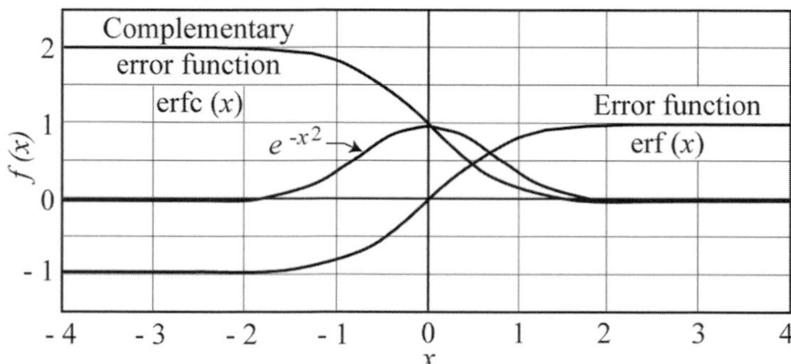

Figure 11.3 Normal, error and complementary error functions

Example 11.2: Analytical solution for a flow pulse

Calculate the propagation of a single flow pulse lasting $T = 6$ h at an initial discharge $Q_0 = 1{,}000$ m^3/s given the mean flow celerity $c = 1$ m/s and dispersion coefficient $K = 1{,}000$ m^2/s, as shown in Fig. E-11.2. Find the discharge at $x = 75$ km after 1 day.

Solution: Consider $x = 75{,}000$ m, $t = 86{,}400$ s and pulse duration $T = 6 \times 3{,}600 = 21{,}600$ s. From Eq. (11.6), we obtain

$$Q(75 \text{ km}, 1 \text{ day}) = \frac{1{,}000}{2}\left\{erfc\left[\frac{-11{,}400}{18{,}590}\right] - erfc\left[\frac{10{,}200}{16{,}099}\right]\right\} + \frac{1{,}000}{2} \times 3.733 \times 10^{32}\left\{erfc\left[\frac{161{,}400}{18{,}590}\right] - erfc\left[\frac{139{,}800}{16{,}099}\right]\right\}$$

$$Q(75 \text{ km}, 1 \text{ day}) = [500(1.6142 - 0.3703)] + \left[500 \times 3.733 \times 10^{32}\left(1.1873 \times 10^{-34} - 1.1574 \times 10^{-34}\right)\right]$$

$$Q(75 \text{ km}, 1 \text{ day}) = 622 + 0.56 = 622.5 \text{ cms}$$

The main characteristics of floodwave propagation are clearly visible from Fig. E-11.2: (1) translation of the floodwave moving downstream at the celerity $c = 1 \text{ m/s} = 86.4 \text{ km/day}$; and (2) floodwave attenuation through the parameter $K = \frac{\Omega Q}{2WS_f} \simeq \frac{(1-Fr^2)Q}{2WS_f}$ as the flood propagates downstream.

It is noted that the dispersion of the floodwave is due to the momentum equation Eq. (11.3) because $K = 0$ when $\Omega = 0$. Also, the principle of superposition can be applied

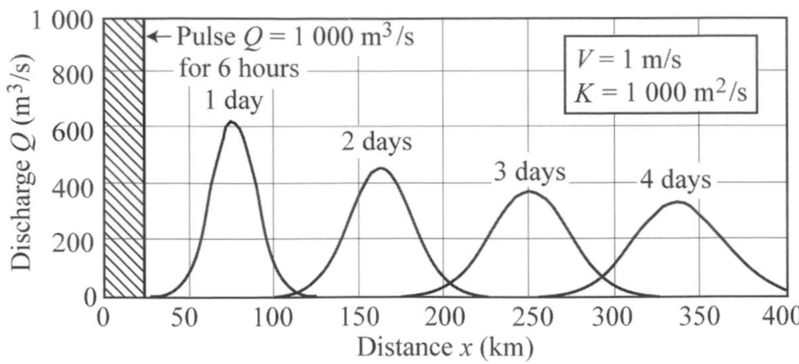

Fig. E-11.2 Analytical advection–diffusion example

to a sequence of step functions because Eq. (11.5) is linear. The advantage of the analytical solution is that we can directly calculate the values of discharge at any time and space value. However, the analytical solution becomes less practical for long hydrographs where discharge varies rapidly with time. To handle large variability in discharge, the numerical method of Section 11.2.2 is usually more convenient.

11.2.2 Numerical Solution for Floodwave Propagation

The numerical solution to the advection–diffusion equation can contaminate the results by adding numerical diffusion, which artificially attenuates the floodwave. Higher-order numerical schemes can eliminate numerical diffusion (Abbott and Basco 1989). Considering the Leonard (1979) numerical scheme in Julien (2018), the grid size Δx and time step Δt are determined from the flood celerity c and diffusion coefficient K as $\Delta x = 10K/c$ and $\Delta t = 10K/c^2$. A practical finite-difference numerical scheme of Eq. (11.5) without numerical diffusion is

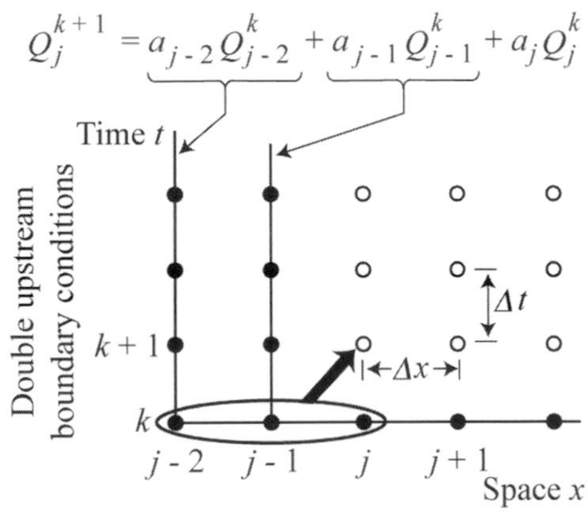

Figure 11.4 Double upstream boundary condition

$$Q_j^{k+1} = 0.1\, Q_{j-2}^k + 0.8\, Q_{j-1}^k + 0.1\, Q_j^k. \tag{11.7}$$

The subscript from $j-2$ to j refers to space, the superscript refers to time from k to $k+1$. This algorithm requires two upstream boundary conditions at $j-2$ and $j-1$ as shown in Figure 11.4, and the initial condition at $k=0$ describes the flow discharge along the channel reach at the beginning of the flood.

Example 11.3 shows calculations for a double pulse, and Case Study 11.1 presents the case study of a dam-break event in the Doce River.

Example 11.3: Numerical solution for triangular pulses

Simulate the propagation of a double triangular flow pulse in a river where the celerity is $c = 1$ m/s and the diffusion coefficient $K = 1{,}000$ m²/s. Calculate the hydrograph at a distance $x = 100$ km downstream of the double pulse shown in Fig. E-11.3.

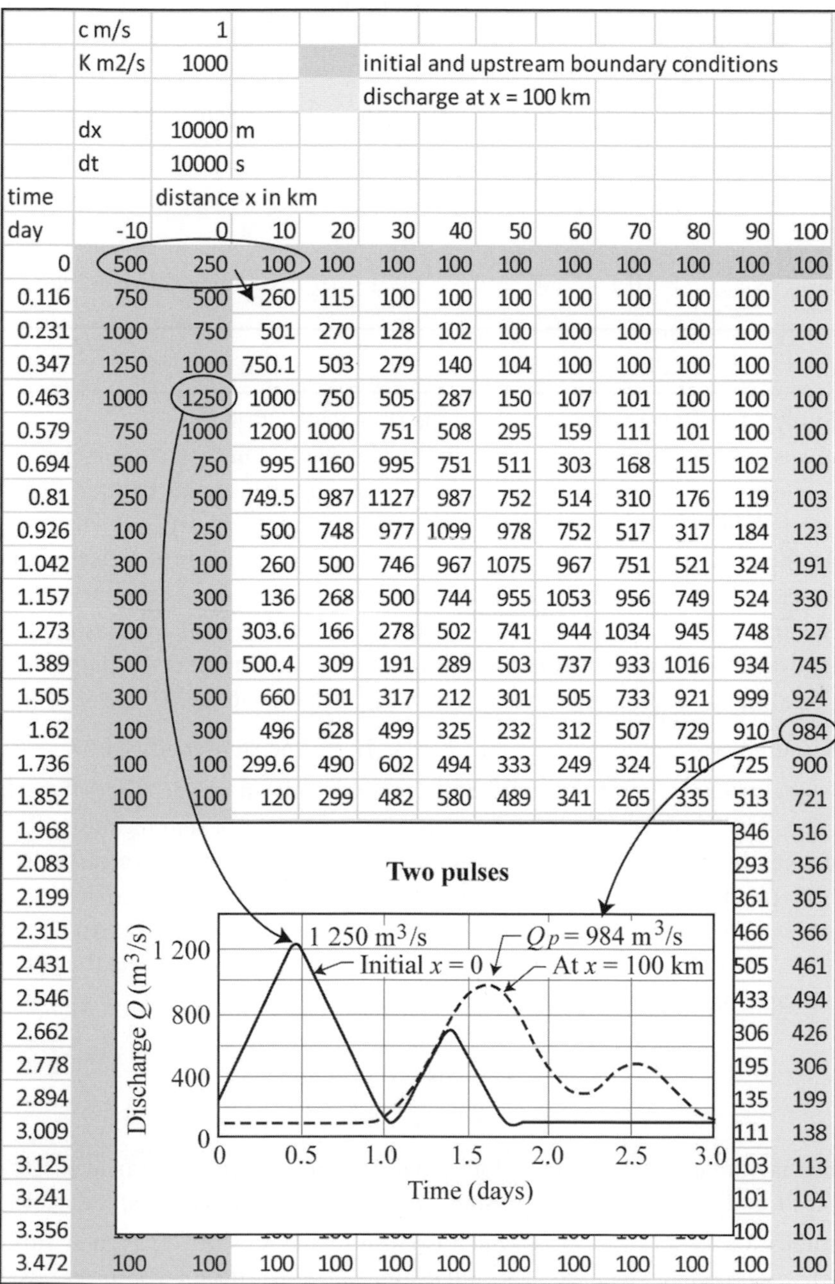

	c m/s	1									
	K m2/s	1000		initial and upstream boundary conditions							
				discharge at x = 100 km							
	dx	10000	m								
	dt	10000	s								
time		distance x in km									
day	-10	0	10	20	30	40	50	60	70	80	90	100
0	500	250	100	100	100	100	100	100	100	100	100	100
0.116	750	500	260	115	100	100	100	100	100	100	100	100
0.231	1000	750	501	270	128	102	100	100	100	100	100	100
0.347	1250	1000	750.1	503	279	140	104	100	100	100	100	100
0.463	1000	1250	1000	750	505	287	150	107	101	100	100	100
0.579	750	1000	1200	1000	751	508	295	159	111	101	100	100
0.694	500	750	995	1160	995	751	511	303	168	115	102	100
0.81	250	500	749.5	987	1127	987	752	514	310	176	119	103
0.926	100	250	500	748	977	1099	978	752	517	317	184	123
1.042	300	100	260	500	746	967	1075	967	751	521	324	191
1.157	500	300	136	268	500	744	955	1053	956	749	524	330
1.273	700	500	303.6	166	278	502	741	944	1034	945	748	527
1.389	500	700	500.4	309	191	289	503	737	933	1016	934	745
1.505	300	500	660	501	317	212	301	505	733	921	999	924
1.62	100	300	496	628	499	325	232	312	507	729	910	984
1.736	100	100	299.6	490	602	494	333	249	324	510	725	900
1.852	100	100	120	299	482	580	489	341	265	335	513	721
1.968											346	516
2.083											293	356
2.199											361	305
2.315											466	366
2.431											505	461
2.546											433	494
2.662											306	426
2.778											195	306
2.894											135	199
3.009											111	138
3.125											103	113
3.241											101	104
3.356											100	101
3.472	100	100	100	100	100	100	100	100	100	100	100	100

Fig. E-11.3 Numerical calculation table for floodwave propagation

Solution: First, the grid spacing is $\Delta x = 10K/c = 10{,}000$ m $= 10$ km and the time step $\Delta t = 10K/c^2 = 10{,}000\,\mathrm{s} = 0.116$ day. The algorithm is $Q_j^{k+1} = 0.1\,Q_{j-2}^k + 0.8\,Q_{j-1}^k + 0.1\,Q_j^k$ and we can develop the marching procedure shown in Fig. E-11.3. The boundary conditions are in the first two columns. Note that we offset the upstream boundary condition with a one time-step lag (because $c\Delta t = \Delta x$) in the downstream direction (see the table at $x = -10$ km and $x = 0$). For example, the discharge at successive times where $x = 10$ km is calculated as

At $t = 0.116$ day,

$$Q_{10\,\mathrm{km}}^{0.116\,\mathrm{day}} = (0.1 \times 500) + (0.8 \times 250) + (0.1 \times 100) = 260\ \mathrm{cms},$$

At $t = 0.231$ day,

$$Q_{10\,\mathrm{km}}^{0.231\,\mathrm{day}} = (0.1 \times 750) + (0.8 \times 500) + (0.1 \times 260) = 501\ \mathrm{cms,\ etc.}$$

Case Study 11.1: Dam-break floodwave of the Doce River, Brazil

The collapse of the Fundão tailings dam in Brazil spilled 32 million cubic meters of mine waste, causing a severe socio-economic and environmental impact in the Doce River (Palu and Julien 2019, 2020). The main river parameters are the channel slope $S_f = 0.0005$, channel width $W = 150$ m, Manning's coefficient $n = 0.044$ and Froude number Fr $= 0.17$. The one-dimensional advection–diffusion equation was applied to the discharge measurements at station G6 listed in Problem 11.1 to calculate the flow discharge at station G5, located 73.5 km downstream. Starting from the initial discharge of 60 m^3/s, the peak discharges reached $Q = 871$ m^3/s at G6 and $Q = 704$ m^3/s at G5. The measured wave celerity is $c = 1.17$ m/s. The dispersion coefficient $K \simeq \frac{(1-\mathrm{Fr}^2)Q}{2WS_f}$ varies widely ($400 < K < 5{,}500$ m^2/s) because it depends on a wide range of discharges ($75 < Q < 850$ m^3/s). This traditional approach stimulated research regarding the question of what discharge should be used for the analysis of floodwaves when the discharge varies so rapidly. After considering that the floodwave celerity also has a unique measurable value, despite the fact that celerity does also vary with discharge according to $c = \beta V$, Palu and Julien (2020) demonstrated that the dispersion coefficient can be approximated as

$$K \simeq \left[\frac{1 - 0.444\mathrm{Fr}^2}{2n\sqrt{S_0}}\right]\left[\frac{0.6c}{\sqrt{g}\mathrm{Fr}}\right]^{10/3}.$$

The main advantage of this formulation is that c can be obtained from field observations while n, Fr and S_0 are relatively constant. With the above parameters, the reader finds $K \simeq [502][1.32]^{10/3} = 1{,}260$ m^2/s, which is close to the $K = 975$ m^2/s, best fitting the field observations in Fig. CS-11.1. Interestingly, the diffusion coefficient is better estimated from the low values of discharge rather than based on peak discharges during floods.

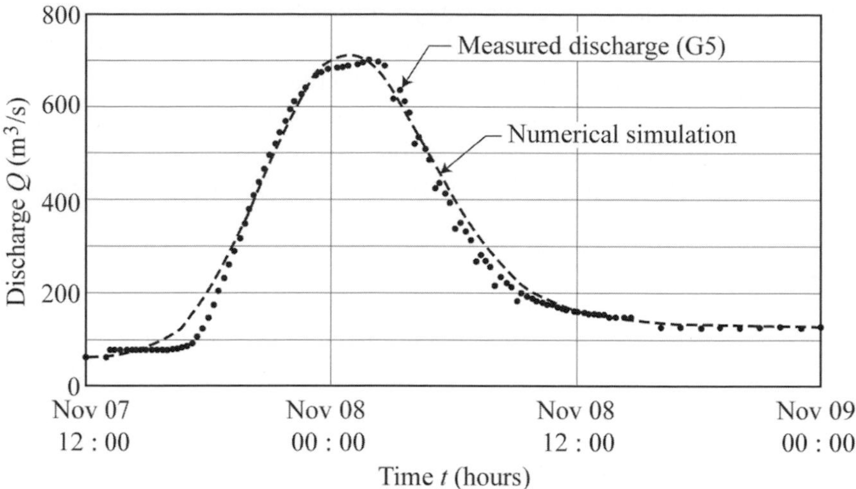

Fig. CS-11.1 Comparison of the discharge calculations and measurements

Additional Resources

Additional information on the topic of unsteady flow is available in Liggett and Cunge (1975), Leonard (1979), Abbott and Basco (1989), Fennema and Chaudhry (1990), Singh (1997), Sturm (2001), Ponce (2014) and Battjes and Labeur (2017). Regarding floodwave propagation, useful references include Woolhiser (1975), Chapra (1997), Chanson (2004), Chaudhry (2008) and Woo et al. (2015). For more details on the analysis of floodwaves from dam break, the reader is referred to Palu and Julien (2019, 2020).

EXERCISES

These exercises review the essential concepts from this chapter.

1. Which three most important equations are combined to define unsteady flow in open channels?
2. What does the Saint-Venant equation describe?
3. What is flow conveyance?
4. How many terms describe the diffusive-wave approximation of the Saint-Venant equation?
5. What physically causes the attenuation of floodwaves?
6. Which of mountain streams or large rivers have a better ability to attenuate floods?
7. What is the difference between the *erf* and *erfc* functions?
8. Why would a numerical scheme have two upstream boundary conditions?
9. What is the difference between physical and numerical diffusion and which should be eliminated?
10. True or false?
    (a) The continuity equation links space and time.
    (b) A kinematic wave does not deform in the downstream direction.
    (c) It is impossible to get a negative diffusion coefficient.

(d) A floodwave moves faster than the river flow.
(e) The celerity increases with flow depth.
(f) Floodwaves attenuate when the flow is subcritical.
(g) Error functions are linked to the Gaussian normal distribution.
(h) Numerical diffusion cannot be avoided.
(i) Resistance to flow causes the attenuation of floodwaves.

PROBLEMS

1. ♦♦ Doce River Flood. With reference to the Doce River in Case Study 11.1, the observed hydrographs measured at stations G6 (located 95 km downstream of the source) and station G5 (73.5 km downstream of G6) are shown in Table P-11.1. Find the following:

Table P-11.1. Discharge at G6 and G5

Station G6		Station G5	
Time (h)	Discharge (m^3/s)	Time (h)	Discharge (m^3/s)
16	59.4	34.5	80.7
17	60.2	35.5	107.3
18	84.4	36.5	174.8
19	187.9	37.5	317.2
20	272.3	38.5	439.1
21	382.2	39.5	520.8
22	460.3	40.5	628.9
23	528.9	41.5	676.3
24	681.3	42.5	686.7
25	792.5	43.5	697.2
26	870.8	44.5	693.7
27	803.5	45.5	639.0
28	717.5	46.5	511.7
29	609.0	47.5	414.3
30	439.1	48.5	351.5
31	351.2	49.5	268.9
32	308.8	50.5	224.4
33	253.9	51.5	200.8
34	223.7	52.5	180.2
35	202.5	53.5	171.2
36	185.0	54.5	165.9
37	178.0	55.5	157.3
38	171.1	56.5	153.9
39	157.6	57.5	153.9
40	147.2	58.5	152.2

(a) The diffusion coefficient K for this range of discharge based on a channel width of 150 m, a Manning coefficient $n = 0.044$, slope of 0.0005 and Froude number $\text{Fr} = 0.17$.

(b) Determine the wave celerity by comparing the peak time at these two stations.

2. ◆◆ With reference to Problem 11.1, consider the hydrograph at the source in Fig. P-11.2 where the discharge increases from 60 m^3/s at $t = 0$ to 1,900 m^3/s in four hours. It then decreases to a constant value of 150 m^3/s three hours after the peak discharge. Use the numerical solution to the advection–diffusion equation to calculate the hydrograph with $200 < K < 6{,}000$ m^2/s. Which value of K provides the best agreement with the field measurements? Plot and compare the results for $K = 200$ m^2/s, $K = 3{,}000$ m^2/s and $K = 6{,}000$ m^2/s.

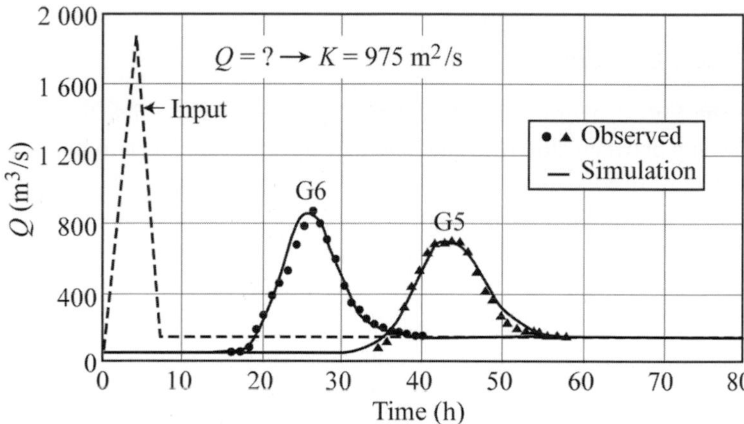

Fig. P-11.2

12 Culverts

Culverts are described in Section 12.1 followed with an analysis of culvert performance curves in Section 12.2 and outlet works in Section 12.3.

12.1 Culvert Characteristics

Originating from the French word *couvert* designating covered conduits, the flow in culverts bridges the gap between closed conduits and open channels. The two main types of culverts are: (1) reinforced concrete pipes (RCPs); and (2) corrugated metal pipes (CMPs). High-density polyethylene (HDPE) pipes are also becoming popular. RCPs have a smooth surface ($n \simeq 0.012$) while CMPs/HDPE pipes with undulated surfaces have a high roughness ($n \simeq 0.024$). The cost per linear foot $C_{\$/\mathrm{ft}}$ is higher for RCPs ($C_{\$/\mathrm{ft}} \sim \$30D_{\mathrm{ft}}^{1.8}$ in 2019, where the pipe diameter D_{ft} in feet is raised to the power 1.8) than CMPs/HDPE pipes ($C_{\$/\mathrm{ft}} \sim \$5D_{\mathrm{ft}}^{1.8}$). Various culvert shapes are shown in Figure 12.1 with concrete circular and box culverts used in urban areas. Other features are listed for HDPE pipes in Table 12.1 and RCPs in Table 12.2.

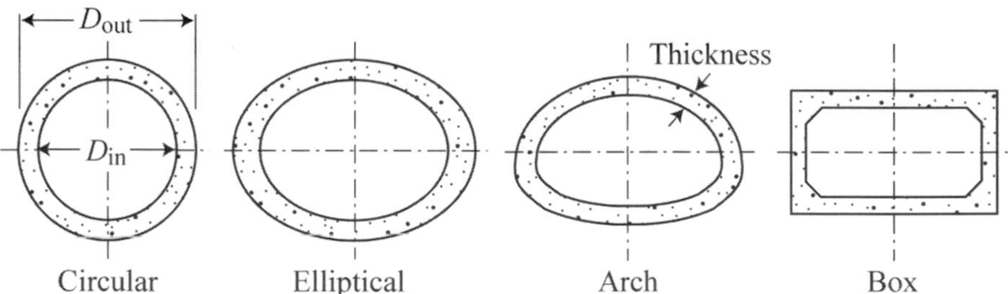

Figure 12.1 Types of culverts

The geometric properties of circular culverts with partial flow are described as a function of the flow depth in Example 8.1. Accordingly, we can calculate the main geometric properties as a function of flow depth as: (1) the angle $\theta = 2\cos^{-1}\left(1 - 2\frac{y}{D}\right)$ where θ is in radians; (2) the flow area $A = 0.5(\theta - \sin\theta)r^2$; (3) the wetted perimeter is $P = \theta r$; and (4) the hydraulic radius is $R_h = \frac{A}{P} = 0.5r\left(1 - \frac{\sin\theta}{\theta}\right)$. When normalized to the value flowing full, we obtain the graphs in Figure 12.2

Table 12.1. Characteristics of HDPE pipes (customary units)

D_{out} (in.)	HDPE 110 psi			HDPE 128 psi			HDPE 200 psi		
	Thickness (in.)	D_{in} (in.)	Weight (lb/ft)	Thickness (in.)	D_{in} (in.)	Weight (lb/ft)	Thickness (in.)	D_{in} (in.)	Weight (lb/ft)
12	0.83	11	13.5	0.94	10.8	15.3	1.4	9.7	22
18	1.16	15.5	26.8	1.33	15.2	30.5	2	13.8	44
24	1.55	20.7	47.7	1.78	20.2	54.2	2.7	18.3	78
30	1.94	25.9	74.6	2.22	25.3	84.7	3.3	22.9	122
36	2.32	31.1	107	2.67	30.3	122			
42	2.71	36.2	146	3.10	35.4	167			
48	3.1	41.4	176	3.56	40.5	218			
54	3.48	46.6	242						

Table 12.2. Characteristics of RCPs (SI units)

D_{in} (mm)	D_{out} (mm)	Thickness (mm)	Mass per m (kg/m)	Truck Load number	In 2019		
					Class 50-D ($/m)	Class 100-D ($/m)	Class 140-D ($/m)
300	445	70	225	74	92	92	92
375	533	76	306	54	113	114	114
450	622	83	381	44	117	117	147
525	711	89	467	35	127	149	176
600	800	95	578	29	172	198	233
675	889	102	689	24	263	301	352
750	978	108	781	21	347	398	465
825	1,067	114	912	18	403	513	539
900	1,156	121	1,039	16	484	554	647
975	1,245	127	1,195	14	531	636	745
1,200	1,511	146	1,561	10	762	916	1,068
1,500	1,829	152	2,123	7	1,199	1,439	1,679
1,800	2,184	178	2,865	5	1,735	2,083	2,431
2,100	2,540	203	3,807	4	2,309	2,770	3,231
2,400	2,896	229	4,869	2	3,071	3,684	4,298
2,700	3,251	254	5,752	2	3,839	4,608	5,376
3,000	3,607	279	7,043	1	4,701	5,645	6,586

12.2 Culvert Performance Curves

The performance curve of a culvert defines the discharge at a given flow depth. We recognize four types of performance curves: barrel control (Section 12.2.1), inlet control (Section 12.2.2), outlet control (Section 12.2.3) and submerged outlets (Section 12.2.4).

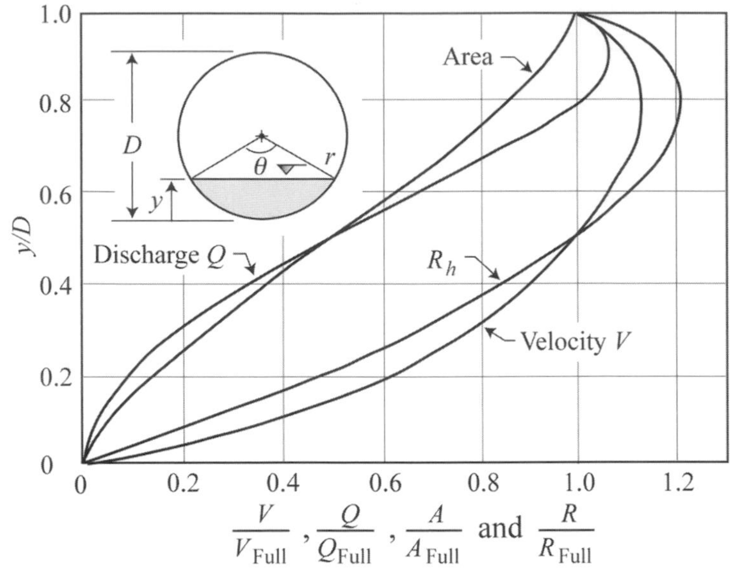

Figure 12.2 Relative area, hydraulic radius, velocity and discharge for culverts

12.2.1 Barrel Control ($H < D$)

As shown in Figure 12.3, the maximum flow in a circular culvert is obtained when the flow depth is slightly below flowing full. To make sure that a culvert will not cause backwater, the maximum discharge is calculated from

$$Q = 0.44\sqrt{g}D^{2.5}, \tag{12.1}$$

which reduces to $Q_{\mathrm{cfs}} = 2.5D_{\mathrm{ft}}^{2.5}$ where Q_{cfs} in ft^3/s and the culvert diameter D_{ft} in ft.

Under very particular conditions (rarely met in practice), the slope of the barrel is perfectly adjusted to maintain a critical flow depth across the entire culvert length. This corresponds to the point of maximum discharge $Q = 0.93\sqrt{g}D^{2.5}$, which only occurs at an optimum slope $S_{\mathrm{opt}} = 111\, n^2/D_{\mathrm{ft}}^{1/3}$. Note that $S_{\mathrm{opt}} \approx 1\%$ for smooth pipes and $4\% < S_{\mathrm{opt}} < 7\%$ for corrugated surfaces.

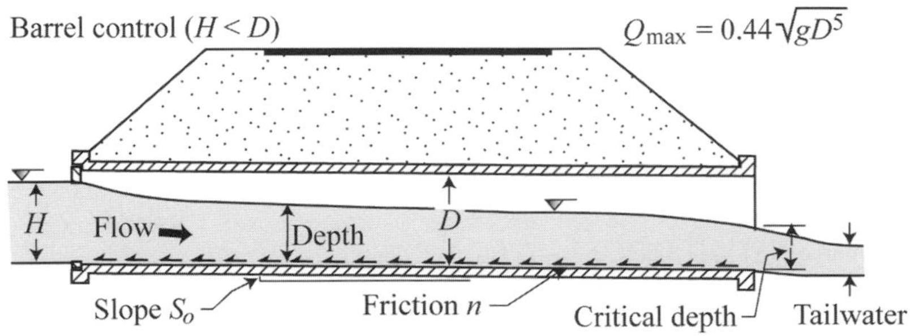

Figure 12.3 Barrel control without causing backwater

In Example 12.1, we seek first approximations of the culvert size required to pass a certain discharge without causing backwater with the formula $D = 1.39g^{-0.2}Q^{0.4}$, which corresponds to $D_{\text{ft}} = 0.69Q_{\text{cfs}}^{0.4}$, or $D_{\text{m}} = 0.89Q_{\text{cms}}^{0.4}$ in SI.

♦♦ **Example 12.1:** Culvert without backwater

What is the culvert size that would pass a discharge of 3 m^3/s without causing backwater?

Solution: The diameter is $D_{\text{m}} = 0.89 \times 3^{0.4} = 1.38$ m, or 54 in.

12.2.2 Inlet Control ($H > D$)

The case of inlet control for culverts with a submerged inlet is the most common case. It is particularly applicable to short and smooth culverts, e.g. RCPs. Various inlet types are sketched in Figure 12.4. In general, streamlined entrances will increase the flow discharge at a given upstream water level.

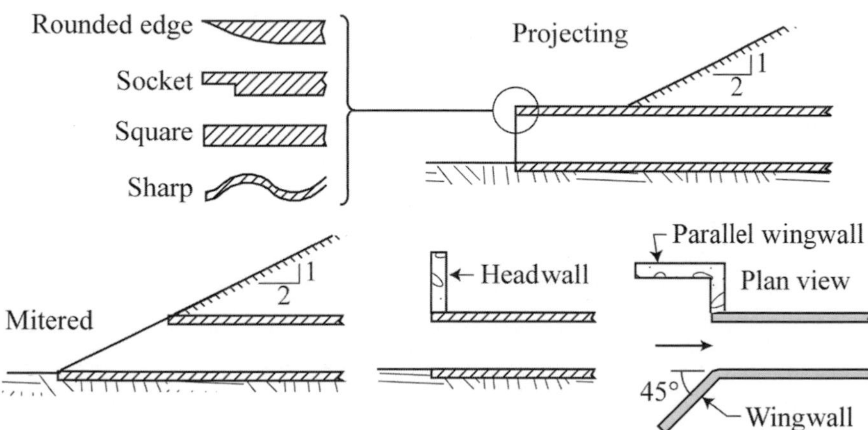

Figure 12.4 Various types of culvert inlets

As sketched in Figure 12.5, inlet control has a free surface inside the pipe.

Two parameters (h_1/D and k_1 depending on the inlet type in Table 12.3) are needed to calculate the relationship between the water level H and flow discharge:

$$\frac{H}{D} + 0.5S_0 = \frac{h_1}{D} + k_1\frac{Q^2}{gD^5}. \qquad (12.2)$$

The relationship between H and Q is called a culvert performance curve.

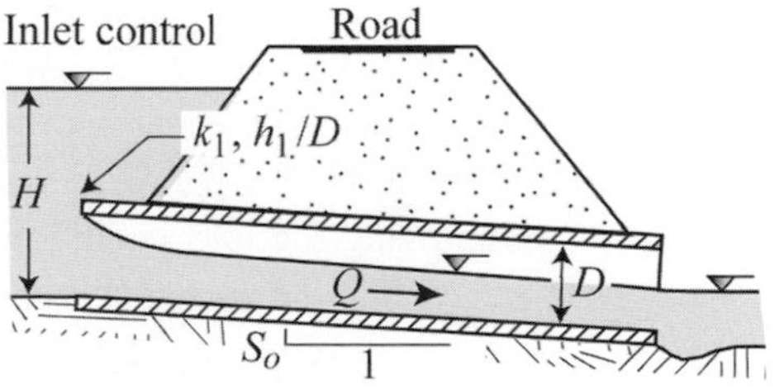

Figure 12.5 Inlet control for culverts

Table 12.3. Inlet-control coefficients

Entrance type		h_1/D	k_1	K_E
Headwall				
	Round	0.74	1.35	0.10
	Socket	0.74	1.51	0.19
	Square	0.67	2.08	0.43
Headwall and 45-degree wingwalls				
	Socket	0.73	1.52	0.20
	Square	0.7	1.91	0.35
Headwall and parallel wingwalls				
	Socket	0.74	1.70	0.30
Mitered entrance				
	2H:1V	0.74	2.42	0.62
Projecting entrance				
	Socket	0.7	1.66	0.25
	Square edge	0.64	2.15	0.46
	Sharp edge	0.53	2.98	0.92

Example 12.2 illustrates how to calculate the performance curve for inlet control.

♦♦ Example 12.2: Culvert with inlet control

Part I. Calculate, plot and compare the dimensionless performance curves $H/D = f\left(Q/\sqrt{gD^5}\right)$ for a culvert at a 1% slope under inlet control given: (1) an RCP with a headwall and a rounded entrance; and (2) a CMP with a projecting sharp edge.

Part II. Calculate the discharge in m^3/s for a 1-m-diameter pipe at a slope of 1% for these two culvert conditions when the water level is 5 m above the culvert invert.

Solution:

Part I. For a culvert at a slope $S_0 = 0.01$ from Eq. (12.2) for an RCP with headwall and rounded entrance (coefficients $h_1/D = 0.74, k_1 = 1.35$ from Table 12.3), $\frac{H}{D} = 0.74 - 0.005 + 1.35\frac{Q^2}{gD^5}$. For the case of maximum roughness with a projecting CMP with a sharp entrance $(h_1/D = 0.53, k_1 = 2.98)$, $\frac{H}{D} = 0.53 - 0.005 + 2.98\frac{Q^2}{gD^5}$. Both curves are plotted on Fig. E-12.2. These represent the extreme conditions for different culvert types. At a given water level, the figure shows how much of an increase in discharge can be gained by improving the inlet conditions. When the culvert is flowing full ($H = D$), the figure also shows that the discharge is approximately $Q = 0.44\sqrt{gD^5}$, as expected from Eq. (12.1).

Part II. When $H = 5$ m and $D = 1$ m, the corresponding discharges are calculated from

$$Q = \sqrt{\left[\left(\frac{H}{D} + 0.5S_0 - \frac{h_1}{D}\right)\frac{1}{k_1}\right]}\sqrt{gD^5},$$

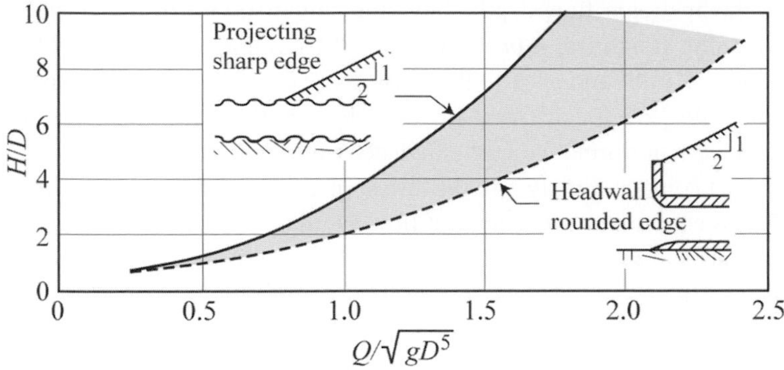

Fig. E-12.2 Culvert performance curve for inlet control

resulting in $Q = 1.77\sqrt{9.81 \times 1^5} = 5.54\ \text{m}^3/\text{s}$ for the rounded-edge pipe with a headwall, and $Q = 1.22\sqrt{9.81 \times 1^5} = 3.82\ \text{m}^3/\text{s}$ for the projecting sharp-edged pipe. Both discharges exceed the discharge that would not cause backwater from Eq. (12.1), i.e. $Q = 0.44\sqrt{9.81 \times 1^5} = 1.38\ \text{m}^3/\text{s}$.

12.2.3 Outlet Control ($H > D$)

Outlet control occurs when the pipe is long and/or when Manning's coefficient n is high, e.g. CMPs. For outlet control, the pipe is flowing full except at the downstream end. There are two main components in the analysis of outlet control: the inlet losses depending on the entrance condition and the friction losses along the barrel. For outlet control, the culvert is flowing full and the coefficients K_E are listed in Table 12.3. The energy losses in the culvert with outlet control are sketched in Figure 12.6.

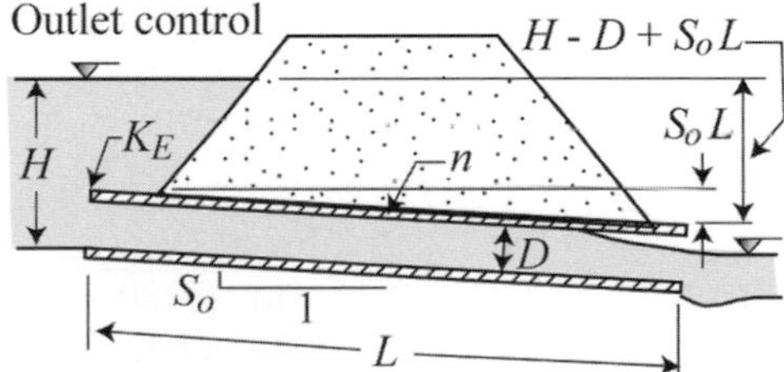

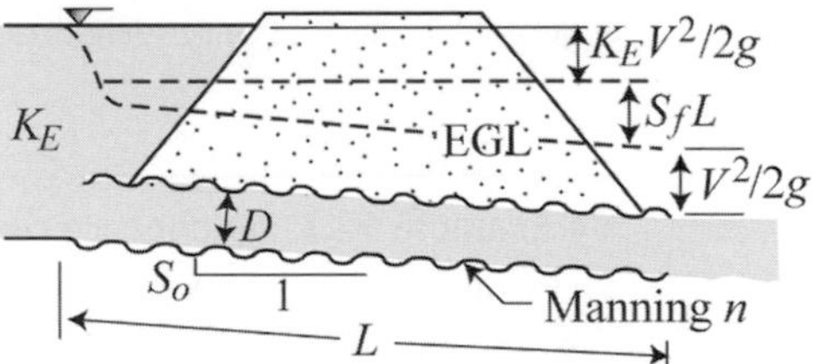

Figure 12.6 Culvert with outlet control

The available energy between both ends of the culvert is $H - D + S_0 L$. The velocity head is simply $V^2/2g = (8/g\pi^2)Q^2/D^4$ and the friction losses from Manning's equation in the culvert flowing full are $S_f L = 10.3\left(\frac{n^2 L}{m^2 D^{4/3}}\right)\frac{Q^2}{D^4}$.

The main equation for culverts with outlet control is therefore rearranged in dimensionless form after dividing all terms by the culvert diameter:

$$\frac{H}{D} - 1 + \frac{S_0 L}{D} = \left[(1 + K_E)\frac{8}{g\pi^2} + \frac{10.3 n^2 L}{m^2 D^{4/3}}\right]\frac{Q^2}{D^5}, \tag{12.3}$$

where H is the upstream flow depth above the culvert invert of diameter D, length L, slope S_0, Manning roughness n ($n = 0.012$ for RCPs and $n = 0.024$ for CMPs) and also considering $m = 1$ in SI and $m = 1.49$ in customary units. The entrance coefficients K_E are listed in Table 12.3 for different inlet conditions. Example 12.3 illustrates how to establish a culvert performance curve and determine whether the flow is under inlet control or outlet control. Simply put, the lowest discharge at a given flow depth controls the flow. This is equivalent to the highest flow depth at a given discharge. The effects of culvert length are shown in Figure 12.7a, with a short pipe ($L = 100$ ft) usually under inlet control, because the curve for inlet control is to the left of the curve for outlet control. For the long culvert, $L = 400$ ft, we get inlet control when $H < 10$ ft and outlet control when $H > 10$ ft. Similarly, on Figure 12.7b, this RCP is always under inlet control, but the CMP ($n = 0.024$) would turn to outlet control at a flow depth above 7 ft.

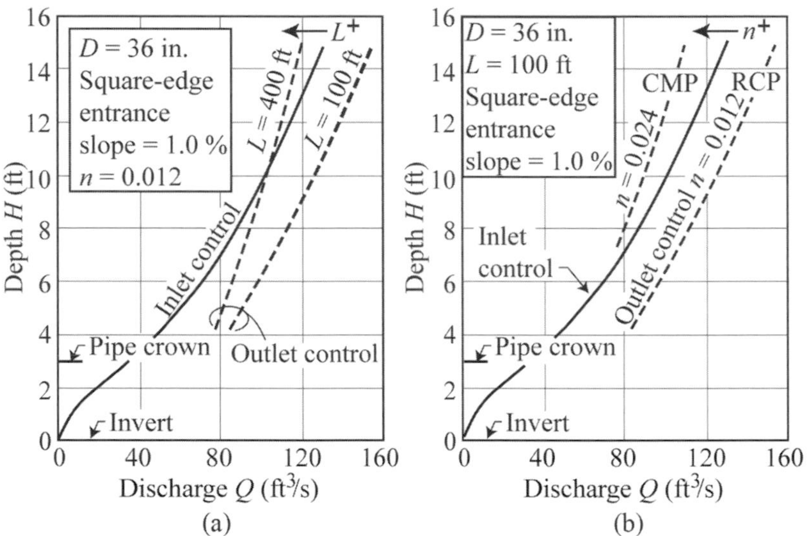

Figure 12.7 Effects of culvert length and roughness on inlet/outlet control

---

♦♦ Example 12.3: Performance curves with inlet/outlet control

Develop a culvert curve for a 200-ft-long 48-in. CMP with a sharp-edged projecting entrance at a 2% slope. Use the performance curve to find the range of discharges under inlet and outlet control.

Solution: In this case, $L = 200$ ft, $D = 48$ in. $= 4$ ft at a slope $S_0 = 0.02$, with $n = 0.024$. We have a sharp-edged projecting entrance ($h_1/D = 0.53, k_1 = 2.98, K_E = 0.92$).

For inlet control in customary units from Eq. (12.2),

$$\frac{H}{D} + 0.5S_0 = \frac{h_1}{D} + k_1 \frac{Q^2}{gD^5},$$

which gives $H_{\text{ft}} = 2.08 + 0.000362Q^2_{\text{cfs}}$.

For outlet control, from Eq. (12.3), with $g = 32.2$ ft/s^2 and $m = 1.49$ in customary units

$$\frac{H}{D} - 1 + \frac{S_0 L}{D} = \left[0.0252(1 + K_E) + \frac{4.63 n^2 L_{\text{ft}}}{D_{\text{ft}}^{4/3}}\right] \frac{Q_{\text{cfs}}^2}{D_{\text{ft}}^5}$$

or

$$H_{\text{ft}} = 0.000517 Q_{\text{cfs}}^2.$$

As shown in Fig. E-12.3, we have inlet control when $H < 7$ ft and outlet control when $H > 7$ ft. The change in control occurs at a discharge Q of 116 ft^3/s.

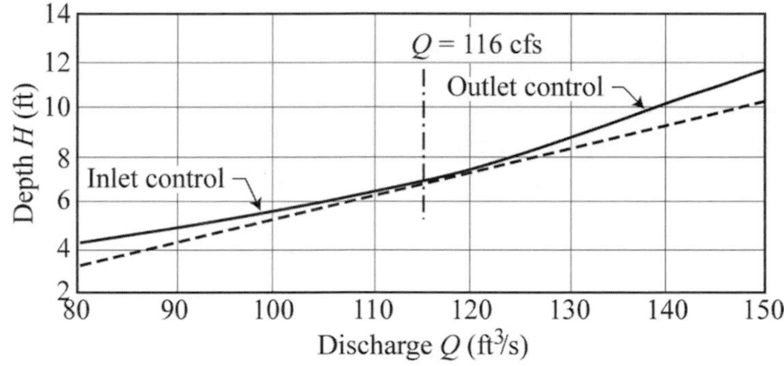

Fig. E-12.3 Example of culvert performance curve with inlet and outlet control

12.2.4 Submerged Outlets

This is a particular case where the flow submerges both ends of the culvert as sketched in Figure 12.8. This situation is common during high tides in coastal areas and can also happen when draining fields into flooded areas. The flow discharge depends on entrance and friction losses as described for outlet control except that the elevation difference is ΔH; thus,

$$\frac{\Delta H}{D} = \left[(1 + K_E)\frac{8}{g\pi^2} + \frac{10.3 n^2 L}{m^2 D^{4/3}}\right] \frac{Q^2}{D^5}. \qquad (12.4)$$

Figure 12.9 shows three main approaches to improve the performance curve of culverts with inlet control: (1) beveling the entrance with rounded edges and avoiding sharp edges; (2) broader sides with wingwalls for side-tapered inlets at angles between 6:1 and 4:1; and (3) steepened entrance slope up to 3:1 or 2:1 with a drop less than a quarter of the main barrel diameter for slope-tapered inlets.

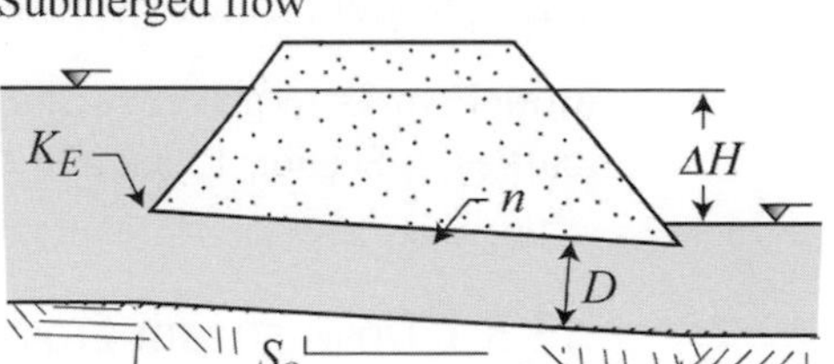

Figure 12.8 Culvert with submerged outlet

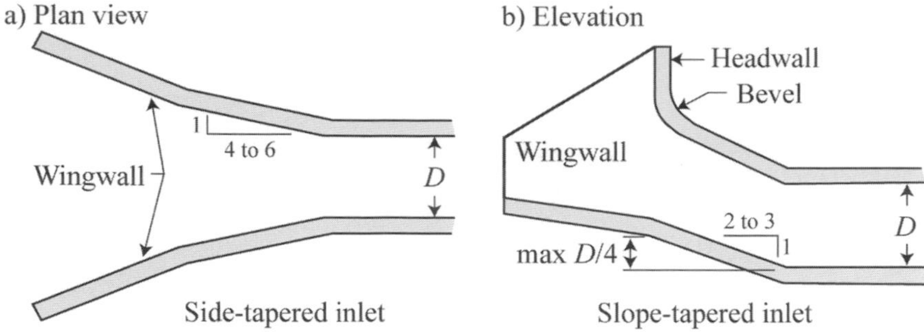

Figure 12.9 Improved culvert inlets

Example 12.4 calculates the culvert performance for submerged flow conditions.

♦ **Example 12.4:** Culverts with submerged outlets
Develop performance curves for a 1-m-diameter and 100-m-long CMP culvert with a projecting sharp edge in a coastal area when the head differential is between 1 m and 2 m.

Solution: For this case, $L = 100$ m, $D = 1$ m, $n = 0.024$ for a CMP, $K_E = 0.92$ from Table 12.3, $g = 9.81\ \text{m/s}^2$ and $m = 1$ in SI. The calculations from Eq. (12.4) reduce to:

$$\Delta H = \left[(1 + 0.92)\frac{8}{9.81\pi^2} + \left(10.3 \times 0.024^2 \times 100\right)\right]Q^2 = 0.752Q^2$$

or

$$Q = 1.15\sqrt{\Delta H}.$$

Accordingly, the discharge becomes $Q = 1.15\ \text{m}^3/\text{s}$ when $\Delta H = 1$ m, and $Q = 1.62\ \text{m}^3/\text{s}$ when $\Delta H = 2$ m.

12.3 Culvert Outlet Works

Outlet works to prevent scour are described in Section 12.3.1, energy dissipation with baffles in Section 12.3.2 and the bearing capacity of pipes in Section 12.3.3.

12.3.1 Downstream Erosion Control

The outflow of clear water downstream of culverts can form large scour holes. In the design of outlet structures, the use of large rocks called riprap is very common in engineering practice, as sketched in Figure 12.10.

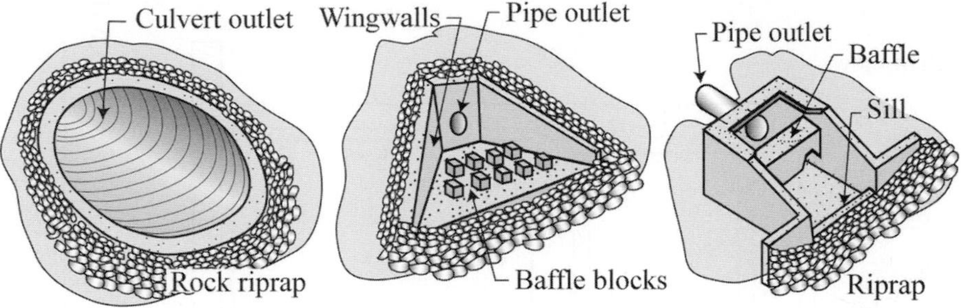

Figure 12.10 Erosion control at culvert outlets

In the design of channels downstream of culvert outlets, the maximum permissible velocity to prevent significant erosion can be determined from Table 12.4.

Table 12.4. Maximum permissible velocities for canals at flow depth < 1 m

Soil type	Manning n	Clear water, no detritus	Water transporting colloidal silt	Water transporting silts, sands/gravels
Stiff clay (very colloidal)	0.025	1.14	1.52	0.91
Alluvial silt when colloidal	0.025	1.14	1.52	0.91
Alluvial silt when noncolloidal	0.02	0.61	1.07	0.61
Volcanic ash	0.02	0.76	1.07	0.61
Silt loam (noncolloidal)	0.02	0.61	0.91	0.61
Ordinary firm loam	0.02	0.76	1.07	0.69
Sandy loam (noncolloidal)	0.02	0.53	0.76	0.61
Fine sand (colloidal)	0.02	0.46	0.76	0.46
Fine gravel	0.02	0.76	1.52	1.14
Coarse gravel (noncolloidal)	0.025	1.22	1.83	1.98
Cobbles and shingles	0.035	1.52	1.68	1.98
Shales and hardpans	0.025	1.83	1.83	1.52

When the kinetic energy in the culvert is high, the design of an outlet box with baffles will dissipate the excess energy. The design of a culvert pipe outlet with a baffle and stilling basin is shown in Figure 12.11, from Young (1983a and 1983b) and Hernandez (1984), with approximate dimensions in Table 12.5.

12.3.2 Trenches and Pipe-Bearing Capacity

Loading factors L_F of culverts and pipes can be defined from the type of bedding in trenches shown in Figure 12.12. Concrete beddings with compacted backfill (Class A) are

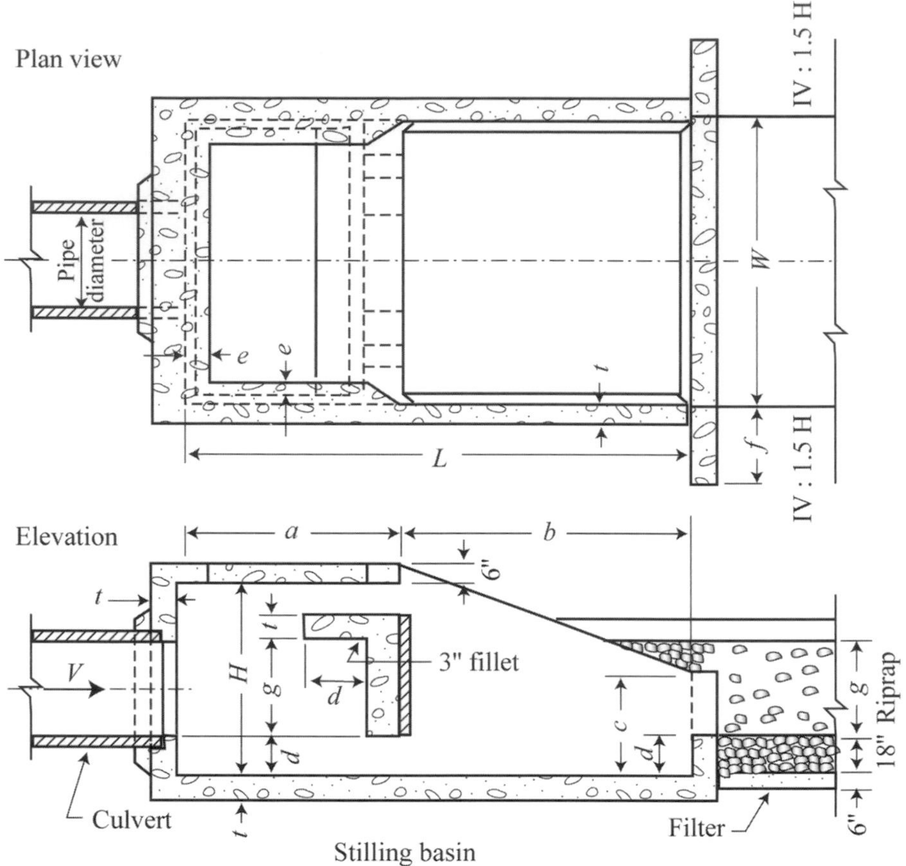

Figure 12.11 Culvert stilling basin with baffle

Table 12.5. Culvert outlet stilling basin (modified after Hernandez 1984)

Pipe diam. D (in.)	Flow ratc Q (cfs)	W (ft-in.)	H (ft-in.)	L (ft-in.)	a (ft-in.)	b (ft-in.)	c (ft-in.)	d (ft-in.)	e (ft-in.)	f (ft-in.)	g (ft-in.)	T (in.)	Riprap diam. D_s (in.)
18	21	5-6	4-3	7-4	3-3	4-1	2-4	0-11	0-6	1-6	2-1	6	4
24	38	6-9	5-3	9-0	3-11	5-1	2-10	1-2	0-6	2-0	2-6	6	7
30	59	8-0	6-3	10-8	4-7	5-1	3-4	1-4	0-8	2-6	3-0	7	9
36	85	9-3	7-3	12-4	5-3	7-1	3-10	1-7	0-8	3-0	3-6	8	9
42	115	10-6	8-0	14-0	6-0	8-0	4-5	1-9	0-10	3-0	3-11	9	10
48	151	11-9	9-0	15-8	6-9	8-11	4-11	2-0	0-10	3-0	4-5	10	11
54	191	13-0	9-9	17-4	7-4	10-0	5-5	2-2	1-0	3-0	4-11	10	12
60	236	14-3	10-9	19-0	8-0	11-0	5-11	2-5	1-0	3-0	5-4	11	13
72	339	16-6	12-3	22-0	9-3	12-9	6-11	2-9	1-3	3-0	6-2	12	14

most suitable with $L_F = 2.2$, while Class D is simply not recommended (Spangler 1962). The three-edged bearing capacity of concrete pipes (ASTM C76) is summarized in Table 12.6 as a function of the pipe diameter D.

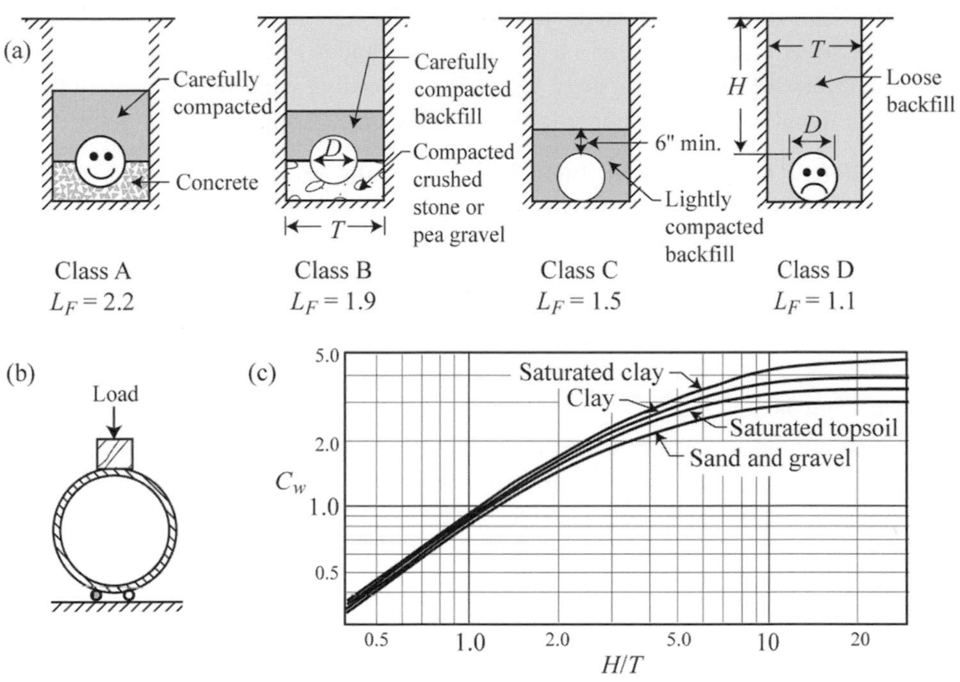

Figure 12.12 (a) Trenches, (b) three-edge bearing test (b) and (c) weight coefficient

Table 12.6. Three-edged bearing strengths of concrete pipes (in 1,000 lb/ft)

D/in. (cm)	Nonreinforced class			Reinforced class			
	I	II	III	II	III	IV	V
12 (30)	1.8	2.25	2.6	1.5	2.0	3.0	3.75
18 (45)	2.2	3.0	3.3	2.25	3.0	4.5	5.62
24 (61)	2.6	3.6	4.4	3.0	4.0	6.0	7.50
30 (76)	3.0	4.3	4.75	3.75	5.0	7.5	9.38
36 (91)	3.3	4.5	5.0	4.5	6.0	9.0	11.25
48 (122)				6.0	8.0	12.0	15.00
60 (152)				7.5	10.0	15.0	18.75
72 (183)				9.0	12.0	18.0	22.50

Note: multiply table numbers by 0.0146 for values in kN/m.

The applied external load W on a buried pipe is calculated from the soil specific weight γ_{soil}, the pipe diameter D, trench width T and the weight coefficient C_W from Figure 12.12c. The Marston formulas are $W = C_W \gamma_{soil} T^2$ for rigid pipes and $W = C_W \gamma_{soil} TD$ for flexible pipes.

Example 12.5 shows basic pipe-trench calculations. The three-edge bearing strength S_3 of a pipe is $S_3 = WS_F/L_F$, where S_F is the safety factor.

Example 12.5: Concrete pipe strength in a trench

An 18-in. sewer pipe is buried in a 3-ft-wide trench under 15 ft of a clay at 120 lb/ft^3. Find the three-edge bearing capacity for a Class C bedding considering a safety factor $S_F = 1.25$. Would a Class II nonreinforced concrete pipe be appropriate?

Solution: Here, $T = 3$ ft, $H = 15$ ft and $H/T = 5$. Figure 12.12c gives $C_W = 3$ and the applied weight is $W = C_W \gamma_{soil} T^2 = 3 \times 120 \times 3^2 = 3{,}240$ lb/ft. For Class C bedding (Figure. 12.12a), the load factor is $L_F = 1.5$. The minimum three-edge bearing strength of the pipe $S_3 = WS_F/L_F = 3{,}240 \times 1.25/1.5 = 2{,}700$ lb/ft is the minimum required and a nonreinforced Class II pipe would work.

Additional Resources

Additional information can be found about culverts in PCA (1964), urban hydraulics in Mays (2019) and channel stabilization with riprap in USDA (2007), ASCE (2008, 2013), FHWA (2009, 2012), USBR (2015), USBR–ASCE (2015).

EXERCISES

These exercises review the essential concepts from this chapter.

1. What are the two main culvert types?
2. What is the most common culvert shape?
3. What are the four main types of controls on culvert performance curves?
4. What flow type does not cause backwater?
5. Which inlet type causes maximum energy loss?
6. Why does the culvert slope affect the performance under inlet control?
7. What are the two main sources of energy loss in outlet control?
8. What is the lowest point of a culvert inlet called?
9. When do you expect a submerged outlet to control the flow?
10. True or false?
    (a) The maximum flow through a culvert is always when the culvert is flowing full.
    (b) Headwalls help reduce the entrance losses of culverts.
    (c) The entrance coefficient is the sole factor affecting inlet control.
    (d) Barrel losses for outlet control are calculated using Manning's formula.
    (e) RCPs are most likely to flow under outlet control.
    (f) Long culverts are more likely to flow under outlet control.
    (g) When comparing inlet and outlet performance curves the line with maximum discharge at a given stage determines the control.
    (h) Submerged outlets always increase the performance of a culvert.
    (i) Culverts are best placed flat at the bottom of a trench.
    (j) Riprap is the name used for baffles at culvert outlets.

SEARCHING THE WEB

Can you find more information about the following?

1. ♦ RCPs, HDPE pipes and CMPs.
2. ♦ Circular, elliptical, arch and box culverts.
3. ♦ Culverts with mitered inlets, wingwalls and headwalls.
4. ♦ Culvert outlets and downstream erosion control.

PROBLEMS

1. ♦ Calculate the discharge in m^3/s flowing through a 1-m-diameter culvert without causing backwater.
2. ♦ Calculate the maximum discharge in m^3/s flowing through a 1-m-diameter culvert under critical flow conditions.
3. ♦ In the 1.2-m-diameter circular pipe in Fig. P-12.3, the water level is $y = 0.36d$. Determine the flow discharge if the flow velocity is 2 m/s.
4. ♦ In a circular pipe, flow is allowed up to 85% of the maximum depth. Determine the required pipe diameter to carry a flow rate of 2.25 m^3/s if the flow velocity is 0.8 m/s.
5. ♦♦ A 100-ft-long, 36-in. concrete culvert at a 1% slope has a sharp-edged projecting entrance. At an upstream flow depth of 5 ft, find the flow discharge, inlet or outlet control.
6. ♦♦ Recalculate Problem 12.5 for a headwall and rounded entrance. Would the discharge increase at a 5 ft flow depth?

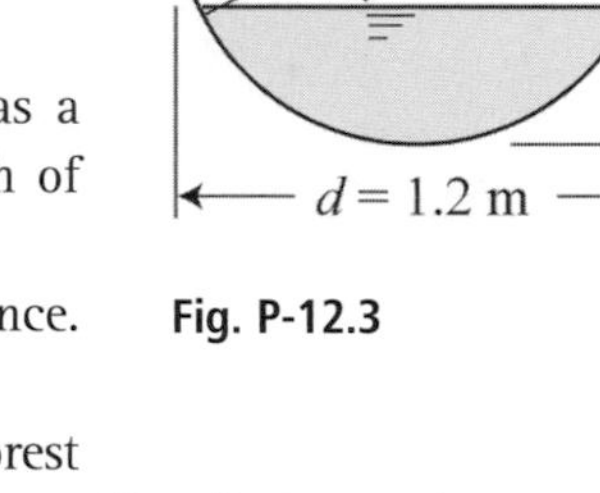

Fig. P-12.3

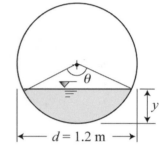

7. ♦♦♦ **Design problem.** Design a culvert across a 20-ft-wide forest road over a stream with a 50-hectare drainage basin. The specific discharge is 500 cfs per square mile. The culvert invert is 15 ft below the road surface and the embankment slopes are 2H:1V. Suggest the culvert type and size, the Manning coefficient n and slope *with a sketch*. Plot the performance curve and find at what flow depth will the culvert operate for this design discharge.
8. ♦♦♦ **Design problem.** Determine the culvert capacity curve for a 500-ft-long concrete culvert under a highway. The use of a square edge with a 45° wingwall at the entrance is considered. You need to keep the upstream water level lower than 8 ft above the culvert invert and the design discharge is 75 cfs. The soil type of the area is a silt loam with 10% clay. Sketch an appropriate outlet energy dissipation structure if the level drops 3 ft under the highway. Provide a sketch for your proposed design.
9. ♦ Reconsider Problem 12.7. Design a culvert across a forest road. If you have measurements after a flood that the upstream flow depth reached 8 ft above the culvert invert, can you find the maximum discharge in that culvert?
10. ♦ Recalculate Problem 12.8, with a 500-ft-long, 36-in. concrete culvert under a highway with a square edge and 45° wingwall. At the same upstream flow depth, would the discharge increase in the culvert if you design a very nice rounded-edge entrance?

Fig. P-12.11

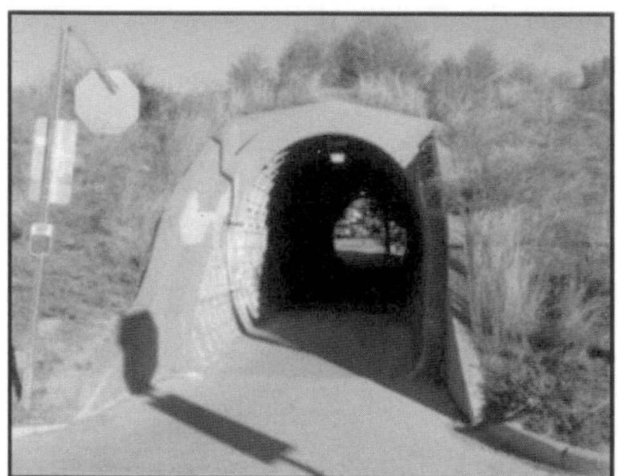

Fig. P-12.12

11. ♦♦♦ Fort Collins Flood 1997. Examine the design of the culvert flows for Spring Creek under the Burlington–Northern Santa Fe (BNSF) railroad track in Fort Collins (Fig. P-12.11). As a first approximation, consider a 100 ft length and 1% slope with 3 ft of freeboard for three 36-in. concrete culverts with the invert 12 ft below the track level – assume all three at the same level. During the Fort Collins Flood in 1997, the flow overtopped the embankment. Estimate the flow discharge for that condition.

12. ♦♦♦ Fort Collins Flood 1997. Examine the design of the culvert flows for Spring Creek under the BNSF railroad track in Fort Collins (Fig. P-12.12). As a first approximation, consider the 100-ft-long and 10-ft-diameter circular CMP at a 1% slope with the invert 12 ft below the track level. During the Fort Collins Flood in 1997, the flow overtopped the railroad track. Estimate the flow discharge for that condition.

13 Spillways and Gates

Hydraulic structures provide human-made control of flow depth and discharge. This chapter provides a broad overview of this complex topic with a few calculation examples for two main types of structures: spillways in Section 13.1 and gates in Section 13.2.

13.1 Spillways

The two main types of spillways are morning glory (Section 13.1.1) and ogee (Section 13.1.2). Gates can also be used to control the spillway discharge (Section 13.1.3). Downstream erosion can be prevented with flip buckets and plunge pools (Section 13.1.4) and energy dissipators with stilling basins (Section 13.1.5). Finally, we will look at stepped spillways (Section 13.1.6) and give a detailed design example for a baffle-chute drop structure (Section 13.1.7).

13.1.1 Morning Glory Spillways

The name of the morning glory spillway stems from its circular funnel shape like the flower. The circular intake is streamlined and control gates can be placed at the crest. The advantage of the morning glory spillway is that it can be located upstream of the dam inside the reservoir. The water can be conveyed in a pipe further away from the base of the dam. The performance curve of spillways in Figure 13.1 resembles culvert performance curves with inlet and outlet control. The inlet and outlet control with $Q \propto \sqrt{H}$ can be a disadvantage at high stages.

13.1.2 Ogee Spillways

The word "ogee" originates from the French word *ogive* (meaning a rounded or arch-like profile), and ogee spillways are used for large dams. The spillway shape matches the profile of free-flowing water at the design discharge. The design discharge Q_D and the spillway length L define the unit discharge $q = Q_D/L$. The critical depth y_c is two-thirds of the head H above the spillway and the critical velocity is $V_c = \sqrt{gy_c}$. The velocity and area give the discharge $Q = AV = Ly\sqrt{gy} = L\sqrt{g}y^{3/2}$. Since the flow depth varies on the spillway, the fixed design head H_D designating the energy grade line (EGL) above the spillway crest is used with a design discharge coefficient C_D to give the design discharge Q_D as

$$Q_D = C_D\sqrt{2g}LH_D^{3/2}, \qquad (13.1)$$

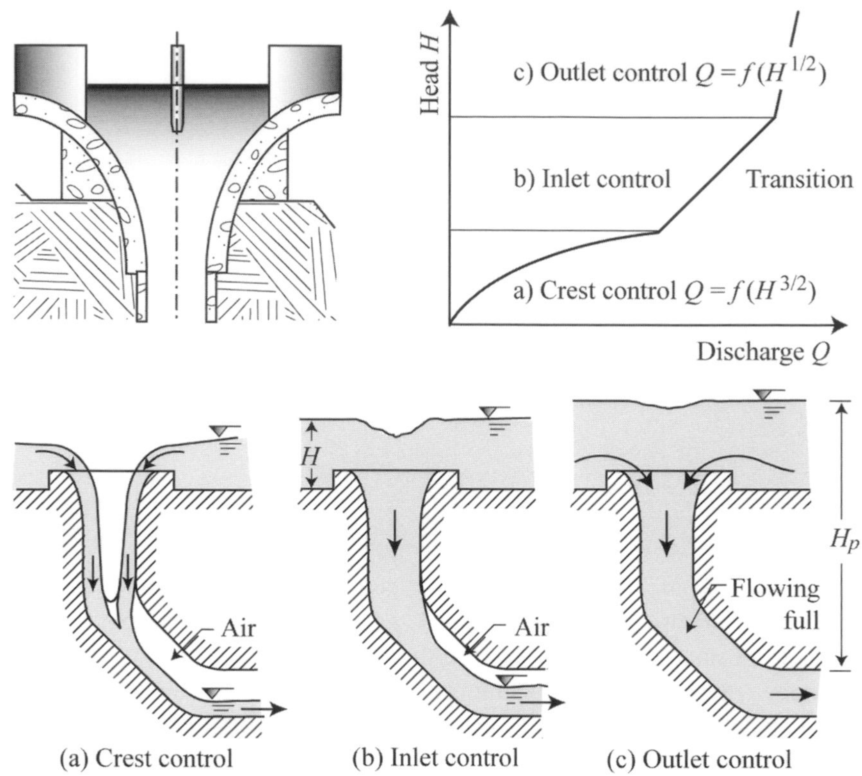

Figure 13.1 Morning glory spillway

where the design discharge coefficient C_D depends on the reservoir flow depth H_D above and P below the crest as shown in Figure 13.2. The value of $C_D = 0.492$ is reached for deep reservoirs ($P >> H_D$). The flow discharge at any head H is calculated in Example 13.1 from $Q = C\sqrt{2g}LH^{3/2}$, with the value of C depending on the ratio H/H_D.

♦♦Example 13.1: Ogee spillway rating curve

Define the rating curve for a 100-ft-wide ogee spillway at a crest elevation of 1,700 ft with an 80-ft-deep large reservoir with a design discharge of 35,300 cfs. Also, can you explain why the discharge coefficient increases with discharge?

Solution: The design head is obtained from Eq. (13.1) with $C_D = 0.492$ using $P/H_D = 80/20 = 4$ in Figure 13.2:

$$H_D = \left(\frac{Q}{C_D L\sqrt{2g}}\right)^{2/3} = \left(\frac{35{,}300}{0.492 \times 100\sqrt{2 \times 32.2}}\right)^{2/3} = 20\,\text{ft}.$$

The flow discharge for various flow depth ratios H/H_D is calculated with C from Figure 13.2:

$$Q_D = C\sqrt{2g}LH^{3/2} = C\sqrt{2 \times 32.2} \times 100\, H^{3/2}. \qquad (13.2)$$

The rating curve calculated in Table E-13.1 is also plotted in Fig. E-13.1

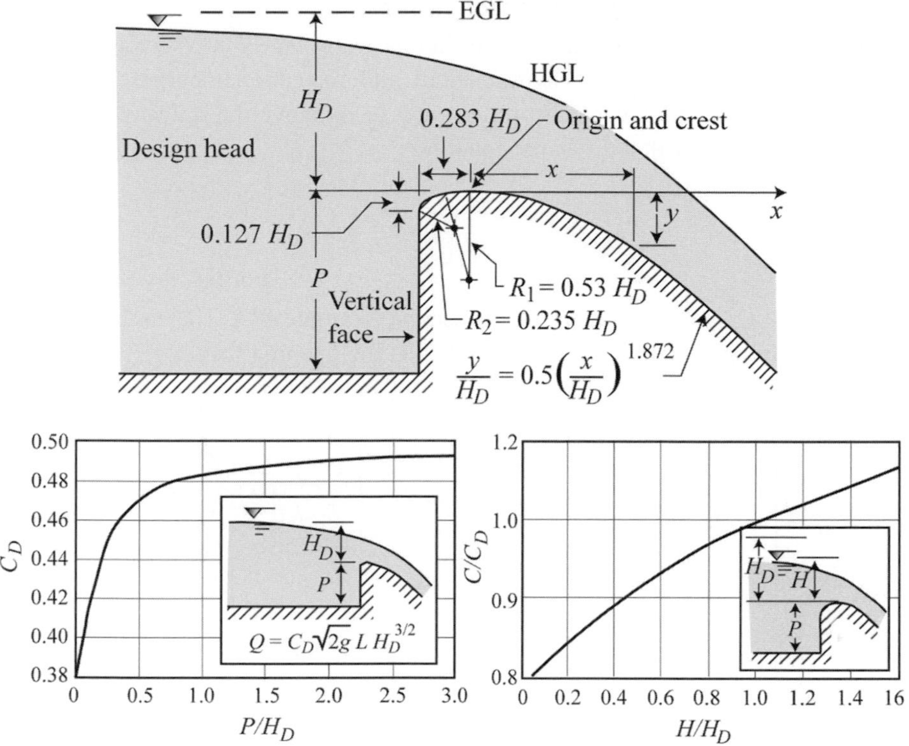

Figure 13.2 Ogee spillway shape and discharge coefficients (USBR 1977)

Table E-13.1. Ogee spillway curve

Elevation (ft)	Head H (ft)	H/H_D	C/C_D	C	Q (ft^3/s)
1,700	0	0	0.8	0.394	0
1,705	5	0.25	0.85	0.418	4,414
1,710	10	0.5	0.91	0.448	12,486
1,715	15	0.75	0.96	0.472	22,937
1,720	20	1	1	0.492	35,314
1,725	25	1.25	1.04	0.512	49,354
1,730	30	1.5	1.06	0.522	64,877

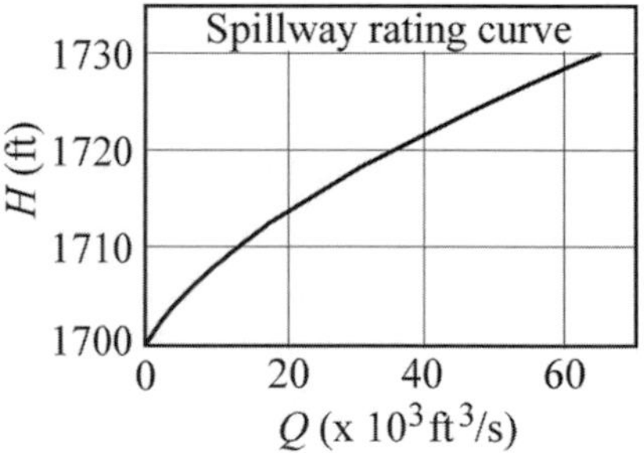

Fig. E-13.1 Spillway rating curve

In the second part, the reason why the discharge coefficient increases beyond the design value ($C > C_D$) when ($Q > Q_D$) can be explained by the negative pressure distribution that develops on the spillway crest when the discharge is larger than the design discharge. Naturally, the streamlines at higher discharges would be higher than the streamline of lower discharges. Therefore, a negative pressure (suction) develops on the crest of the spillway when $Q > Q_D$, as discussed in Section 9.3.5. This negative pressure distribution increases the flow rate near the spillway crest. Conversely, a positive pressure distribution develops when $Q < Q_D$ and this decreases the discharge coefficient $C < C_D$ accordingly.

13.1.3 Ogee Spillways with Control Gates

Spillways with control gates can maintain higher water levels in reservoirs to increase hydropower production or for environmental and recreational purposes. The rating curve of the spillway then depends on the water level above the spillway crest H_1 and the gate opening H_2 with a discharge equation:

$$Q = \frac{2}{3}\sqrt{2g}CL\left(H_1^{3/2} - H_2^{3/2}\right).$$

The coefficient C plotted in Figure 13.3 is used in Example 13.2. The discharge can also be estimated from $Q \simeq 0.72Ld\sqrt{2gH_1}$ where d is the opening height.

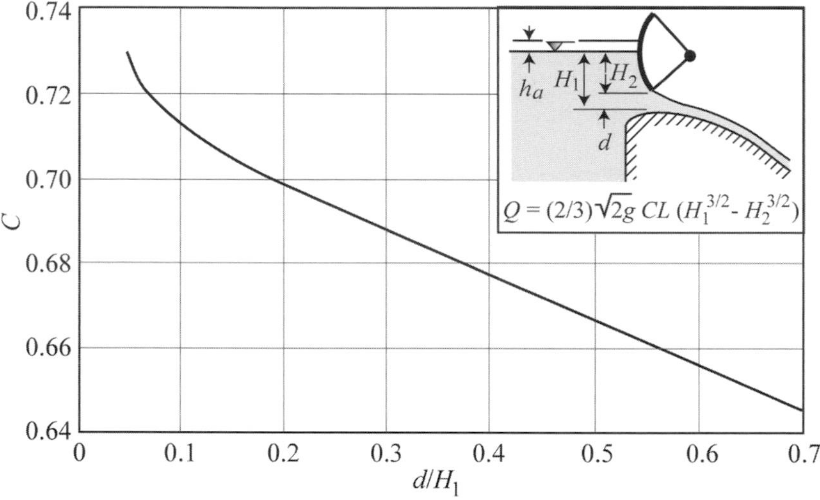

Figure 13.3 Discharge coefficient for spillways with gates

♦♦ **Example 13.2:** Spillway with control gates

The water level in a deep reservoir with control gates is maintained at an elevation 10 m above the ogee spillway crest. What is the flow discharge over the 35-m-wide spillway if the opening at the bottom of the gate is 1 m?

Solution: The flow discharge is obtained from the following formula, with $H_1 = 10$ m, $d = 1$ m, $H_2 = 9$ m, $d/H_1 = 1/10 = 0.1$, and from Figure 13.3 we get $C = 0.715$, thus giving

$$Q = \frac{2}{3}\sqrt{2g}CL\left(H_1^{3/2} - H_2^{3/2}\right) = \frac{2}{3}\sqrt{2 \times 9.81} \times 0.715 \times 35\left(10^{3/2} - 9^{3/2}\right) = 341 \text{ m}^3/\text{s}.$$

Alternatively, $Q \simeq 0.72Ld\sqrt{2gH_1} = 0.72 \times 35\sqrt{2 \times 9.81 \times 10} = 353 \text{ m}^3/\text{s}$.
The difference between these two methods is less than 4%.

13.1.4 Flip-Bucket Spillways with Plunge Pools

The extreme water velocities at the base of a very high and smooth spillway chute can cause significant erosion at the toe of the dam and threaten the stability of the entire structure. As sketched in Figure 13.4, the concept behind the flip bucket is to push the fast-flowing waters as far away as possible from the spillway toe.

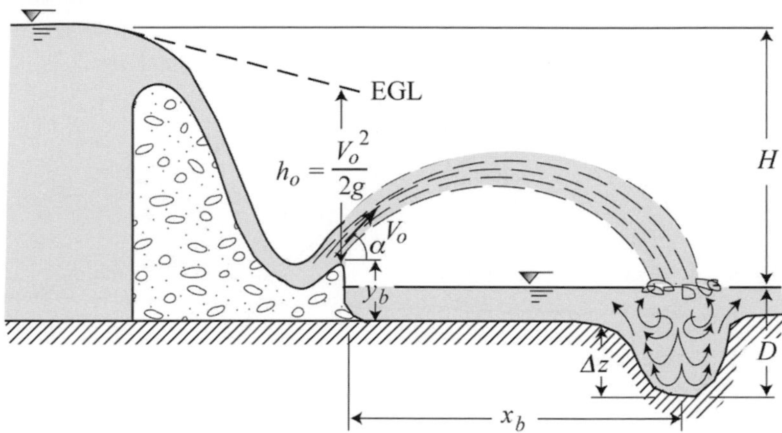

Figure 13.4 Flip-bucket spillway and plunge pool scour

Like a ski jump, the horizontal downstream distance x_b from the launching point to the point of impact on the riverbed is computed from

$$\frac{x_b}{h_0} = \sin(2\alpha) + 2\cos(\alpha)\sqrt{\frac{y_b}{h_0} + \sin^2(\alpha)}, \tag{13.3}$$

where y_b is the height of the launching point above the bed, h_0 is the velocity head at the launching point and α is the angle of the jet at the launching point with the horizontal. For the case where $y_b = 0$, this equation simplifies to $x_b = 2h_0 \sin(2\alpha)$, with a maximum downstream distance reached when $\alpha = 45°$.

Among many formulas, the scour depth can be estimated from the Veronese equation,

$$D_{\mathrm{m}} = 1.9 H_{\mathrm{m}}^{0.225} q_{\mathrm{m^2/s}}^{0.54}, \tag{13.4}$$

where the scour depth D in m includes the tailwater level, H in m is the elevation drop from the reservoir to the tailwater and q in $\mathrm{m^2/s}$ is the unit discharge, with calculations in Example 13.3.

For low head structures, or when the jet is submerged, a complex flow configuration in the plunge pool can develop as sketched in Figure 13.5. A hydraulic jump close to the structure is a great source of concern as the high flow velocities can cause significant scour. In some cases, slotted-bucket spillways can break up the jet and induce a hydraulic jump close to the structure, and therefore slotted-bucket spillways can only be considered for small structures and low Froude numbers.

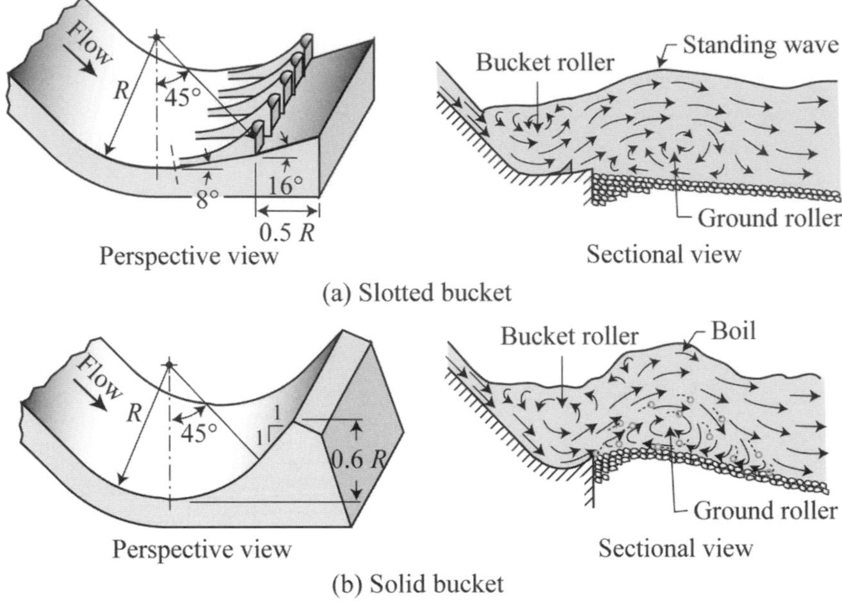

Figure 13.5 Submerged bucket spillways (USBR 1976, 1977)

♦♦ **Example 13.3:** Flip-bucket spillways

A flip-bucket spillway launches a 35-cm-thick water jet at 30 m/s starting at a 4-m elevation above the riverbed at an angle of 45°. The reservoir level located 80 m above the river is maintained at an elevation 10 m above the ogee spillway crest. If the water depth in the river is 2 m, estimate the depth and location of the scour hole.

Solution: Consider, $y_b = 4$ m, $V_0 = 30$ m/s, $h_0 = V_o^2/2g = 30^2/2 \times 9.81 = 46$ m, $H = 80$ m and $q = V_0 h_0 = 30 \times 0.35 = 10.5$ m²/s.

The horizontal distance of the scour hole is located $x_b = h_0[\sin(90°) + 2\cos(45°)\sqrt{\frac{4}{46} + \sin^2(45°)}] = 2.09 \times 46 = 96$ m away from the launching point.

From the Veronese equation, the total scour depth from the free surface in the river is $D = 1.9 H_{\text{m}}^{0.225} q_{\text{m}^2/\text{s}}^{0.54} = 1.9 \times 80^{0.225} 10.5^{0.54} = 18.1$ m.

Given the flow depth of 2 m, the expected scour depth would reach $\Delta z = 16$ m below the riverbed elevation unless bedrock is reached closer to the riverbed.

13.1.5 Energy Dissipation through Stilling Basins

Standard stilling basins were developed by the US Bureau of Reclamation (USBR 1977, 1983) to dissipate energy below drop structures and high dams. As sketched in Figure 13.6, there are three useful stilling basin designs (Type II, III and IV) using the dimensions D_1 and D_2 from the upstream and downstream flow depths, while w and s refer to the width and spacing of chute and baffle blocks. Most dimensions are functions of the upstream Froude number Fr_1.

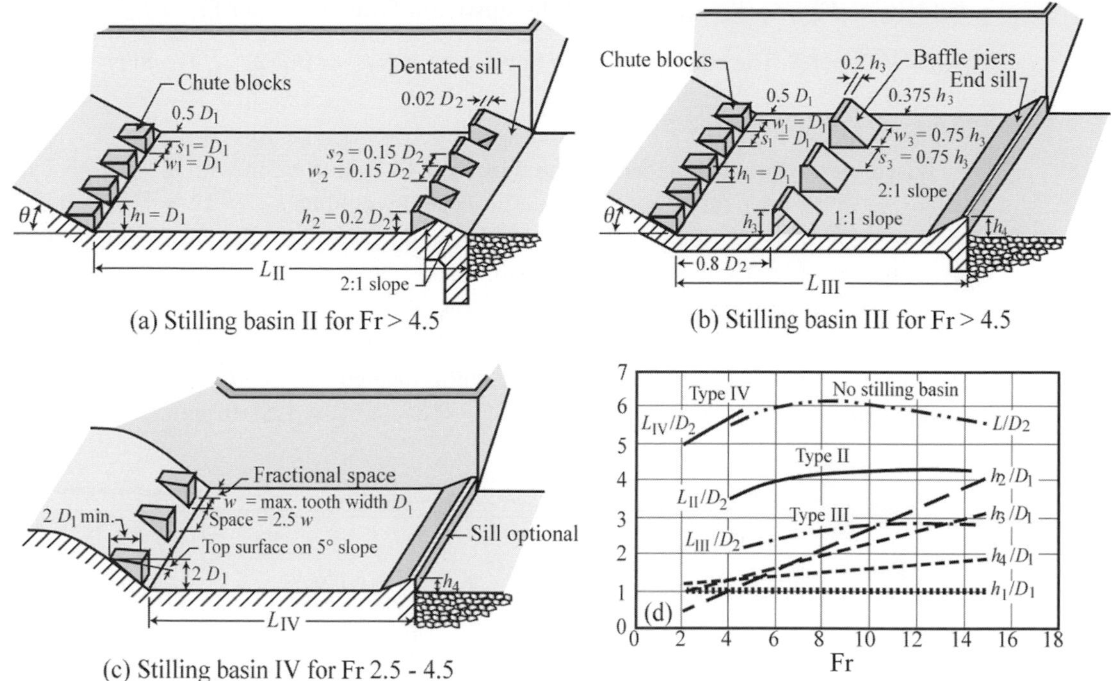

Figure 13.6 Stilling basins (USBR 1976, 1977)

Canal structures and low head diversion dams ($2.5 < \text{Fr}_1 < 4.5$) use Type IV stilling basins with length $L_{IV} = D_2[5.2 + 0.4\,(\text{Fr}_1 - 2.5)]$ close to the length of regular hydraulic jumps.

For large spillways ($\text{Fr}_1 > 4.5$), the Type II stilling basin has a length $L_{II} = D_2[4.0 + 0.055\,(\text{Fr}_1 - 4.5)]$, with a maximum length $L_{II\,max} = 4.35D_2$. Type III stilling basins are limited to flow velocities less than 50–60 ft/s (15–18 m/s). They were developed to reduce the length by adding baffle blocks and forcing the hydraulic jump as far upstream as possible. The length of Type III stilling basins is $L_{III} = D_2[2.4 + 0.073\,(\text{Fr}_1 - 4.5)]$, with a maximum $L_{III\,max} = 2.8D_2$. The Type III basin dimensions are: (1) chute block height $h_1 = D_1$; (2) dentated sill height $h_2 = 0.2D_2$; (3) baffle pier height $h_3 = D_1[1.3 + 0.164\,(\text{Fr}_1 - 4.0)]$; and (4) end sill height $h_4 = D_1[1.25 + 0.056\,(\text{Fr}_1 - 4.0)]$. The dimensions of Type III stilling basins are illustrated in Example 13.4

♦♦ Example 13.4: Stilling-basin dimensions

Select the stilling basin type and find appropriate dimensions to dissipate the energy for a discharge of 3,000 cfs on a long and smooth 60-ft-wide spillway at a 17% slope.

Solution: Here, the unit discharge is $q = Q/W = 3{,}000/60 = 50$ ft²/s and the critical depth is $h_c = (q^2/g)^{1/3} = (50^2/32.2)^{1/3} = 4.27$ ft. The normal depth for a smooth surface ($n = 0.012$) is obtained from $q = (1.49/n)h_n[Wh_n/(W + 2h_n)]^{2/3}S^{1/2} = (1.49/0.012)h_n$ $[60h_n/(60 + 2h_n)]^{2/3}0.17^{1/2} = 50$ ft²/s for which a solver gives $h_n = 1$ ft. The upstream velocity is $V_n = q/h_n = 50$ ft/s and the upstream Froude number $\mathrm{Fr}_1 = V_n/\sqrt{gh_n} = 50/\sqrt{32.2} = 8.8$. The conjugate depth downstream is $h_2 = (h_1/2)\left(\sqrt{1+8\mathrm{Fr}_1^2} - 1\right) = 0.5\left(-1 + \sqrt{1+8\times 8.8^2}\right) = 12$ ft.

Based on the Froude number, a Type II or Type III stilling basin with $h_1 = D_1 = 1$ ft, $D_2 = h_2 = 12$ ft would be appropriate. The Type II would be $L_{II} = 12 \times [4.0 + 0.055\,(8.8 - 4.5)] = 51$-ft long without baffles and with a dentated sill height $h_2 = 0.2 \times 12 = 2.4$ ft at the downstream end.

A Type III stilling basin with baffle blocks sketched in Fig. E-13.4 would be significantly shorter: $L_{III} = 12 \times [2.4 + 0.073(8.8 - 4.5)] = 32.5$ ft, with baffles $h_3 = 1 \times [1.3 + 0.164\,(8.8 - 4.0)] = 2.1$-ft high located $X = 0.8D_2 = 9.6$ ft between blocks. The end sill is $h_4 = 1 \times [1.25 + 0.056\,(8.8 - 4.0)] = 1.51$-ft high.

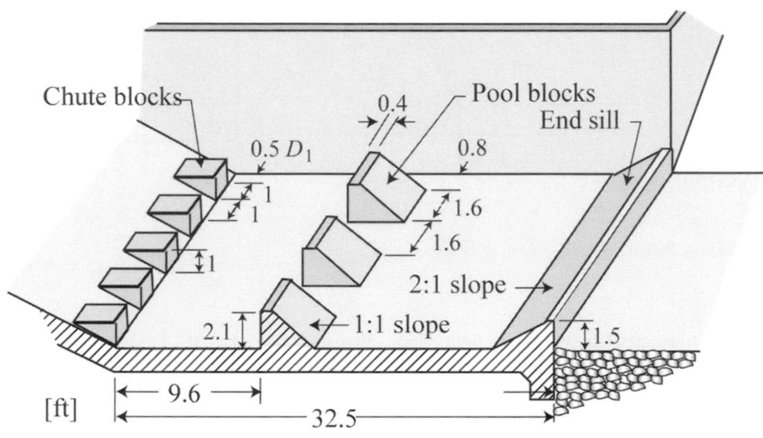

Fig. E-13.4 Type III stilling basin

13.1.6 Stepped Spillways

Stepped spillways are designed to increase surface roughness, aerate the flow and dissipate as much energy as possible on the face of the spillway (Chanson 2004). Figure 13.7 shows a recent physical model of a stepped spillway from C. Thornton and R. Ettema at the Colorado State University (CSU) Hydraulics Laboratory. The amount of air entrainment is paramount to alleviate the negative pressure and prevent cavitation under high velocities on spillway faces. Velocity profiles are examined in Case Study 13.1.

Figure 13.7 CSU stepped spillway

♦♦ **Case Study 13.1:** Stepped spillway

In the 1990s, a large physical stepped spillway model was designed by J. Ruff at CSU. It had a total length of 30.5 m and height 15.2 m, comprising 25 steps (height = 0.61 m and length = 1.22 m) on a 1V:2H slope. At a model discharge of 2.83 m^3/s, the measured velocity profiles (Ward 2002) reached 15 m/s. More recently, Kositgittiwong et al. (2012 and 2013) carried out computational fluid dynamics (CFD) simulations with five different turbulence models: the standard $\kappa-\varepsilon$, the realizable $\kappa-\varepsilon$, the renormalized group $\kappa-\varepsilon$, the standard $\kappa-\omega$ and the shear-stress transport $\kappa-\omega$ models. Details of these complex turbulence models are found in Kositgittiwong et al. (2013), and Fig. CS-13.1 shows that the calculated profiles compare very well with these large-scale velocity measurements.

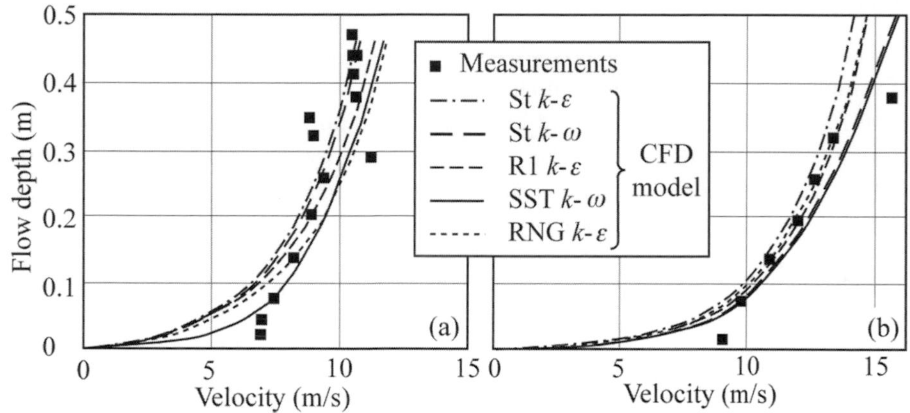

Fig. CS-13.1 Velocity profiles on a stepped spillway (Ward 2002 and Kositgittiwong et al. 2012)

13.1.7 Baffle Chutes

As sketched in Figure 13.8, baffle chutes can also be designed to dissipate a lot of energy on very steep spillways. Their design uses baffle blocks which break the flow to dissipate

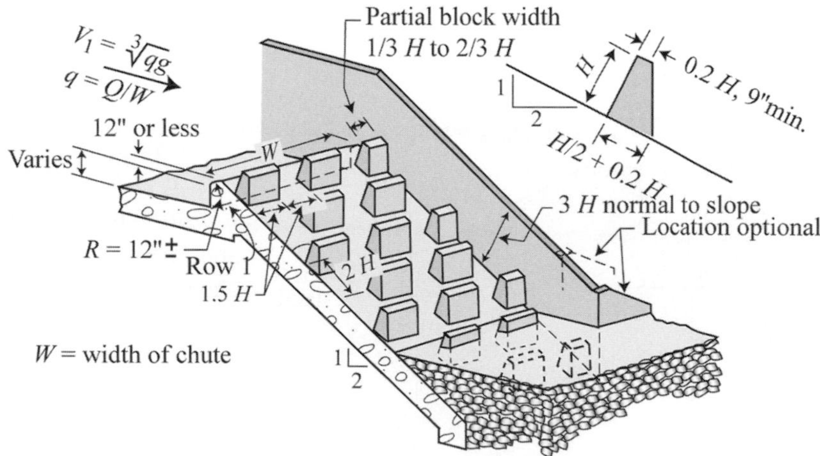

Figure 13.8 Baffle-chute sketch

Table 13.1. Typical size of baffle chutes

H (ft)	q (ft^2/s)	N	w_b (ft)	W (ft)	Q (cfs)
1	6.6	2–4	1–1.5	4–12	25–80
1.5	12.2	2–3	1.5–2	6–12	75–150
2	19	2–3	2–3	8–18	150–340
2.5	26	2–3	2.5–3.5	10–21	260–540
3	35	2–3	3–4.5	12–27	420–940

energy. Typical scales for baffle chutes are listed in Table 13.1, with design details in Example 13.5.

♦♦ Example 13.5: Baffle-chute dimensions

Hayes (1983) spells out the details of baffle-chute design with a discharge $Q = 120$ ft^3/s and an $F = 6$-ft elevation drop. The flow leaves a trapezoidal channel 8 ft at the base and 1.5H:1V side slope. The channel slope is 0.00035, Manning's coefficient n is 0.025, with an approach flow depth of 4.1 ft.

Solution: This design details: (1) the baffles; (2) the entrance; (3) the crest; (4) the outlet; and (5) others.

(1) To dimension the baffles, the unit discharge over an 8-ft-wide chute is $q = 120/8 = 15$ cfs/ft.

The corresponding critical depth is $h_c = (q^2/g)^{1/3} = (15^2/32.2)^{1/3} = 1.91$ ft, and the block height is $h_b = 0.9h_c = 0.9 \times 1.91 = 1.72$ ft. The baffle width is $h_b < w_b < 1.5h_b$, and $w_b = 2$ ft is selected with $N = 2$ baffles per row (plus one spacing and two side halves) for a total chute width $W = 4w_b = 8$ ft. The minimum baffle thickness is $T = 9$ in. $= 0.75$ ft. The minimum spacing between each row is 6 ft or $2h_b$, i.e. $S = 6$ ft. The chute slope is 2H:1V, i.e. the angle $\phi = 26.5°$.

(2) To dimension the entrance in Fig. E-13.5a, the upstream conditions include the flow depth $d_1 = 4.1$ ft, area $A_1 = 58$ ft^2, velocity $V_1 = Q/A_1 = 120/58 = 2.08$ ft/s with velocity head $V_1^2/2g = 2.08^2/2 \times 32.2 = 0.067$ ft and energy level $E_1 = 4.1 + 0.07 = 4.17$ ft over an entrance length $L_1 = 2d_1 = 8.2$ ft.

The flow over the sill is critical at $h_c = 1.91$ ft with $V_c^2/2g = 0.5\ h_c = 0.97$ ft and $E_c = 1.91 + 0.97 = 2.9$ ft. The energy loss over the sill is $\Delta E = 0.5(V_c^2 - V_1^2)/2g = 0.5(0.97 - 0.07) = 0.45$ ft. The step height is therefore $h_s = E_1 - E_c - \Delta E = 4.17 - (2.92 + 0.45) = 0.8$ ft.

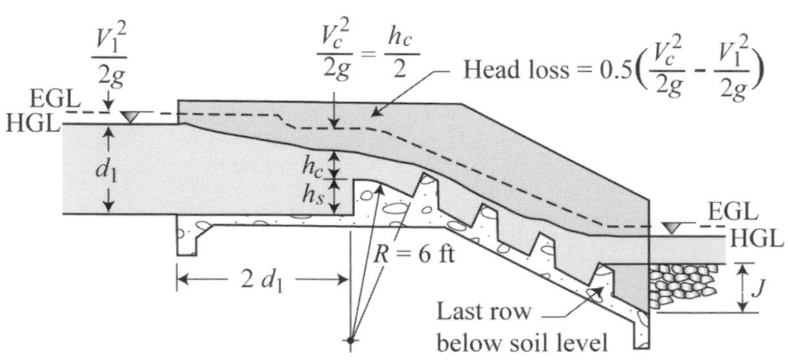

Fig. E-13.5a Baffle-chute intake

(3) To dimension the crest of the baffle chute in Fig. E-13.5b, the distances x, y and z are calculated with a radius $R = 6$ ft such that $z = R\tan(\phi/2) = 6\tan(26.5°/2) = 1.41$ ft, $y = R(1 - \cos\phi) = 6(1 - \cos 26.5°) = 0.63$ ft and $x = y/\tan\phi = 0.63 \times 2 = 1.26$ ft, and the baffle will be placed at an elevation $e = h_s - y = 0.8 - 0.63 = 0.17$ ft above the channel floor.

(4) At the downstream end, the last row of baffles needs to be below the downstream channel floor and, therefore, $J = S\sin\phi + h_b\cos\phi = 4.25$ ft. Given the elevation drop F between upstream and downstream levels, the structural drop in elevation is $L_y = e + F + J = 0.2 + 6 + 4.25 = 10.45$ ft. The elevation drop from each row of baffles is $S_y = S\sin\phi = 6\sin 26.5° = 2.68$ ft. The minimum number of rows is $L_y/S_y = 10.45/2.68 = 3.9$, and we need four rows of baffles. Note that a minimum of four rows is always required.

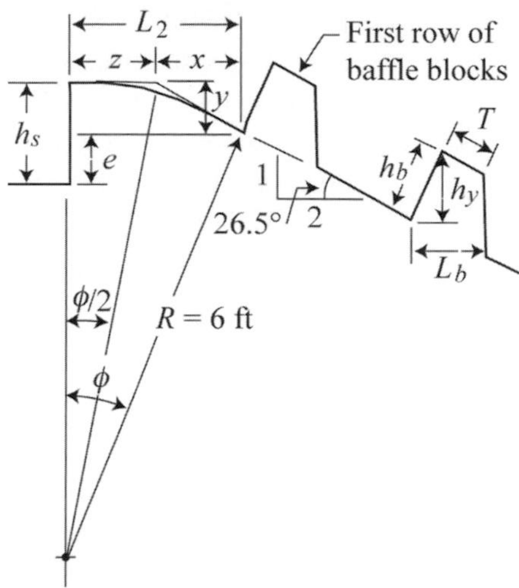

Fig. E-13.5b Baffle-chute entrance

(5) In Fig. E-13.5c, other dimensions include the upstream wall height $h_1 = d_1 + 1 = 5.1$ ft, $h_2 = h_1 - h_s = 5.1 - 0.8 = 4.3$ ft and $h_3 = 3h_b = 3 \times 1.72 = 5.16$ ft, or $h_3^{'} = h_3\cos\phi = 5.77$ ft. The cutoff depths are selected as $C_1 = C_3 = 2.5$ ft.

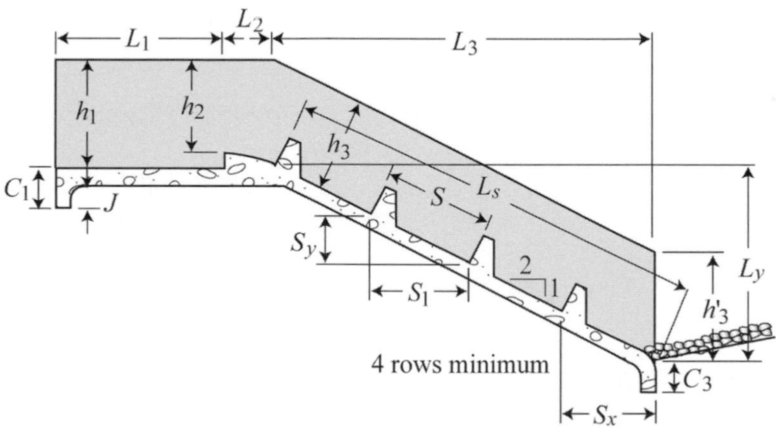

Fig. E-13.5c Baffle-chute profile

13.2 Gates and Weirs

Flow control structures include gates and weirs. Gates (Section 13.2.1) typically raise the water level while allowing the flow under the structure. Weirs (Section 13.2.2) choke the flow and elevate surface flow conditions at a critical flow depth above the structure. Parshall flumes (Section 13.2.3) measure the flow discharge in agricultural areas.

13.2.1 Gates

Gates block the flow and control the upstream water level. As sketched in Figure 13.9, sluice gates slide vertically and open at the bottom with an opening height w. Sluice gates are very commonly used at the intake of irrigation canals. To find the flow discharge, the specific energy is the same on both sides of the gate $y_1 + (V_1^2/2g) = y_2 + (V_2^2/2g)$. Due to the vena contracta, the downstream flow depth is less than the opening height w and the contraction coefficient C_c is used as $y_2 = C_c w$. This equation is solved for the unit discharge as

$$
\begin{aligned}
q &= C_c w \sqrt{\frac{y_1}{y_1 + y_2}} \sqrt{2gy_1}, \\
q &\simeq 0.6 w \sqrt{2gy_1}.
\end{aligned} \tag{13.5}
$$

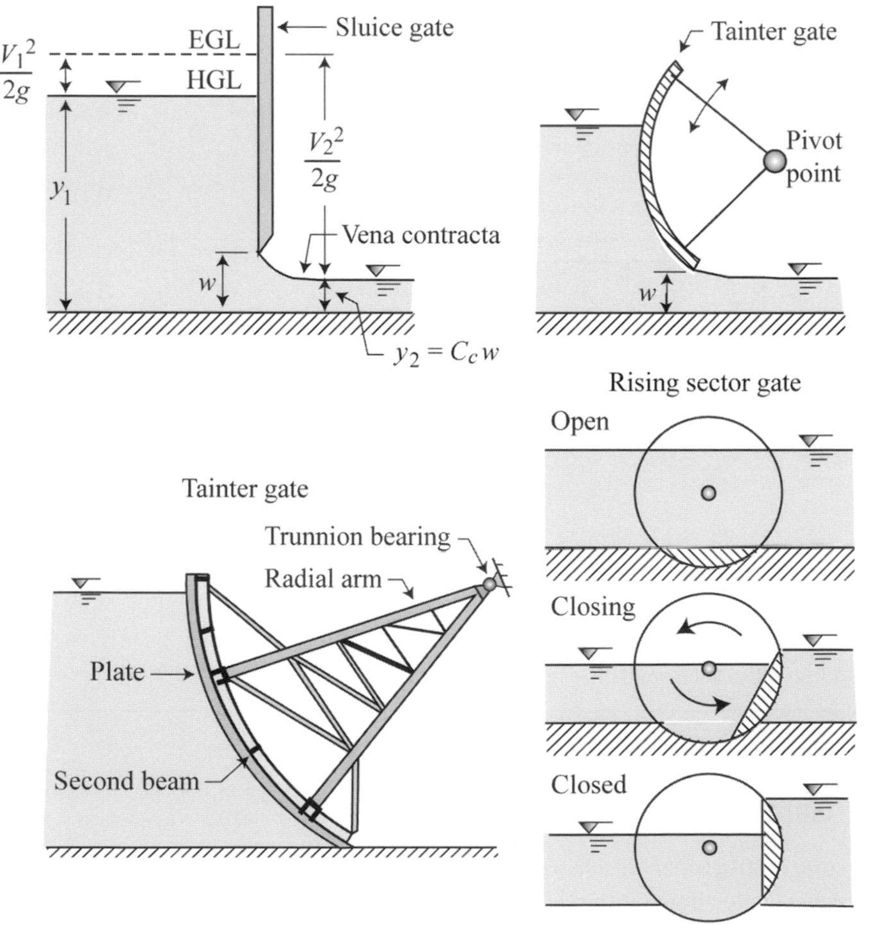

Figure 13.9 Sluice gate, Tainter gate and rising sector gate

Tainter gates, named after Wisconsin engineer Jeremiah Tainter, are circular and allow flow underneath to be better streamlined than for vertical sluice gates. They are often used on larger rivers and in navigable waterways.

Rising sector gates rotate around a pivot point on the sides of the gates. The water can flow above the rising sector, and they effectively serve as a weir with a variable height.

13.2.2 Weirs

Weirs allow the flow to overtop the structure. The water level will rise until the critical flow depth is reached on top of the structure. There are three main types of weirs, depending on the weir thickness ℓ: broad-crested weirs, short-crested weirs and sharp-crested weirs sketched in Figure 13.10. Broad-crested weirs include overtopped roadways, levees and embankments. Sharp-crested weirs are thin structures, typically made of concrete, steel or sheet piles. The common characteristic of all weirs is that critical flow is reached in the opening, which can be used to determine the flow discharge.

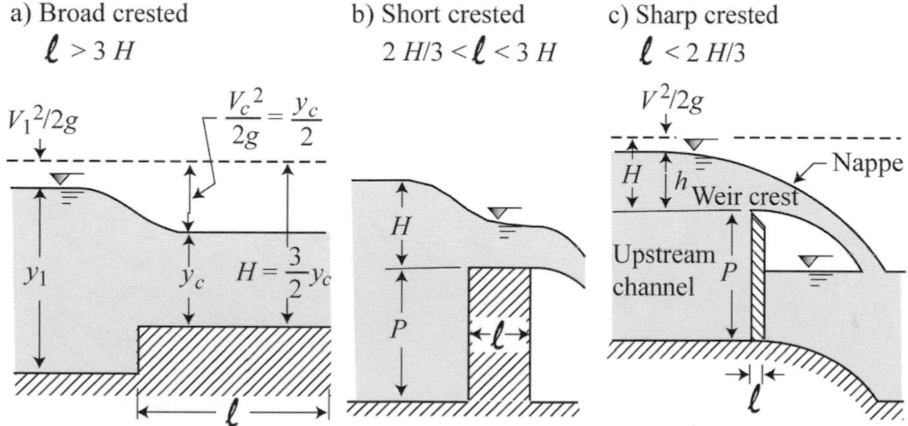

Figure 13.10 Broad-crested weir, short-crested weir and sharp-crested weir

For rectangular weirs, the cross-sectional area is proportional to the critical depth while the velocity is proportional to the square root of the critical depth. Thus, with the use of a discharge coefficient $C \simeq 0.54\sqrt{g}$, or ($2.9 < C < 3.1$ customary units), the flow discharge of broad-crested rectangular weirs of length L is

$$Q = L\frac{2}{3}H\sqrt{g\frac{2}{3}H} \simeq 0.54\sqrt{g}LH^{3/2} = CLH^{3/2}. \tag{13.6}$$

Without the additional friction at the top of the weir, sharp-crested rectangular weirs have a slightly higher coefficient, i.e. $C \simeq 1.8$ in SI and $C \simeq 3.3$ in customary units.

For triangular sharp-crested weirs (V-notch weirs), Example 13.6 illustrates how to analytically derive the following flow discharge relationship

$$Q = C_\Delta \frac{8}{15}\sqrt{2g}\tan\left(\frac{\theta}{2}\right)H^{5/2}, \tag{13.7}$$

where θ is the angle at the base and $C_\Delta \simeq 0.6$. Additionally, the discharge of trapezoidal sharp-crested weirs can be calculated from a combination of the triangular and rectangular formulas in Example 13.7. Finally, a Cipolletti weir, named after the Italian Cesare Cipolletti, is the particular case of a trapezoidal channel when $\tan(\theta/2) = 1/4$, or $\theta = 28.1°$.

Other types of weirs include labyrinth and piano key weirs, as sketched in Figure 13.11. The labyrinth and piano key weirs are designed to increase the effective length of a spillway. The design is more complicated, and they can be subjected to debris and ice problems. Nevertheless, they are quite attractive in urban water parks.

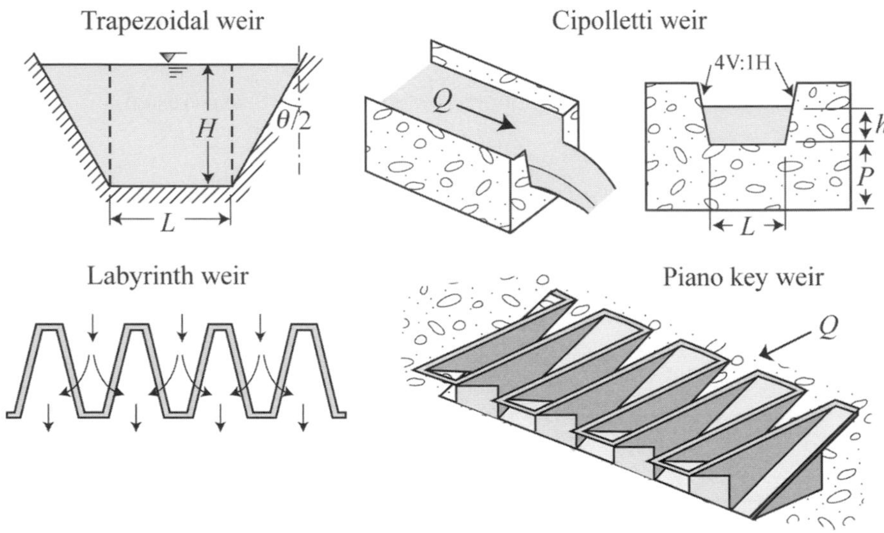

Figure 13.11 Trapezoidal, Cipolletti, labyrinth and piano key weirs

---

♦ **Example 13.6:** Triangular sharp-crested weir

Define the discharge relationship for the triangular sharp-crested weir sketched in Fig. E-13.6 with a base angle θ.

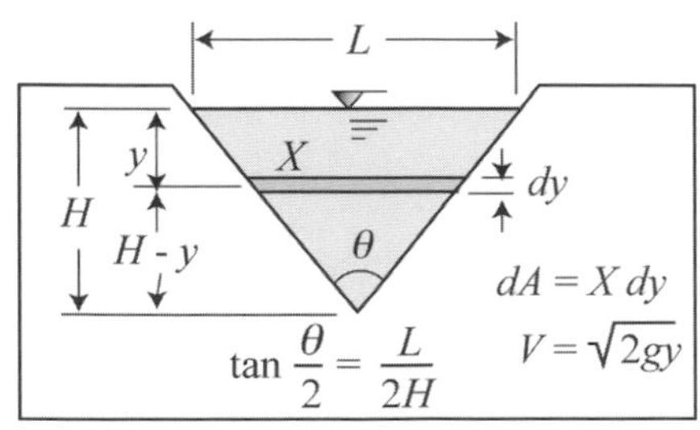

Fig. E-13.6 Triangular weir

Solution: We can define the width X as a function of elevation y below the free surface as

$$\frac{X}{L} = \frac{H-y}{H}, \text{or } X = \frac{L}{H}(H-y) = 2\tan\left(\frac{\theta}{2}\right)(H-y),$$

and the flow discharge dQ in the elementary area $dA = Xdy$ is

$$dQ = VdA = \sqrt{2gy}Xdy = \sqrt{2gy}2\tan\left(\frac{\theta}{2}\right)(H-y)dy.$$

After integration we obtain

$$Q = \int dQ = 2\sqrt{2g}\tan\left(\frac{\theta}{2}\right)\int_0^H \left(H\sqrt{y} - y^{3/2}\right)dy$$
$$= 2\sqrt{2g}\tan\left(\frac{\theta}{2}\right)\left[\left(\frac{2Hy^{3/2}}{3}\right) - \left(\frac{2}{5}y^{5/2}\right)\right]_0^H.$$

The free surface is below the upstream energy level H, and we introduce a discharge coefficient C_Δ:

$$Q = 2\sqrt{2g}\tan\left(\frac{\theta}{2}\right)H^{5/2}\left(\frac{2}{3}-\frac{2}{5}\right)C_\Delta = C_\Delta\frac{8}{15}\sqrt{2g}\tan\left(\frac{\theta}{2}\right)H^{5/2}. \quad (13.7)$$

In practice, we typically use $C_\Delta \simeq 0.6$.

♦♦ Example 13.7: Trapezoidal sharp-crested weir

Calculate the flow discharge in SI units for a sharp-crested trapezoidal weir of length $L = 5$ m and $\theta \simeq 45°$ when the head upstream is $H = 2$ m. Repeat calculations in customary units for a Cipolletti weir ($\theta = 28.1°$) of the same length and height.

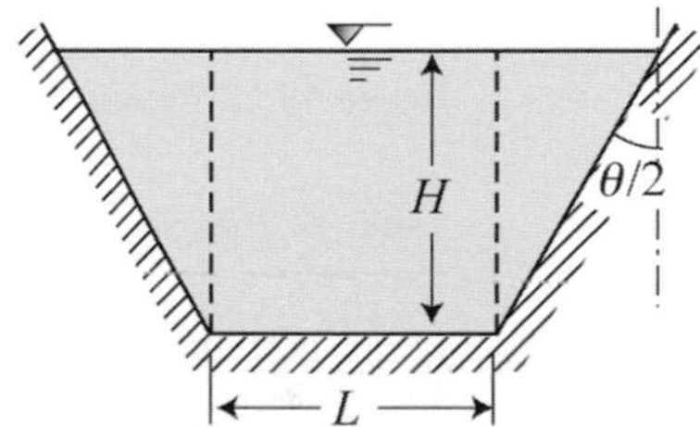

Fig. E-13.7 Trapezoidal weir

Solution: Combining the rectangular and triangular weir equations gives

$$Q = 1.8 \times 5 \times 2^{3/2} + 0.6 \times \frac{8}{15}\sqrt{2 \times 9.81}\tan\left(\frac{45°}{2}\right)2^{5/2} = 25.4 + 3.32 = 28.8\,\text{m}^3/\text{s},$$

$$Q = 1.8 \times 5 \times 2^{3/2} + 0.6 \times \frac{8}{15}\sqrt{2 \times 9.81}\tan\left(\frac{28.1°}{2}\right)2^{5/2} = 35.32 \times (25.4 + 2.0)$$
$$= 970\,\text{ft}^3/\text{s}.$$

13.2.3 Parshall Flumes

Developed by Ralph Parshall at CSU, Parshall flumes are commonly used to measure the discharge in irrigation channels. The flume has three main sections: a contracted area, a throat and a downstream area, as sketched in Figure 13.12.

For flumes with a discharge capacity up to 100 cfs, the main geometric characteristics are the following: $L_T = 2$ ft, $L_D = 3$ ft , $D = 3$ ft, $N = 0.75$ ft, $K = 0.25$ ft, $a = 0.167$ ft, $b = 0.25$ ft, and the other dimensions are listed in Table 13.2. Based on flow depth measurements at h_0 and h_T, the flow discharge (unsubmerged when $h_T < 0.7h_0$) is calculated in Example 13.8 from

$$Q_f = \sqrt{gW_T^5}\left(\frac{h_0^{1.5504}}{1.31 \times W_T^{1.4738} L^{0.0766}}\right). \tag{13.8}$$

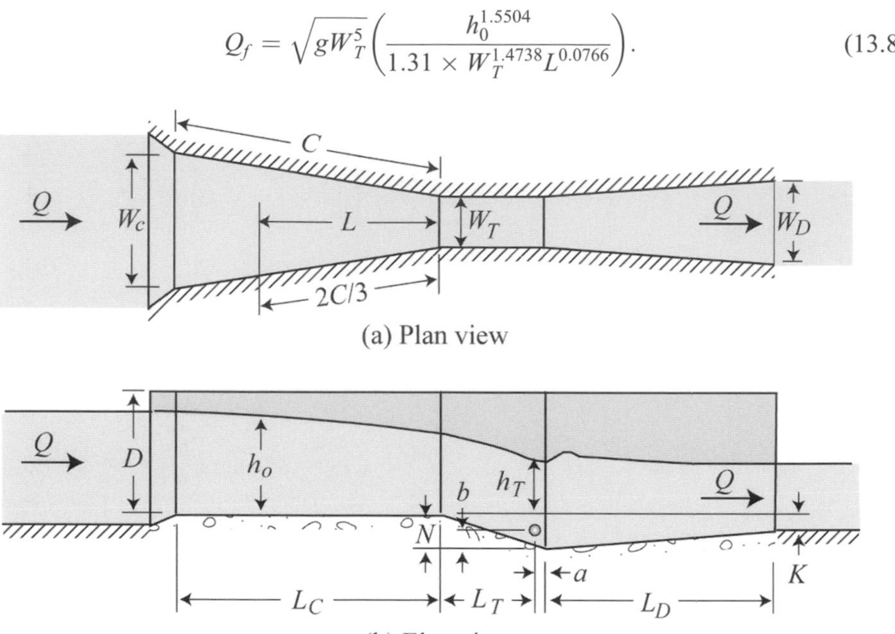

Figure 13.12 Parshall flume

Table 13.2. Parshall flume dimensions

W_T (ft)	W_C (ft)	W_D (ft)	L_C (ft)	C (ft)	L (ft)	Q_{min} (ft^3/s)	Q_{max} (ft^3/s)
1.0	2.77	2.0	4.41	4.50	3.00	0.11	16.1
1.5	3.36	2.5	4.66	4.75	3.17	0.15	24.6
2.0	3.96	3.0	4.91	5.00	3.33	0.42	33.1
3.0	5.17	4.0	5.40	5.50	3.67	0.61	50.4
4.0	6.35	5.0	5.88	6.00	4.00	1.30	67.9
5.0	7.55	6.0	6.38	6.50	4.33	1.60	85.6
6.0	8.75	7.0	6.86	7.00	4.67	2.60	103.5

♦ **Example 13.8:** Parshall flume dimensions

Calculate: (1) the flow in a Parshall flume with a 4-ft throat width, $L = 4$ ft, and a measured flow depth $h_0 = 2.4$ ft; and (2) determine the dimensions of a Parshall flume to measure discharges from 1 to 50 cfs.

Solution:

(1) For $W_T = 4$ ft and $L = 4$ ft from Table 13.2, the unsubmerged discharge formula gives

$$Q_f = \sqrt{gW_T^5}\left(\frac{h_0^{1.5504}}{1.31 \times W_T^{1.4738} L^{0.0766}}\right) = \sqrt{32.2 \times 4^5}\left(\frac{2.4^{1.5504}}{1.31 \times 4^{1.4738} \times 4^{0.0766}}\right)$$
$$= 62.8 \text{ cfs}.$$

(2) The dimensions for $Q = 50$ cfs and $W_T = 3$ ft from Table 13.2 are shown in Fig. E-13.8.

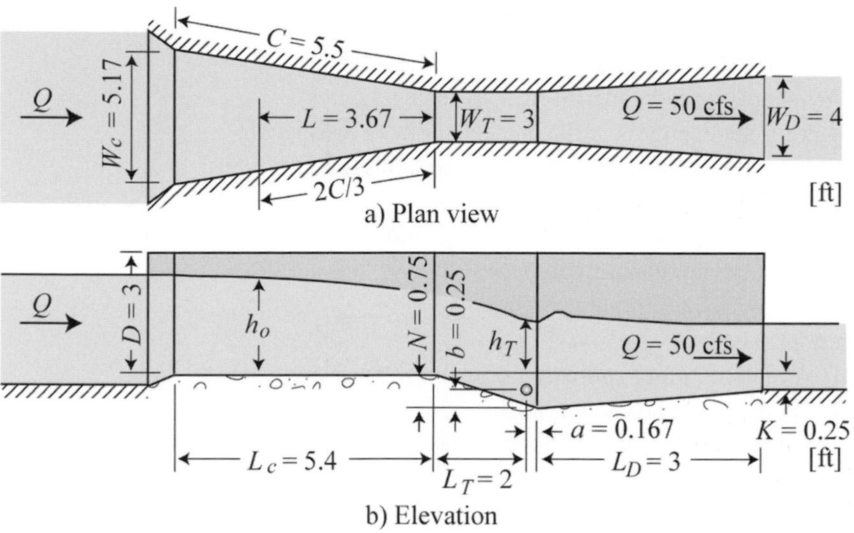

Fig. E-13.8 Parshall flume dimensions

Additional Resources

There is ample information on hydraulic structures in the US Bureau of Reclamation reports (USBR 1976, 1977, 1983). Other helpful references include Davis and Sorensen (1984), Smith (1985), Senturk (1994), Simon and Korom (1997), Novak et al. (2001) and Hager and Schleiss (2009). More recently, Lee et al. (2019b) introduced an additional correction factor to describe the interference between multiple spillways in close proximity. More information on spillway crest gates is available in Mayer and Bowman (1984). Knauss (1987) and Hite (1992) discuss intake swirling and cavitation around hydraulic structures.

Useful references on scour from jet diffusion include Bauer and Beck (1984), Young (1983b), Bormann and Julien (1991), Stein et al. (1993), Vischer and Hager (1995), Hoffmans and Verheij (1997), An et al. (2015) and Julien (2018). Also, regarding river degradation below dams, Richard et al. (2005), Shin and Julien (2011), Ji et al. (2011) and Kim and Julien (2018) review detailed case studies. More details on baffle chutes in steep channels are found in Hayes (1983) and Young (1983a). On milder slopes, rock ramps made of riprap can be designed based on Abt and Johnson (1991) and USBR (2007).

Numerous references for water discharge measurements are valuable: (1) USBR (1983) for small canal structures; (2) Muller (1988), Bos (1989), USBR (1997) and WMO (2010) for discharge measurements; (3) Hager and Schleiss (2009) for mountain streams; (4) Derbyshire (2006) and USBR (2006) for fish-friendly structures; (5) Kim and Julien (2018) and Kim et al. (2018) for low head dams; and (6) Ji et al. (2011) for estuary barrages. Irrigation structures are well described in Warren (1983), Hernandez (1984) and USBR (1997).

EXERCISES

These exercises review the essential concepts from this chapter.

1. When does a morning glory spillway act like a culvert?
2. How is the shape of an ogee spillway determined?
3. Why should gates be placed near the spillway crest?
4. Where is the energy of a flip-bucket spillway dissipated?
5. What are the three most common USBR stilling basin types?
6. What is the purpose of a stilling basin?
7. What is an effective way to prevent cavitation on the face of a spillway?
8. What are the three main gate types?
9. What is a Parshall flume used for?
10. True or false?
    (a) A morning glory spillway keeps the water away from the dam.
    (b) There cannot be suction at the crest of an ogee spillway.
    (c) A flip-bucket spillway dissipates energy far from the spillway toe.
    (d) Type III Stilling basins use baffles to reduce the basin length.
    (e) Stepped spillways dissipate energy at the base of the spillway.
    (f) The height of baffle blocks is slightly less than the critical depth.
    (g) Broad-crested weirs have a slightly better performance than sharp-crested weirs.
    (h) Labyrinth weirs increase the effective length of a spillway.

SEARCHING THE WEB

Find photos of the following features, study them carefully and write down your observations.

1. ♦ Pick three of these structures: morning glory spillways, flip-bucket spillways, ogee spillways, Tainter gates, sluice gates and rising sector gates.
2. ♦ Pick three of these structures: sharp-crested, broad-crested, Cipolletti weirs, stepped spillways and baffled chutes.

PROBLEMS

1. ♦ A broad-crested weir of a height h is placed in a rectangular channel of width b. If the upstream depth is y_1 and the upstream velocity head and frictional losses are neglected, develop an equation for the discharge in terms of the upstream flow depth. Assume the flow over the broad-crested weir to be critical.
2. ♦ A sharp-crested rectangular weir crosses a 10-m-wide channel. The weir height is 7 m. If the head on the weir is measured to be 1.5 m, what is the discharge in the channel?
3. ♦ A 50-m-wide Tainter gate holds 8 m of water above the crest of an ogee spillway. If the opening at the bottom of the gate is 1 m high, calculate the discharge over the spillway.

4. ♦ A 1-m-thick water jet reaches 25 m/s at the tip of a flip-bucket spillway angled 30° above the horizontal. If the tip elevation is 8 m above the downstream water elevation, determine the expected scour depth from the Veronese equation.
5. ♦ Determine the rating curve of a 31-ft-wide broad-crested weir in Fig. P-13.5 and a crest elevation of 2,906 ft.
6. ♦ Design a stilling basin for the following conditions: $S = 30\%$, $V = 58$ ft/s, $Q = 3{,}590$ ft^3/s and width = 31 ft.

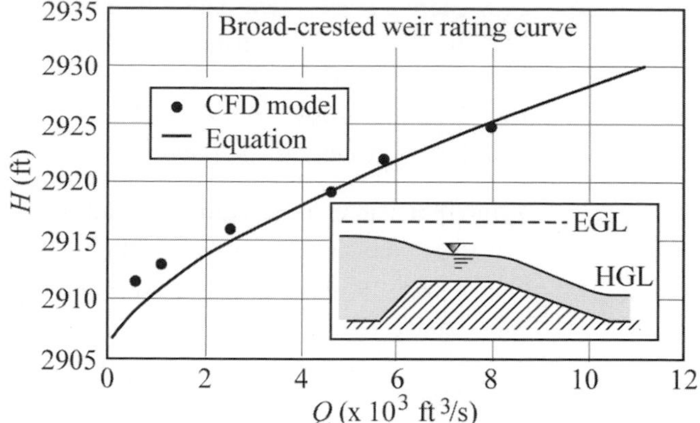

Fig. P-13.5

7. ♦♦ Design a Parshall flume to measure the discharge in the range 0.5–30 cfs.
8. ♦♦ Dimension a 25-ft-drop baffle chute for $Q = 375$ cfs if the entrance channel is 20 ft wide.
9. ♦♦♦ Let's revisit the problem first posed in Problem 4.9 in Chapter 4; part (c) was Problem 8.9 in Chapter 8. This important problem focuses on design adaption, communication and professionalism.
   (d) Joe (Joe is Jan's boss and your company's president) seeks the best outcome for your district, where the population increases steadily at 4% per year. Since the water demand increases with population, Joe wants to meet the projected water demand for the next 25 years. What pipe diameter would be optimal with the expected discharge 25 years from now? Prepare an adaptation plan (a maximum of one page with bullet items) on how you would adapt your proposed design to meet the rapidly increasing water demand? For instance, will you need to build a second pipeline in the future? Would it be preferable to use a larger pipe from the start and replace the pump as the future demand grows? How does your plan impact the project cost?
   (e) You worked hard lately and feel ready to ask Joe for a promotion, which your supervisor Jan heartily supports. At the meeting, however, Joe is unsettled since Jack called him with bad news. At the last public hearing, Jill could hardly explain why the cost of your design has increased so drastically (from ~\$300 thousand to \$2 million). She is up for re-election and is increasingly concerned. Jack also said that your analysis keeps increasing the pipe diameter while the cost skyrocketed. Jack still thinks that his proposal (dual 6-in. pipelines) is the cheapest option over the duration of Jill's elected term (next four years) – is it true? Your company may lose this contract to your competitor, a company headed by Jim. How did this happen? Can you sort this out? What would you do?

Prepare a revised two-page summary of your analysis. One page on your adaptation plan (part d), and one page on how you would sort out the office situation (part e). Write down bullet items (a maximum of five items) of the sequence of events that led to this situation? Also write down bullet items (a maximum of five items) of what you could do to remedy the situation.

[Hint: Your adaptation plan may look like: (1) year 2020, initial construction Q = xx, D = yy, cost = $zz; and (2) year 20XX, Q = YY, build second pipeline for an additional cost = $ZZ. There is no unique solution to this important assignment. You may have great ideas for your adaptation plan and to untangle office situations in a very professional manner. Appendices B and C should be considered.]

14 Hydrology

Hydrology keeps a focus on flood magnitude and frequency to better design hydraulic structures. This chapter reviews hydrologic processes in Section 14.1, flood discharge in Section 14.2 and extreme floods in Section 14.3.

14.1 Hydrologic Processes

This section covers the hydrologic cycle (Section 14.1.1), precipitation (Section 14.1.2), river basins (Section 14.1.3) and infiltration (Section 14.1.4).

14.1.1 Hydrologic Cycle

The hydrologic cycle in Figure 14.1 illustrates processes contributing to the source and yield of both water and sediment from upland areas to the fluvial system. The cycle includes the hydrologic processes of condensation, precipitation, interception, evaporation, transpiration, infiltration, subsurface flow, exfiltration, deep percolation, groundwater flow, surface runoff and surface-detention storage. These processes play a role in hydrology, but precipitation, infiltration, surface runoff and upland erosion are the most important in the analysis of river flows.

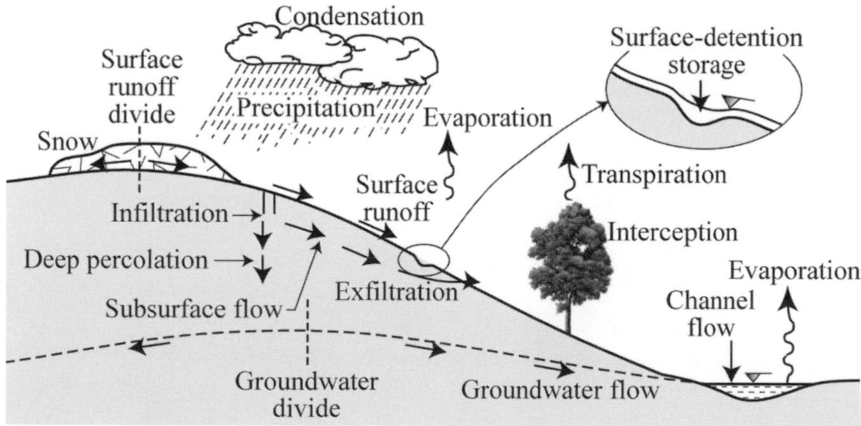

Figure 14.1 Hydrologic cycle

Floods and surface runoff are calculated from the excess rainfall. The excess rainfall precipitation is obtained by subtracting the interception, evapotranspiration, surface storage and infiltration losses from the rainfall precipitation. Some rainfall is intercepted by vegetation before it reaches the ground. The amount of interception varies with the type, density and stage of vegetation growth, rainfall intensity and wind speed. On a single storm basis, the interception storage capacity is generally less than 4 mm, and will typically reach ~1 mm for coniferous and broadleaf forests. Interception storage becomes negligible under heavy and extreme rainfall events.

Evapotranspiration is the combination of evaporation and transpiration. Evaporation refers to the phase change from liquid to vapor from wet surfaces. Water evaporation from plant surfaces is termed transpiration. The evaporation from a leaf surface includes both transpiration and intercepted water. The potential evapotranspiration is mostly (~80%) related to the air temperature and solar radiation. Wind also contributes (~20%) to the evapotranspiration potential. Over long periods of time, evapotranspiration can return large fractions of the total precipitation to the atmosphere. When calculating runoff from single rainstorms, however, evapotranspiration is usually negligible.

Surface storage is the volume of water filling land depressions before surface runoff begins. It is estimated at 0.2–0.6 in. (5–15 mm) in pervious areas such as open fields, woodlands and grasslands. Values of 0.05–0.3 in. (1–8 mm) are typical for paved surfaces and rooftops.

14.1.2 Precipitation

Rainstorms result from the cooling of saturated air masses. Figure 14.2 illustrates the cases of cold and warm fronts, orographic precipitation and hurricanes.

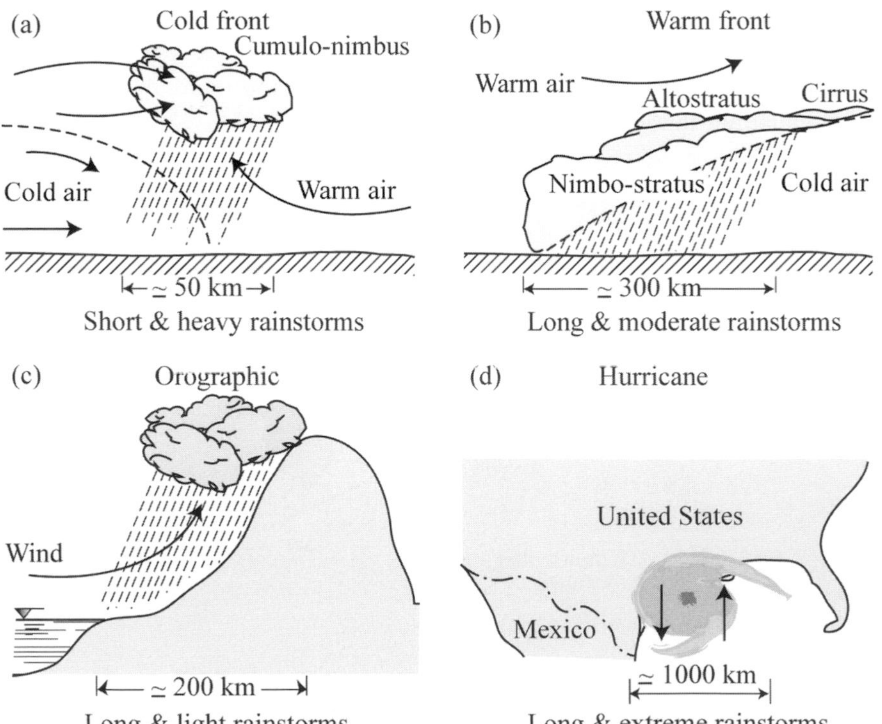

Figure 14.2 Precipitation from cold and warm fronts, mountains and hurricanes

Light rain describes rainfall with a precipitation rate of less than 1 mm/h; moderate rain, 1–4 mm/h; heavy rain, 4–16 mm/h; very heavy rain, 16–50 mm/h; and extreme rain, >50 mm/h. Cold fronts will typically produce heavy but short thunderstorms while warm fronts will produce long and light to moderate rainstorms. Orographic precipitation is generated from winds pushing saturated air masses up a mountain range, and the resulting precipitation is generally a light and steady drizzle. Hurricanes and typhoons generate heavy to extreme rainstorms on large surface areas with strong winds and long durations. Most devastating floods result from hurricanes and tropical storm precipitation.

The mean annual rainfall precipitation in the United States increases southward from 50 mm in the north to over 4,000 mm near the Mississippi River delta as shown in Figure 14.3a. The spatial variability in climate over very large river basins contrasts with the relative climatic homogeneity over small watersheds. The corresponding spatial distribution of a three-hour rainstorm with a period of return of 100 years is shown in Figure 14.3b. Large precipitation events in the United States are typically determined by the magnitude and path of hurricanes. Hurricanes form when sea surface temperatures are higher than 26 °C (>79 °F). Figure 14.3c illustrates the prevailing path line of the hurricanes near the Gulf of Mexico. Formed at sea, they are most devastating when they move inland, where they lose their strength but cause extreme windstorms and rainstorms. Note the spatial variability of large rainstorms in relation to the mean annual precipitation and the hurricane paths.

14.1.3 River Basins

River-basin characteristics include topography and physiography, geology and pedology, forestry and climatology. Very small watersheds typically cover less than 10 km^2, small watersheds less than 100 km^2 and large watersheds exceed 1,000 km^2. Watershed boundaries are separated by basin divides located at high points between watersheds, as shown in Figure 14.4. Lines of constant elevation define contour lines. Hypsometric curves give the relative basin area higher than a given elevation. The hypsometric curves of streams typically plot higher than rivers (Kang et al. 2021).

Geographic information systems (GIS) are helpful for watershed delineation, topography and mapping slopes, soil types and land use. The surface slope is a dominant parameter in the calculation of surface runoff and sediment transport. Digital elevation models (DEMs) display watershed information in a raster-based GIS format with grid sizes typically finer than 30 m. Figure 14.5 shows typical storm/radar resolution scales alongside watershed/model scales.

Rivers follow the low points along the watershed topographic profiles. Drainage networks are delineated using the Strahler channel numbering system illustrated in Figure 14.5. The river length is measured from the remotest point on the watershed to the outlet and increases with drainage area. In Figure 14.6, river length increases roughly with the square root of the drainage area ($L_{\mathrm{km}} \cong 2\sqrt{A_{\mathrm{km}^2}}$).

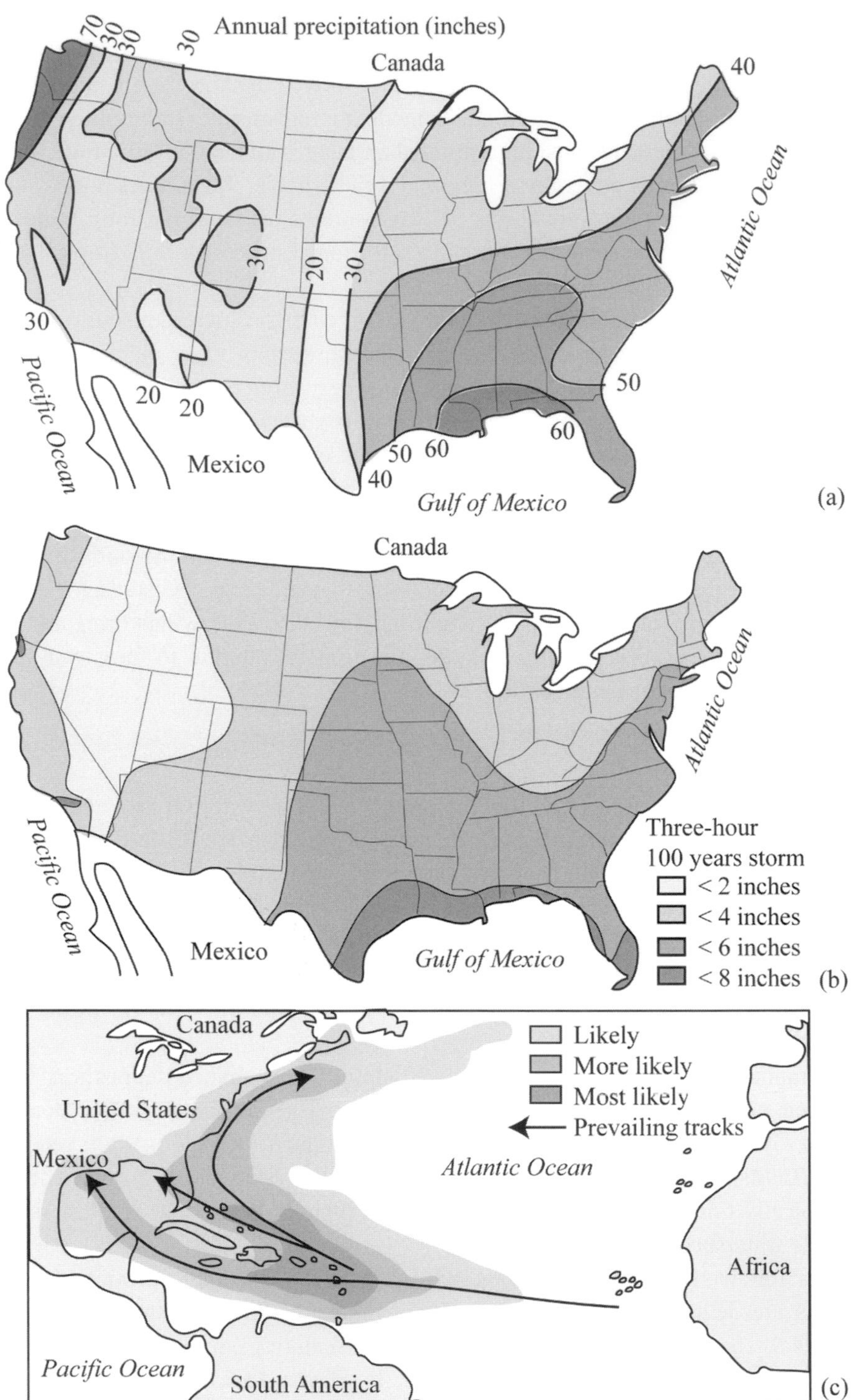

Figure 14.3 (a) Mean annual rainfall in the USA [1 in. = 25.4 mm]; (b) three-hour rainstorm with 100-hour return period (after NOAA 2006); and (c) prevailing hurricane tracks

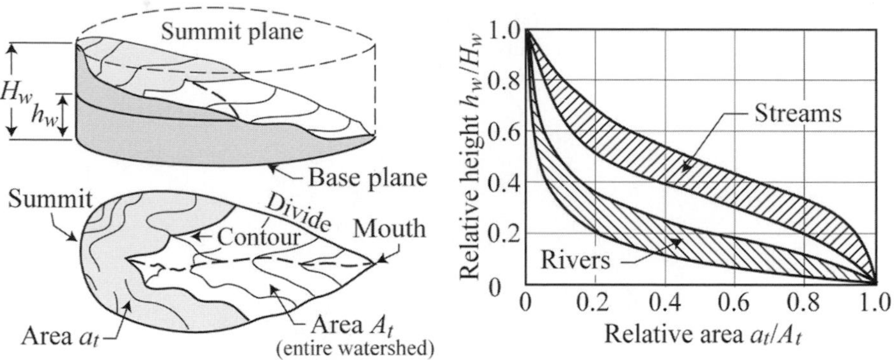

Figure 14.4 River basin and hypsometric curve

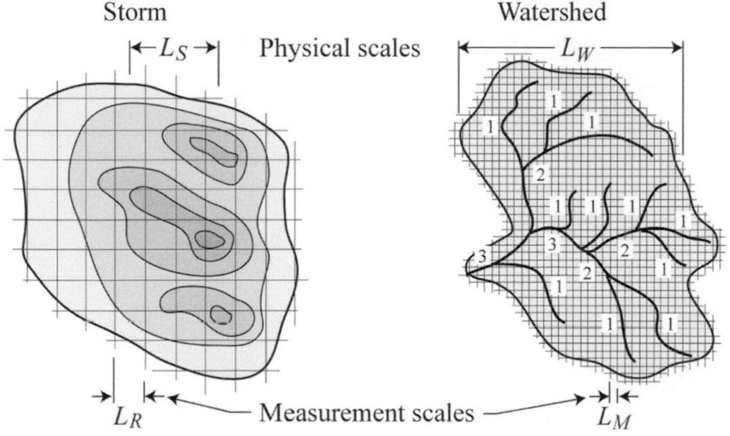

Figure 14.5 Rainstorm and river-basin scales

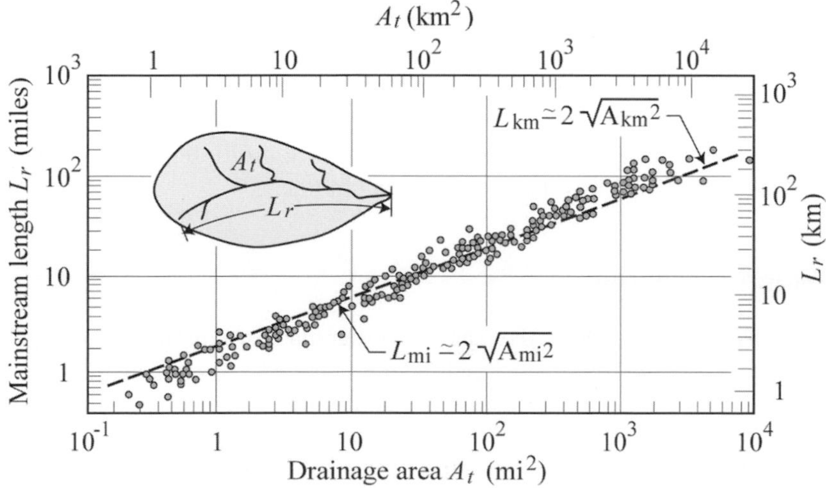

Figure 14.6 River length vs drainage area (modified after Eagleson 1970)

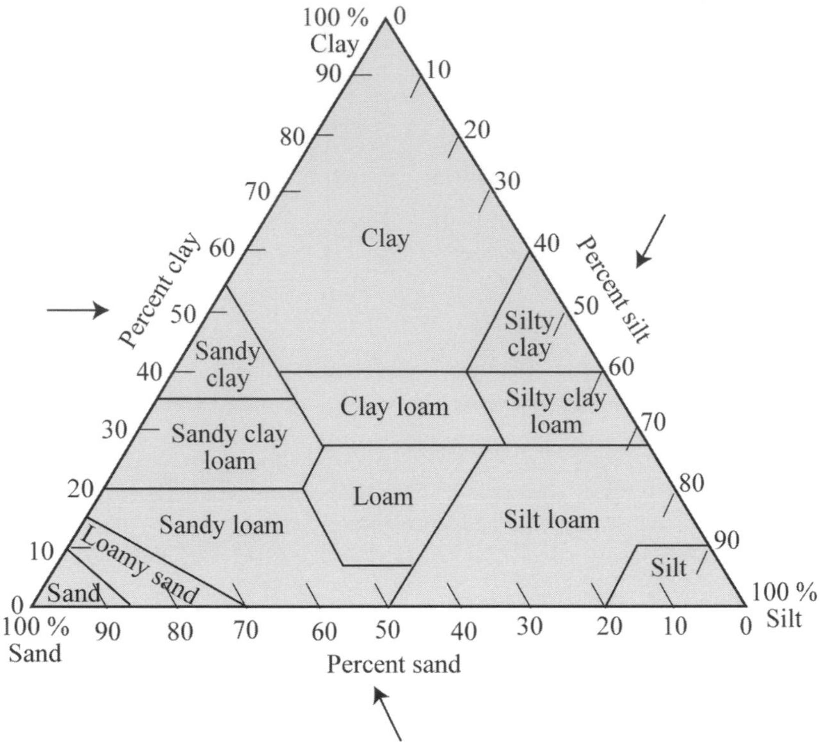

Figure 14.7 Triangular soil classification

14.1.4 Infiltration

Infiltration is the process of water permeating into soil pores. Dry soils absorb water rapidly and over time, the infiltration rate decreases until it reaches the saturated hydraulic conductivity of the soil. According to the Soil Conservation Service classification shown in Figure 14.7, soil types depend on the percentage of sand, silt and clay. For instance, a soil with 60% sand, 30% silt, and 10% clay is a sandy loam.

The potential infiltration curves for initially dry soils in terms of infiltration rate and cumulative infiltration are plotted in Figure 14.8, respectively. In general, soils containing more than 90% sand are highly permeable and do not generate surface runoff. For soils containing at least 20% clay, the cumulative infiltration is less than 2.5 cm (1 in.) in three hours, and less than 10 cm (4 in.) in a day.

14.2 Flood Discharge

The design discharge depends on the type of hydrograph based on the time to equilibrium (Section 14.2.1). We briefly discuss small watersheds and urban areas (Section 14.2.2), dynamic watershed modeling (Section 14.2.3) and flood frequency (Section 14.2.4).

14.2.1 Time to Equilibrium

The amount of rainfall in excess of infiltration generates surface runoff. The time to equilibrium is the time for the floodwave celerity to reach the outlet of a watershed.

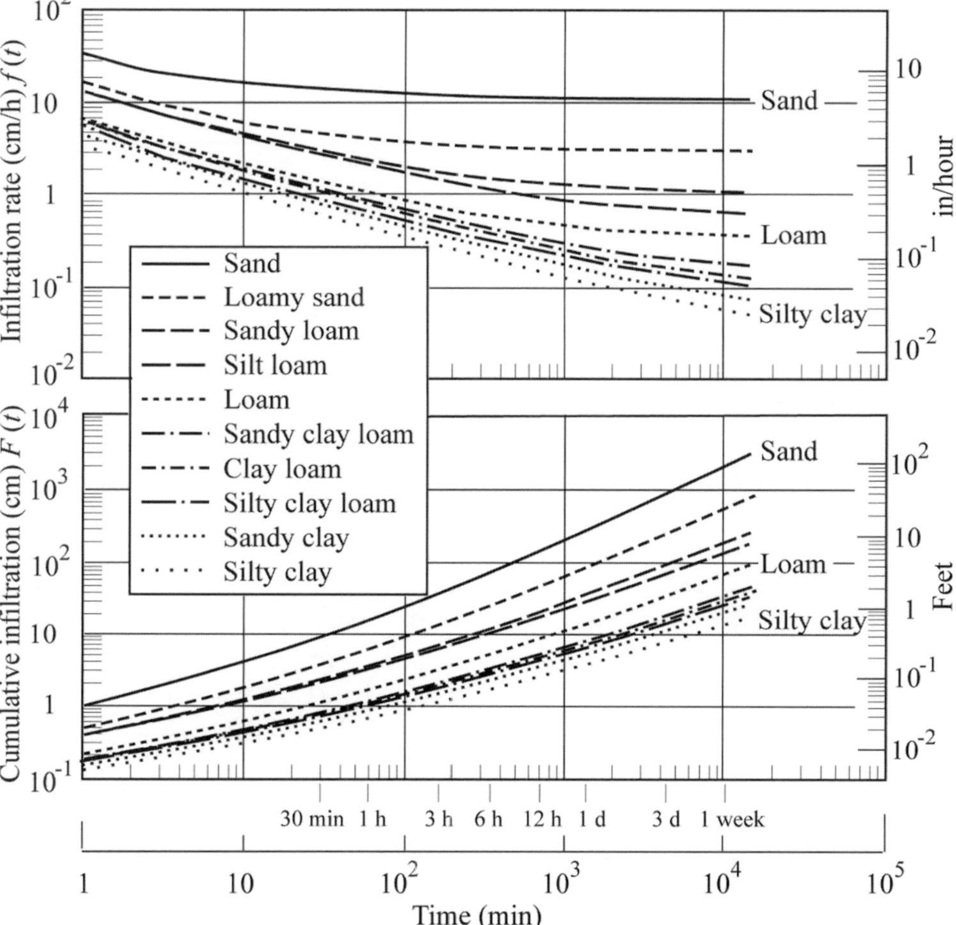

Figure 14.8 Infiltration curves for dry soils (from Julien 2018)

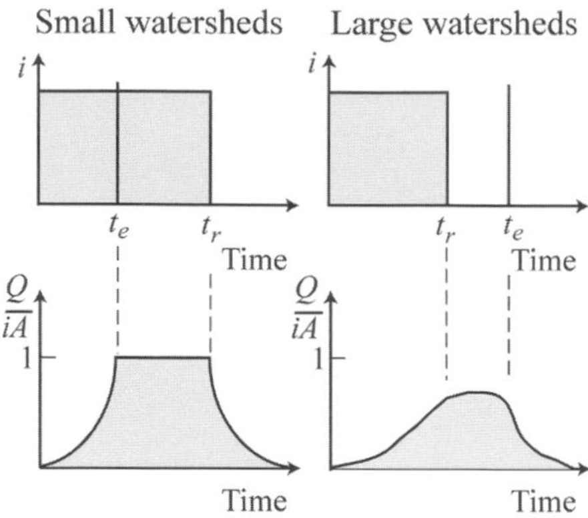

Figure 14.9 Dimensionless hydrographs

Figure 14.9 captures the essential features of the rainfall–runoff relationship (Woolhiser 1975). There are two main types of hydrographs: (1) complete hydrographs (small watersheds with $t_r > t_e$) where the rainfall duration t_r exceeds the time to equilibrium t_e; and (2) partial hydrographs (large watersheds with $t_r < t_e$) where the time to equilibrium exceeds the rainfall duration.

The time to equilibrium can be estimated as a function of drainage area, e.g. from the relationship between the length of runoff in Figure 14.6 and floodwave celerity c: $t_{e(\text{hour})} = L/c \approx 2{,}000\sqrt{A_{\text{km}^2}}/3{,}600 c_{\text{m/s}} \cong (0.5\sqrt{A_{\text{km}^2}})/c_{\text{m/s}}$

Figure 14.10 essentially shows that short (order of 15 minutes) rainfall durations are indicated for watersheds smaller than 1 km^2. Hourly rainfall data are suitable for

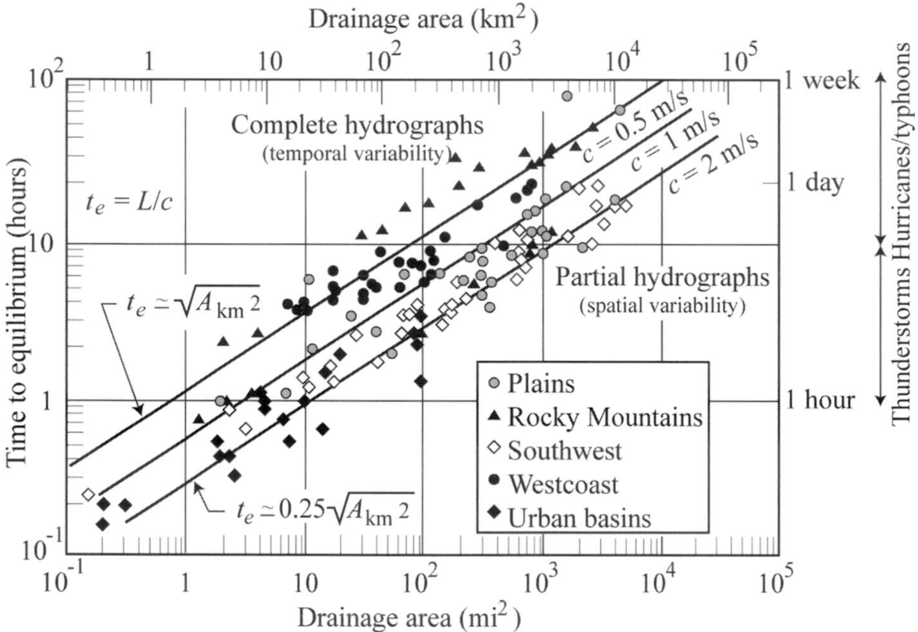

Figure 14.10 Time to equilibrium for small and large watersheds

small and medium watersheds. Daily rainfall precipitation data will be suitable for watersheds larger than 10,000 km^2. For small watersheds, temporal variability becomes most important because the runoff adjusts rapidly to changes in rainfall precipitation. For large watersheds, the spatial variability of rainfall becomes predominant and all hydrological processes need to be carefully modeled.

14.2.2 Rational Method for Small Watersheds

For small watersheds ($A < 10$ mi^2), and more specifically urban areas, the design discharge is determined from the rational method ($Q = CiA$) based on the runoff coefficient C, the design rainfall intensity i and the drainage area A. Because of the short basin response time, high rainfall intensities over short time intervals are used to design urban drainage systems. The runoff coefficient depends on the permeability of the land surface, between $C = 1$ on impervious surfaces and $C \simeq 0.1$ on sandy soils, with a range of practical values in Table 14.1. The upper range can be used on steep surfaces and lower values on flatter surfaces (see Example 14.1).

Table 14.1. Typical ranges of runoff coefficients

Land use		Runoff coefficient C
Parking lots, roads, roofs		0.85–0.95
Commercial areas		0.75–0.95
Industrial		0.5–0.9
Residential	Complex	0.6–0.8
	Single houses	0.3–0.5
Croplands	Clay soil	0.45–0.55
	Loam	0.35–0.45
	Sandy soil	0.25–0.35
Forested areas		0.2–0.4
Parks, open space		0.15–0.35
Lawn	Clay soil	0.25–0.45
	Sandy soil	0.1–0.25

Example 14.1: Urban runoff

A land developer in Fort Collins wants to a change 120 acres of flat loamy grassland into commercial real estate. Consider a 50-year rainstorm bringing 2 inches of rain in 30 minutes. Estimate the peak discharge before and after development.

Solution: Before development, $C \simeq 0.4$ is a representative value from Table 14.1. The discharge

$$Q = 0.4 \times \frac{4 \text{ in.}}{\text{h}} \times 120 \text{ ac.} = 192 \text{ ft}^3/\text{s}.$$

(Note: 1 acre = 43,560 ft^2 and 1 ac. × in./h $\cong$ 1 cfs.)

After commercial development with mostly paved surfaces, $C \cong 0.95$, and the design discharge would increase to

$$Q = 0.95 \times \frac{4 \text{ in.}}{\text{h}} \times 120 \text{ ac.} = 460 \text{ ft}^3/\text{s}.$$

The volumetric difference between these two discharges over 30 minutes requires a detention storage of

$$\mathit{Volume} = (460 - 192) \text{ cfs} \times 30 \times 60 \text{ s} = 480{,}000 \text{ ft}^3 \text{ or } 11 \text{ ac} \cdot \text{ft}.$$

See Problem 14.2 for more rainfall data in Fort Collins and FCSCM (2018) for the Fort Collins Stormwater Criteria Manual.

14.2.3 Dynamic Watershed Modeling

The analysis of large watersheds (partial hydrographs) is far more complex and requires detailed watershed modeling. Besides the popular Hydraulic Engineering Center's Hydrologic Modeling System (HEC–HMS) approach and the CASC2D–SED model (Johnson et al. 2000), fully distributed two-dimensional models can be used. Two powerful methods include dynamic watershed modeling in Case Study 14.1 and flood routing through reservoirs in Case Study 14.2.

Case Study 14.1: Large watershed modeling of Naesung Stream, South Korea

Dynamic watershed modeling calculates runoff and sediment transport over large watersheds. TREX is an event-based model that simulates overland flow and soil erosion at the watershed scale. A model is sketched in Fig. CS-14.1a with a detailed model description in Velleux et al. (2006, 2008).

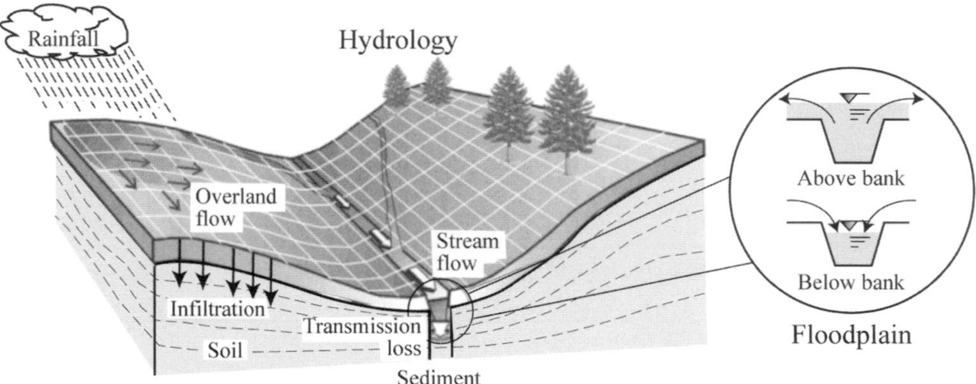

Fig. CS-14.1a TREX model schematic (after Velleux et al. 2008)

Input data are prepared using Colorado State University's TREX (two-dimensional, runoff, Erosion, and export) tool and the surface topography of a watershed is discretized at a fine grid scale. From the precipitation at each pixel, TREX calculates the excess rainfall precipitation from the Green–Ampt infiltration equation, uses the Manning resistance equation and solves the

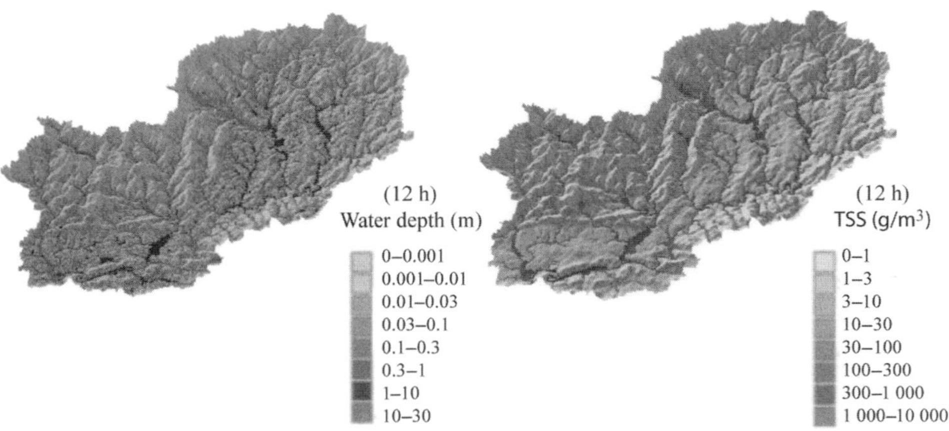

Fig. CS-14.1b TREX simulation of flow depth and TSS at Naesung stream

diffusive-wave approximation of the Saint-Venant equations for surface runoff. The saturated hydraulic conductivity K_h and Manning's coefficient n are the most important calibration parameters. Several applications can be found on small mountain watersheds (Velleux et al. 2006, 2008, 2012, Julien and Halgren 2014, Steininger 2014), on a large 12,000 km^2 watershed (England et al. 2014), as well as applications in South Korea (Kim 2012, Ji et al. 2014) and under tropical monsoons (Abdullah et al. 2013, 2014, 2018, 2019). Dynamic video simulations of storm events are usual for this model. Fig. CS-14.1b illustrates the TREX model simulation results for flow depth and total suspended solids (TSS) at 12 h for a 300-mm design

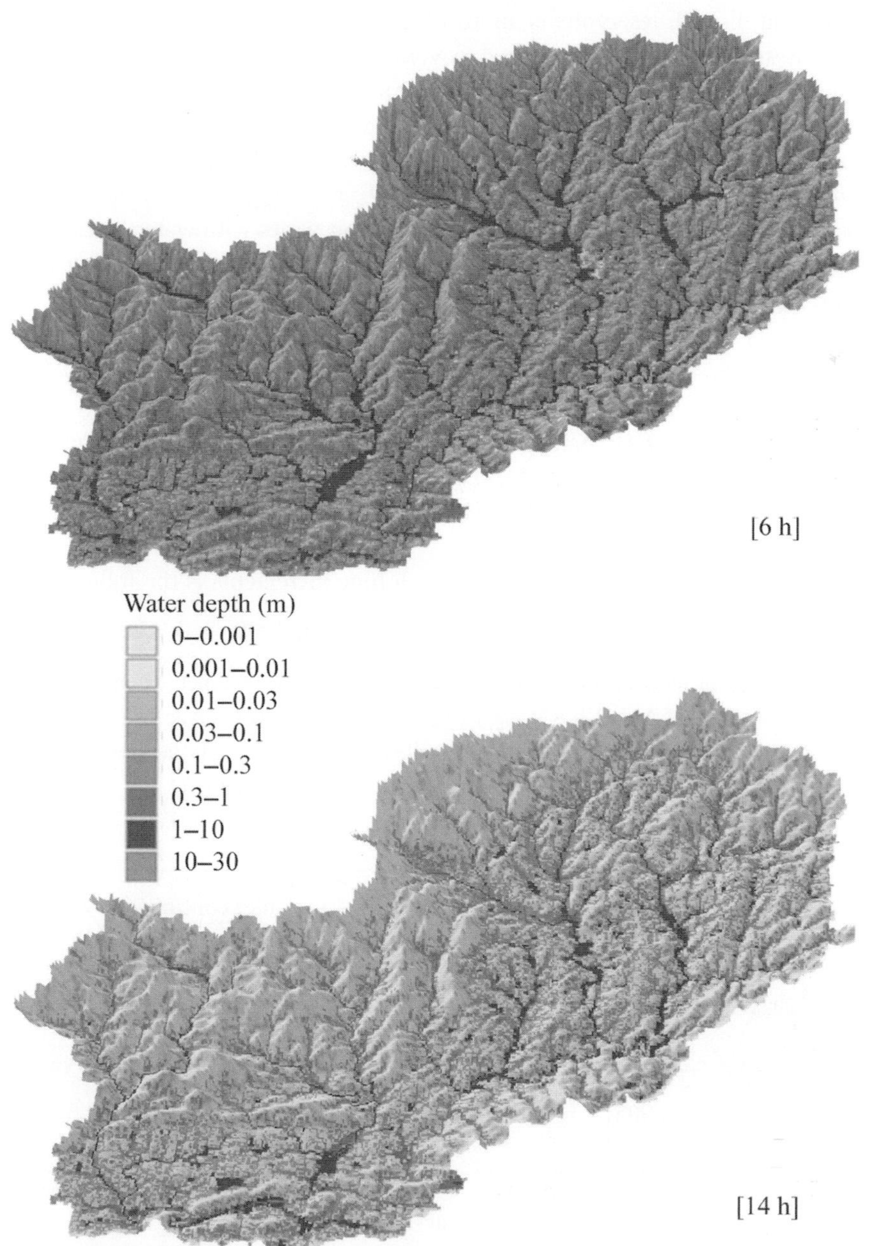

Fig. CS-14.1c TREX simulation of flow depth after 6 h and 14 h

rainstorm lasting 6 h (i.e. 50 mm/h). The calculations at a time step of ~1 s were performed at a 150-m grid scale on Naesung stream (South Korea) covering 1,815 km^2, thus a total of 80,690 pixels. Fig. CS-14.1c illustrates the complex interaction between upland areas, floodplains and main channels. The simulation shows that after 6 h, upland areas are still draining into the main streams. After 14 h, the upland areas are drying out and the flow is fully concentrated in the main channel network and on the floodplains.

Case Study 14.2: Flood routing through a reservoir in Montana

Flood routing through reservoirs is another very important hydraulic engineering tool. Consider a steep mountain watershed covering 9.4 mi^2 and draining into a small reservoir with a 1,200 ac. · ft capacity. The reservoir outlet consists of a culvert with plenty of freeboard. We have to calculate the maximum water level in the reservoir and the maximum outlet discharge from a very large storm. The basic relationship to be solved in reservoir routing is $Q_{in} - Q_{out} = storage$ where Q_{in} and Q_{out} are the in and outflowing discharges, and the volumetric storage in the reservoir varies with water level.

More specifically, the initial water volume is 118.6 ac · ft at a water level in the reservoir at El. 2,909.4 ft. The reservoir storage curve is approximated by $\forall_{\text{ac.·ft}} = 15\,(\text{El.} - 2{,}905)^{1.4}$ and the culvert curve has inlet control as discussed in Section 12.2.2 with the following conditions: (1) the invert is at El. 2,909.4 ft; (2) the culvert diameter is $D = 3.2$ ft; (3) at an elevation below El. 2,912.6 ft, the discharge is approximately linear as $Q_{out\,\text{cfs}} = 9.4(\text{El.} - 2{,}909.4)$; and (4) at an elevation above El. 2,912.6 ft, the culvert curve applies $k_1 Q_{out}^2/gD^5 = [(\text{El.} - 2{,}909.4)/D] + (0.5S_0) - (h_1/D)$, where $k_1 = 1.7$, $S_0 = 0.1$, $h_1/D = 0.74$, and $D = 3.2$ ft. The reservoir capacity curve and the culvert rating curve are plotted in Fig. CS-14.2a.

The sample flood lasting 30 h with $\Delta t = 2$-h time increments is the following:

$$Q_{in} = 10,\ 50,\ 70,\ 90,\ 120,\ 140,\ 250,\ 420,\ 830,\ 1800,\ 1200,\ 360,\ 120,\ 40 \text{ and } 10 \text{ cfs}.$$

The flood routing calculations are shown in Table CS-14.2.1. The volumetric conversion to ac · ft from cfs over a period of $\Delta t = 2$ h is $\forall_{\text{ac.·ft}} = Q\Delta t = 1\ \text{cfs} \times 2 \times 3{,}600\ \text{s} \times \text{ac.}/43{,}560\ \text{ft}^2$, or 1 cfs for 2 h = 0.1653 ac · ft. From the calculations detailed in Table CS-14.2.1, the maximum reservoir elevation is El.$_{max}$ = 2,924 ft at $t_{max} = 26$ h with an outflow discharge $Q_{out\,max} = 157$ cfs. The corresponding graphics of the routed flood and water level are shown in Fig. CS-14.2b.

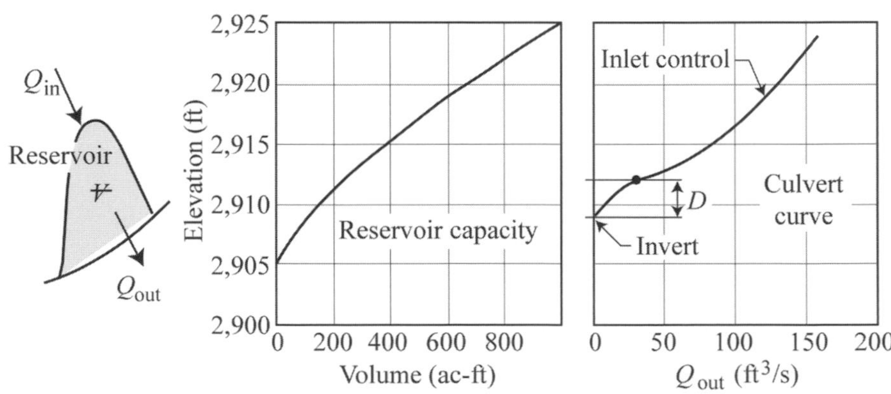

Fig. CS-14.2a Reservoir capacity curve and culvert rating curve

Table CS-14.2.1. Flood routing calculations through the reservoir

Time (h)	Q_{in} (cfs)	$\forall_{in}$ (ac · ft)	$\forall_{res}$ (ac · ft)	El. (ft)	Q_{out} (cfs)	$\forall_{out}$ (ac · ft)
2	10	1.7	118.6	2,909	0	0.0
4	50	8.3	120.3	2,909	0	0.1
6	70	11.6	128.5	2,910	2	0.4
8	90	15.0	139.8	2,910	5	0.9
10	120	20.0	153.9	2,910	8	1.4
12	140	23.3	172.5	2,911	13	2.1
14	250	41.6	193.7	2,911	17	2.9
16	420	69.9	232.4	2,912	25	4.2
18	830	138.1	298.1	2,913	61	10.1
20	1,800	299.6	426.1	2,916	93	15.4
22	1,200	199.7	710.2	2,921	135	22.4
24	360	59.9	887.5	2,923	153	25.5
26	**120**	**20.0**	**921.9**	**2,924**	**157**	**26.1**
28	40	6.7	915.8	2,924	156	26.0
30	10	1.7	896.4	2,924	154	25.7
32	0	0.0	872.4	2,923	152	25.3
34	0	0.0	847.1	2,923	149	24.9
36	0	0.0	822.3	2,922	147	24.5

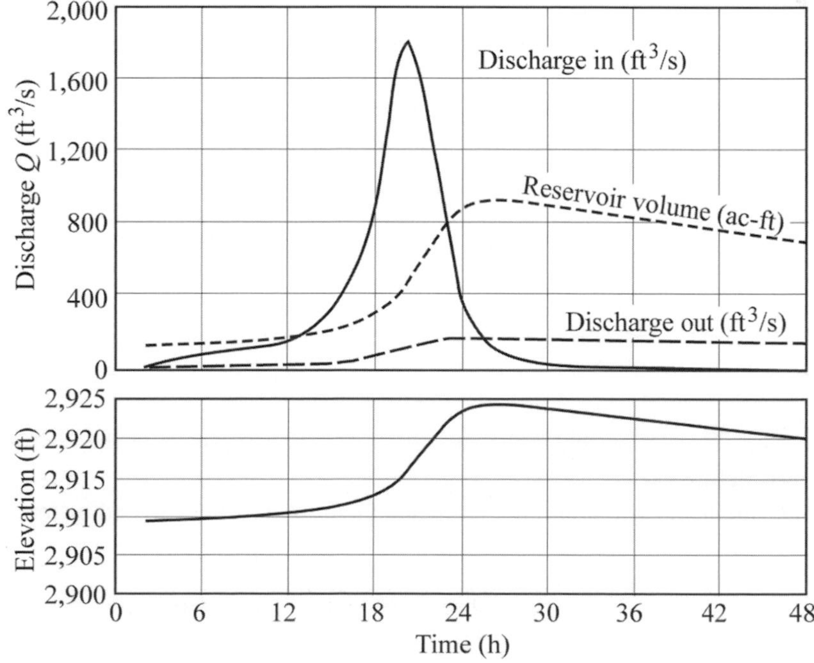

Fig. CS-14.2b Flood routing through the reservoir

14.2.4 Flood Frequency Analysis

For large watersheds with average daily discharge measurements, the design discharge can be obtained from the flood frequency analysis. Accordingly, the probability p for a flood to exceed a certain discharge Q is the exceedance probability $p = 1/T$, where T is the period of return. The non-exceedance probability is $F = 1 - p = 1 - (1/T)$. Figure 14.11 compares the data series for flood frequency (solid symbols) with flow duration (all circles).

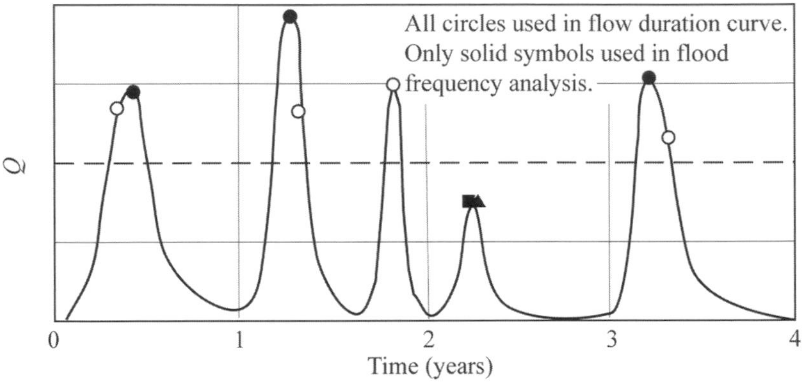

Figure 14.11 Sketch for flow duration curve and flood frequency

From the sample of n independent values (e.g. annual peak daily discharge values), the mean $\bar{x}$, variance σ^2 and skewness γ are first obtained from

$$\bar{x} = \sum_1^n \frac{x_i}{n},$$

$$\sigma^2 = \frac{\sum_1^n (x_i - \bar{x})^2}{n - 1}, \tag{14.1}$$

$$\gamma = \frac{n \sum_1^n (x_i - \bar{x})^3}{(n - 1)(n - 2)\sigma^3}.$$

The standard deviation σ equals the square root of the variance and the coefficient of variation is defined as $C_v = \sigma/\bar{x}$.

For the Gumbel distribution, the frequency factor $K_G(T)$ for a period of return T is

$$K_G(T) = -\frac{\sqrt{6}}{\pi}\left\{0.5772 + \ln\left[\ln\left(\frac{T}{T-1}\right)\right]\right\}. \tag{14.2}$$

It is used to calculate the discharge corresponding to the period of return T from

$$x_p = \bar{x} + \sigma K_G(T). \tag{14.3}$$

The Gumbel method has two parameters $(\bar{x}, \sigma)$ and it is relatively simple.

The normalized value of the exceedance probability z_p is approximately

$$z_p \cong \frac{F^{0.135} - (1-F)^{0.135}}{0.1975}. \tag{14.4}$$

Exact values can also be obtained from math packages (e.g. the norm.inv function in Excel with mean $\mu = 0$ and standard deviation $\sigma = 1$). For instance, according to the normal distribution, the value of z exceeded 5% of the time ($T = 20$) is $z_p = 1.649$ from Eq. (14.4) with $F = 0.95$, compared to the exact $z_p = 1.645$.

The log-Pearson type III distribution has three parameters $(\bar{x}, \sigma, \gamma)$ and is particularly useful to describe the frequency of flood flows. For example, annual peak discharges are first log transformed as $x = \ln Q$. The three distribution parameters are calculated from the mean $\bar{x}$, variance σ^2 and skewness coefficient γ of the log-transformed variables. When $|\gamma| < 2$, the frequency factors K_p can then be calculated with the Wilson–Hilferty formula,

$$K_p(\gamma) = \frac{2}{\gamma}\left(1 + \frac{\gamma z_p}{6} - \frac{\gamma^2}{36}\right)^3 - \frac{2}{\gamma}, \tag{14.5}$$

where z_p is the normalized variable at a certain probability level from Eq. (14.4). The discharge is then obtained:

$$Q_p = e^{\left(\bar{x} + \sigma K_p\right)}. \tag{14.6}$$

The log Pearson III method has three parameters $(\bar{x}, \sigma, \gamma)$ and should be better than the Gumbel method. Detailed calculations are shown in Case Study 14.3.

Case Study 14.3: Flood frequency analysis of the Muda River, Malaysia

We estimate the 100-year discharge of the Muda River in Malaysia from a long record of daily discharge measurements from 1961 to 2005 ($n = 44$ years). The first step is to extract the maximum daily discharge for each year of the data set. These 44 values are then ranked from the highest to lowest: $Q =$ 1,340, 1,225, 1,200, 1,100, 980, 912, 861, 789, 781, 706, 661, 640, 626, 626, 612, 602, 572, 572, 565, 565, 549, 546, 542, 539, 516, 500, 480, 450, 449, 436, 433, 399, 393, 382, 377, 375, 374, 340, 332, 326, 319, 315, 268 and 264 m^3/s (data from Julien et al. 2010). Using $x = Q$ in Eq. (14.1), the mean discharge $\bar{Q} = 587\ m^3/s$ and standard deviation $\sigma = 263\ m^3/s$ are obtained from this observation set.

The Gumbel distribution uses $Q_p = \bar{Q} + \sigma K_G$. For instance, for a period of return $T = 50$ years, Eq. (14.2) gives $K_G = 2.592$, and the 50-year discharge is calculated from Eq. (14.3) as $Q_{50} = 587 + 2.592(263) = 1{,}268\ m^3/s$.

For the log Pearson III method, after logarithmic transformation with $x = \ln Q$, the three parameters from Eqs. (14.1) are $\bar{x} = 6.29$, $\sigma = 0.408$ and $\gamma = 0.418$. For a $T = 100$-year discharge, $p = 1/T = 0.01$ and $F = 0.99$ and the approximation $z_p = 2.337$ from Eq. (14.4) is very close to the exact value $z_p = 2.326$ obtained from the norm.inv function in Excel. The corresponding value of $K_p = 2.63$ is obtained from Eq. (14.5). Finally, the 100-year discharge from the log Pearson III method is then $e^{\bar{Q}+K_p\sigma} = e^{6.29+(2.63\times 0.408)} = 1{,}577\ \mathrm{m^3/s}$.

Table CS-14.3 lists the Gumbel and log Pearson III calculations.

Table CS-14.3. Muda River flood frequency calculations

Return period T in years	$F(Q)$ $(=1-(1/T))$	K_G	Q_G (m^3/s)	z_p	K_P	$Q_{LP\,III}$ (m^3/s)
20	0.95	1.866	1,077	1.645	1.754	1,103
50	0.98	2.592	1,268	2.054	2.271	1,362
100	0.99	3.137	1,410	2.326	2.630	1,577
200	1.995	3.679	1,553	2.576	2.969	1,811

For comparison with the observations, the exceedance probability of the highest discharge ($Q = 1{,}340\ \mathrm{m^3/s}$) is $p = 1/(N+1) = 1/45 = 0.0222$, which obviously corresponds to a period of return of $T = 45$ years. This process is repeated for the second highest discharge [$Q = 1{,}225\ \mathrm{m^3/s}$ with $p = 2/(N+1) = 2/45$], and $T = 45/2 = 22.5$ years, and so on. Both methods agree well with the measurements in Fig. CS-14.3. As expected, the log Pearson III method performs slightly better at high discharges because it has three parameters instead of two. The differences become increasingly pronounced for long return periods. The 100-year discharge ranges between 1,400 m^3/s and 1,600 m^3/s depending on the preferred method.

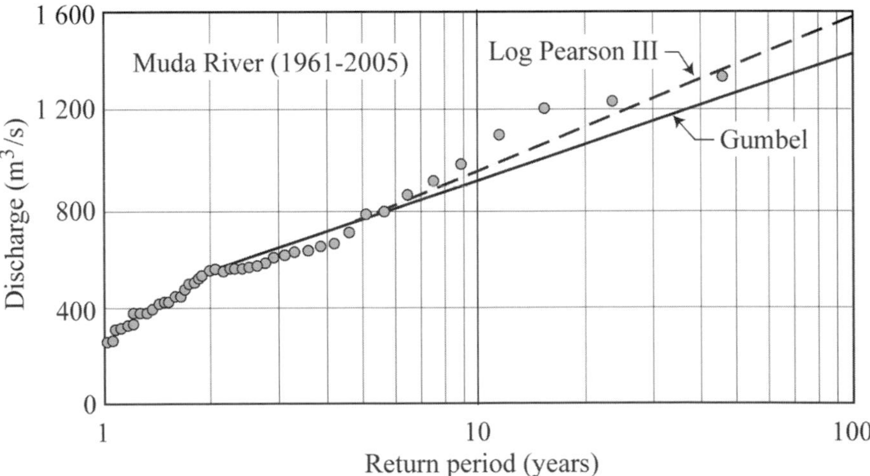

Fig. CS-14.3 Flood frequency of the Muda River

14.3 Extreme Floods

We now consider extreme events starting with risk analysis (Section 14.3.1), then extreme precipitation (Section 14.3.2) and, finally, extreme floods (Section 14.3.3).

14.3.1 Risk Analysis

Risk is defined as the product of the probability of occurrence times the consequences, or damage. The hydrologic risk is the exceedance probability $p = 1 - F(x)$ of an event, which describes how often a certain value will be exceeded during a given time interval. The return period $T = 1/p$ is often preferred to the exceedance probability p.

The binomial distribution defines the probability P_k of having k occurrences in n independent trials given the probability p of an event (nonoccurrence $1 - p$) as

$$P_k = \left[\frac{!n}{!k!(n-k)}\right] p^k (1-p)^{n-k}, \tag{14.7}$$

where $!4 = 4 \times 3 \times 2 \times 1 = 24$. From Eq. (14.7), the probability of having two floods exceeding the 10-year flood in the next 5 years is obtained with $p = 0.1$, $n = 5$, $k = 2$ as $P_2 = [!5/(!2!3)]0.1^2(0.9)^3 = (5 \times 4/2)0.01 \times 0.729 = 0.0729$.

This concept can be extended to determine the probability that there is not any large flood $k = 0$ in the next n years as $P_0 = (1-p)^n$. The probability for at least one event with a period of return T to occur in the next n years is the hydrologic risk R:

$$R = 1 - P_0 = 1 - (1-p)^n = 1 - \left(1 - \frac{1}{T}\right)^n. \tag{14.8}$$

For instance, consider a dam designed to withstand a 100-year flood with a useful life of 30 years. The probability that the 100-year flood may be exceeded at least once in 30 years is $R = 1 - 0.99^{30} = 0.26$. Table 14.2 illustrates the relationship between risk and period of return over the useful life of a project from Eq. (14.8) as $T = 1/\left[1 - (1-R)^{1/n}\right]$. For instance, at an $R = 1\%$ probability of failure in 100 years, the design needs to be based on a 9,950-year flood.

Table 14.2. Return period T in years at a risk level R over a useful life n

Risk R (%)	**Project useful life n (in years)**						
	2	5	10	20	25	50	100
25%	7.46	17.9	35.3	70.0	87.4	174	348
10%	19.5	48.1	95.4	190	238	475	950
5%	39.5	98.0	196	390	488	976	1,950
2%	99.5	248	496	990	1,250	2,480	4,950
1%	199	498	996	1,990	2,490	4,980	9,950

14.3.2 Extreme Precipitation

For small watersheds ($t_r > t_e$), the extreme discharge can use the rational formula with extreme rainfall with short durations ($t_r < 3$ h) from Figure 14.12.

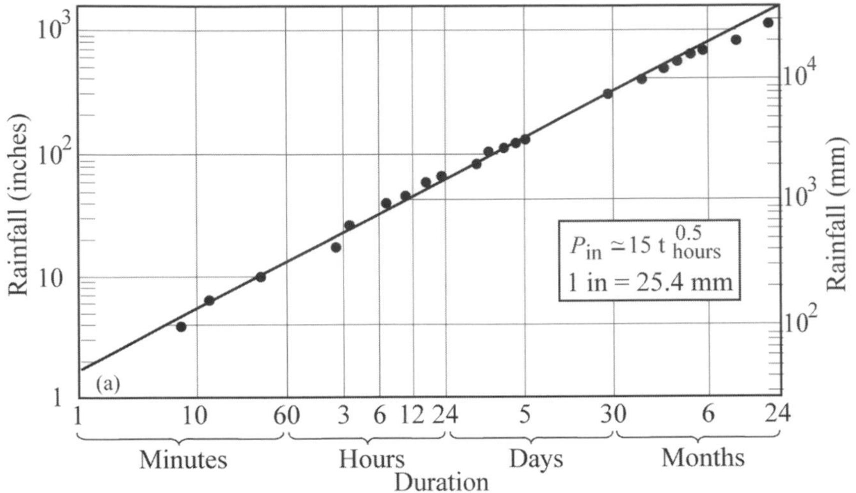

Figure 14.12 World maximum precipitation

On large watersheds ($t_r < t_e$) the two main approaches are: (1) to use hydrologic models to transform a maximum precipitation into a maximum discharge; and (2) use the largest measured floods for a given drainage area. The maximum observed rainfall depth depends on rainfall duration and the drainage area. The maximum observed rainfall depths in the United States are listed in Table 14.3. For instance, on a drainage area of 10 mi^2, the maximum 6-h precipitation is 627 mm which is less than the world maximum daily point precipitation of 933 mm from Figure 14.12. The bold values roughly correspond to $t_r = t_e = 0.5\sqrt{A_{\text{km}^2}}$.

Table 14.3. Maximum observed rainfall depth (in mm) as function of duration and surface area for the United States (after WMO, 1986)

Area (mi^2 [km^2])	Duration (h)						
	6	12	18	24	36	48	72
10 (26)	627	757	922	983	1,062	1,095	1,148
100 (260)	**498**	668	826	894	963	988	1,031
200 (518)	455	**650**	798	869	932	958	996
500 (1,295)	391	625	**754**	831	889	914	947
1,000 (2,590)	340	574	696	**767**	836	856	886
2,000 (5,180)	284	450	572	630	**693**	721	754
5,000 (12,950)	206	282	358	394	475	**526**	620
10,000 (25,900)	145	201	257	307	384	442	**541**
100,000 (259,000)	43	64	89	109	152	170	226

14.3.3 Extreme Floods

With risk being equal to the probability times the damage, the risk of failure of very large infrastructures can quickly become excessively large. For instance, assume an estimated damage cost from a dam break to be \$1 billion (\$1 $\times$ 10^9). A design based on a 100-year flood has a 1% annual probability of failure. The annualized cost of the failure becomes 0.01 $\times$ \$1 billion = \$ 10 million/year, which is far from negligible. Nowadays, the trend is to design large infrastructures, like dams and spillways, to withstand extreme floods. Two concepts can be considered: (1) the maximum flood (or rainfall) ever recorded; and (2) the probable maximum flood (PMF; based on the probable maximum precipitation [PMP]). The concept of PMF is based on an exceedance probability small enough that it would result in a manageable annualized cost when multiplied by the cost of damage. Recent work on extreme floods include England et al. (2019) and Salas et al. (2014, 2018, 2020).

The specific discharge of a watershed is the peak discharge divided by the drainage area. We can thus get a first estimate of specific discharge for the largest measured floods based on the drainage area in Figure 14.13. The dashed line shows the highest possible discharge from the largest US precipitation and $C = 1$ at $t_r = t_e = 0.5\sqrt{A_{\mathrm{km}^2}}$ from Table 14.3.

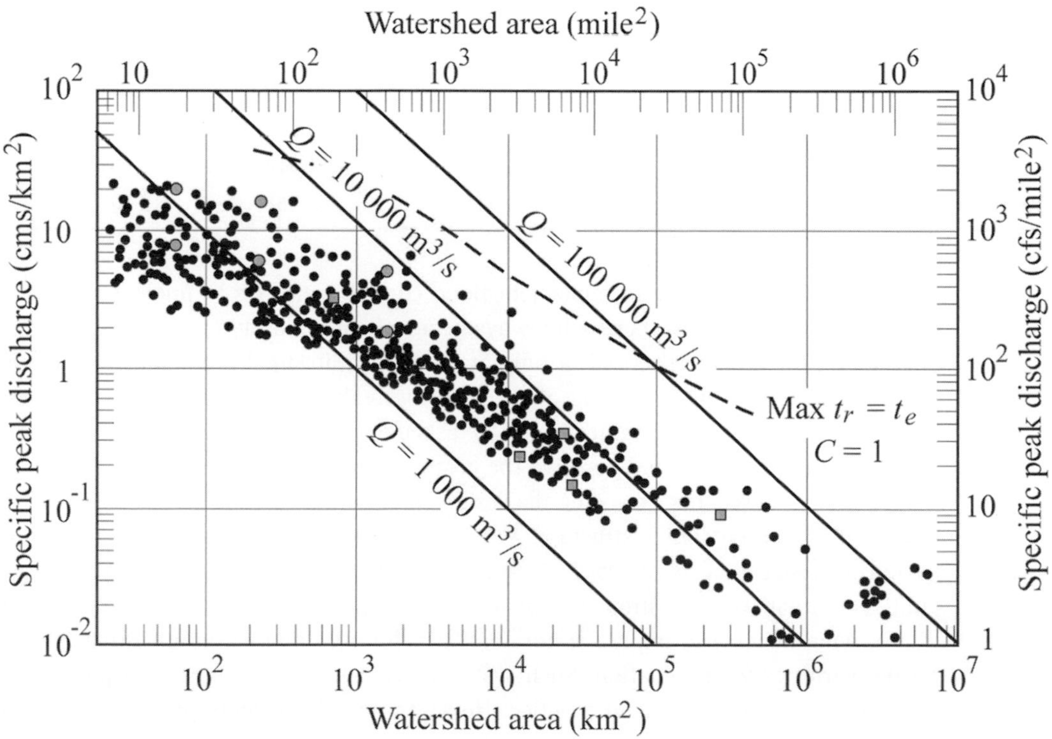

Figure 14.13 Specific discharge of measured large floods as a function of drainage area (modified after Creager et al. 1945 and Abdullah et al. 2013)

The sample of European reservoir characteristics in Table 14.4 shows an increase in design flood discharge with drainage area.

Table 14.4. Sample of European reservoir design floods

Site	Country	Reservoir			Drainage area (km^2)	Design flood discharge (m^3/s)
		Height (m)	Area (km^2)	Volume ($\times 10^6$ m^3)		
Oschenik	Austria	107	0.43	34	1.7	10
Zillergrundl	Austria	186	1.41	89.5	30	165
Alpe Gera	Italy	174	1.15	62.7	36	248
Ridracoli	Italy	101	1.04	33	37	600
Ancipa	Italy	111	1.39	27.8	51	1,100
Gura Raului	Romania	73	0.65	17.5	147	800
Vidraru	Romania	166	10	465	286	660
Pollaphuca	Ireland	32	20	173	309	1,045
Inniscarra	Ireland	45	5	30	793	1,157
Sylvenstein	Germany	48	6.6	124	1,138	2,012
Kozyak	Macedonia	114	14	550	1,850	1,200
Thissavros	Greece	172	20	705	3,690	7,650
Platanovryssi	Greece	95	3.3	57	4,090	7,400
Fala	Slovenia	34	1	4.2	13,257	5,600
JM de Oriol	Spain	130	104	3,162	51,916	10,775
Aldeadavila	Spain	139	3.7	114	73,458	11,670

As engineers continue to implement flood control measures to prevent the damage from improbable floods, there will always be a finite probability to experience an event exceeding the design criteria. Not only are these infinitesimal probabilities quite challenging to define, but engineers often enter the realm of what may be termed social hydrology. For instance, engineers may be tasked to design a system to prevent floods so large that the public would perceive the cost as excessive. If the system was built and failed under a discharge exceeding the design condition, it then becomes ever so difficult to justify an even larger structure. When facing huge floods (e.g. category IV and V hurricanes), engineers can easily be asked: Why do you build a structure so large? And, in the aftermath of a disastrous failure, being asked again: Why did you build it so small?

Extreme floods also imply the notion of developing flood warning systems that can issue warnings for evacuation. Such systems, typically based on rainfall precipitation forecasts, work quite well in practice. However, many residents prefer to stay home when the warnings are issued because they never experienced anything like it before and they feel safest in their homes. The perspective of coping with larger floods in the perspective of climate change will also continue to challenge engineers for generations. In conclusion, Case Study 14.4 looks at a devastating flood in Colorado.

Case Study 14.4: The Big Thompson Flood, Colorado

Two reports of Grozier et al. (1976) and McCain et al. (1979) document the Big Thompson River flood of July 31–August 1, 1976 (Fig. CS-14.4a). On July 31, several violent thunderstorms released large volumes of rain from Estes Park to the Wyoming border. Larimer County reported 139 lives lost and property damage of $16.5 million. The isohyetal map of the total precipitation from July 31 to August 2, 1976 is shown in Fig. CS-14.4a. Eastern Colorado was under conditions favorable for heavy rain on July 31. A slowly moving cold front in the state combined with easterly winds, moving moist air upslope. The large thunderstorms near Estes Park moved very slowly. Heavy rainfall started around 18:30 MDT and ended around 23:30 MDT (Fig. CS-14.4b). Most of the rain fell in one hour and precipitation totals reached 10 in. (254 mm) between Estes Park and Drake and more than 12 in. (305 mm) in the Glen Haven area. Very little rainfall contributed to the flood east of Drake and west of Estes Park.

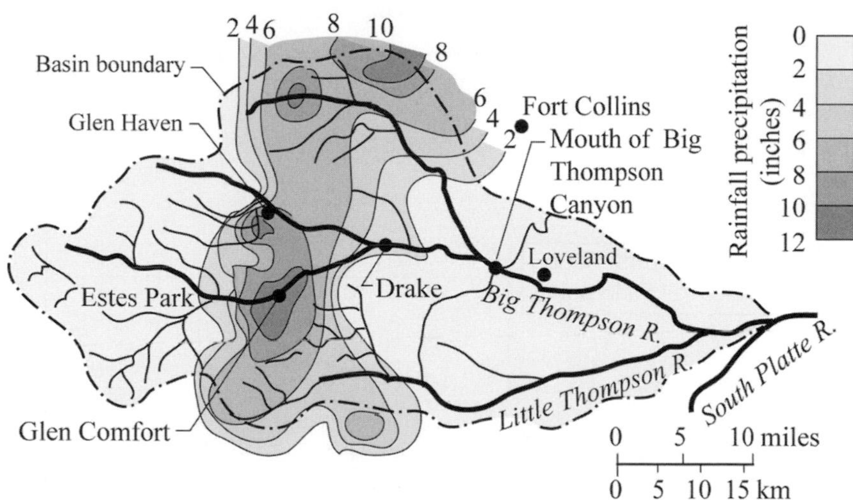

Fig. CS-14.4a Big Thompson flood

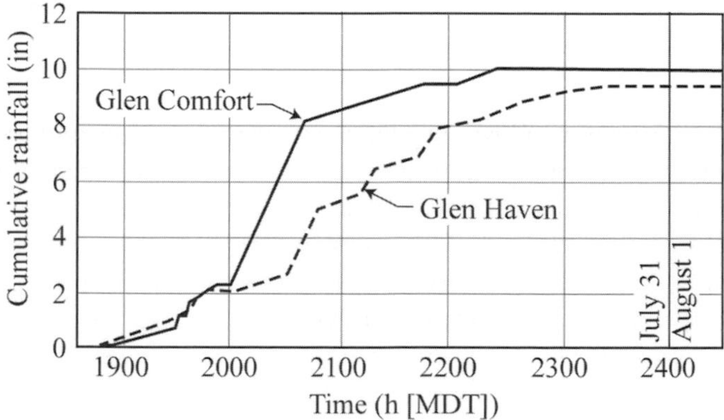

Fig. CS-14.4b Hyetographs of the 1976 flood

Flood runoff in the Big Thompson basin derived from an area of 60 square miles (155 km^2) centered on the Big Thompson River between Estes Park and Drake. The topography of the area has very steep north- and south-facing slopes with rugged rock faces and a thin soil mantle. The storm runoff quickly reached surface channels and the flash flood only lasted a few hours. The reported peak stages on the Big Thompson River occurred as follows: 21:00 at Drake, 21:30 at the Loveland power plant, and approximately 23:00 at the mouth of the canyon ~8 miles (13 km) west of Loveland. The flood peak moved through the 7.3-mile (11.7-km) reach between Drake and the canyon mouth in ~2 h. This corresponds to a floodwave celerity $c \approx 11{,}700/(2 \times 3{,}600) = 1.62$ m/s in this canyon. East of the canyon mouth, the valley widens substantially and the flood discharge receded quickly. The discharge at the mouth of the Big Thompson Canyon peaked at 31,200 ft^3/s (883 m^3/s), and the peak decreased to ~2,500 ft^3/s (71 m^3/s) at noon on August 1 near LaSalle. The peak discharges at various locations are shown as a function of drainage area in Fig. CS-14.4c.

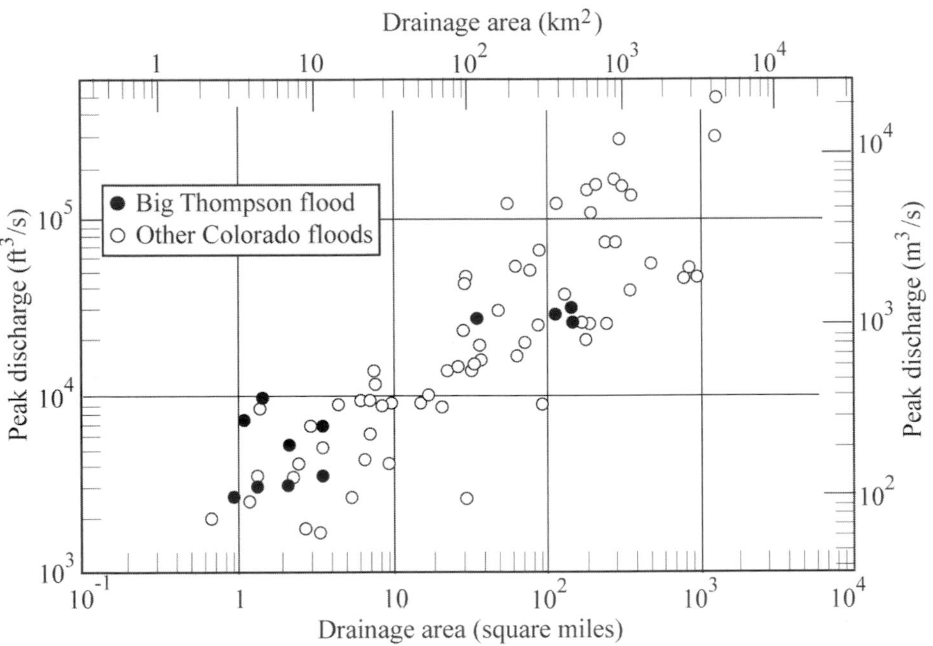

Fig. CS-14.4c Big Thompson peak discharge vs drainage area

Additional Resources

Additional resources include hydrology textbooks like Eagleson (1970), Anctil et al. (2012), Ponce (2014) and McCuen (2016). A detailed analysis of small watersheds is available in Haan et al. (1993) with land-use-change effects in Calder (1992) and McCuen (2016). Restoration efforts are discussed in Brooks and Shields (1996), USDA (2007) and USBR (2015). Hillslope stability problems are discussed in Lu and Godt (2013) with landslide applications in Batista and Julien (2019). Climatic teleconnections are discussed in Lee et al. (2018, 2019a).

An in-depth analysis of flood frequency is found in England et al. (2019). The reader is also referred to Stedinger et al. (1992) and Anctil et al. (2012) for L-moments and for the Weibull, Pareto and generalized extreme value distributions. For nonstationary conditions such as climate change, the reader is referred to Salas and Obeysekera (2014). A stochastic analysis of monsoon rainfall is also covered in Muhammad et al. (2015).

Recent model applications using the PMP to calculate the PMF include: (1) semi-arid areas in the United States by England et al. (2007, 2010, 2014); and (2) tropical monsoon watersheds by Abdullah et al. (2014, 2018, 2019, 2020). Novelties include: (1) a stochastic storm transposition approach by England et al. (2014); (2) the field application of autoregressive precipitation models like DARMA by Muhammad et al. (2015); (3) field verification with paleoflood records by England et al. (2010 and 2014); and (4) climate preparedness on hydraulic infrastructure by Friedman et al. (2016) and O'Brien (2017).

EXERCISES

These exercises review the essential concepts from this chapter.

1. What is the difference between evaporation and evapotranspiration?

2. What is a cold front?

3. What is a hypsometric curve?

4. When do you get a complete hydrograph?

5. What is the main difference between hurricanes and thunderstorms?

6. What is a reservoir capacity curve?

7. What is the difference between a flow duration curve and a flood frequency analysis?

8. What is risk?

9. How does the world maximum precipitation vary with the storm duration?

10. True or false?

- (a) Altostratus clouds generate intense precipitation.
- (b) Hurricanes generate intense and short rainstorms.
- (c) One-hundred-year storms are influenced by hurricanes.
- (d) The channel length is linearly proportional to the drainage area.
- (e) Intense storms rarely generate runoff on sandy soils.
- (f) Hurricanes can bring larger watersheds under complete equilibrium than thunderstorms.
- (g) The rational method is best used for large watersheds.
- (h) In reservoir routing, the maximum outflow discharge always occurs when the reservoir level is at the maximum.
- (i) Gumbel's method can be used to predict 100-year floods.
- (j) The maximum observed rainfall for 14-day rainfall is observed in monsoon climates.

PROBLEMS

Rainfall and Runoff

1. ♦ From the data for the Big Thompson flood in Case Study 14.4, compare these extreme flood discharges with those of Figure 14.13. Also compare the data from Table 14.4 with Figure 14.13.
2. ♦ The rainfall precipitation in Fort Collins is given as a function of exceedance probability and rainfall duration in Fig. P-14.2. Examine ratios of the 100-year to 2-year rainstorm? Compare the 1,000-year data with the world maximum precipitation in Figure 14.12. What is the period of return of a rainstorm in Fort Collins reaching 6 in. of rain in 2 h?

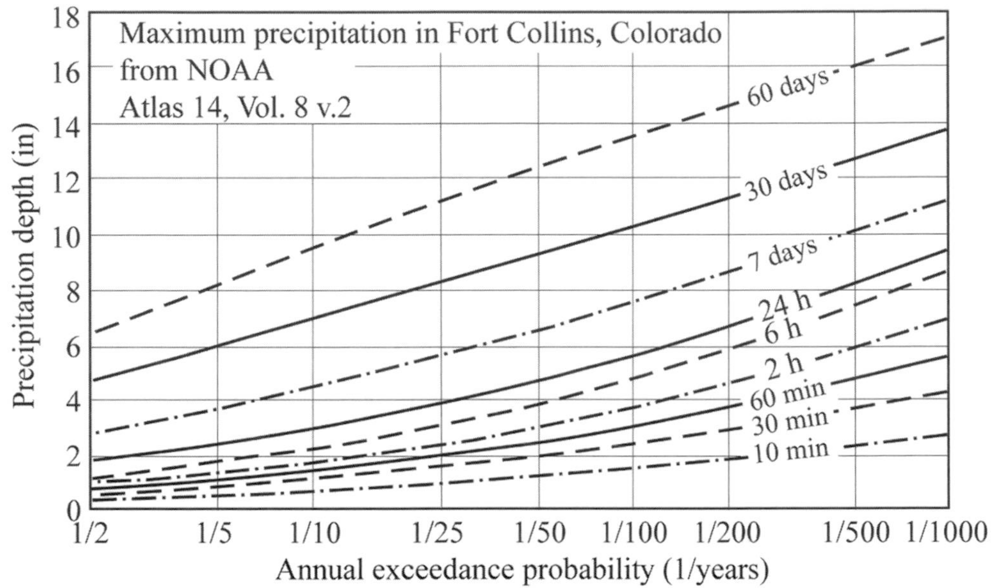

Fig. P-14.2 Maximum rainfall precipitation in Fort Collins, CO

3. ♦ Estimate the time to equilibrium for a three-hour rainstorm with a period of return of 100 years in Tennessee on a rectangular farmland covering 150 acres. Estimate the infiltration on bare soil for a silty clay loam. Can you estimate the maximum discharge for this storm?
4. ♦♦ Duksan Creek covers 33.1 km^2 and the channel length is 12 km (Kim 2012). This steep mountain watershed has an average slope of 25% and a Manning coefficient $n \simeq 0.03$. Estimate the time to equilibrium for a 3-h rainfall intensity of 60 mm/h. Determine the maximum discharge for two conditions: (1) the surface is impervious; and (2) the soils are primarily a sandy loam.
5. ♦ A 9.4-mi^2 drainage area is expected to receive 6.4 in. of rain in 6 h. If the runoff coefficient is 0.9, estimate the peak runoff discharge in cfs and the total runoff volume in ac · ft.
6. ♦ A 9.4-mi^2 drainage area is expected to receive 13.7 in. of rain in 72 h. If the runoff coefficient is 0.8, estimate the peak runoff discharge in cfs and the total runoff volume in ac · ft.

Risk Analysis

7. ♦♦ A dam is designed to withstand a 100-year flood and its failure would result in \$1 billion in damage. A permanent engineering solution will be implemented six years from now. In the meantime, a temporary structure could be built at a cost of \$40 million to reduce the hydrologic failure probability to a 1,000-year flood. Is the temporary structure worth building?
8. ♦♦ A dam is designed to withstand a 100-year flood and its failure would result in \$1 billion in damage. A permanent engineering solution will be implemented six years from now. In the meantime, a temporary structure could be built at a cost of \$40 million to reduce the hydrologic failure probability to a 250-year flood. Is the temporary structure worth building?

Rational Method

9. ♦♦ With reference to Case Study 14.2, consider flood routing based on the reservoir curve in Fig. CS-14.2a. If the runoff coefficient is $C = 0.8$, what is the rainfall depth on a 24.5-km^2 watershed that can be contained in a reservoir with a capacity of 1.23×10^6 m^3?
10. ♦♦ With reference to Case Study 14.2, consider flood routing based on the culvert performance curve in Fig. CS-14.2a. If the runoff coefficient is $C = 0.64$, what is the rainfall intensity on this 25-km^2 watershed that corresponds to the maximum culvert capacity of 4.25 m^3/s?

Routing through Reservoirs

11. ♦ For the conditions detailed in Example 14.2, and referring to Fig. P-14.11, run a triangular inflow discharge increasing 50 cfs every 2-h period for 24 h and (inflow $Q_{max} = 600$ cfs at $t = 24$ h) and decreasing by 50 cfs hourly until $t = 48$ h. Determine the maximum outflow discharge and reservoir elevation and capacity.

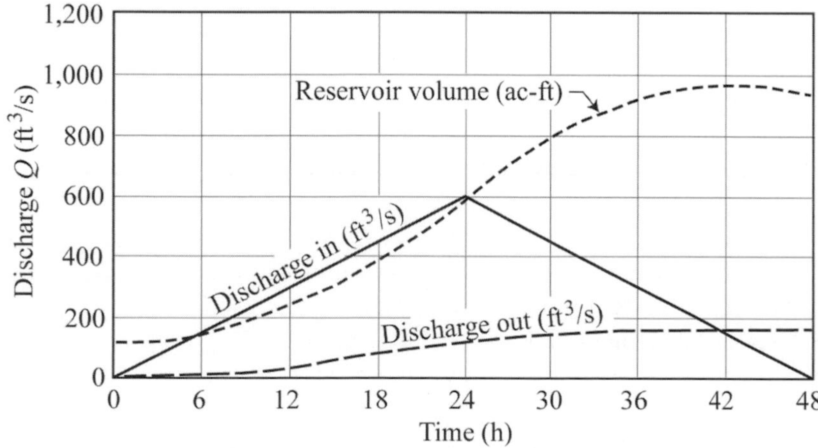

Fig. P-14.11

12. ♦ With reference to Example 14.2, run a triangular inflow discharge, increasing 150 cfs every 2-h period for 12 h and (inflow $Q_{max} = 900$ cfs at $t = 12$ h) and decreasing by 150 cfs hourly until $t = 48$ h. Referring to Table P-14.12, determine the maximum outflow discharge and reservoir elevation and capacity.

Table P-14.12. Flood routing in a reservoir

Time hrs	Q_{in} cfs	$\forall_{in}$ ac. · ft	$\forall_{res}$ ac. · ft	Elev. ft	Q_{out} cfs	$\forall_{out}$ ac. · ft
0	0	0.0	118.6	2,909	0	0.0
2	150	25.0	118.6	2,909	0	0.0
4	300	49.9	143.6	2,910	6	1.0
6	450	74.9	192.5	2,911	17	2.8
8	600	99.9	264.6	2,913	48	8.1
10	750	124.8	356.4	2,915	77	12.9
12	900	149.8	468.3	2,917	101	16.7
14	750	124.8	601.3	2,919	121	20.1
16	600	99.9	706.0	2,921	134	22.3
18	450	74.9	783.5	2,922	143	23.8
20	300	49.9	834.6	2,923	148	24.7
22	150	25.0	859.9	2,923	151	25.1
24	0	0.0	859.8	2,923	151	25.1
26	0	0.0	834.7	2,923	148	24.7
28	0	0.0	810.0	2,922	146	24.2
30	0	0.0	785.8	2,922	143	23.8

15 Geohydrology

This chapter provides a brief overview of the physical properties of wet soils in Section 15.1 and discussion of processes associated with high water content in Section 15.2.

15.1 Soil Properties

This section briefly discusses soil granulometry and mineralogy (Section 15.1.1) and physical properties of wet soils (Section 15.1.2).

15.1.1 Soil Granulometry and Mineralogy

The particle size distribution is the main characteristic of a soil with gravels (2 mm $< d_s <$ 64 mm), sands (0.0625 mm $< d_s <$ 2 mm), silts (2 μm $< d_s <$ 62.5 μm) and clays ($d_s <$ 4 μm). The percentage of sand, silt and clay has been used for the soil classification in Section 14.1.4. The US standard sieves include No. 10 at 2 mm to separate sands from gravels and No. 200 at 0.075 mm to separate sands from silts. From the particle size distribution in Figure 15.1, the grain size d_{10} designates that 10% of the material by weight is finer. Well-graded mixtures include nearly impervious glacial till and hardpans that experienced consolidation under high pressure.

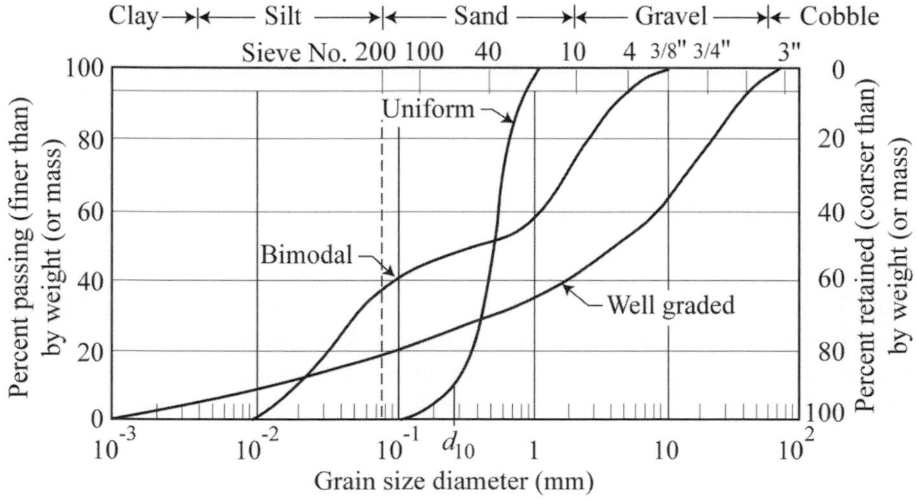

Figure 15.1 Particle size distribution

Clays are very complex given their property to adsorb water. As shown in Figure 15.2, one basic clay sheet is the tetrahedral silica sheet sketched as a trapezoid with a layer of silicon atoms between a layer of oxygen atoms with one minus charge per silicon atom ($Si_2O_5^{-2}$ structure). The second basic sheet represented by a rectangle is the octahedral sheet or layer of aluminum (or iron/magnesium) atoms between layers of densely packed hydroxyls. Combinations of these basic sheets define the main types of clays. For simplicity, only three types of clay particles are considered: kaolinite, illite and montmorillonite.

Kaolinite has a 1:1 structure, meaning that it superposes a silica sheet to an octahedral aluminum sheet to form a structural unit. The unit thickness is $\cong 7$ Å (1 Å = angstrom = 10^{-10} m). The bonds between the oxygens of the silica sheet and the hydroxyls of the aluminum sheet are strong and many units can be stacked up to form very stable minerals that cannot easily be penetrated by water (low swelling). Kaolinite crystals are typically compact 300–4,000 nm (a nanometer is 1 nm = 10^{-9} m) wide and 50–2,000 nm thick, with a low specific surface (surface per unit mass) of 10–20 m^2/g.

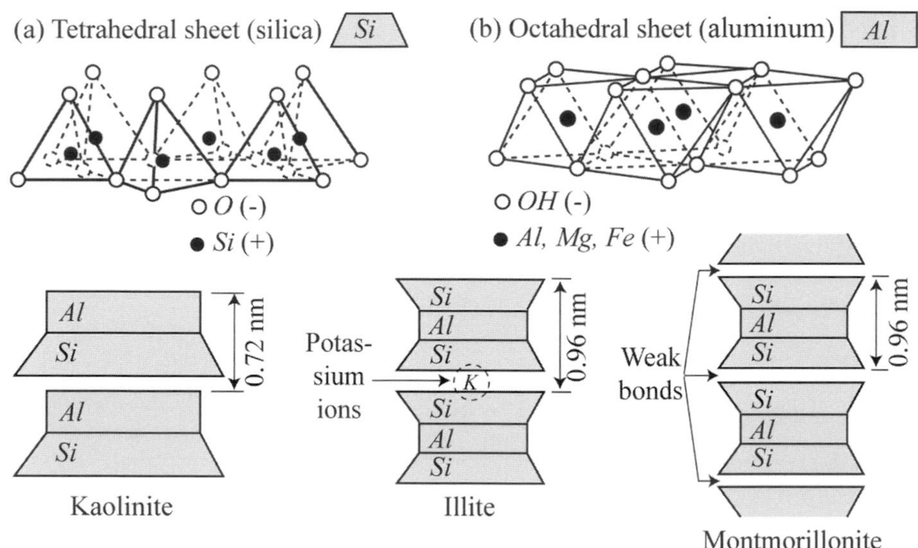

Figure 15.2 Tetrahedral and octahedral sheet, kaolinite, illite and montmorillonite

Illite has a 2:1 structure, meaning that a unit is composed of an octahedral sheet sandwiched between two silica sheets. The unit thickness is about 1 nm = 10 Å. The units are linked by nonexchangeable potassium, and illite will not swell as water cannot easily penetrate between the units. Illite crystals are typically flat, 10,000 nm wide and 30 nm thick, with a moderate specific surface of 80–100 m^2/g.

Montmorillonite also has a 2:1 structure, but the bonds between the units are much weaker than the potassium bonds of illite. As a result, water can easily enter between the units and cause rapid swelling and shrinkage of montmorillonite clays. Montmorillonite crystals are typically very thin, 100–1,000 nm wide and 3 nm thick, with a very high specific surface of 800 m^2/g. Soils with montmorillonite also tend to have higher plasticity.

The cation exchange capacity represents the capacity of a clay to accept cations. Cations in water can be attracted to clay surfaces and higher valence cations can easily replace cations of lower valence. Replacement reactions in soils follow the approximate order

$$Li^+ < Na^+ < H^+ < K^+ < NH_4^+ < Mg^{+2} < Ca^{+2} < Al^{+3} < Zn^{+2} < Fe^{+3} < Ni^{+2} < Cu^{+2}.$$

For instance, the swelling of sodium montmorillonite clays can be significantly reduced by the addition of lime (CaOH) which replaces Na^+ by stronger Ca^{+2} cations.

Atterberg limits define the water content at certain critical stages of the soil behavior. The water content w is measured as the ratio of the weight of water to the weight of solids in a soil sample. The liquid limit LL is the water content (in %) at which the soil behaves like a fluid. The plastic limit PL is the water content (in %) at which a soil reaches a plastic state and the plasticity index $PI = LL - PL$ is the difference in water content between the plastic and liquid limits. Figure 15.3 provides a plasticity chart for different types of clays.

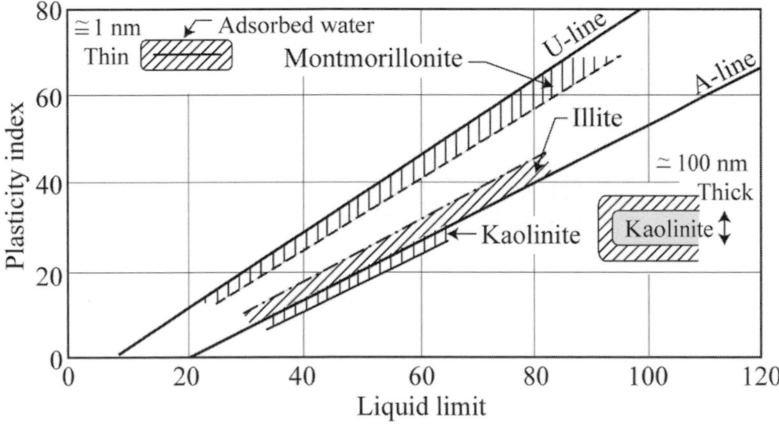

Figure 15.3 Plasticity index chart for different clays (after Holtz and Kovacs, 1981)

The activity of a clay $A = PI/clay\ fraction$ is the ratio of the plasticity index to the clay fraction of a soil. Clays are considered inactive when $A < 0.75$ (e.g. kaolinite), normal when $0.75 < A < 1.25$ (e.g. illite) and active when $A > 1.25$ (e.g. montmorillonite). For instance, calcium montmorillonite is active ($A = 1.5$) while sodium montmorillonite is very active ($4 < A < 7$). This means that, with only 5% montmorillonite, a soil would have a very high plasticity index (e.g. $20 < PI < 35$). Montmorillonite increases swelling, shrinkage and plasticity of soils. Table 15.1 compares the main characteristics of clays and Table 15.2 lists some of the main properties of saturated soils.

Table 15.1. Typical properties of clays

	Specific gravity	Thickness (nm)	Diameter (nm)	Specific surface (m^2/g)	Exchange capacity (Meq/100 g)	Cation	Liquid limit (%)	Plastic limit (%)	Plasticity index (%)	Activity A	Shrinkage limit (%)
Kaolinite	2.62–2.68	50–2,000	100–1,000	10–20	3–8		38–60	27–37	11–23	0.4	25–29
Illite	2.60–2.86	30	10,000	80–100	25–40		95–120	45–60	50–70	0.9	14–18
Montmorillonite	2.75–2.78	3	300–4,000	800–1,000	80–100		290–710	55–100	215–650	7.2	9.3–15
						Na	710	55	650		
						K	660	100	560		
						Ca	510	80	430		
						Mg	410	60	350		
						Fe	290	75	215		

Table 15.2. Typical properties of saturated soils

	Porosity n (%)	Void ratio e (%)	Water content w (%)	ρ_d dry (kg/m^3)	ρ_t total (kg/m^3)	γ_d dry (lb/ft^3)	γ_t total (lb/ft^3)
Glacial till	20	25	9	2,120	2,320	132	145
Dense mixed sand	30	43	16	1,860	2,160	116	135
Dense sand	34	51	19	1,750	2,090	109	130
Stiff glacial clay	37	60	22	1,700	2,070	106	129
Loose mixed sand	40	67	25	1,590	1,990	99	124
Loose sand	46	85	32	1,430	1,890	90	118
Soft glacial clay	55	120	45	1,220	1,770	76	110
Lightly organic clay	66	190	70	930	1,580	58	98
Very organic clay	75	300	110	680	1,430	42	89
Soft bentonite	84	520	194	430	1,270	27	80

Compiled from Lambe and Whitman (1969) and Holtz and Kovacs (1981).

15.1.2 Properties of Wet Soils

Soils divide the total volume between solids, water and air as sketched in Figure 15.4. The void space $\forall_v$ can be filled by water $\forall_w$, and the degree of saturation $S = \forall_w/\forall_v$. For saturated soils, the volume of voids is entirely filled with water ($S = 100\%$). The void ratio $e = \forall_v/\forall_s$ is the ratio of the volume of voids to the volume of solids $\forall_s$. Alternatively, we can use the porosity $n = \forall_v/\forall_t$ as the ratio of the volume of voids to the total volume $\forall_t$. In summary, $e = \forall_v/\forall_s$ and $n = \forall_v/\forall_t$, such that $n = e/(1+e)$ and $e = n/(1-n)$. Porosity n in this chapter should not be confused with Manning's coefficient n for open-channel flows.

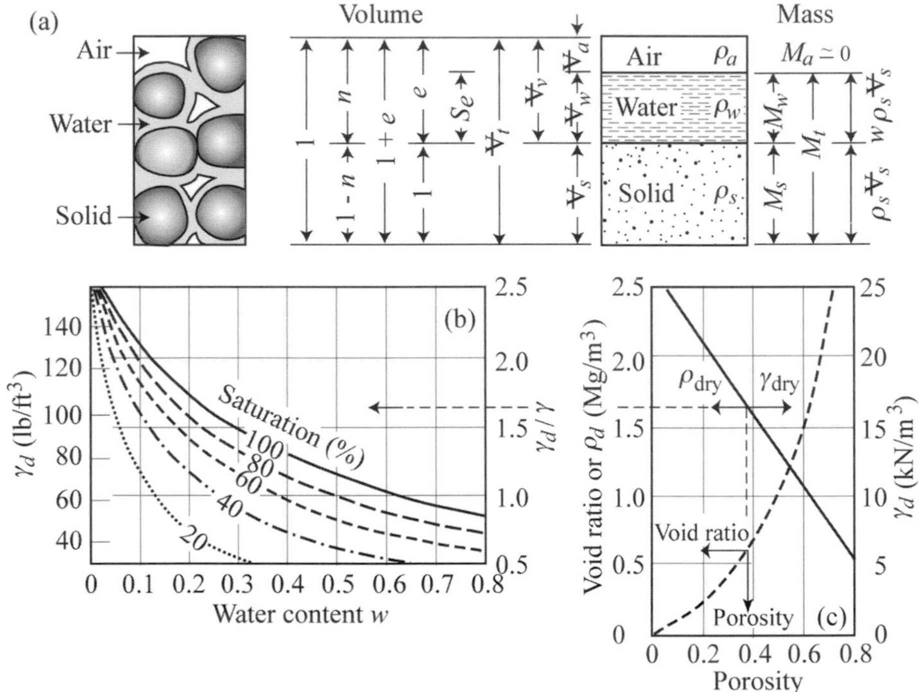

Figure 15.4 Volumetric and mass properties of wet soils

In terms of mass, the density of water is $\rho = M_w/\forall_w$ and the mass density of solids $\rho_s = M_s/\forall_s$. This ratio defines the specific gravity of solids $G = \rho_s/\rho$. The water content is the ratio of the mass (or weight) of water to solids $w = M_w/M_s = W_w/W_s$. The total mass density $\rho_t = M_t/\forall_t = (M_s + M_w)/(\forall_s + \forall_w)$. The specific weight ($\gamma = \rho g$) is the product of mass density and gravity. For unsaturated soils, the mass to volume relationship is $Gw = Se$ (as demonstrated in Example 5.2 below).

The dry mass density of a soil is $\rho_d = M_s/\forall_t = \rho_s/(1+e) = \rho_t/(1+w)$.

The mass density for saturated soils ($\rho_{sat} = \rho_t$ for $S = 100\%$) is $\rho_{sat} = M_t/\forall_t = \rho_s(1+w)/(1+e)$.

The submerged mass density is $\rho' = \rho_{sat} - \rho$.

Typical mass densities are listed in Table 15.3. The basic soil properties for a saturated soil (Example 15.1) and unsaturated soil (Example 15.2) stem from laboratory measurements.

Table 15.3. Typical mass densities and specific weights of soils

Soil type	ρ_{sat} (Mg/m^3)	ρ_d (Mg/m^3)	ρ' (Mg/m^3)		e (%)	n (%)	γ_d (lb/ft^3)
Glacial till	2.1–2.4	1.7–2.3	1.1–1.4	Silty sand and gravel	14–85	12–89	89–146
Sand and gravel	1.9–2.4	1.5–2.3	0.9–1.4	Fine to coarse sand	20–95	17–49	85–138
Crushed rock	1.9–2.2	1.5–2.0	0.9–1.2	Silty sand	30–90	23–47	87–127
Silt and clay	1.4–2.1	0.6–1.8	0.4–1.1	Inorganic silt	40–110	29–52	80–118
Organic silt and clay	1.3–1.8	0.5–1.5	0.3–0.8	Sand	40–100	29–50	83–118
Peat	1.0–1.1	0.1–0.3	0–0.1	Mica sand	40–120	29–55	76–120

♦♦ Example 15.1: Saturated soil properties

Determine the soil properties from the following laboratory measurements: the wet soil in a dish has a mass of 460 g; the sample is dried and the dry sample + dish has a mass of 360 g; and the dish has a mass of 60 g. Assume a solid mass density $G = 2.65$.

Solution:

(1) Mass of water $M_w = 460 - 360 = 100$ g, $\forall_w = 100$ cc, or 100 cm^3.
(2) Mass of dry soil $M_s = 360 - 60 = 300$ g, $\forall_s = M_s/\rho G = 300/2.65 = 113$ cc.
(3) Water content $w = M_w/M_s = 33$ %.
(4) Total mass of wet soil $M_t = 400$ g, $\forall_t = 213$ cc when saturated (no air).
(5) Total mass density $\rho_{sat} = \rho_t = 400/213 = 1.88$ g/cc $= 1.88$ Mg/m^3, $\gamma_t = 1.88(62.4) = 117$ lb/ft^3.
(6) Dry mass density $\rho_d = 300/213 = 1.41$ g/cc $= 1.41$ Mg/m^3, $\gamma_d = 1.41(62.4) = 88$ lb/ft^3.
(7) Submerged mass density $\rho' = 1.88 - 1 = 0.88$ g/cc $= 0.88$ Mg/m^3, $\gamma' = 0.88(62.4) = 55$ lb/ft^3.
(8) Void ratio $e = 100/113 = 0.88$.
(9) Porosity $n = 100/213 = 0.47$; also, $n = e/(1+e) = 0.88/1.88 = 0.47$.
(10) Saturation $S = Gw/e = 2.65(0.33)/0.88 = 100\%$, and $\rho_{sat} = \rho_t$.

♦ Example 15.2: Unsaturated soil properties

Derive the relationship between S, e, w and G and define ρ_t and ρ_{sat}.

Solution: Referring to Fig. E-15.2, we start with a volume of solids $\forall_s$; the volumes of void and water are $\forall_v = e\forall_s$ and $\forall_w = Se\forall_s$.

The total volume is $\forall_t = \forall_s + \forall_w = \forall_s(1+e)$.

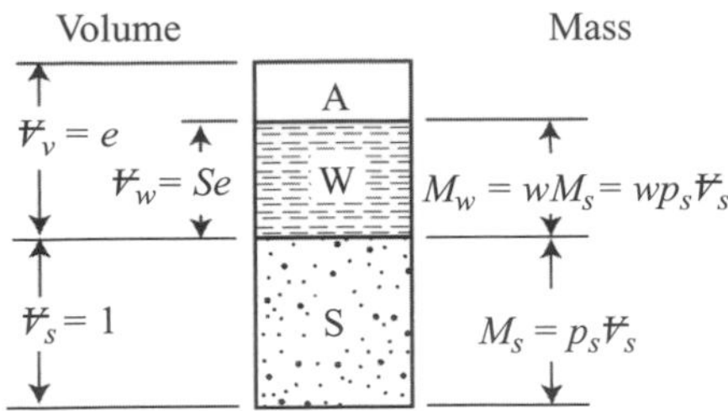

Fig. E-15.2 Unsaturated soils

On the mass side, the mass of solids is $M_s = \rho_s \forall_s$. The mass of water is $M_w = \rho \forall_w = w M_s = w \rho_s \forall_s$.

The total mass is $M_t = M_s + M_w = \rho_s \forall_s (1 + w)$.

Thus, $\forall_w / \forall_s = Se = w\rho_s / \rho = Gw$ from which we convert volume to mass with $Gw = Se$.

The total mass density is $\rho_t = M_t / \forall_t = \rho_s (1 + w) / (1 + e) = \rho(G + Se)/(1 + e)$.

The saturated mass density (when $S = 100\%$) becomes $\rho_{sat} = \rho(G + e)/(1 + e)$.

The dry mass density $\rho_d = M_s / \forall_t = \rho_s \forall_s / \forall_t = \rho G/(1+e) = \rho_t/(1+w)$.

15.2 Wet Soil Processes

This section describes processes such as capillarity (Section 15.2.1), boiling sands (Section 15.2.2), static liquefaction (Section 15.2.3), consolidation and subsidence (Section 15.2.4) and seismic surveys (Section 15.2.5).

15.2.1 Capillarity

Capillarity is a property of the interaction between the surface tension σ_t of water and solid surfaces. For instance, the surface tension of water is $\sigma_t \cong 0.073$ N/m ($\sigma_t \cong 0.005$ lb/ft) at 20 °C. In a capillary tube of diameter d in Figure 15.5, the surface tension exerts an upward force on the water column equal to $F_{cap} = \sigma_t \pi d \cos(\alpha)$. This force is equal to the weight W in the column of height h_c, or $W = \rho g h_c \pi d^2/4$. When $W = F_{cap}$, the capillary rise h_c becomes $h_{c\ cm} = 4\sigma_t \cos(\alpha)/\rho g d \approx 3/d_{mm}$. The capillary pressure at the top of a capillary tube is $u = -\rho g h_c$. Note that the hydrostatic pore pressure in soils is denoted as u instead of p.

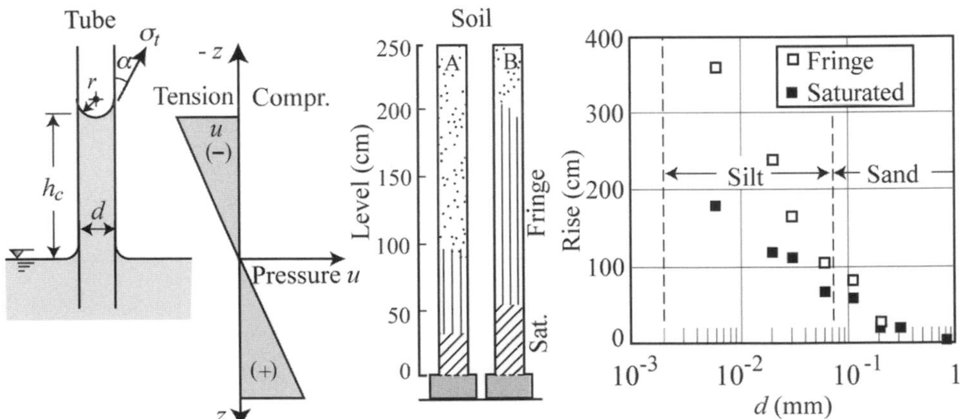

Figure 15.5 Capillary rise in a tube and granular soils

For example, the capillary rise in tubes with diameters of 0.0625 mm and 2 μm (= 2 microns) are, respectively, $h_c = 3/0.0625 = 48$ cm, and $h_c = 3/0.002 = 15$ m. It is interesting to note that the capillary pressure would be respectively $p_c = -4{,}710$ Pa, and $p_c = -147$ kPa for these two diameters. The capillary rise in the smallest tube is impossible due to the maximum height ≈ 10 m before water turns to vapor. Capillarity thus becomes significant in very fine soils (silts and clays).

In granular material, the concept is similar except that the pore diameter varies. The zone below the groundwater table is saturated. The capillary rise is a function of the effective grain diameter d_{10} as shown in Figure 15.5. Two heights can be considered: the saturated layer and the unsaturated capillary fringe (Lambe and Whitman 1969). The saturated zone between the capillary fringe and groundwater table has a negative pore pressure.

15.2.2 Effective Stress and Quick Condition

The concept of effective stress is very important as illustrated in Figure 15.6. Under hydrostatic conditions, the fluid pressure increases linearly with depth (e.g. $u = \rho g h_w$ above the soil sample). The total pressure in the saturated soil sample of length L increases by $\Delta\sigma = \rho_{sat} g L$, such that the total pressure σ on the support screen is $\sigma = (\rho h_w + \rho_{sat} L) g$. The hydrostatic pressure at the base of the screen is $u = \rho g (h_w + L)$. The effective pressure σ' is the intergranular pressure given by the difference between the total pressure and the pore pressure, thus $\sigma' = \sigma - u = (\rho_{sat} - \rho) g L$.

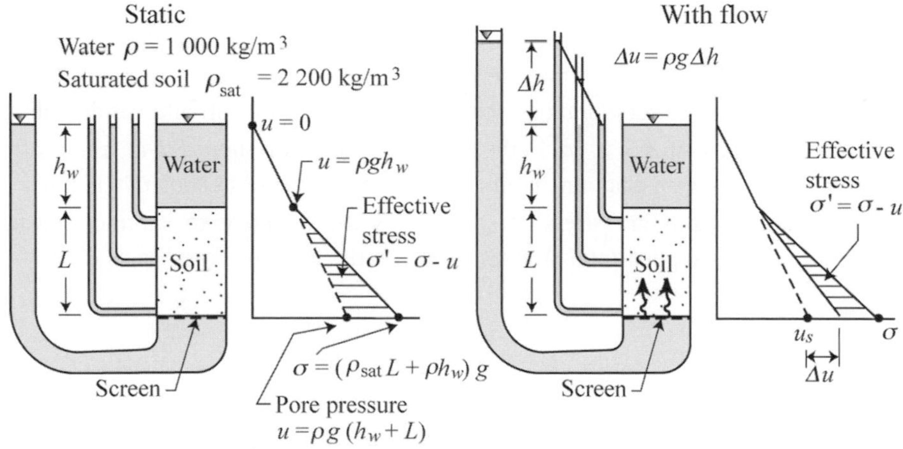

Figure 15.6 Concept of effective stress with groundwater flow

In the case with groundwater flow, the pore pressure increases due to the upward pressure gradient. The piezometric head becomes larger at the screen than at the top of the soil sample. At the screen, the piezometric head increased by Δh up to $L + h_w + \Delta h$, while the piezometric head at the top of the sample is unchanged. The total downward pressure from the weight on the screen is unchanged from hydrostatic conditions. With the flow, the screen pore pressure increases by $\Delta u = \rho g \Delta h$ to

$u = \rho g(L + h_w) + \Delta u = \rho g(L + h_w + \Delta h)$. Therefore, the effective stress in the soil ($\sigma' = \sigma - u$) decreases due to the upward groundwater flow. From this effective stress relationship, the stress between the grains vanishes when $\sigma' = 0$, i.e. $\sigma' = \sigma - u = [(\rho h_w + \rho_{sat} L)g] - [\rho g(L + h_w + \Delta h)] = (\rho_{sat} - \rho)gL - \rho g \Delta h = 0$.

The condition when the force on the screen is zero defines the critical hydraulic gradient i_c from $i_c = \Delta h_c / L = [(\rho_{sat}/\rho) - 1] = (G - 1)/(1 + e)$. We observe that when $G = 2.65$ with $1 > e > 0.5$, we obtain $0.82 < i_c < 1.1$. In practice, we need to avoid $i_c = \Delta h / L \approx 1$, see Example 15.3. This critical gradient causes a boiling, or quick condition, where soils lose their entire bearing capacity.

♦♦ **Example 15.3:** Quick condition

For the case in Figure 15.6 with ρ_{sat} = 2,200 kg/m^3, what is the critical gradient corresponding to boiling condition?

Solution: The answer is simply $i_c = (\rho_{sat}/\rho) - 1 = 1.2$. Alternatively, with G = 2.65, $\rho_{sat} = \rho(G + e)/(1 + e) = 2{,}200$ kg/m^3 yields $e = (G\rho - \rho_{sat})/(\rho_{sat} - \rho) = (2.65 - 2.2)/(2.2 - 1) = 0.375$, and $i_c = \Delta h / L = (G - 1)/(1 + e) = 1.65/1.375 = 1.2$.

15.2.3 Static Liquefaction

Static liquefaction refers to the sudden strength loss of loaded soils under undrained conditions. Essentially, the resistance of noncohesive soils can be examined in Figure 15.7 where the shear stress τ is a function of the soil pressure along the vertical σ_v and horizontal σ_h axes. The plane of failure of a Mohr–Coulomb diagram is at an angle ϕ. We can also define the average pressure $p = 0.5\,(\sigma_h + \sigma_v)$ and half difference $q = 0.5\,(\sigma_v - \sigma_h)$. At the condition of failure, p_f and q_f, the plane of failure is described at an angle ϕ defined as $\sin\phi = q_f/p_f$ in the τ–σ plane. This corresponds to an angle α from $\tan\alpha = q_f/p_f$ on the p–q diagram. In other words, $\tan\alpha = \sin\phi$ as shown in Figure 15.7.

To describe intergranular stress in saturated soils, we have to subtract the pore pressure u from the applied total stresses to define the effective stresses as $\sigma' = \sigma - u$,

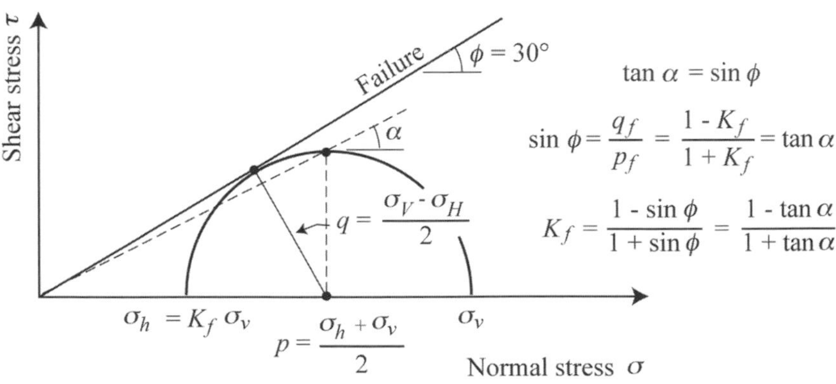

Figure 15.7 Mohr–Coulomb diagram

as shown in Figure 15.8a. The plane of failure for effective stresses is now defined by the angle of failure ϕ' in the Mohr–Coulomb diagram or by the angle α' in the $p' - q'$ diagram given $\tan\alpha' = \sin\phi'$. According to Olson (1974), typical values for α' are $25° < \alpha' < 35°$ for quartz, $20° < \alpha' < 25°$ for kaolinite, $15° < \alpha' < 25°$ for illite, and $5° < \alpha' < 15°$ for montmorillonite. Other useful relationships include the ratio $K_f{}' = \sigma_h{}'/\sigma_v{}'$ at failure such that $K_f{}' = (1 - \tan\alpha')/(1 + \tan\alpha')$, which can also be rewritten as $\tan\alpha' = (1 - K_f{}')/(1 + K_f{}')$.

This means that a saturated soil is more likely to fail because the effective stresses are much closer to the plane of failure than a dry soil (see Figures 15.8b and c). Therefore, properly drained soils retain their bearing capacity and are more stable.

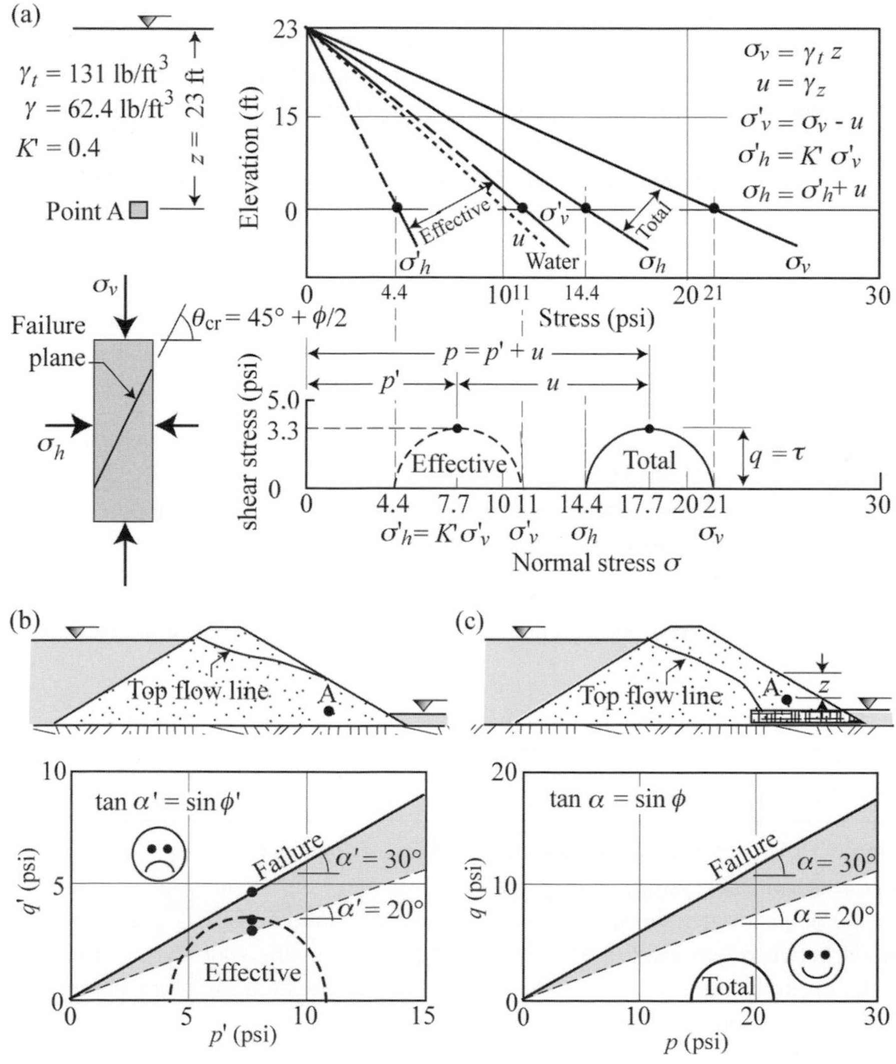

Figure 15.8 Total and effective stresses

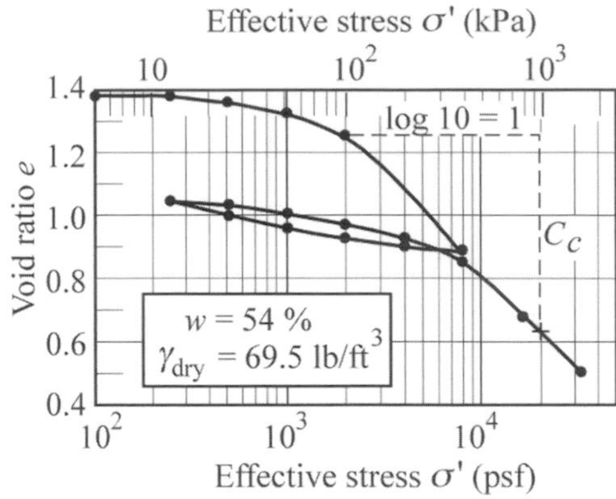

Figure 15.9 Oedometer test

15.2.4 Consolidation and Subsidence

Soil consolidation refers to the decrease in soil volume by the drainage of pore water under pressure. Soils under an increase in effective stress from $\sigma_1{}'$ to $\sigma_2{}'$ consolidate over time due to the decrease in void ratio from e_1 to e_2. From a consolidation test (called oedometer test), the change in void ratio over the log of the ratio of effective stresses defines the compression index $C_C = (e_1 - e_2)/\log(\sigma_2{}'/\sigma_1{}')$.

For the oedometer test in Figure 15.9 starting at $\sigma_1{}' = 2{,}000$ psf and $e_1 = 1.25$, the void ratio would decrease to $e_2 = 0.63$ as the pressure increases to $\sigma_2{}' = 20{,}000$ psf. The compression index over one log cycle in pressure is thus $C_C = (1.25 - 0.63)/\log(10) = 0.62$; see Example 15.4.

When consolidation tests are not available, an estimate from the formula of Terzaghi and Peck (1967) where LL is the liquid limit (%) is $C_C \cong 0.009(LL - 10)$.

To calculate the ultimate settlement s_u from an increase in pressure from $\sigma_1{}'$ to $\sigma_2{}'$, we can use $s_u = H\Delta e/(1+e) = HC_C{}'\log(\sigma_2{}'/\sigma_1{}')$, where the modified compression index $C_C{}' = C_C/(1+e_1)$ from the initial void ratio e_1 and the layer thickness H.

Subsidence refers to lowering of the ground surface from groundwater pumping. It is the result of an increase in effective stress dues to the reduction of pore pressure. Example 15.4 illustrates settlement calculations and Example 15.5 calculates subsidence from long-term pumping.

♦ **Example 15.4:** Long-term consolidation

Estimate the consolidation settlement of a 30-ft layer of soil from Figure 15.9 if the effective pressure increases from 2,000 to 8,000.

Solution: $s_u = 30\,[0.62/(1+1.25)]\log(8{,}000/2{,}000) = 5$ ft.

Alternatively, we consider the change in dry density as the void ratio decreases from 1.25 to 0.88

$$s_u = H[(e_1 - e_2)/(1+e_1)] = 30(1.25 - 0.88)/(1+1.25) = 4.9 \text{ ft}.$$

Example 15.5: Long-term subsidence from groundwater pumping

An $H = 20$-m thick saturated clay layer has a water content $w = 40\%$, a liquid limit $LL = 60\%$, a saturated density $\rho_{sat} = 1.82$ g/cc and a solid density $G = 2.7$. The layer is drained between two sand layers. The initial pressure in the clay layer is $\sigma = 728$ kPa and the groundwater level in this area will be lowered by 30 m for a long period of time. Estimate the ultimate settlement for this layer.

Solution: The saturated void ratio is $e_1 = Gw = 2.7 \times 0.4 = 1.08$, and the compression index is $C_C \cong 0.009(60 - 10) = 0.45$ with a decrease in pore pressure

$\Delta u = \gamma \Delta h = 9{,}810 \times 30 = 294$ kPa. This corresponds to an increase in effective stress and the settlement is

$$s_u \cong H[C_C/(1+e_1)] \log(\sigma_2'/\sigma_1') = 20[0.45/(2.08)] \log[(728+294)/728] = 0.64 \text{ m}.$$

The timescale for the consolidation of soils is now examined. The consolidation process for the settlement s as a function of time follows the diffusion equation $ds/dt = c_v d^2s/dz^2$ where c_v is the consolidation coefficient – note that c_v is different from C_c. Therefore, the consolidation time becomes of interest for clayey soils with very low permeability. From the consolidation theory named after Arthur Casagrande, a saturated soil layer of thickness Z drains on one side ($H = Z$). For a layer of thickness Z between two highly pervious surfaces, we use $H = Z/2$. The layer drains with time t, as shown in Figure 15.10, where the dimensionless consolidation time is $T = c_v t/H^2$.

Casagrande and Taylor defined the consolidation percentage $U_{\%} = 100\,U$ as a function of time $T = \pi U^2/4$ when $U < 0.6$ (or $U_{\%} < 60\%$) and $T = 1.781 - 0.933 \log(100 - U_{\%})$ when $U_{\%} > 60\%$.

As shown in Figure 15.10b, the consolidation is almost complete ($U_{\%} = 93\%$) when $T = 1$, and we also notice that half the consolidation ($U_{\%} = 50\%$) takes place when $T = 0.2$.

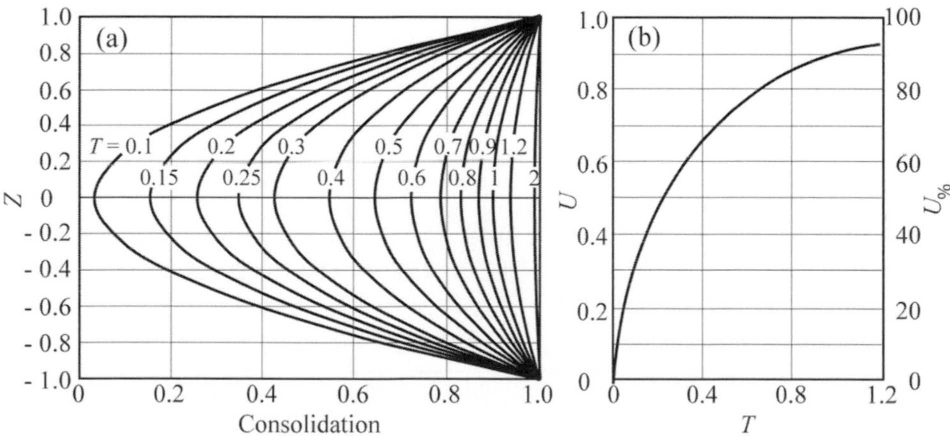

Figure 15.10 Consolidation percentage as a function of time

The consolidation coefficient c_v is most important for clays and consolidation times become very significant for soils with high plasticity indices and high liquid limits. Some empirical relationships from the US Navy (1971) are: (1) $c_v\ (\text{m}^2/\text{s}) \cong PI^{-5}$; and (2) $c_v\ (\text{cm}^2/\text{s}) \cong 200LL^{-3}$ where $PI(\%) = LL - PL$ is the plasticity index (in %), the liquid limit (LL%) and the plastic limit (PL%). For instance, at a liquid limit of $LL = 60\%$ we obtain $c_v \cong 200 \times 60^{-3} = 0.001\ \text{cm}^2/\text{s}$.

Consolidation becomes particularly important when thick clay layers with high plasticity and liquid limit consolidate very slowly, as found in Example 15.6 and Case Study 15.1.

♦ **Example 15.6:** Subsidence time for clay layers

Estimate the 93% subsidence time for the clay layer in Example 15.5.

Solution: A first estimate of c_v from the liquid limit is obtained from $c_v \cong 200 \times 60^{-3} = 0.001\ \text{cm}^2/\text{s} = 2.9\ \text{m}^2/\text{year}$.

From $H = Z/2 = 10$ m between two sand layers, we reach 93% of the subsidence when $t_{93} \cong H^2/c_v = 100/2.9 = 34$ years, and half the subsidence $t_{50} \cong 0.2H^2/C_v = 0.2 \times 34 = 6.8$ years. The time is sufficiently long to justify a consolidation test with a sample from this clay.

Case Study 15.1: Consolidation properties of clays

Consider the clay properties in Table CS-15.1 assembled from Holtz and Kovacs (1981). From this sample, plot the void ratio e and the measured compression index C_c (and C_c') as a function of the liquid limit. The values of c_v and t_{93} for an $H = 10$-m layer are estimated from the above relationships.

Table CS-15.1. Sample of clay properties

	LL (%)	PL (%)	PI (%)	w_n (%)	e_0	σ_p' (kPa)	C_c	C_c'	c_v (m^2/year)	t_{93} (years)
Sandy clay	24	12	12	11	0.37	370	0.08	0.06	> 45	1–2
Sandy clay	27	14	13	11	0.32	900	0.11	0.08	31–84	1–3
ML sandy silt	31	17	14	30	0.83	350	0.16	0.09	21–60	2–5
Clay till	34	20	14	23	0.65	420	0.19	0.12	16–60	2–6
CL soft clay	41	24	17	34	0.94	200	0.34	0.18	9–20	4–12
CL firm clay	50	23	27	36	1	250	0.44	0.22	2–5	20–40
CH clay+silt	71	28	43	43	1.17	290	0.52	0.24	0.2–2	> 50
CH soft clay	81	25	56	51	1.35	350	0.84	0.36	0.1–1	> 100

Here, M is for silt and C is for clay. The L is for low, and H for high, plasticity. Thus, an ML soil denotes a low-plasticity silt.

Solution: Table CS-15.1 is quite instructive, and results are plotted in Fig. CS-15.1. The void ratio and the compression indices (C_c and C_c') increase with the liquid limit of a soil. It also turns out that the parameter $C_c' = C_c/(1 + e_0)$ is most interesting since it simply equals the unit consolidation $C_c' = s_u/H$ when the effective stress increases by one order of magnitude. As an example, for the CH soft clay in Table CS-15.1, if the effective stress increases from 350 kPa to 3,500 kPa, a 1-m layer of soil would ultimately consolidate by 36 cm. Consequently, soils with high liquid limits will experience the largest consolidation. We also learn that thick layers of soils with high liquid limits will consolidate very slowly. The time t_{93} to achieve 93% of the consolidation increases with the liquid limit. In this regard, it is interesting that the value $C_c' \simeq w_n(\%)/200$ from this sample is similar to the observations of Fadum (1941).

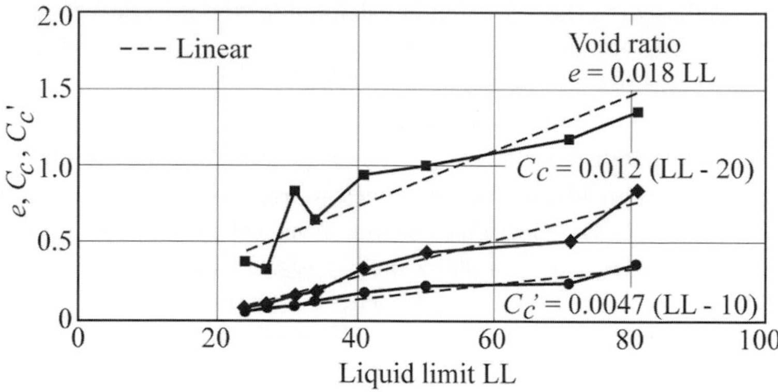

Fig. CS-15.1 Compression indices for various clays

15.2.5 Seismic Surveys

Seismic surveys provide a nonintrusive estimate of the groundwater-table elevation. Seismic waves propagate through the ground as pressure waves (P-waves) or shear waves (S-waves). S-waves do not propagate well in gas or fluids because the motion is perpendicular to the direction of the wave propagation. The propagation velocity of primary waves increases with soil density and water content with a range from 250 m/s in loose unsaturated material up to 5 km/s in dense crystalline rock. For instance, the sound velocity in air is ~300 m/s, ~500 m/s in loose soils and up to 1 km/s in dry sand and gravel mixtures. The velocity reaches 1.4 km/s in clear water, up to 1.7 km/s in saturated sands and gravels, 1.5–1.8 km/s in clays, 1.8–2.6 km/s in sandstones and shales, and 3.0–7.0 km/s in solid bedrock. Figure 15.11 locates a groundwater table from a seismic survey with a geophone line (A–H) spaced 25 ft apart, starting at the point source at $x = 0$.

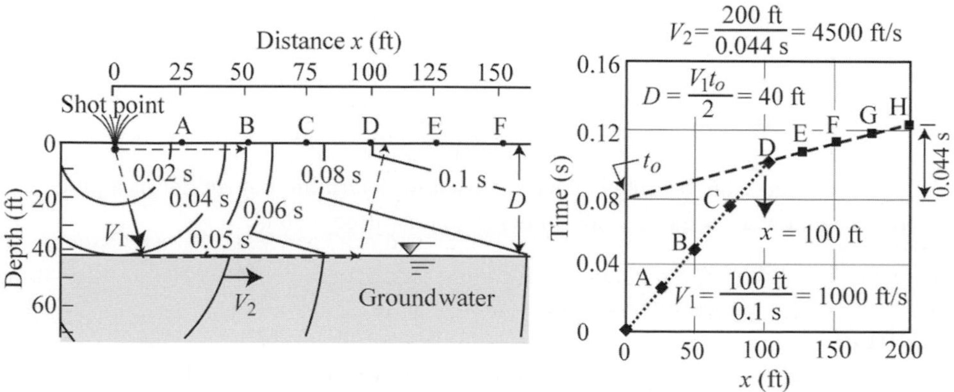

Figure 15.11 Seismic survey for detecting groundwater level

The approach based on the refraction law determines the thickness D of a homogeneous layer. The following formula gives D from the distance X at which the two lines intersect at two velocities V_1 and V_2, as $D = 0.5x\sqrt{(V_2 - V_1)/(V_2 + V_1)}$.

From the example in Figure 15.11 where $x = 100$ ft, $V_1 = 100/0.1 = 1{,}000$ ft/s and $V_2 = 200/0.044 = 4{,}500$ft/s yields $D = 0.5 \times 100\sqrt{3{,}500/5{,}500} = 39.9$ft.

Two other methods are also available: (1) we can use $D = 0.5x[1 - (V_1/V_2)] = 0.5\times 100\,[1 - (1/4.5)] = 38.9$ ft or (2) from the time intercept t_0 at $x = 0$ in Figure 15.11. In this case, $D = 0.5V_1t_0 = 0.5 \times 1{,}000 \times 0.08 = 40$ ft.

The difference between these three approaches is less than 5% when $V_1 < 0.33V_2$.

Additional Resources

There is ample more information about the material covered in this chapter in geotechnical engineering textbooks such as Terzaghi and Peck (1967), Lambe and Whitman (1969), Holtz and Kovacs (1981), Das (1983) and Lu and Godt (2013).

EXERCISES

These exercises review the essential concepts from this chapter.

1. Why is the plasticity of montmorillonite higher than for kaolinite at a given liquid limit?
2. What is the main relationship to convert mass to volume in soils?
3. Why is it possible to have a negative pore pressure?
4. Why should engineers avoid the quick condition?
5. Why is the drainage of permeable dams beneficial?
6. Does the compression index increase or decrease with the liquid limit?
7. Does the rate of consolidation increase or decrease with the liquid limit?
8. Does the plasticity index of a soil normally increase or decrease with grain size?
9. Do seismic waves travel faster in bedrock or in loose soils?
10. True or false?
    (a) Kaolinite is more prone to swelling than montmorillonite.
    (b) Soil plasticity increases with the montmorillonite clay content.
    (c) In saturated soils, the void ratio equals Gw.
    (d) The dry specific weight is the same as the submerged specific weight.
    (e) Capillarity is at a maximum in very fine-grained soils.
    (f) Quick sands happen when the effective stress of a soil reduces to zero.
    (g) Groundwater pumping causes subsidence.
    (h) Clayey soils consolidate faster than silty soils.
    (i) The three-edge bearing capacity is used to determine the weight of a pipe.
    (j) Seismic waves travel faster in dense and saturated soils.

PROBLEMS

1. Classify the soils from the particle size distributions in Fig. P-15.1.
    (a) Uniform silt loam, $LL = 30$ and $PI = 3$
    (b) Silt loam, $LL = 34$ and $PI = 3$, from Holtz and Kovacs (1981).

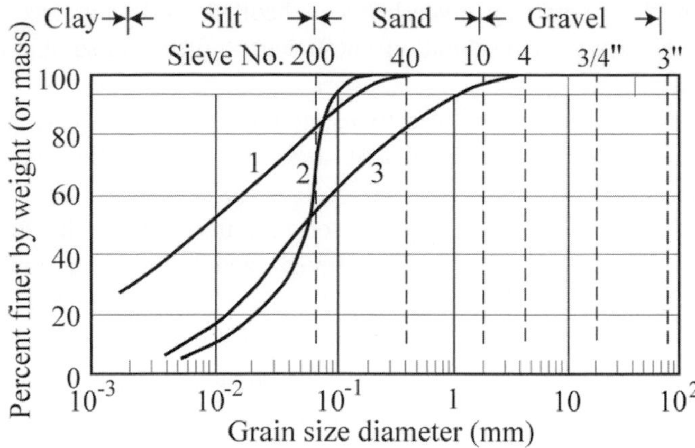

Fig. P-15.1

2. What is the soil type for the particle sizes in Fig. P-15.2 given $LL = 33$, $PL = 23$ and $PI = 10$.

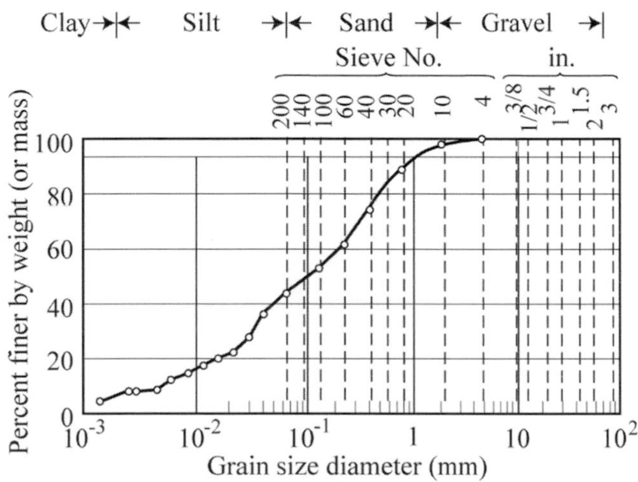

Fig. P-15.2

3. ♦♦ A soil is being mechanically compacted and its dry density increases from 102 to a maximum 109 lb/ft^3 as the water content increases from 13% to 18%. If the specific gravity of solids is 2.65, determine the corresponding void ratios and porosities, as well as total unit weights and degrees of saturation. Remember, $\gamma_t = (1 + w)\gamma_d$, $e = (G\gamma/\gamma_d) - 1$, $n = e/(1 + e)$ and $Se = Gw$.
4. ♦♦ The dry density of a soil decreases from a maximum of 1.75 Mg/m^3 to 1.65 Mg/m^3 as the water content increases from 18% to 22%. If the specific gravity of solids is 2.65, determine the corresponding total unit weights, void ratios, porosities and degrees of saturation. Remember, $\rho_t = (1 + w)\rho_d$, $e = (G\rho/\rho_d) - 1$, $n = e/(1 + e)$ and $Se = Gw$.

5. ♦♦ A 12-m-thick soft clay layer between bedrock and sand has a liquid limit $LL = 71$. Estimate the consolidation coefficient and the consolidation percentage after 10 years.
6. ♦♦ A 36-ft-thick soil layer between two sand layers has a plasticity index $PI = 26$. Estimate the consolidation coefficient and the consolidation percentage after $t = 5$ years.
7. ♦ A seismic survey gives the following times for seven geophones separated by 20 m, with $T = 57, 114, 129, 143, 157, 171, 186$ milliseconds. Find the velocities V_1 and V_2 as well as the depth at which the groundwater table is located.
8. ♦ A seismic survey in Fig. P-15.8 gives the following times for 9 geophones separated by 25 ft, with $T = 25, 50, 75, 100, 108, 113, 119, 124, 130$ milliseconds. Find the velocities V_1 and V_2 and the groundwater table depth.

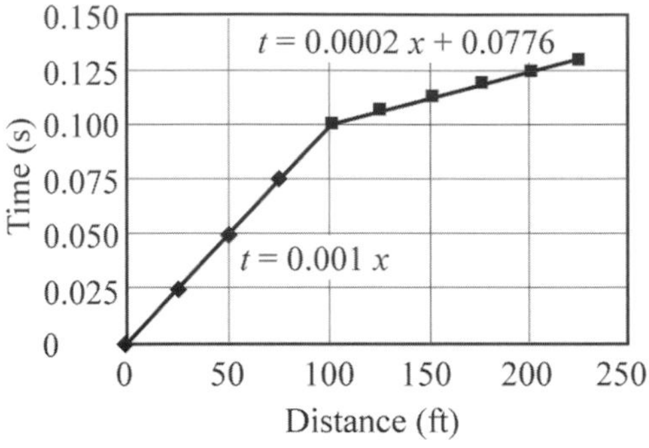

Fig. P-15.8

16 Groundwater

This chapter reviews groundwater flows in terms of permeability in Section 16.1, steady flow in Section 16.2 and unsteady groundwater flow in Section 16.3.

As sketched in Figure 16.1a, infiltrated water permeates into soil pores and the deep percolation reaches the groundwater table to form unconfined aquifers. Confined aquifers are found between two impervious layers (aquicludes) or near-impervious layers called aquitards. The flow in confined aquifers is pressurized.

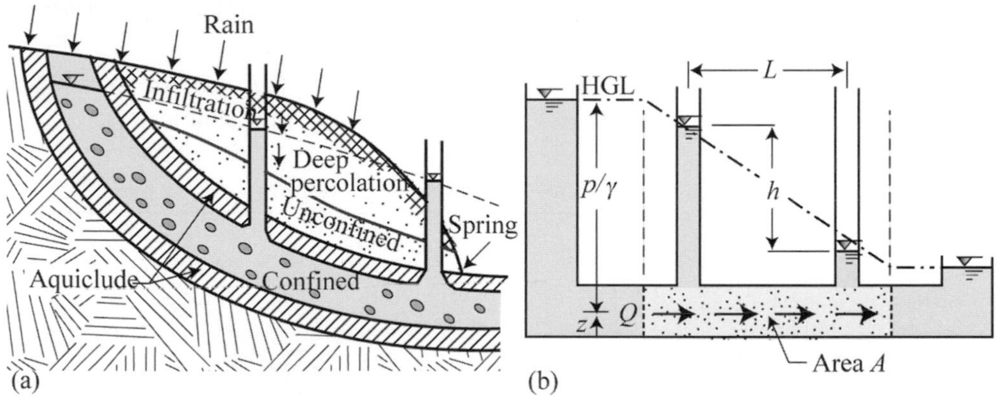

Figure 16.1 Sketch of aquifers and Darcy's experiment

16.1 Permeability

In 1856, Henry Darcy carried out experiments on the permeability of saturated sands. As shown in Figure 16.1b, the permeability coefficient K can be defined experimentally from the flow rate Q, the cross-sectional area A and the drop in piezometric elevation h over the distance L. Darcy's law is

$$V = \frac{Q}{A} = K\left(\frac{-h}{L}\right) = K\left(\frac{-dh}{dx}\right) = Ki. \tag{16.1}$$

The slope of the piezometric line defines the gradient $i = -dh/dx$ in the flow direction x. The negative sign shows a velocity in the direction of decreasing piezometric head. Most groundwater flows are laminar ($\mathrm{Re} < 10$), with exceptions for flows with large pore spaces such as rock aquifers. The velocity V is the mean flow velocity over the entire cross-sectional area while the velocity V' in the pores is $V' = V/n$ where n is the porosity.

A closer look at permeability allows us to separate the permeability coefficient K from the intrinsic permeability k, such that $k = Cd^2$, and

$$K = gk/\upsilon = gCd^2/\upsilon, \tag{16.2}$$

as a function of pore size d, gravity g, fluid kinematic viscosity υ and a dimensionless coefficient C. The two permeability coefficients are: (1) the intrinsic permeability k in L^2; and (2) the permeability coefficient K in L/T.

The typical unit used for the intrinsic permeability k is the darcy, 1 darcy $= 0.987 \times 10^{-8}$ cm^2 $= 1.062 \times 10^{-11}$ ft^2. Converting the intrinsic permeability k to a permeability coefficient K at 20 °C ($\upsilon = 1 \times 10^{-6}$ m^2/s and $g = 9.81$ m/s^2), gives $K = 1$ cm/s $= 40.7$ l/m$^2 \cdot$ d when $k \simeq 1 \times 10^{-5}$ cm^2. Also, 1 darcy gives 0.84 m/d $=$ 1×10^{-3} cm/s $=$ 2.7 ft/d $=$ 20 gal/ft$^2 \cdot$ d. For the US, 1 ft/d $=$ 7.48 gal/ft$^2 \cdot$ d; 1 ft^3 $=$ 7.48 US gallons and 1 US gallon $=$ 3.785 liters.

The permeability coefficients vary widely as a function of grain size and substrate as indicated in Figure 16.2. Example 16.1 shows permeability conversion factors.

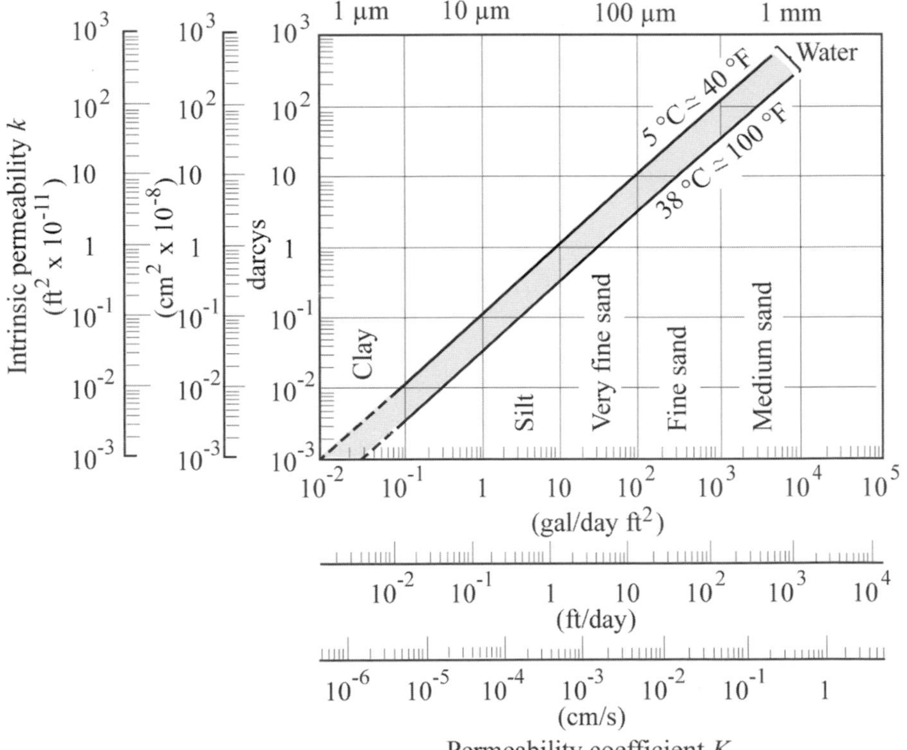

Figure 16.2 Intrinsic permeability and permeability coefficients

---

♦ **Example 16.1:** Permeability coefficients

A uniform coarse sand ($d_s = 1$ mm) has a permeability coefficient $K \simeq 1$ cm/s. Calculate K in m/d and in gal/ft^2 · d, and the intrinsic permeability in cm^2 and in darcy units at 20 °C.

Solution:

$K = 1$ cm/s $= 86{,}400/100 = 864$ m/d $= 2{,}830$ ft/d
$K = 864 \times 3.28 \times 7.48 = 21{,}200$ gal/ft^2 · d
$k = K\upsilon/g = 10^4 \times 0.01 \times 10^{-6}/9.81 = 1.02 \times 10^{-5}$ cm^2 $= 1.1 \times 10^{-8}$ ft^2
$k = 1.02 \times 10^{-5}/0.987 \times 10^{-8} = 1{,}030$ darcy.

It has been known since Jean Poiseuille in the mid-nineteenth century that for laminar flow in small tubes, the velocity is proportional to the square of the diameter. We owe to Allen Hazen the corresponding approximate formulation for sandy soils that the permeability coefficient is proportional to the square of the effective diameter d_{10} of the particle size distribution for which 10% by weight is finer, or

$$K\,(\text{cm/s}) \cong 100 d_{10}^2\,(\text{cm}). \tag{16.3}$$

As a useful reference in Figure 16.3a, the permeability of very coarse sand (1 mm) is approximately 1 cm/s; medium sand, 0.1 cm/s; fine sand, 0.02 cm/s; and very fine sand, 0.005 cm/s. The permeability of sandy river deposits is typically 0.02 cm/s $< K <$ 0.2 cm/s. For silts, the permeability coefficient is typically 0.00001 cm/s $< K <$ 0.001 cm/s, while clays are nearly impervious: 10^{-6} cm/s $< K < 10^{-5}$ cm/s for kaolinite and $10^{-10} < K < 10^{-7}$ cm/s for montmorillonite.

In a modification of the Kozeny–Carman equation, Irmay (Todd, 1964) formulated permeability to include the effects of grain size, porosity and the degree of saturation. The permeability coefficient for unsaturated soils is defined as

$$K = C d_{10}^2 (S - S_0)^3 \left[\frac{n^3}{(1-n)^2}\right] = C d_{10}^2 (S - S_0)^3 \left[\frac{e^3}{(1+e)}\right],$$

where C is a constant, d_{10} is the effective particle diameter, S is the degree of saturation, S_0 is the degree of saturation for nonmoving water (i.e. adsorbed to the solids) and n is the porosity or void ratio e as $n^3/(1-n)^2 = e^3/(1+e)$. The interest in this formulation, shown in Figure 16.3, is to show the relative effect on permeability of: (1) particle size d_{10}; (2) net degree of saturation $(S - S_0)$; and (3) soil porosity n. We learn that the particle diameter is the primary factor to determine the permeability coefficient of saturated soils while permeability also increases with porosity. For unsaturated soils, the permeability coefficient decreases very rapidly with the degree of saturation. It is important to realize that at 50% saturation the permeability coefficient will decrease by at least one order of magnitude. The effects of water temperature from Eq. (16.2) are less pronounced with permeability increasing slightly with temperature. For instance, the effects of porosity on permeability are examined in Example 16.2.

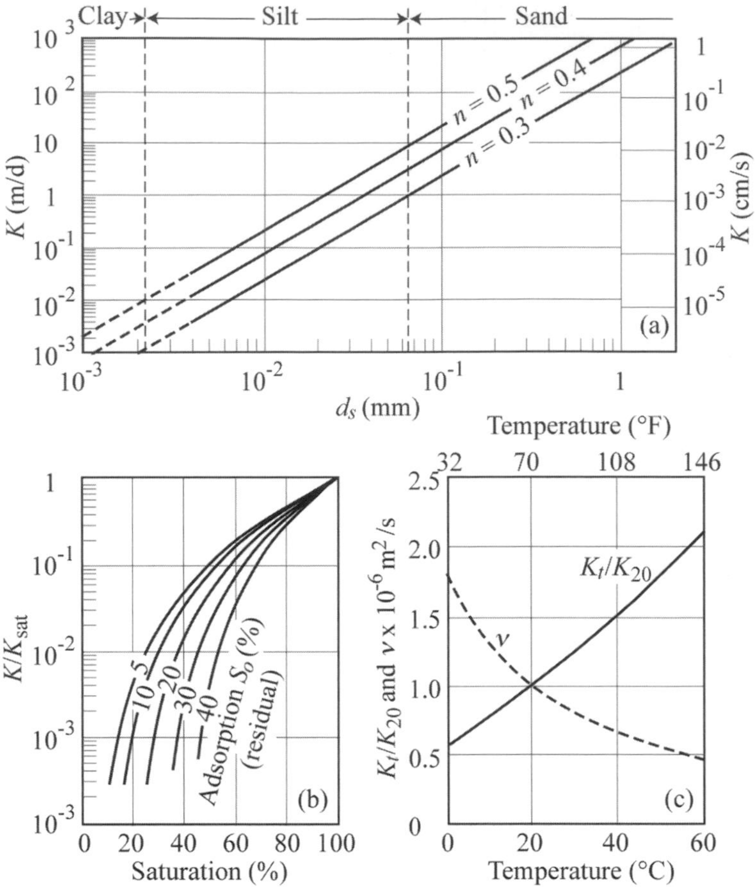

Figure 16.3 Effects of diameter, saturation and viscosity on permeability

---

♦ **Example 16.2:** Porosity effects on permeability coefficients

Estimate the permeability coefficient K in cm/s for a soil given $d_{10} = 0.0625$ mm. If the void ratio $e = 0.89$ decreases by 30% through compaction, what would be the effect on K?

Solution: From Hazen's formula, $K \approx 100 \times 0.00625^2 = 0.004$ cm/s or 3.4 m/d. With $e = 0.89$, we obtain $e^3/(1+e) = 0.89^3/1.89 = 0.37$.

A decrease in void ratio to $e = 0.7 \times 0.89 = 0.623$ would give $e^3/(1+e) = 0.623^3/1.623 = 0.149$. The compaction would result in a $[(0.37 - 0.149)/0.37] = 60\%$ decrease in permeability coefficient from $K = 3.4$ m/d to $K \cong 3.4 \times 0.149/0.37 = 1.4$ m/d.

16.2 Steady Groundwater Flow

This section covers steady one-dimensional flow in aquifers (Section 16.2.1) and seepage through embankments (Section 16.2.2). For two-dimensional flows (Section 16.2.3), we use flow nets for deep foundations (Section 16.2.4) and shallow foundations (Section 16.2.5).

16.2.1 Steady Flow in Aquifers

The concept of aquifers is sketched in Figure 16.4. Impervious layers are called aquicludes when essentially impermeable and aquitards when they have a very low permeability compared to the permeable layers. A confined aquifer is a saturated layer of pervious material between two impervious layers, for example a sand layer between two clay layers. The water level in unconfined aquifers corresponds to the groundwater table, while the piezometric line defines confined aquifers.

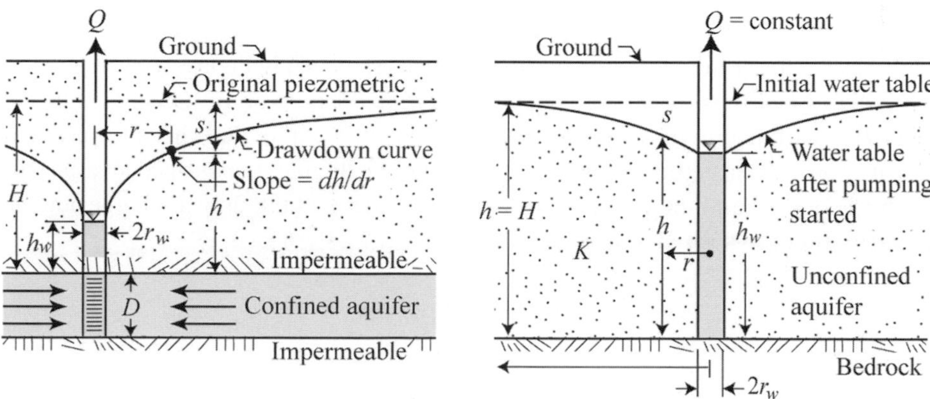

Figure 16.4 Confined and unconfined aquifers

Steady flow from a pumped well in a confined aquifer can be analyzed by assuming that Darcy's law is applicable and vertical accelerations are neglected, which is called the Dupuit–Forchheimer assumption to honor Jules Dupuit in France and Philipp Forchheimer in Austria. For the case of an aquifer of constant depth D, the cylindrical flow area at a radius r from the well is $A = 2\pi rD$. The flow velocity from Darcy's law is $V = Ki = Kdh/dr$ and the steady-state flow discharge becomes $Q = AV = 2\pi rDKdh/dr$. Separating the variables r and h and integrating $Qdr/r = 2\pi KDdh$ gives

$$Q = 2\pi T(h_2 - h_1)/\ln(r_2/r_1). \tag{16.4}$$

The aquifer transmissivity $T = KD$ with dimensions L^2/T, see Example 16.3.

♦♦ **Example 16.3:** Confined aquifer

A pumping well in a 20-ft-deep confined aquifer yields 450 gallons per minute. If the piezometric lines at a distance of 100 ft and 200 ft from the well indicate a piezometric difference of 2 ft, determine T and K.

Solution:

$Q = 450/(60 \times 7.48) = 1$ cfs $= 0.0284$ m^3/s,
$T = Q\ln(r_2/r_1)/[2\pi(h_2 - h_1)] = \ln(2)/4\pi = 0.055$ ft^2/s $= 0.0051$ m^2/s $= 440$ m^2/d,
$K = T/D = 0.055/20 = 0.00275$ ft/s $= 72$ m/d.

In the case of unconfined aquifers, the flow area is $A = 2\pi rh$ and the velocity is $V = Kdh/dr$. Separating the variables in $Q = 2\pi rhK(dh/dr)$ yields

$$Q = \pi K\left(h_2^2 - h_1^2\right)/\ln\left(r_2/r_1\right), \tag{16.5}$$

This approximation is valid for horizontal flow (Dupuit–Forchheimer condition).

16.2.2 Seepage through Shallow Embankments

In the case of linear flow through the embankments sketched in Figure 16.5, a correction of the free discharge surface h_s can be estimated from the ratio of the length L to the flow depth H, as a function of the downstream flow depth h_0; note that the exit flow depth $h_e = h_0 + h_s$. When vertical flow components are negligible, the surface area per unit width b is $A = bh$ and the velocity is $V = Kdh/dx$ such that the flow discharge obtained by integration gives the Dupuit parabola:

$$Q = (Kb/2L)\left(H^2 - h_e^2\right). \tag{16.6}$$

An application of embankment seepage is shown in Example 16.4.

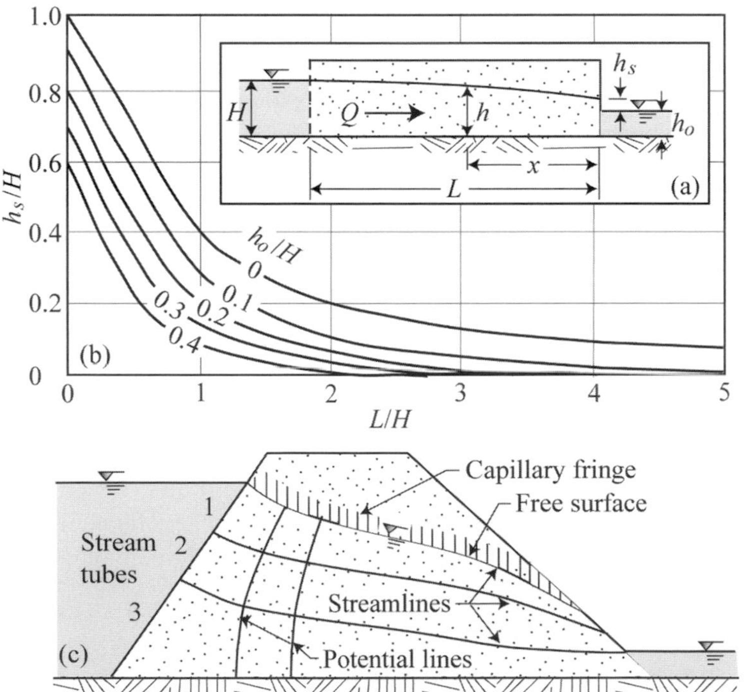

Figure 16.5 Dupuit parabola (modified after Mansur and Kaufman 1962)

♦ **Example 16.4:** Seepage through a levee

A 400-m-long levee is 20 m long at the base with permeability $K = 0.002$ cm/s. Find the seepage flow rate if the upstream flow depth is 6 m and the downstream end is dry.

Solution: With $L = 20$ m and $H = 6$ m, or $L/H = 3.33$, we obtain $h_s = h_e = 0.12 \times 6 = 0.72$ m, and $Q = (Kb/2L)(H^2 - h_e^2) = [2 \times 10^{-5} \times 400/(2 \times 20)](6^2 - 0.72^2) = 0.007\,\text{m}^3/\text{s}$.

16.2.3 Two-Dimensional Groundwater Flow

The seepage through deep earth embankments leads to a two-dimensional analysis of flow conditions (Polubarinova-Kochina 1962). A flow net is a network of orthogonal lines describing streamlines and potential lines shown in Figure 16.6.

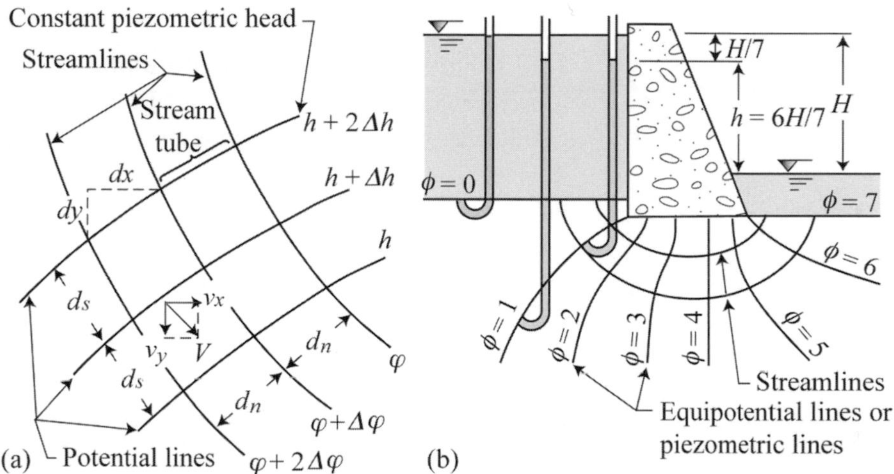

Figure 16.6 Two-dimensional flow nets

Water flows along streamlines φ such that $v_x = -\partial\varphi/\partial y$ and $v_y = \partial\varphi/\partial x$ in the direction of the decreasing potential ϕ, i.e. $v_x = -\partial\phi/\partial x$ and $v_y = -\partial\phi/\partial y$. In the case of groundwater flow, a potential line is equivalent to the piezometric line as shown in Figure 16.6. In two dimensions, the conservation of mass is combined with Darcy's law, i.e. $v_x = -K\partial h/\partial x$ and $v_y = -K\partial h/\partial y$ to give the Laplace equation:

$$\frac{\partial v_x}{\partial x} + \frac{\partial v_y}{\partial y} = -K\left(\frac{\partial^2 h}{\partial x^2} + \frac{\partial^2 h}{\partial y^2}\right) = 0,$$

where h is the piezometric level and K is the permeability. The total head drop H is divided into N_p potential lines with the head drop between two potential lines equal to $dh = H/N_p$. A stream tube has a discharge q per unit width, or $dq = Kdh$ and the total discharge per unit width increases with the number of stream tubes N_s. Therefore, the flow discharge under a structure of width b is

$$Q = KbHN_s/N_p \tag{16.7}$$

Seepage flow through a dam can be calculated as shown in Example 16.5.

♦♦ **Example 16.5:** Seepage through a dam

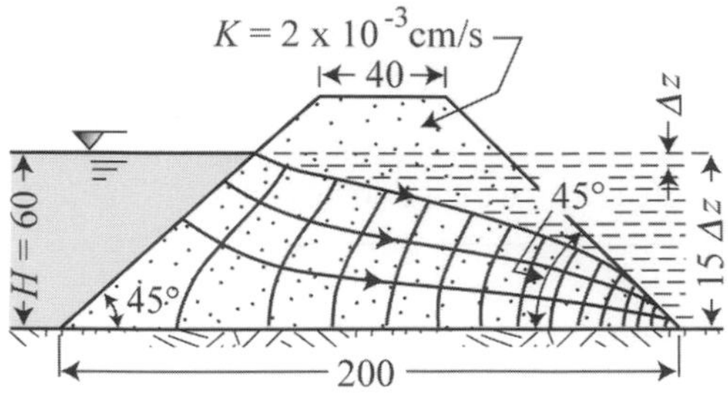

Fig. E-16.5 Seepage through a dam

A 400-m-long dam has a seepage length of 200 m, a flow depth $H = 60$ m and a permeability coefficient $K = 0.002$ cm/s. Use a flow net to calculate the seepage flow per unit width through the dam.

Solution: There are $N_s = 3$ stream tubes and $N_p = 15$ potential lines in Fig. E-16.5.

$$Q = 2 \times 10^{-5} \times 400 \times 60 \times 3/15 = 0.096\ \text{m}^3/\text{s}.$$

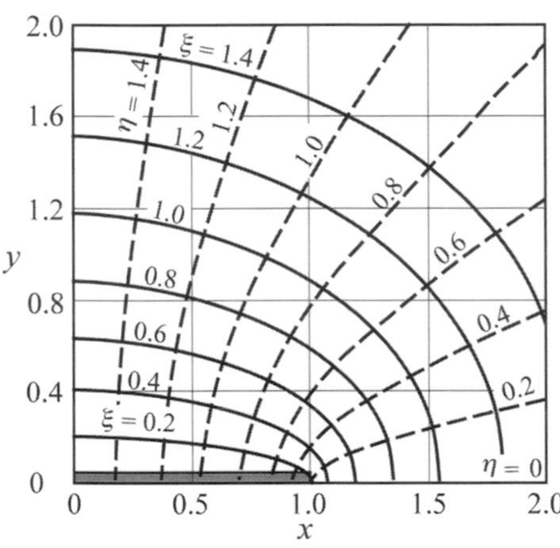

Figure 16.7 Flow net around a flat surface

16.2.4 Flow Nets for Deep Foundations

A useful solution to the flow net problem describes the coordinates x and y as a function of orthogonal lines (ξ, η) as

$$x = \cosh(\xi)\cos(\eta), \tag{16.8}$$

$$y = \sinh(\xi)\sin(\eta). \tag{16.9}$$

With knowledge of hyperbolic functions, interesting applications of this flow net shown in Figure 16.7 include the pressure distribution under a dam (Example 16.6) and the flow around sheet piles (Example 16.7).

♦ **Example 16.6:** Pressure distribution under a dam

Find the pressure distribution under a straight dam.

Solution: This is obtained from the potential lines η along the axis x at $y = 0$, i.e. a streamline value $\xi = 0$ and $\cosh(0) = 1$. The relative distance $-1 < X < 1$ is centered around the mid-base of the dam and the pressure is $\frac{\Delta p}{\gamma H_t} = \left[\frac{\cos^{-1}(X)}{\pi}\right]$, where the argument X is in radians.

From Fig. E-16.6, if a dam base is $L = 100$ ft long $(L/2 = 50)$, $H_1 = 60$ ft and $H_2 = 0$ ft. The pressure at point A, located 5 ft from the downstream end, is $X = (0.5L - x)/\ 0.5L = (50 - 5)/50 = 0.9$ and $\Delta p = \gamma H_t[\cos^{-1}(0.9)]/\pi = 62.4 \times (60 - 0) \times 0.451/\pi = 537$ psf.

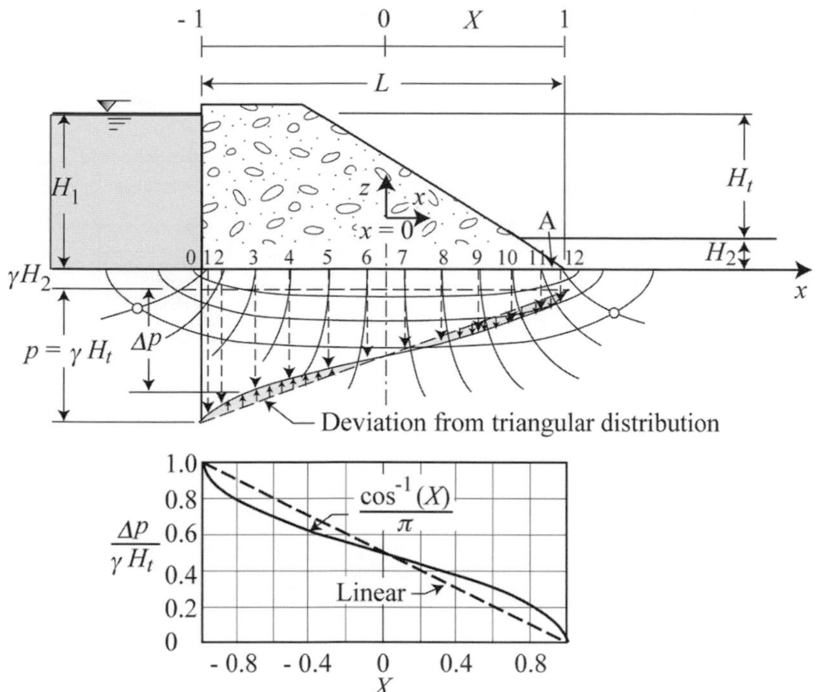

Fig. E-16.6 Pressure distribution under a dam

Example 16.7: Pressure distribution along a sheet pile

Find the saturated pore pressure distribution along a vertical sheet pile.

Solution: This is given from two components: (1) hydrostatic and (2) hydrodynamic from groundwater flow. Consider the flow around a sheet pile with length D into the ground and a head drop H, as sketched in Fig. E-16.7.

The pore pressure distribution on both sides of the sheet pile is

$$p_{up} = \gamma\{(x + H/2) + (H/\pi)\left[\cos^{-1}(x/D)\right]\},$$

$$p_{down} = \gamma\{(x + H/2) - (H/\pi)\left[\cos^{-1}(x/D)\right]\},$$

where $x > 0$ is the distance below the ground and $X = x/D$ is in radians.

For instance, considering a head drop $H = 10$ m and $D = 15$ m, we determine the pressure on both sides of the sheet pile at a depth $x = 14$ m as

$$p_{up} = 9{,}810\{(14 + 5) + (10/\pi)\left[\cos^{-1}(14/15)\right]\} = 198 \text{ kPa},$$

$$p_{down} = 9{,}810\{(14 + 5) - (10/\pi)\left[\cos^{-1}(14/15)\right]\} = 175 \text{ kPa}.$$

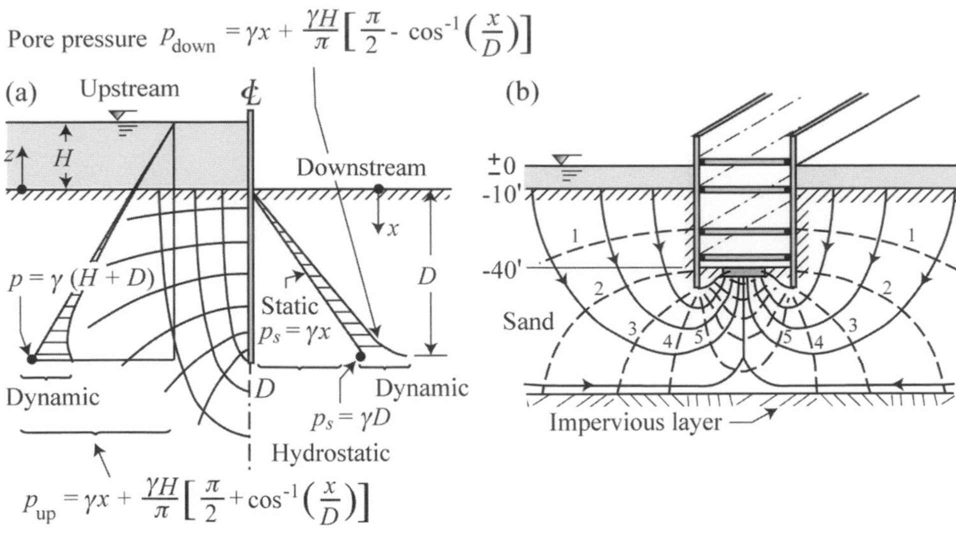

Fig. E-16.7 Flow and pressure distribution along a sheet pile

We can also calculate the pressure on both sides at the ground surface $x = 0$ as

$$p_{up} = 62.4 \times 3.28\{(0 + 5) + (10/\pi)[\cos^{-1}(0)]\} = 2{,}046 \text{ psf},$$

$$p_{down} = 62.4 \times 3.28\{(0 + 5) - (10/\pi)[\cos^{-1}(0)]\} = 0.$$

At the ground surface, the pressure difference between the upstream and downstream sides of the sheet pile must equal the hydrostatic pressure $p_{static} = 62.4 \times 10 \times 3.28 = 2{,}046$ psf.

There is always a need to check against the boiling condition (Section 15.2.2). This means that the pore pressure on the downstream side should be much lower than the submerged weight of the soil in the dry area. Typically, the sheet pile depth D will have to be deeper into the ground than $H/2$.

16.2.5 Flow Nets for Shallow Foundations

In the case of a flat dam on shallow foundations, an approximation for the number of stream tubes N_s to the number of potential lines N_p from Eq. (16.9) becomes

$$N_s/N_p = (1/\pi)\sinh^{-1}(2D/L), \tag{16.10}$$

as shown in Figure 16.8a.

Similarly, the case of sheet piles of length l over shallow foundations of depth D, the number of potential lines over the number of stream tubes is obtained from Eq. (16.8):

$$N_s/N_p = (2/\pi)\cosh^{-1}(D/l) \tag{16.11}$$

for flow on one side in Figure 16.8b.

Different configurations are also illustrated in Example E-16.8.

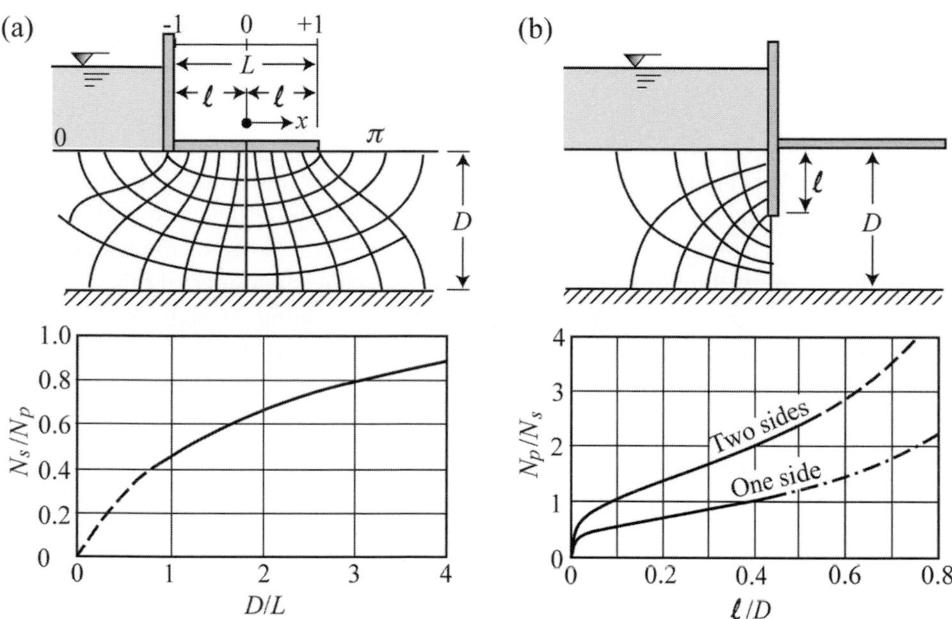

Figure 16.8 Ratio of stream to potential lines for a flat surface and sheet piles

---

♦ **Example 16.8:** Seepage for shallow foundations

(1) A 250-m-long and 50-m-wide dam rests on a 50-m-thick soil with $K = 0.0001$ cm/s. Find the seepage under the impervious dam holding 15 m of water. (2) Consider flow around a sheet pile 25 ft into a 60-ft-thick permeable layer at $K = 0.00001$ ft/s. If the water depth upstream of the sheet pile is 10 ft above the level in the construction area, estimate the seepage per unit width.

Solution:

(1) In this case, $D/L = 50/50 = 1$ and we obtain $N_s/N_p = (1/\pi)\sinh^{-1}(2) = 0.459$, and $Q \cong WKHN_s/N_p = 250 \times 0.0001 \times 0.01 \times 15 \times 0.459 \times 86{,}400 = 149\ \text{m}^3/\text{d}$.

(2) Here, $D/l = 60/25 = 2.4$ and both sides $N_s/N_p = 0.5 \times (2/\pi)\cosh^{-1}(2.4) = 0.48$. The unit discharge is $q \cong KHN_s/N_p = 1 \times 10^{-5} \times 10 \times 0.48 \times 86{,}400 = 4.15\ \text{ft}^3/\text{ft} \cdot \text{d}$.

16.3 Unsteady Groundwater Flow

The continuity equation describing unsteady water flow of mass density ρ with the velocity components v_x, v_y, v_z in the x, y, z directions is given by

$$\frac{\partial \rho}{\partial t} + \frac{\partial \rho v_x}{\partial x} + \frac{\partial \rho v_y}{\partial y} + \frac{\partial \rho v_z}{\partial z} = 0. \tag{16.12}$$

For porous soils, $\partial\rho/\rho = -\partial\forall/\forall$ where $\forall$ designates the volume of fluid.

Combining with Darcy's law, $V = Kdh/ds$ where h is the piezometric head, yields

$$K\left(\frac{\partial^2 h}{\partial x^2} + \frac{\partial^2 h}{\partial y^2} + \frac{\partial^2 h}{\partial z^2}\right) = -\frac{\partial \rho}{\rho \partial t} = \frac{\partial \forall}{\forall \partial t}. \tag{16.13}$$

This volumetric change $d\forall$ in groundwater can take different forms: (1) consolidation (Section 15.2.4); and (2) filling and draining. It is possible to fill the entire soil pores p_0 from initially dry conditions. When draining soils, however, some of the water in the pores will be adsorbed to the solids (also called the specific retention) and this will cause some hysteresis between the cycles of wetting and draining of soils. We consider unsteady flow in a floodplain (Section 16.3.1), in pumping wells (Section 16.3.2) and tidal fluctuations (Section 16.3.3).

16.3.1 Unsteady Flow in Stratified Floodplains

Let us consider an application of unsteady flow through a stratified floodplain with a highly permeable gravel lens as shown in Figure 16.9. If the water level is suddenly raised by H above the initial level, we need to determine the rate of rise of water in the alluvium of permeability K. In this case, we have one-dimensional vertical flow. The porosity of the dry soil is n, the thickness of the porous layer is z, the available pore volume is nAz and the difference in piezometric head is $H - z$. The vertical flow through surface area A has a gradient $(H - z)/z$. The governing equation is $Q = AV = AK(H - z)/z = nAdz/dt$. After separating the variables, we obtain $Kdt/n = zdz/(H - z)$, which is integrated to give

$$t = \frac{nH}{K}\left\{-z^* + \ln\left[\frac{1}{(1 - z^*)}\right]\right\}. \tag{16.14}$$

Note that $t^* = Kt/nH = 1$ when $z^* = z/H = 0.841$, as calculated in Example 16.9.

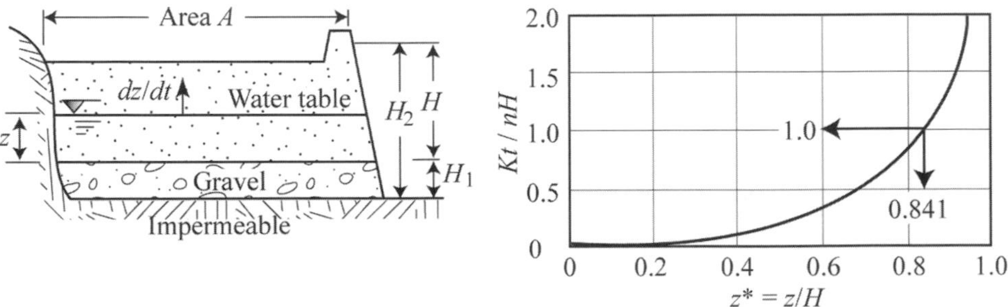

Figure 16.9 Unsteady floodplain flow

♦ **Example 16.9:** Unsteady flow in a stratified floodplain

Consider the flow through a stratified floodplain deposit with porosity $n = 0.4$ and permeability $K = 0.03$ cm/s. If the water level is suddenly raised by 10 m starting at the top of the gravel layer, how long will it take for the groundwater table to rise by 8 m? At what value of $z^* = z/H$ is $t^* = Kt/nH = 1$?

Solution:

$$t = \frac{nH}{K}\left\{-z^* + \ln\left[\frac{1}{(1-z^*)}\right]\right\} = \frac{0.4 \times 10}{3 \times 10^{-4}}\left\{-0.8 + \ln\left[\frac{1}{(1-0.8)}\right]\right\} = \frac{10{,}800}{3{,}600} = 3 \text{ hours.}$$

$t = nH/K = 0.4 \times 10/\left(3 \times 10^{-4} \times 3{,}600\right) = 3.7$ hours when $z_{84}{}^* = 0.841$ or $z_{84} = 8.4$ m.

16.3.2 Unsteady Flow in Pumping Wells

The water storage coefficient or specific yield S_y represents the volumetric fraction of the soil that can be drained. When draining soils, $S_y = n - n_e$, which is the difference between the total soil porosity n minus the residual porosity (or specific retention) n_e of the drained soil. Typical values of the specific yield and retention of soils based on the d_{10} of the granulometric distribution are shown in Figure 16.10.

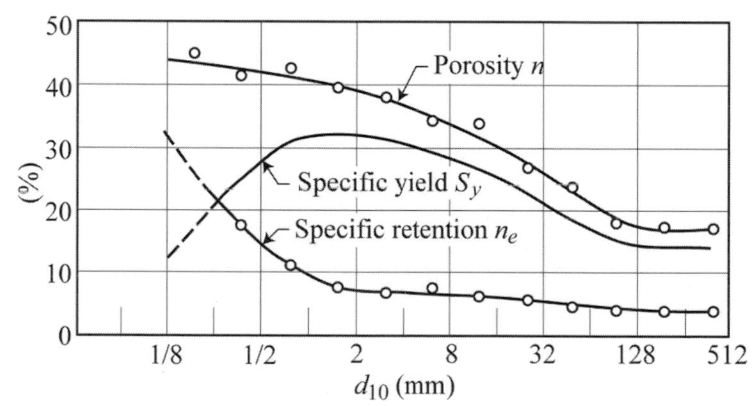

Figure 16.10 Conceptual specific yield based on d_{10} (after Todd 1964)

For pumping wells in confined aquifers of permeability K and thickness D, the transmissivity $T = KD$. The unsteady flow equation can be rewritten in cylindrical coordinates as a function of the piezometric head h and radius r (Bouwer 1978). We obtain in cylindrical coordinates (without derivation here)

$$\frac{\partial^2 h}{\partial r^2} + \frac{\partial h}{r \partial r} = \frac{S_y \partial h}{KD \partial t} = \frac{S_y \partial h}{T \partial t}, \tag{16.15}$$

where S_y is the specific yield of the confined aquifer. This equation can also be applied to unconfined aquifers when the drawdown is small compared to the aquifer thickness.

This unsteady flow equation (also called transient) is simply a diffusion equation equivalent to heat transfer and wave propagation. For flow in a confined aquifer, the solution from Theis in cylindrical coordinates describes the drawdown s as a function of $u = r^2 S_y/4Tt$ as

$$s = \frac{Q}{4\pi T} W(u)$$

where

$$W(u) = \int_u^\infty \frac{e^{-y}}{y} dy = -0.577216 - \ln u + u - \frac{u^2}{2 \times 2} + \cdots$$

The Theis integral $W(u)$ can be solved numerically. However, the following approximation from the Cooper–Jacob method for $W(u)$ based on the first two terms of the series expansion is often preferred:

$$W(u) \simeq -\ln(2.25/4u). \tag{16.16}$$

The drawdown s can then be plotted as a function of the logarithm of time since

$$s = \frac{2.3Q}{4\pi T} \log\left(\frac{2.25Tt}{r^2 S_y}\right). \tag{16.17}$$

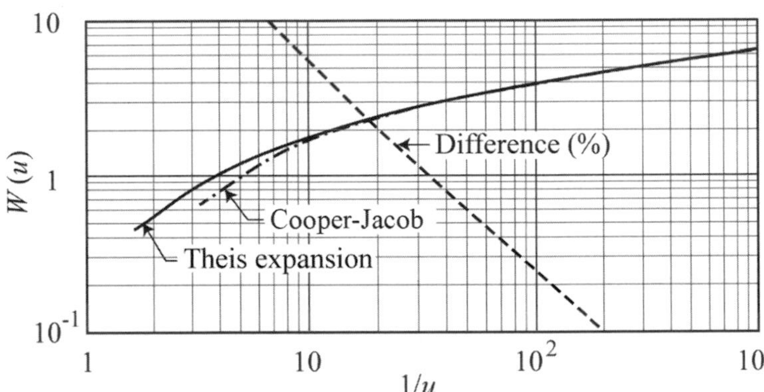

Figure 16.11 $W(u)$ vs $1/u$ from Theis and Cooper and Jacob

Figure 16.11 shows the difference between the Theis infinite series expansion and the simplified logarithmic approximation of Cooper and Jacob. Note that $1/u = 4Tt/r^2 S_y$ increases with time. Also note that the deviation between both solutions is less than 2% after the initial stages of pumping [$u < 0.05$, or $W(u) > 2$]. This condition is usually met after 1 hour of pumping in confined aquifers and after 12 hours for unconfined aquifers. Practical calculations of T and S_y from pumping wells at a constant discharge Q are examined in Example 16.10.

♦♦ Example 16.10: Unsteady flow in pumping wells

Bouwer (1978) lists the following pumping drawdown measurements as a function of time for a borehole located 200 m away from a pumping well drawing 1,000 m^3 of water per day. At times $t = 0.001, 0.005, 0.01, 0.05, 0.1, 0.5, 1, 5, 10$ d, the drawdown levels are $s = 0.017, 0.097, 0.145, 0.267, 0.322, 0.449, 0.504, 0.632, 0.687$ m. Find the transmissivity T and specific yield S_y.

Solution: Fig. E-16.10 shows the drawdown vs time. From the governing equation, Eq. (16.17), we obtain a straight-line fit with intercept at $t_0 = 0.002$ d where $s = 0$.

The transmissivity of the aquifer is calculated from the slope of the line

$$\Delta s = (0.687 - 0)/(\log 10 - \log 0.002) = 0.186:$$

$$T = 2.3Q/(4\pi\Delta s) = 2.3 \times 1{,}000/(4\pi \times 0.186) = 985\,\mathrm{m}^2/\mathrm{d} = 0.0114\mathrm{m}^2/\mathrm{s},$$

and

$$S_y = 2.25Tt_0/r^2 = 2.25 \times 985 \times 0.002/200^2 = 0.00011.$$

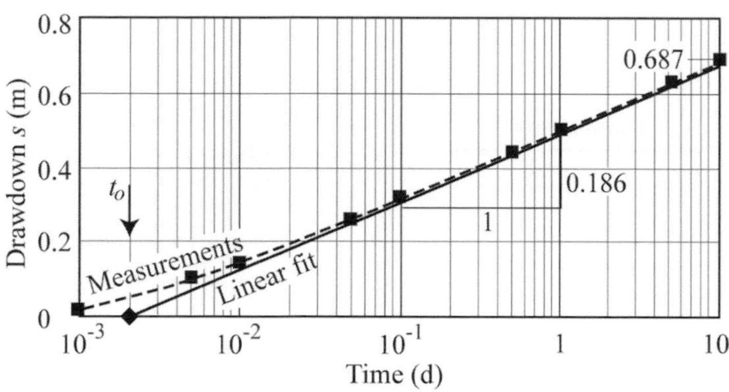

Fig. E-16.10 Pumping test measurements

16.3.3 Unsteady Flow from Tidal Fluctuations

Groundwater-level fluctuations following sinusoidal patterns like tides in coastal areas can be studied by solving the diffusion equation, Eq. (16.15), in Cartesian coordinates, $\partial h/\partial t = (T/S_y)\partial^2 h/\partial x^2$. Todd (1964) provided an elegant estimation for a confined aquifer of permeability K and thickness D described with specific yield S_y and transmissivity $T = KD$. The tides are defined with a half amplitude h_0 and tidal period t_0, which corresponds to an angular velocity $\omega = 2\pi/t_0$. The amplitude of the piezometric level h at a distance x from the coast is obtained from

$$h = h_0 e^{-x\sqrt{\frac{\pi S_y}{Tt_0}}} \sin\left[\left(\frac{2\pi t}{t_0}\right) - \left(x\sqrt{\frac{\pi S_y}{Tt_0}}\right)\right], \tag{16.18}$$

with celerity $c = \sqrt{4\pi T/S_y t_0}$. This formulation can also be used as a first approximation for unconfined aquifers, where the fluctuation is small compared to the aquifer depth. The volume of water $\forall$ that infiltrates over a shore length L during a half period ($t_0/2$) is $\forall = Lh_0\sqrt{2t_0 S_y T/\pi}$. Example 16.11 provides calculations of a tidal wave in coastal groundwater.

Example 16.11: Tidal-wave propagation in groundwater

Consider a 1-m-amplitude tidal wave ($h_0 = 1$ m) in an 80-m-deep coarse sand layer with $K = 1$ cm/s. Assume a specific yield $S_y = 0.2$ and calculate the water-level amplitude as a function of distance at different times.

Solution: We can calculate $T = KD = 0.01 \times 80 \times 3{,}600 = 2{,}880$ m^2/h, $t_0 = 12.4$ h, $\sqrt{t_0 T/\pi S_y} = \sqrt{12.4 \times 2{,}880/(0.2 \times \pi)} = 238$ m and $t_0/2\pi = 2$ h. Therefore,

$$h = h_0 e^{-x\sqrt{\frac{\pi S_y}{T t_0}}} \sin\left[\left(\frac{2\pi t}{t_0}\right) - \left(x\sqrt{\frac{\pi S_y}{T t_0}}\right)\right] = e^{-x/238} \sin\left[(t/2) - (x/238)\right],$$

with the argument in radians, x in meters and t in hours, as shown in Fig. E-16.11.

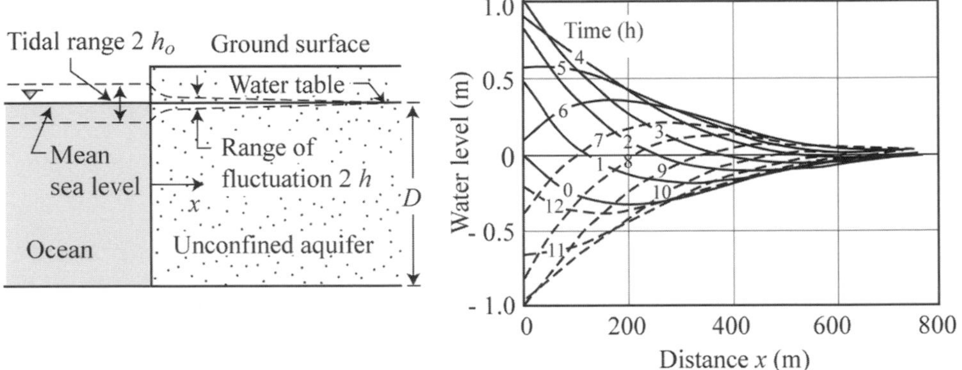

Fig. E-16.11 Groundwater level from tidal variations

Finally, similar applications to rivers during floods are possible. As a first order of magnitude, the volume of water infiltrating into both river banks over the river length L during a flood can be estimated from the flood of duration t_0 and amplitude h_0, the bank permeability K and porosity n, and thickness D as

$$\forall \approx L h_0 \sqrt{\frac{2 t_0 n K D}{\pi}}. \tag{16.19}$$

Consider a river with a $D = 5$-m thick floodplain deposit with $K = 0.01$ cm/s and porosity $n = 0.25$. A flood with a 2-m water-level rise ($h_0 = 2$ m) in 2.5 d, or period $t_0 = 10$ d would infiltrate the following volume per km of length:

$$\forall = L h_0 \sqrt{2 t_0 n K D/\pi} = 2{,}000\sqrt{2 \times 10 \times 86{,}400 \times 0.25 \times 10^{-4} \times 5/\pi}$$

$$= 16{,}500\,\text{m}^3/\text{km}.$$

Note that we consider two river banks but half the infiltration cycle. The estimated average groundwater recharge rate is

$$\bar{Q} \approx \frac{16{,}500 \times 2 \times 2}{10 \times 86{,}400} = 0.08 \text{ m}^3/\text{s per km}$$

during this period.

Case Study 16.1 illustrates groundwater applications.

♦♦ Case Study 16.1: Groundwater applications near Lusaka, Zambia

Baumle (2011) studied groundwater pumping wells near Lusaka in Zambia. The results of more than 150 pumping tests and the regional analysis of the aquifer properties resulted in a better understanding of the relationship between hydraulic parameters and the aquifer properties and water yield. *In situ* aquifer tests were used to determine the transmissivity T, and the specific capacity q was also calculated from pumping rates and dynamic water levels. The results showed a good correlation between the transmissivity and the specific capacity. Understandably, the conversion of $T_{\text{m}^2/\text{d}}$ in m^2/d to $q_{\text{l/m·s}}$ in $\text{l/s} \cdot \text{m}$ using $1 \text{ m}^3 = 1{,}000$ liters and 86,400 s per day gives $T_{\text{m}^2/\text{d}} = 86.4 q_{\text{l/ms}}$, which is in agreement with the field measurements.

Considering an average daily water consumption of 86 liters per capita (23 US gallons per day), a pumping yield of 1 l/s meets the demand of 1,000 people. The groundwater potential can be defined in the following terms:

- high: yield $>$ 10 l/s for water supply to towns and for irrigation;
- moderate: yield $>$ 1 l/s for water supply to smaller communities;
- limited: yield $>$ 0.01 l/s for very local water supply; and
- insufficient: yield $<$ 0.01 l/s for essentially inexistent water supply.

Fig. CS-16.1 shows the relationship between the transmissivity/specific capacity and groundwater yield.

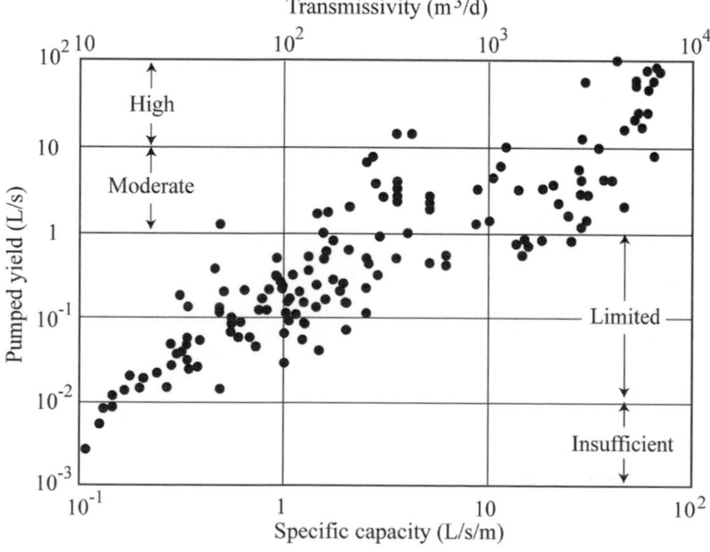

Fig. CS-16.1 Specific transmissivity and yield near Lusaka (Baumle 2011)

As a reference point, the average daily water consumption in the US is 82 gallons per person per day (300 liters per day). The UN considers that the average person is in need of 20–50 liters per day (5–13 US gallons per day) to satisfy basic drinking, cooking, cleaning and sanitation needs. Approximately11% of the world's population (780 million people) still does not have access to clean drinking water. Also, more than two billion people (25% of the world's population) still live without basic sanitation. There is still plenty of work to be done . . .

Additional Resources

Useful references for this chapter include Todd (1964), Davis and DeWiest (1966), Walton (1970), Raudkivi and Callander (1976) and Bouwer (1978).

EXERCISES

These exercises review the essential concepts from this chapter.

1. Why are there two permeability coefficients?
2. Which parameter has the greatest effect on permeability?
3. What is the difference between confined and unconfined aquifers?
4. What is the Dupuit–Forchheimer assumption?
5. What is a flow net?
6. What is the boiling condition?
7. What is the specific retention of a soil?
8. How do you determine the specific yield of a soil?
9. What is a drawdown?
10. True or false?
    (a) The flow velocity in the pores is higher than given by Darcy's law.
    (b) A darcy is a measure of the permeability coefficient.
    (c) The Hazen formula is only valid for saturated soils.
    (d) An increase in porosity corresponds to an increase in permeability.
    (e) Water temperature is the dominant parameter to define soil permeability.
    (f) The Dupuit parabola applies to flow through embankments.
    (g) The Laplace equation combines Darcy's law with the conservation of mass.
    (h) The pressure distribution under a dam is linearly distributed.
    (i) Sheet piles should be driven as deep into the soil as the water level difference.
    (j) The Theis equation solves for unsteady flow in unconfined aquifers.

PROBLEMS

Permeability

1. ♦ A deep well in a 25-m-deep confined aquifer draws 0.05 cms over a long time. The drawdowns at 40 m and 120 m are, respectively, 2.8 m and 1 m. Calculate the permeability coefficient and the transmissivity.
2. ♦ A deep well penetrates a 40-ft-thick confined aquifer and pumps 400 gpm (gallons per minute) for a very long time. Water-level measurements in boreholes 100 ft and and 200 ft from the well have drawdowns of 11 ft and 4 ft, respectively. Find the permeability coefficient and the transmissivity of the aquifer.
3. ♦ A deep well reaches an impervious layer 34 m below the groundwater table and pumps 2 m^3/min over a long period of time. If the drawdowns in two boreholes 20 m and 45 m from the well are 4 m and 2 m, respectively, calculate the permeability coefficient and the transmissivity of the unconfined aquifer.
4. ♦ A deep well reaches an impervious layer 20 m below the groundwater table and pumps 0.03 cms over a long period of time. If the drawdowns in two boreholes 40 m and 120 m from the well are 2.8 m and 1 m, respectively, calculate the permeability coefficient and the transmissivity of the unconfined aquifer.

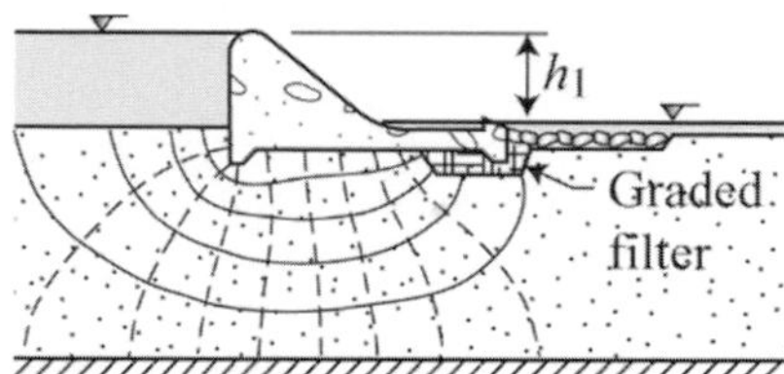

Fig. P-16.5

Seepage

5. ♦ Estimate the seepage discharge per unit width in Fig. P-16.5 if porosity and grain size are $n = 0.25$ and $d_{10} = 0.01$ mm, and the water level drops 20 m.
6. ♦ Estimate the seepage discharge per unit width in Fig. P-16.6 if porosity and grain size are $n = 0.25$ and $d_{10} = 0.01$ mm, and the water level drops 35 ft.

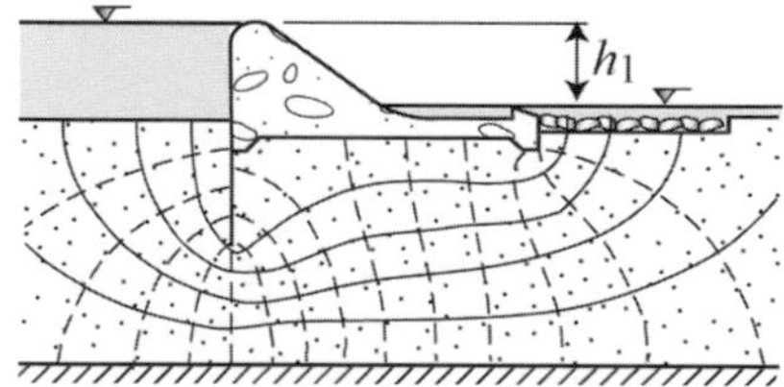

Fig. P-16.6

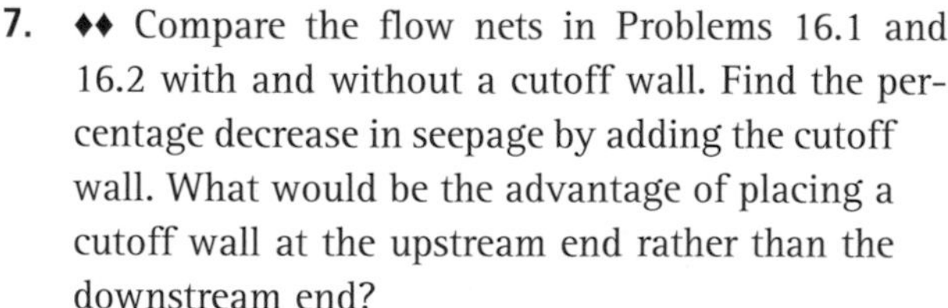

7. ♦♦ Compare the flow nets in Problems 16.1 and 16.2 with and without a cutoff wall. Find the percentage decrease in seepage by adding the cutoff wall. What would be the advantage of placing a cutoff wall at the upstream end rather than the downstream end?
8. ♦♦ Compare the flow net in Fig. P-16.8 with the flow net of Example 16.5. Which one has the largest seepage flow? What is the advantage of the toe filter?

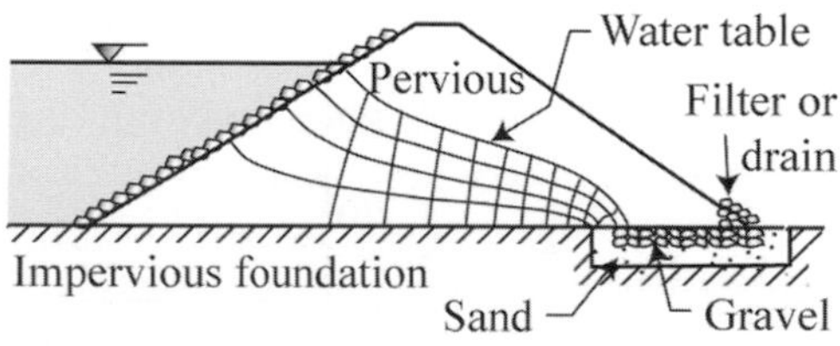

Fig. P-16.8

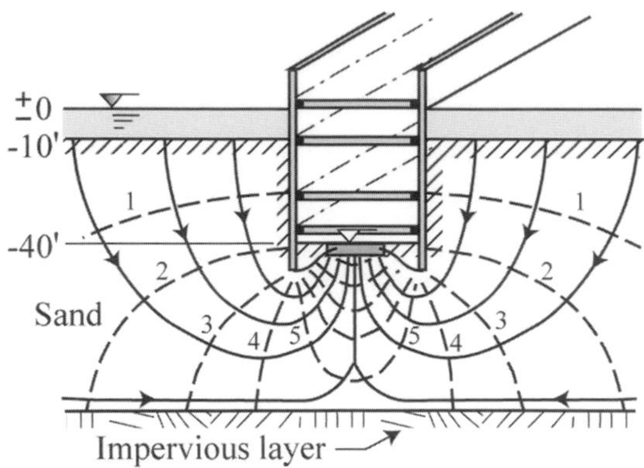

Fig. P-16.9

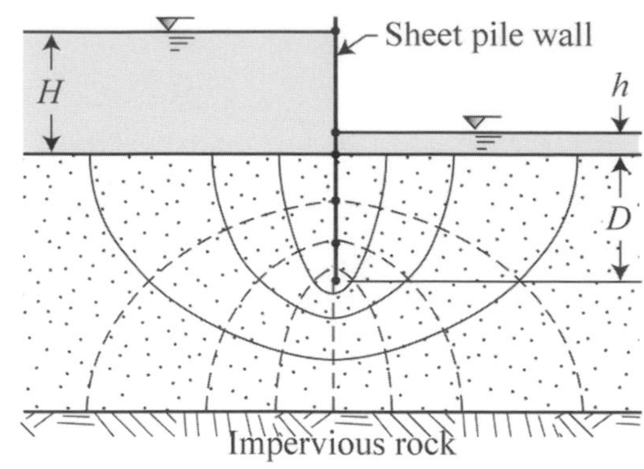

Fig. P-16.11

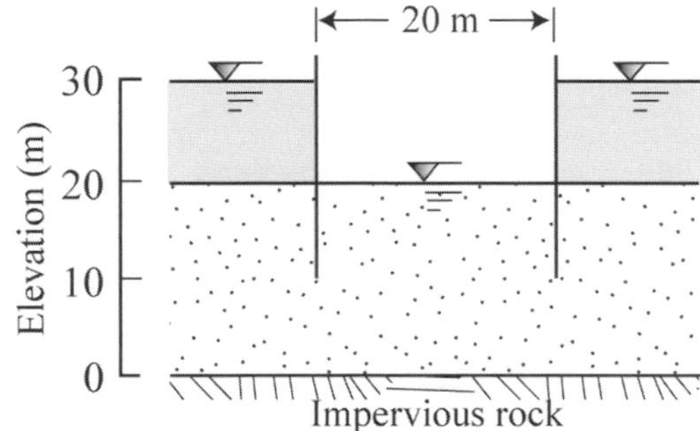

Fig. P-16.10

9. ♦♦ In Fig. P-16.9, estimate the pumping discharge required to drain the 250-ft-long trench in loose fine sand, $e = 0.85$. Is the trench safe against piping if the sheet pile is 10 ft below the floor of the trench?

10. ♦♦ In Fig. P-16.10, estimate the pumping discharge required to drain the 100-m-long trench with $K = 1 \times 10^{-4}$ cm/s. Is the trench safe against piping (boiling sands) if the soil weighs 20 kN/m^3?

Pore Pressure

11. ♦♦ A sheet pile in a 20-m soil layer is 10 m into the ground as shown in Fig. P-16.11. Given $H = 10$ m and $h = 2$ m, calculate the pore pressure distribution and the water force on the sheet pile.

12. ♦♦ A sheet pile in a 60-ft soil layer is 30 ft into the ground (see Fig. P-16.12). Given $H = 30$ ft and $h = 5$ ft, calculate the pressure distribution and the net water force on the sheet pile.

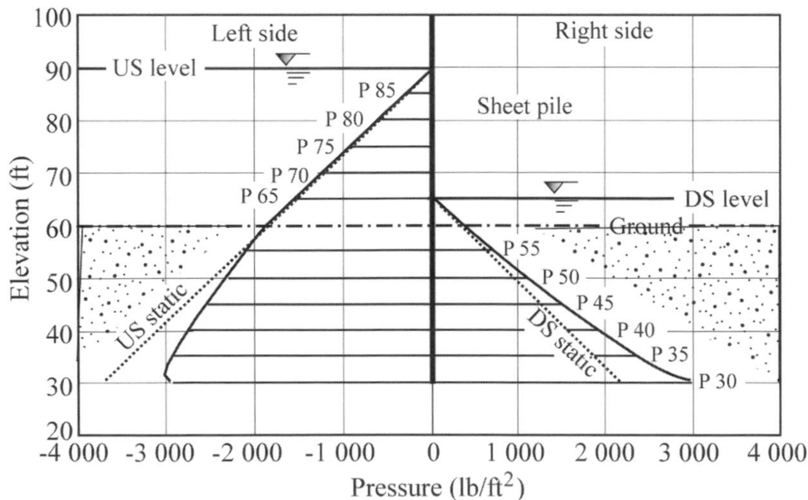

Fig. P-16.12

Unsteady Flow

13. ♦ A pond covers 12 acres alongside a river over 1,000 ft. If the initial water levels in both the pond and river are the same, calculate the rise in water level in the pond in the next three days if the water level in the river suddenly rises by 4 ft for one week. Consider $H = 6$ ft, $L = 200$ ft and $K = 0.003$ ft/s in Fig. P-16.13.

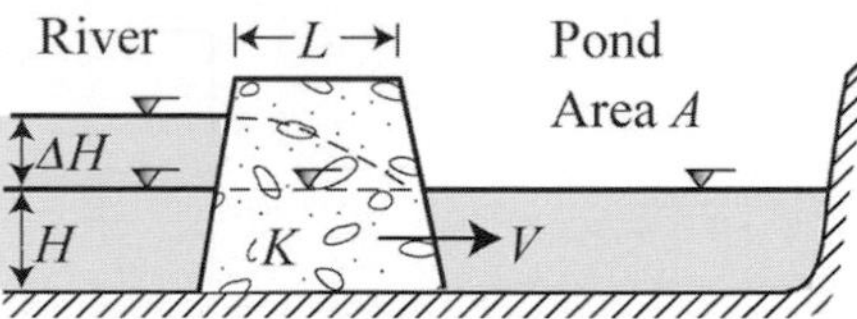

Fig. P-16.13

14. ♦ A pond covering 5 hectares is adjacent to a river over a distance of 300 m. The bottom of the pond has a 2.5-m-thick layer of silt with a permeability $K = 2 \times 10^{-6}$ m/s overlying a highly permeable gravel layer. If the initial water level in both the pond and the river are the same, calculate the rise in water level in the pond in the next three days if the water level in the river suddenly rises by 2 m for one week. Consider that $H = 3$ m, and $L = 100$ m in Fig. P-16.14. If the soil has a void ratio $e = 0.8$ and $G = 2.65$, define the condition leading to piping.

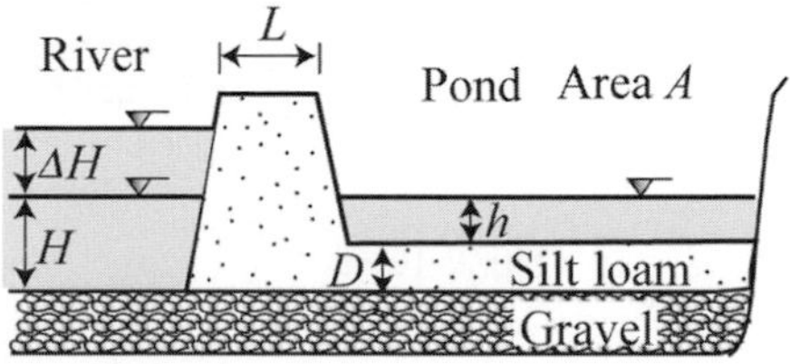

Fig. P-16.14

15. ♦♦ Consider the following confined aquifer pumping drawdown measurements as a function of time for a borehole located 300 ft away from a pumping well drawing 1,500 gpm: $t =$ 1, 2, 3, 4, 6, 8, 10, 30, 40, 50, 60, 80, 100, 200, 400, 600, 800, 1000, 1440 min; and $s =$ 0.45, 0.74, 0.91, 1.04, 1.21, 1.32, 1.45, 2.02, 2.17, 2.3, 2.34, 2.5, 2.67, 2.96, 3.25, 3.41, 3.5, 3.6, 3.81 ft, respectively. Find the transmissivity T and the specific yield S_y. Fig. P-16.15 shows the drawdown s as a function of the log of time.

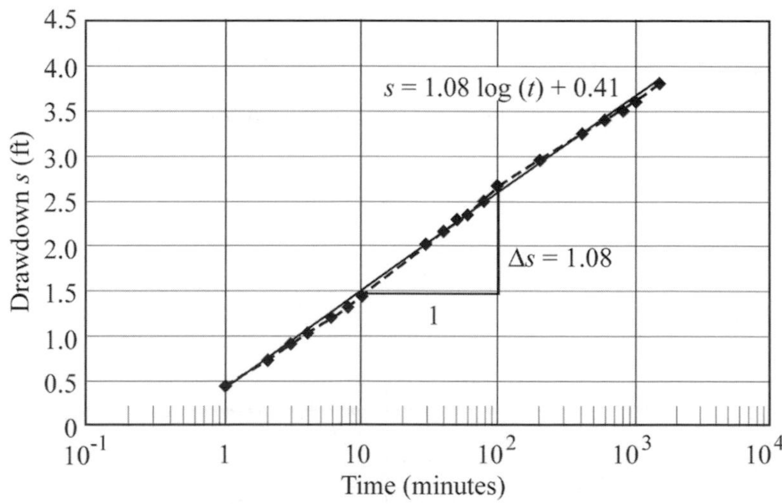

Fig. P-16.15

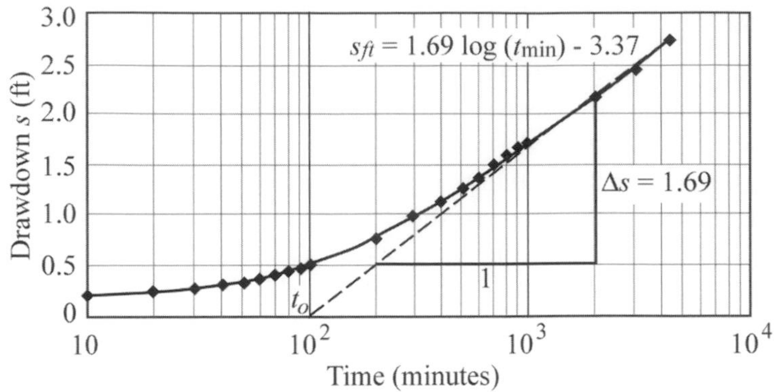

Fig. P-16.16

16. ♦♦ Consider the following pumping drawdown measurements as a function of time for a borehole located 200 ft away from a pumping well drawing 1,100 gpm from a 75-ft-thick unconfined aquifer: $t =$ 10, 20, 30, 40, 50, 60, 70, 80, 90, 100, 200, 300, 400, 500, 600, 700, 800, 900, 1000, 2000, 3000, 4320 min; $s =$ 0.21, 0.24, 0.28, 0.31, 0.34, 0.37, 0.40, 0.43, 0.46, 0.5, 0.77, 0.99, 1.12, 1.27, 1.38, 1.45, 1.56, 1.62, 1.70, 2.15, 2.45, 2.76 ft respectively. Find the transmissivity T, the permeability coefficient K and the specific yield. Fig. P-16.16 shows the drawdown s as a function of the log of time.

APPENDIX A

Basic Cost Analysis

This appendix covers useful relationships for the cost analysis of hydraulic engineering projects. Figure A-1 shows basic parameters: the constant annual interest rate i and the number of years n.

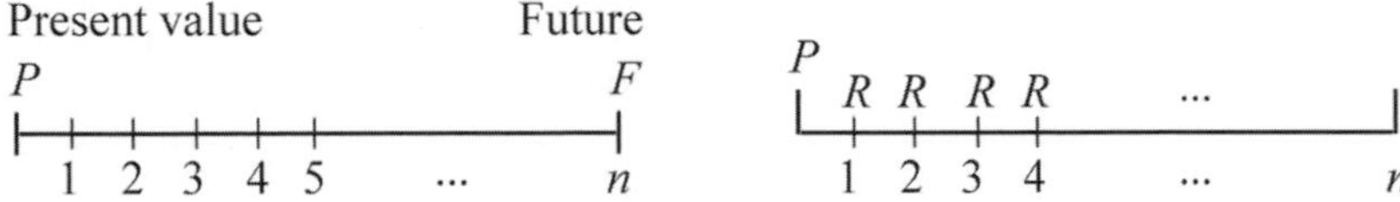

Figure A-1 Present, future and repeated annual values

The additional parameters include: the present value P, the future value F and the end-of-year repeated uniform annual amount R. Useful formulas between future and present-worth factors are simply

$$F = P(1+i)^n$$

and

$$P = F(1+i)^{-n},$$

and the series present-worth P and capital-recovery R factors are, respectively,

$$P = R\left[\frac{(1+i)^n - 1}{i(1+i)^n}\right]$$

and

$$R = P\left[\frac{i(1+i)^n}{(1+i)^n - 1}\right].$$

Examples A-1 to A-3 follow.

♦ **Example A-1**: Eight years ago, you bought a used car for \$5,000 and sell it now for \$1,300. What is the annual rate of return on your investment?

Solution:

$$i = \left(\frac{F}{P}\right)^{1/n} - 1 = \left(\frac{1{,}300}{5{,}000}\right)^{1/8} - 1 = -15.5\%,$$

but it only cost you \$462/year to run.

Compared with a car your parents bought 8 years ago for $25,000 with resale now for $10,000, the loss is now −10.8% per year, but the average cost to run their car was $1,875/year.

♦♦ **Example A-2**: You save $1,200/year ($100/month) to build up a college fund for your child. What will be the amount available for college 20 years from now if you invest at 5% per year?

Solution:

$$F = R\left[\frac{(1+i)^n - 1}{i(1+i)^n}\right](1+i)^n = \$1{,}200\left[\frac{(1.05)^{20} - 1}{0.05(1.05)^{20}}\right](1.05)^{20} = \$39{,}679.$$

♦ **Example A-3**: You saved $400,000 during your working life for your retirement. What annuity can you expect over the next 20 years if the interest rate is 4%.

Solution:

$$R = \$400{,}000\left[\frac{0.04(1+0.04)^{20}}{(1+0.04)^{20} - 1}\right] = \$29{,}432/\text{year}.$$

APPENDIX B
Society and Sustainability

Hydraulic engineers need to balance theory and practice. As sketched in Figure B-1, we readily understand that engineers seek practical applications but also need to understand the underlying concepts and theory behind engineering design. New technical problems experienced in practice define the needs for additional research and better theoretical understanding. Likewise, sound theoretical understanding can help the engineer solve problems beyond standard methods.

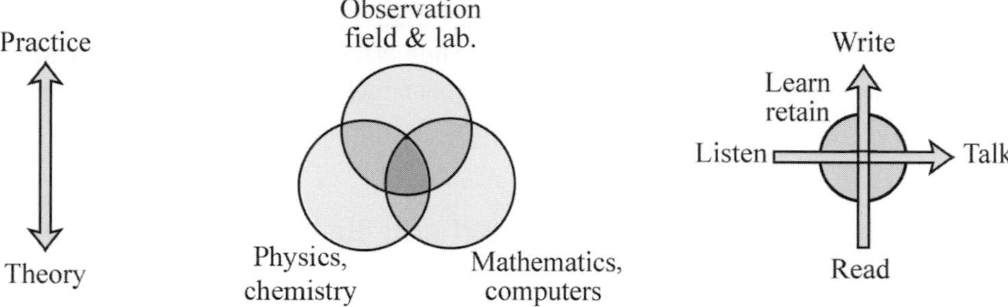

Figure B-1 Basic skills for hydraulic engineers

In the second diagram, a triangular diagram in a counterclockwise loop starts with an observation of the facts. Good engineers collect and extract useful facts from a wealth of information. It is essential to sort out useful from useless information. The second step requires physical understanding of the problem we are trying to solve. In hydraulics, this step usually requires a combination of the first principles of conservation of mass, energy and momentum to define an analytical solution. It is only when the problem is physically understood that we can seek a mathematical solution using algebraic formulas, simple equations or computers. With a quantified solution at hand, we can then return where we started and compare the results of our analytical understanding and numerical models with the field observations. We often start with simple models and cycle back with added complexity until satisfactory results are obtained.

The last diagram illustrates the importance of communication skills along two main poles. Verbal communication involves listening and talking. Written communication requires reading and writing. Written communication skills can be as short as a text message and as long as a thesis or dissertation. The ability to communicate does not replace the technical engineering skills, but it is an essential part of engineering life. It is the ability to process and retain information through communication that contributes to a successful learning experience.

These concepts are explored further with examples below in terms of communication skills (Section B-1.1), the common good (Section B-1.2), a sustainable environment (Section B-1.3) and short-term vs long-term solutions (Section B-1.4).

B-1.1 Communication Skills

Some unexpected circumstances always seem to come up and communication skills become of foremost importance. In conflicting situations, the ability to place ourselves in someone else's situation provides understanding of the viewpoint and interest of others. Empathy helps frame a situation in a broader perspective and we can seek understanding without judging.

The ability to explain a design in simple terms to the public is also very important for engineers. Communication skills can be developed along two main lines: (1) the ability to read and write; and (2) the ability to listen and speak. At times, communication skills become more important than the ability to solve complex technical problems. Besides the ability to listen and read, young engineers should take every opportunity to write and speak. The benefits from public-speaking skills largely outweigh the fear of imperfection. Public speaking gets better with practice.

B-1.2 Technology, the Common Good and Sustainability

The development of technology enables the application of scientific knowledge to solve problems to the benefit of society and raise the living standard of healthy communities. This usually requires a direct interaction with the environment. Engineers make recommendations (and even very important decisions) for the common good of the entire society. It is important to minimize the adverse impact of projects in terms of energy waste and pollution.

While interacting with communities, engineers are seeking the best interest of the entire society, and not solely the interest of the majority, ethnic minorities, fringe groups and lobbying entities, let alone influential individuals. Some methods seeking society's best interests can be quantified. For instance, a benefit–cost analysis can indicate which option in a design would be the most affordable to your community. The lowest-cost approach is a convincing tool to demonstrate the viability of various alternatives. For instance, in the pipe and pump problem from Problem 4.9 in Chapter 4, the determination of an optimal pump size and pipe diameter can be done in two different ways. The lowest cost of materials can be first considered. The cost analysis can also be extended to include energy costs.

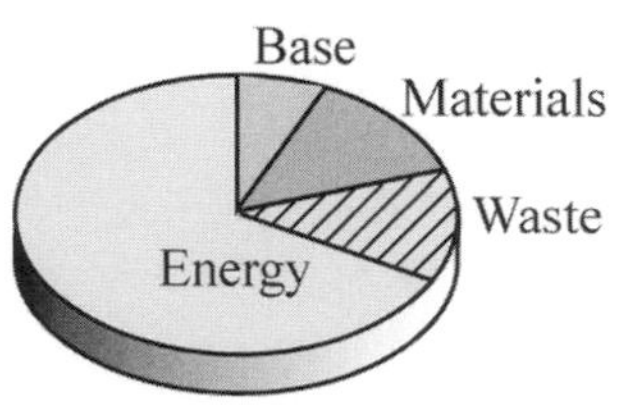

Figure B-2 Cost distribution

In order to cope with unexpected changes, it is important to keep in mind which component contributes to the main portion of the total cost. For the example in Figure B-2, the energy cost covers an overwhelming fraction of the total cost. Keeping a sense of proportions is helpful in making decisions. For instance, doubling base construction costs would have a small impact on the total cost compared to doubling the energy costs. Good engineers develop the ability to identify and eliminate wasted resources. For instance, in the pump and pipe problem (Problem 8.9 in Chapter 8), the cost distribution can be examined by separating (pump, pipes, useful and wasted energy, etc.).

In the analysis of unexpected circumstances, a cost increment ΔC can be perceived as a source of conflict. In many cases such as Example B-1, several solutions can be explored in terms of seeking ways to reduce this added cost increment ΔC.

♦♦ **Example B-1**: After designing a project assuming a pump efficiency of 75%, the pump installed may have a much lower efficiency (e.g. 50%). Energy costs 50% higher than anticipated may not be well perceived by your client. What could be done in this situation?

Solution: If the energy cost is the main fraction of the total cost, considering an additional cost may still lead to a favorable outcome. In this case, finding a more efficient pump (e.g. $e = 70\%$) would still be beneficial despite the additional cost ΔC of a new pump because it would reduce the total cost in the long run.

B-1.3 Sustainability and Global Warming

There is an increasing concern with long-term sustainability, and wasted energy contributing to global warming. Once again, the pump and pipe system should consider a plan to dispose of the pump and pipe materials at the end of their service life. Are there contaminants involved in your project? Can you recycle the materials? Is it possible to consider the amount of energy that would only contribute to global warming? For instance, a pump with a 65% efficiency would waste 35% of the total energy to global warming. Engineers should eliminate waste and achieve long-term sustainability. In looking at future scenarios (e.g. Problem 13.9 in Chapter 13) it is possible to retrofit an aging infrastructure and prepare an adaptation plan to meet future conditions (demographic expansion, climate change . . .). Further reading regarding climate preparedness can be found in Friedman et al. (2016).

B-1.4 Short Term vs Long Term

Most engineering situations involve interaction with others. The best option for a community may imply different perceptions from different stakeholders. Even when the problems can be quantified, the perception for different groups and individuals can be completely different from your own. For instance, if you look at the two options shown in Fig. B-3, it is quite obvious that option B has the lowest long-term cost and should definitely be preferable for the community sharing this cost over the coming decades. But what if the decision is made by individuals with a short-term interest in mind? For instance, consider an executive who has been

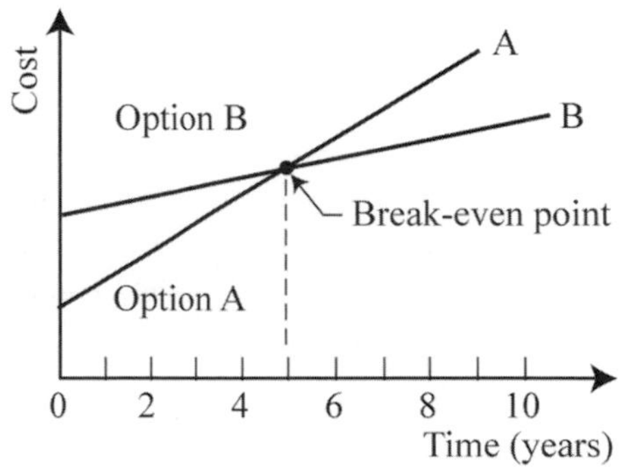

Figure B-3 Short- vs long-term objective

appointed for three years and believes that option A is preferable because it has the lowest cost during an individual mandate. Would this be the best option for the community? Remembering obligations to serve the public interest, the answer is clearly NO! This leads us into ethics and professional obligations, with more details in Appendix C.

APPENDIX C
Professional Engineering Obligations

The following stems from the National Society of Professional Engineers (www.nspe.org/resources/ethics/code-ethics), where some key statements are in italics. Useful other references regarding ethics and the professional engineering obligations can be found in Baura (2006) and Robinson et al. (2007) and Cahn (2020). For a focus on hydraulic engineering, the reader is referred to Julien (2017) and Ettema et al. (2020). Guidelines for engineering education are promoted by the Academy of Engineering (NAE 2005).

1. Engineers shall be guided in all their relations by the highest standards of *honesty and integrity.*
   (a) Engineers shall acknowledge their errors and shall not distort or alter the facts.
   (b) Engineers shall *advise their clients or employers when they believe a project will not be successful.*
   (c) Engineers shall not accept outside employment to the detriment of their regular work or interest. *Before accepting any outside engineering employment, they will notify their employers.*
   (d) Engineers shall not attempt to attract an engineer from another employer by false or misleading pretenses.
   (e) Engineers *shall not promote their own interest at the expense of the dignity and integrity of the profession.*
2. Engineers shall at all times strive to *serve the public interest.*
   (a) Engineers are encouraged to participate in civic affairs; career guidance for youths; and work for the advancement of the safety, health and well-being of their community.
   (b) Engineers *shall not complete, sign or seal plans and/or specifications that are not in conformity with applicable engineering standards.* If the client or employer insists on such unprofessional conduct, they shall notify the proper authorities and withdraw from further service on the project.
   (c) Engineers are encouraged to extend public knowledge and appreciation of engineering and its achievements.
   (d) Engineers are encouraged to *adhere to the principles of sustainable development in order to protect the environment for future generations.*
3. Engineers *shall avoid all conduct or practice that deceives the public.* (See Example C-1).
   (a) Engineers *shall avoid the use of statements containing a material misrepresentation of fact or omitting a material fact.*
   (b) Consistent with the foregoing, engineers may advertise for recruitment of personnel.
   (c) Consistent with the foregoing, engineers may prepare articles for the lay or technical press, but such articles shall not imply credit to the author for work performed by others.

4. Engineers shall *not disclose, without consent, confidential information* concerning the business affairs or technical processes of any present or former client or employer, or public body on which they serve.
   (a) Engineers shall not, without the consent of all interested parties, promote or arrange for new employment or practice in connection with a specific project for which the engineer has gained particular and specialized knowledge.
   (b) Engineers *shall not*, without the consent of all interested parties, *participate in or represent an adversary interest in connection with a specific project* or proceeding in which the engineer has gained particular specialized knowledge on behalf of a former client or employer.
5. Engineers *shall not be influenced in their professional duties by conflicting interests.*
   (a) Engineers shall not accept financial or other considerations, including free engineering designs, from material or equipment suppliers for specifying their product.
   (b) Engineers shall not accept commissions or allowances, directly or indirectly, from contractors or other parties dealing with clients or employers of the engineer in connection with work for which the engineer is responsible.
6. Engineers *shall not attempt to obtain employment or advancement or professional engagements by untruthfully criticizing other engineers*, or by other improper or questionable methods.
   (a) Engineers shall not request, propose or accept a commission on a contingent basis under circumstances in which their judgment may be compromised.
   (b) Engineers in salaried positions shall accept part-time engineering work only to the extent consistent with policies of the employer and in accordance with ethical considerations.
   (c) Engineers shall not, without consent, use equipment, supplies, laboratory or office facilities of an employer to carry on outside private practice.
7. Engineers shall not attempt to injure, maliciously or falsely, directly or indirectly, the professional reputation, prospects, practice or employment of other engineers. Engineers who believe others are guilty of unethical or illegal practice shall present such information to the proper authority for action.
   (a) Engineers in private practice shall not review the work of another engineer for the same client, except with the knowledge of that engineer, or unless the connection of that engineer with the work has been terminated.
   (b) Engineers in governmental, industrial, or educational employ are entitled to review and evaluate the work of other engineers when so required by their employment duties.
   (c) Engineers in sales or industrial employ are entitled to make engineering comparisons of represented products with products of other suppliers.
8. Engineers shall *accept personal responsibility for their professional activities*, provided, however, that engineers may seek indemnification for services arising out of their practice for other than gross negligence, where the engineer's interests cannot otherwise be protected.
   (a) Engineers shall conform with state registration laws in the practice of engineering.
   (b) Engineers shall not use association with a nonengineer, a corporation or partnership as a "cloak" for unethical acts.

9. Engineers shall *give credit for engineering work to those to whom credit is due*, and will recognize the proprietary interests of others.
   (a) Engineers shall, whenever possible, name the person or persons who may be individually responsible for designs, inventions, writings or other accomplishments.
   (b) Engineers using designs supplied by a client recognize that *the designs remain the property of the client* and may not be duplicated by the engineer for others without express permission.
   (c) Engineers, before undertaking work for others in connection with which the engineer may make improvements, plans, designs, inventions or other records that may justify copyrights or patents, should enter into a positive agreement regarding ownership.
   (d) Engineers' designs, data, records and notes referring exclusively to an employer's work are the employer's property. The employer should indemnify the engineer for use of the information for any purpose other than the original purpose.
   (e) *Engineers shall continue their professional development throughout their careers* and should *keep current in their specialty fields by engaging in professional practice*, participating in continuing education courses, reading in the technical literature and attending professional meetings and seminars.

♦ **Example C-1:** In a pump and pipe optimization problem, the preliminary cost estimate included only the cost of materials (Chapter 4, Problem 9). Perhaps this was not specified in earlier communication. Since the energy cost turns out to be significant over time (Chapter 8, Problem 9), it could have been perceived as an attempt to deceive the public. What could be done?

Solution: While great communication skills can be used to clarify diffuse situations, the highest standards of honesty and integrity are required of all engineers at all times.

As an epilog towards a healthy engineering philosophy, let me suggest the following three main pillars of our profession: (1) a commitment to professionalism through service and duty in the pursuit of excellence. For example, in the United States, the Order of the Engineer ceremony stipulates asking "not what we get in performance of duties, but rather what we can give." This does not mean that we should be poorly treated, to the contrary, but clearly our endeavors must be reached without greed. (2) Hydraulic engineers strive for the development of water resources for a better society and a healthy environment, and to alleviate hardship of living populations. (3) The added value of our profession is not in the mere knowledge of hydraulic systems, but in the application of our knowledge for the common good. In hydraulic engineering, this is reflected in the areas of flood control, clean water supply and sanitation, irrigated agriculture, hydropower and renewable energy, disaster mitigation and pollution control, with more details in Julien (2017). It is the development of hydraulic resources in harmony with the aquatic environment that enlightens the spirit of achieving success through fulfilling engineering work. In closing, our profession will continue to pose interesting new challenges for generations to come and one thing is sure: the future of hydraulic engineering will be exciting.

References

Abbott, M. B. and D. R. Basco (1989). *Computational Fluid Dynamics*. Addison-Wesley, London.

Abdullah, J. and P. Y. Julien (2014). Distributed flood simulations on a small tropical watershed with the TREX model. *J. Flood Eng.*, 5, 17–37.

Abdullah, J., S. N. Muhammad, P. Y. Julien, J. Ariffin and A. Shafie (2018). Flood flow simulations and return period calculation for the Kota Tinggi watershed, Malaysia. *J. Flood Risk Manag.*, 11, S766–S782; doi: https://doi.org/10.1111/jfr3.12256.

Abdullah, J., N. S. Muhammad, S. A. Muhammad and P. Y. Julien (2019). Envelope curves for the specific discharge of extreme floods in Malaysia. *J. Hydro-Environ. Res.*, 25, 1–11.

Abdullah, J., J. Kim and P. Y. Julien (2020). Hydrologic modeling of extreme events. In *Fresh Water and Watersheds*, W. Yang (ed.). CRC Press, Boca Raton, FL, ch. 11.

Abt, S. R. and T. L. Johnson (1991). Riprap design for overtopping flow. *J. Hydraul. Eng.*, 117, 959–972.

Akan, O. A. (2006). *Open Channel Hydraulics*. Elsevier, Burlington, MA.

Albertson, M. L., J. R. Barton and D. B. Simons (1960). *Fluid Mechanics for Engineers*. Prentice Hall, Englewood Cliffs, NJ.

An, S. D. and P. Y. Julien (2014). Three-dimensional modeling of turbid density currents in Imha Reservoir, South Korea. *J. Hydraul. Eng.*, 140, doi: https://doi.org/10.1061/(ASCE)HY.1943-7900.0000851.

An, S. D., H. Ku and P. Y. Julien (2015). Numerical modelling of local scour caused by submerged jets. *Maejo Intl. J. Sci. Tech.*, 9, 328–343.

Anctil, F., J. Rousselle and N. Lauzon (2012). *Hydrologie, Cheminement de l'Eau*, 2nd ed. Presses Internationales Polytechnique, Montréal.

Arcement, G. K. and V. R. Schneider (1984). *Guide for Selecting Manning's Roughness Coefficients for Natural Channels and Flood Plains. USGS Water Supply Paper 2339*. US Geological Survey, Washington, DC.

Arneson, L. A. and J. O. Shearman (1998). *User's Manual for WSPRO. A Computer Model for Water Surface Profile Computations*. Federal Highway Administration, Washington, DC.

ASCE (American Society of Civil Engineers) (2008). *Sedimentation Engineering: Processes, Measurements, Modeling and Practice. Manual of Practice 110*. American Society of Civil Engineers, New York.

ASCE (2013). *Inland Navigation, Channel Training Works. Manual of Practice 124*, American Society of Civil Engineers, New York.

Ball, J. W. (1981). Gates and valves. In *Closed-Conduit Flow*, Water Resources Publishing, Littleton, CO, 341–366.

Ball, J. W. and D. J. Hebert (1948). *The Development of High Head Outlet Valves. Hydraulic Laboratory Report No. Hyd.-240*, US Bureau of Reclamation, Denver, CO.

Barnes, H. H. (1987). *Roughness Characteristics of Natural Channels: USGS Supply Paper 1849*. US Geological Society, Denver, CO.

Batista, J. N. and P. Y. Julien (2019). Remotely sensed survey of landslide clusters: case study of Itaoca, Brazil. *J. S. Am. Earth Sci.*, 92, 145–150.

Battjes, J. and R. J. Labeur (2017). *Unsteady Flow in Open Channels.* Cambridge University Press, Cambridge.

Bauer, W. J. and E. J. Beck (1984). Spillways and stream-bed protection works. In *Handbook of Applied Hydraulics*, 3rd ed., C. V. Davis and K. E. Sorensen (eds.). McGraw Hill, New York, section 20.

Baumle, R. (2011). *Development of a Groundwater Information & Management Program for the Lusaka Groundwater Systems. Technical Note No. 4.* Ministry of Energy and Water Development, Department of Water Affairs, Lusaka, and German Federal Institute for Geosciences and Natural Resources (BGR), Lusaka, Zambia.

Baura, G. D. (2006). *Engineering Ethics, an Industrial Perspective.* Elsevier Academic Press, Boston, MA.

Beckwith, T. and P. Y. Julien (2020). *Middle Rio Grande Escondida Reach Report: Morpho-dynamic Processes and Silvery Minnow Habitat from Escondida Bridge to US-380 Bridge (1918–2018).* US Bureau of Reclamation, Colorado State University, Fort Collins, CO.

Bender, T. R. and P. Y. Julien (2011). *Bosque Reach. Arroyo de Las Canas to South Boundary Bosque del Apache National Wildlife Refuge. Overbank Flow Analysis 1962–2002.* US Bureau of Reclamation, Colorado State University, Fort Collins, CO.

Bormann, N. E. and P. Y. Julien (1991). Scour downstream of grade-control structures. *J. Hydraul. Eng.*, 117, 579–594.

Bos, M. G. ed. (1989). *Discharge Measurement Structures.* International Institute for Land Reclamation and Improvement, Wageningen.

Bouwer, H. (1978). *Groundwater Hydrology.* McGraw Hill, New York.

Bradley, J. N. (1978). *Hydraulics of Bridge Waterways*, 2nd ed. US Department of Transportation, Federal Highway Administration, Washington, DC.

Bray, D. I. (1979). Estimating average velocity in gravel bed rivers. *J. Hydraul. Eng.*, 105, 1103–1122.

Brookes, A. and F. D. Shields Jr., eds. (1996). *River Channel Restoration.* Wiley, Chichester.

Cahn, S. M. (2020). *Exploring Ethics. An Introductory Anthology*, 5th ed., Oxford University Press, Oxford.

Calder, I. R. (1992). Hydrologic effect of land-use change. In *Handbook of Hydrology*, D. R. Maidment (ed.). McGraw-Hill, New York, ch. 13.

Camp, T. R. and J. C. Lawler (1984). Water distribution. In *Handbook of Applied Hydraulics*, 3rd ed., C. V. Davis and K. E. Sorensen (eds.). McGraw-Hill, New York, Section 37.

Chang, H. H. (2006). *Generalized Computer Program FLUVIAL-12 Mathematical Model for Erodible Channels: User's Manual.* Chang Consultants, San Diego, CA.

Chanson, H. (2004). *The Hydraulics of Open Channel Flow: An Introduction*, 2nd ed. Elsevier, Amsterdam.

Chapra, S. C. (1997). *Surface Water-Quality Modeling.* McGraw-Hill, New York.

Chaudhry, M. H. (2008). *Open-Channel Flow*, 2nd ed. Springer, New York.

Chaudhry, M. H. (2014). *Applied Hydraulic Transients*, 3rd ed. Springer, Berlin.

Chaudhry, M. H. (2020). Innovative strategies for controlling hydraulic transients in pumping systems and hydroelectric powerplants. *Hydrolink*, No. 2, 41–43.

Cheng, N. S. (2015). Resistance coefficients for artificial and natural coarse-bed channels: alternative approach for large scale roughness. *J. Hydraul. Eng.*, 141, 515–529.

Chow, V. T. (1959). *Open-Channel Hydraulics*, McGraw-Hill, New York.

Chwang A. T. (1978). Hydrodynamic pressures on sloping dams during earthquakes, part 2: exact theory. *J. Fluid Mech.*, 87, 343–348.

Chwang A. T. and G. Housner (1977). Hydrodynamic pressures on sloping dams during earthquakes, part 1: momentum methods. *J. Fluid Mech.*, 87, 335–341.

Creager, W. P., J. D. Justin and J. Hinds (1945). *Engineering for Dams: Vol. 1 General Design.* Wiley and Sons, New York.

Cruise, J. F., M. M. Sherif and V.P. Singh (2007). *Elementary Hydraulics.* Cengage, Mason, OH.

Das, B. M. (1983). *Advanced Soil Mechanics.* Hemisphere Pubishing, New York.

Davis, S. N. and R. J. M. DeWiest (1966). *Hydrogeology.* Wiley and Sons, New York.

Davis, C. V. and K. E. Sorensen, eds. (1984). *Handbook of Applied Hydraulics*, 3rd ed., McGraw Hill, New York.

Derbyshire, K. (2006). *Fisheries Guidelines for Fish-Friendly Structures: Guideline FHG 006*, The State of Queensland, Department of Primary Industries and Fisheries, Brisbane.

Deukmejian, G., G. K. van Vleck and D. N. Kennedy (1985). *A.D. Edmonston Pumping Plant and Tehachapi Crossing.* California Department of Water Resources, Sacramento, California.

Doidge, S. (2019). *Middle Rio Grande San Acacia Reach: Morphology and Silvery Minnow Habitat Analysis. Plan B Report.* Colorado State University, Fort Collins, CO.

Duan, J. G. and P. Y. Julien (2005). Numerical simulation of the inception of channel meandering. *Earth Surf. Process. Landf.*, 30, 1093–1110.

Duan, J. G. and P. Y. Julien (2010). Numerical simulation of meandering evolution. *J. Hydrol.*, 391, 34–46.

Eagleson, P. S. (1970). *Dynamic Hydrology.* McGraw-Hill, New York.

England, J. F. Jr., M. L. Velleux and P. Y. Julien (2007). Two-dimensional simulations of extreme floods on a large watershed. *J. Hydrol.*, 347, 229–241.

England Jr., J. F., J. E. Godaire, R. E. Klinger, T. R. Bauer and P. Y. Julien (2010). Paleohydrologic bounds and extreme flood frequency of the Upper Arkansas River, Colorado, USA. *J. Geomorphol.*, 124, 1–16.

England, J. F., P. Y. Julien and M. L. Velleux (2014). Physically based extreme flood frequency analysis using stochastic storm transposition and paleoflood data. *J. Hydrol.*, 510, 228–245.

England, J. F. Jr., T. A. Cohn, B. A. Faber, J. R. Stedinger, W. O. Thomas Jr., A. G. Velleux, J. E. Kiang and R. R. Mason Jr. (2019). Guidelines for determining flood flow frequency. Bulletin 17C. In *Surface Water Book 4, Hydrologic Analysis and Interpretation.* US Geological Survey, Reston, VI, section B, ch. 5.

ESHA (European Small Hydropower Association) (1998). *Layman's Guidebook on How to Develop a Small Hydro Site.* European Small Hydropower Association. Brussels.

Ettema, R., C. Thornton, P. Julien and T. Hogan (2020). Applied research can enhance hydraulic engineering education. *J. Hydraul. Eng.*, 146, doi: https://doi.org/10.1061/(ASCE)HY.1943-7900.0001730.

Fadum, R. E. (1941). *Observations and Analysis of Building Settlements in Boston.* D.Sc. Thesis, Harvard University, Boston, MA.

FCSCM (2018). *Fort Collins Stormwater Criteria Manual.* City Council of the City of Fort Collins, Fort Collins, CO.

Fennema, R. J. and M. H. Chaudhry (1990). Numerical solution of two-dimensional transient free-surface flows. *J. Hydraul. Eng.*, 116, 1013–1034.

Ferguson, R. (2007). Flow resistance equations for gravel- and boulder-bed streams. *Wat. Res. Res.*, 43, W05427.

Ferras, D., G. DeCesare, D. I. C. Covas and A. J. Schleiss (2020). Hydraulic transients in hydropower systems: from theory to practice, *Hydrolink*, No. 2, 44–47.

FHWA (Federal Highway Administration) (1984). *Guide for Selecting Manning's Roughness Coefficients for Natural Channels and Flood Plains. FHWA-TS-84-204.* Turner-Fairbank Highway Research Center, McLean, VA.

FHWA (2001). *River Engineering for Highway Encroachments: Highways in the River Environment. FHWA NHI 01-004.* Federal Highway Administration, Washington, DC.

FHWA (2009). *Bridge Scour and Stream Instability Countermeasures: Experience, Selection, and Design Guidance. 3rd ed. Volumes 1 and 2, Hydraulic Engineering Circular 23, FHWA-NHI-09-111.* Federal Highway Administration, Washington, DC.

FHWA (2012). *Stream Stability at Highway Structures, 4th ed., Hydraulic Engineering Circular 20, FHWA-HIF-12-004.* Federal Highway Administration, Washington, DC.

Fogarty, C. (2020). *Linking Morphodynamic Processes and Silvery Minnow Habitat Conditions in the Middle Rio Grande – Isleta Reach, New Mexico.* M.S. Thesis, Colorado State University, Fort Collins, CO.

French, R. H. (1985). *Open-Channel Hydraulics.* McGraw-Hill, New York.

Frenette, M. and P. Y. Julien (1980). *Rapport synthèse sur les caractéristiques hydro-physiques du bassin de la rivière Matamec. Report CENTREAU-80-06.* Laval University, Québec.

Friedman, D., J. Schechter, B. Baker, C. Mueller, G. Villarini and K. D. White (2016). *US Army Corps of Engineers Non-stationarity Detection Tool User Guide.* USACE, Climate Preparedness and Resilience Community of Practice, Washington, DC.

Gessler, D., B. Hall and M. Spasojevic (1999). Application of 3D mobile-bed hydrodynamic model. *J. Hydraul. Eng.*, 125, 737–749.

Ghidaoui, M. S., M. Zhao, D. A. McInnis and D. H. Axworthy (2005). A review of water hammer theory and practice. *Appl. Mech. Rev.*, 58, 49–76.

Grozier, R. V., J. F. McCain, L. F. Lang and D. R. Merriman (1976). *The Big Thompson River Flood of July 31–August 1, 1976, Larimer County, Colorado. Flood Information Report.* US Geological Survey and Colorado Water Conservation Board, Denver, CO.

Guo, J., K. Woldeyesus, J. Zhang and X. Ju (2017). Time evolution of water surface oscillations in surge tanks. *J. Hydraul. Res.*, 55, 657–667.

Haan, C. T., B. J. Barfield and J. C. Hayes (1993). *Design Hydrology and Sedimentology for Small Catchment.* Academic Press, San Diego, CA.

Hager, W. H. and A. J. Schleiss (2009). *Constructions Hydrauliques: Ecoulements stationnaires.* EPFL Press, Lausanne.

Hayes, R. B. (1983). Baffled apron drops. In *Design of Small Canal Structures*, US Department of the Interior, Denver Federal Center, Denver, CO, 299–308.

Henderson, F.M. (1966). *Open Channel Flow*, Macmillan, New York.

Hernandez, N. M. (1984). Irrigation structures. In *Handbook of Applied Hydraulics*, 3rd ed., C. V. Davis and K. E. Sorensen (eds.). McGraw-Hill, New York, section 34.

Hite, J. E. (1992). *Vortex Formation and Flow Separation at Hydraulic Intakes.* Ph.D. Dissertation, Washington State University, Pullman, WA.

Hoffmans, G. J. C. M. and H. J. Verheij (1997). *Scour Manual.* Balkema, Rotterdam.

Holste, N. and D. Baird (2020). *One-Dimensional Numerical Modeling of Perched Channels. Technical Report ENV-2020-031*. US Bureau of Reclamation, Colorado State University, Fort Collins, CO.

Holtz, R. D and W. D. Kovacs (1981). *An Introduction to Geotechnical Engineering*. Prentice Hall, Englewood Cliffs, NJ.

Houghtalen, R. J., A. O. Akan and N. H. C. Hwang (1996). *Fundamentals of Hydraulic Engineering Systems*, 4th ed. Prentice Hall, New York.

Idelchik, I. E. (2008). *Handbook of Hydraulic Resistance*, 3rd ed. Jaico Publishing, Mumbai.

Jain, S. C. (2001). *Open-Channel Flow*. Wiley, New York.

Jarrett, R. D. (1985). *Determination of Roughness Coefficients for Streams in Colorado. USGS Water Resources Investigations Report 85-4004*. US Geological Survey, Lakewood, CO.

Ji, U., P. Y. Julien and S. K. Park (2011). Case study: sediment flushing and dredging near the Nakdong river estuary barrage. *J. Hydraul. Eng.*, 137, 1522–1535.

Ji, U., M. Velleux, P. Y. Julien and M. Hwang (2014). Risk assessment of watershed erosion at Naesung Stream, South Korea. *J. Env. Manag.*, 136, 16–26.

Jia, Y. and S. S. Y. Wang (1999). Numerical model for channel flow and morphological change studies. *J. Hydraul. Eng.*, 125, 924–933.

Jia, Y., S. S. Y. Wang and Y. Xu (2009). Validation and application of a 2D model to channels with complex geometry. *Intl. J. Comput. Eng. Sci.*, 3, 57–71.

Johnson, B. E., P. Y. Julien, D. K. Molnar and C. C. Watson (2000). The two-dimensional upland erosion model CASC2D-SED. *J. Am. Wat. Res. Assoc.*, 36, 31–42.

Julien, P. Y. (2010). *Erosion and Sedimentation*, 2nd ed. Cambridge University Press, Cambridge.

Julien, P. Y. (2017). Our hydraulic engineering profession, 2015 Hunter Rouse lecture. Hydraulic engineering lecture, 60th Anniversary State-of-the-Art Reviews. *J. Hydraul. Eng.*, 143, doi: https://doi.org/10.1061/(ASCE)HY.1943-7900.0001267.

Julien, P. Y. (2018). *River Mechanics*, 2nd ed. Cambridge University Press, Cambridge.

Julien, P. Y., A. Ab. Ghani, N. A. Zakaria, R. Abdullah and C. K. Chang (2010). Case study: flood mitigation of the Muda River, Malaysia. *J. Hydraul. Eng.*, 136, 251–261.

Julien, P. Y. and J. S. Halgren (2014). Hybrid hydrologic modeling. In *Handbook of Engineering Hydrology: Modeling Climate Change and Variability*, ed. S. Eslamian. CRC Press, Boca Raton, FL, 331–352.

Kang, W., E. K. Jang, C. Y. Yang and P. Y. Julien (2021). Geospatial analysis and model development for specific degradation in South Korea using model tree data mining. *Catena*, 200, doi: https://doi.org/10.1016/j.catena.2021.105142.

Karney, B. (2020). Hydraulic transients and negative pressures-consequences and risks, *Hydrolink*, No. 2, IAHR, Madrid, 83–40.

Kim, H. Y. and P. Y. Julien (2018). Case study: hydraulic thresholds to mitigate sedimentation problems at Sangju Weir. *J. Hydraul. Eng.*, 144, doi: https://doi.org/10.1061/(ASCE)HY.1943-7900.0001467.

Kim, H. Y., D. G. Fontane, P. Y. Julien and J. H. Lee (2018). Multi-objective analysis of the sedimentation behind Sangju Weir, South Korea. *J. Wat. Res. Plan. Manag.*, 144, doi: https://doi.org/10.1061/(ASCE)WR.1943-5452.0000851.

Kim, J. (2012). *Hazard area mapping during extreme rainstorms in South Korean mountains*. Ph.D. Dissertation, Colorado State University, Fort Collins, CO.

Knauss, J. ed. (1987). *Swirling Flow Problems at Intakes. IAHR, Hydraulic*

Structures Design Manual. Balkema, Rotterdam.

Kohler, W. H. and J. W. Ball (1984). High-pressure outlets, gates and valves. In *Handbook of Applied Hydraulics*, 3rd ed., C. V. Davis and K. E. Sorensen (eds.). McGraw-Hill, New York, section 22.

Kositgittiwong, D., C. Chinnarasri and P. Y. Julien (2012). Two-phase flow over stepped and smooth spillways: numerical and physical models. *Ovidius Univ. Ann. Ser.: Civil Eng*, 14, 147–154.

Kositgittiwong, D., C. Chinnarasri and P. Y. Julien (2013). Numerical simulation of flow velocity profiles along a stepped spillway. *Proc. Inst. Mech. Eng., Part E: J. Proc. Mech. Eng.*, 472171, 227, 327–335.

LaForge, K., C.Y. Yang, S. Doidge, T. Beckwith and C. Fogarty (2020). *Middle Rio Grande Rio Puerco Reach: Rio Puerco to San Acacia Diversion Dam Morphodynamic Processes (1918–2016) and Silvery Minnow Habitat (1992–2012).* US Bureau of Reclamation, Colorado State University, Fort Collins, CO.

Lambe, T. W. and R. V. Whitman (1969). *Soil Mechanics.* Wiley, New York.

Lai, Y. G. (2008). *SRH-2D Version 2: Theory and User's Manual. Sedimentation and Hydraulics – Two-Dimensional River Flow Modeling.* US Department of the Interior, USBR, Technical Service Center, Denver, CO.

Lee, J. S. and P. Y. Julien (2006). Electromagnetic wave surface velocimeter. *J. Hydraul. Eng.*, 132, 146–153.

Lee, J. S. and P. Y. Julien (2012a). Resistance factors and relationships for measurements in fluvial rivers. *J. Korea Contents Assoc.*, 12, 445–452.

Lee, J.S. and P.Y. Julien (2012b). Utilizing the concept of vegetation freeboard equivalence in river restoration. *Intl. J. Contents Assoc.*, 8, 34–41.

Lee, J. S. and P. Y. Julien (2017). Composite flow resistance. *J. Flood Eng.*, 8, 55–75.

Lee, J. S., P. Y. Julien, J. Kim and T. W. Lee (2012). Derivation of roughness coefficient relationships using field data in vegetated rivers. *J. Korean Wat. Res. Assoc.*, 45, 137–149.

Lee, J., J. H. Lee and P. Y. Julien (2018). Global climate teleconnection with rainfall erosivity in South Korea. *Catena*, 167, 28–43.

Lee, J. H., J. A. Ramirez, W. Kim and P. Y. Julien (2019a). Variability, teleconnection, and predictability of Korean precipitation in relation to large scale climate indices. *J. Hydrol.*, 568, 12–25.

Lee, J. H., P. Y. Julien and C. Thornton (2019b). Interference of dual spillway operations. *J. Hydraul. Eng.*, 145, doi: https://doi.org/10.1061/(ASCE)HY.1943-7900.0001593.

Leonard, B. P. (1979). A stable and accurate convective modeling procedure based on quadratic upstream interpolation. *Comp. Meth. Appl. Mech. Eng.*, 19, 59–98.

Liggett, J. A. and J. A. Cunge (1975). Numerical methods of solution of the unsteady flow equations. In *Unsteady Flow in Open Channels*, Water Ressources Publications, Fort Collins, CO, 89–182.

Lu, N. and J. W. Godt (2013). *Hillslope Hydrology and Stability.* Cambridge University Press, Cambridge.

Mansur, C. I. and R. I. Kaufman (1962). Dewatering. In *Foundation Engineering*, G. A. Leonards (ed.). McGraw Hill, New York, 241–350.

Mayer, P. R. and J. R. Bowman (1984). Spillway crest gates. In *Handbook of Applied Hydraulics*, 3rd ed., C. V. Davis and K. E. Sorensen (eds.). McGraw-Hill, New York, section 21.

Mays, L. W. (2019). *Water Resources Engineering*, 3rd ed. Wiley, Hoboken, NJ.

McCain, J. F., L. R. Hoxit, R. A. Maddox, C.F. Chappell, F. Caracena, R.R. Shroba, P.W.

Schmidt, E.J. Crosby, W.R. Hansen and J.M. Soule (1979). *Storm and Flood of July 31–August 1, 1976, in the Big Thompson River and Cache la Poudre River Basin, Larimer and Weld Counties, Colorado. USGS Professional Paper 1115.* US Geological Survey, Washington, DC.

McCuen, R. H. (2016). *Hydrologic Analysis and Design*, 4th ed. Pearson, Boston, MA.

Moglen, G. E. (2015). *Fundamentals of Open Channel Flow.* CRC Press, Taylor and Francis, New York.

Molinas, A. (2000). *User's Manual for BRI-STARS (BRIdge and Stream Tube model for Alluvial River Simulation). FHWA-RD-99-190.* Federal Highway Administration, Washington, DC.

Moody, L. F. and T. Zowski (1984). Hydraulic machinery. In *Handbook of Applied Hydraulics*, 3rd ed., C. V. Davis and K. E. Sorensen (eds.). McGraw-Hill, New York, section 26.

Mortensen, J. G, R. K. Dudley, S. P. Platania, G. C. White, T. F. Turner, P. Y. Julien, S. Doidge, T. Beckwith and C. Fogarty (2020). *Linking Morpho-dynamics and Bio-habitat Conditions on the Middle Rio Grande. Linkage Report I – Isleta Reach Analyses.* US Bureau of Reclamation, Albuquerque, NM.

Muhammad, N. S., P. Y. Julien and J. D. Salas (2015). Probability structure and return periods of multi-day monsoon rainfall. *J. Hydrol. Eng.*, 21, doi: https://doi.org/10.1061/(ASCE)HE.1943-5584.0001253.

Muller, A., ed. (1988). *Discharge and Velocity Measurements.* Balkema, Rotterdam.

NAE (2005). *Educating the Engineer of 2020: Adapting Engineering Education to the New Century.* National Academies Press, Washington, DC.

Naudascher, E. and D. Rockwell (1994). *Flow-Induced Vibrations. IAHR Hydraulic Structures Design Manual.* Balkema, Rotterdam.

Nelson, J. M., R. R. McDonald and P. J. Kinzel (2006). Morphologic evolution in USGS surface water modeling system. In Proceedings of the 8th Federal Interagency Sedimentation Conf.erence, Reno, NV, April 2–6.

NOAA (2006). *Precipitation Frequency Atlas of the United States, Atlas 14, Volume 2.* US Department of Commerce, National Oceanic and Atmospheric Administration and the National Weather Service, Silver Springs, MD.

Novak, P., A. I. B. Moffat, C. Nalluri and R. Narayanan (2001). *Hydraulic Structures*, 3rd ed. Spon Press, London.

O'Brien, P. (2017). *A Framework for the Analysis of Coastal Infrastructure Vulnerability under Global Sea Level Rise.* Ph.D. Dissertation, Colorado State University, Fort Collins, CO.

O'Brien, J. S., P. Y. Julien and W. T. Fullerton (1993). Two-dimensional water flood and mudflow simulation. *J. Hydraul. Eng.*, 119, 244–261.

Olson, R. E. (1974). Shearing strength of kaolinite, illite and montmorillonite. *J. Geotech. Eng. Div.*, 100, 1215–1229.

Padmanabhan, M. (1987). Design recommendations: pump sumps. In *Swirling Flow Problems at Intakes, IAHR, Hydraulic Structures Design Manual*, Balkema, Rotterdam, 101–124.

Palu, M. and P. Y. Julien (2019). Case study: modeling the sediment load of the Doce River after the Fundão tailings dam collapse, Brazil. *J. Hydraul. Eng.*, 145, doi: https://doi.org/10.1061/(ASCE)HY.1943-7900.0001582.

Palu, M. and P. Y. Julien (2020). Test and improvement of 1-D routing algorithms for dam break floods. *J. Hydraul. Eng.*, 146, 13p.

Palu, M. C., R. P. G. Salles and D. D. B. de Souza (2018a). Design of the Cambambe Dam heightening. In Proceedings of the 3rd International Dam World Conference, Foz do Iguacu, Brasil.

Palu, M. C., R. P. G. Salles and D. D. B. de Souza (2018b). Design of the second stage of the Cambambe HPP project. *Dam Eng.*, XXIX, 89–100.

Parmakian, J. (1963). *Waterhammer Analysis.* Dover, New York.

Pezzinga, G. and V. C. Santoro (2017). Unitary framework for hydraulic mathematical models of transient cavitation in pipes: numerical analysis of 1D and 2D flow. *J. Hydraul. Eng.*, 143, doi: https://doi.org/10.1061/(ASCE)HY.1943-7900.0001384.

Polubarinova-Kochina, P. Y. (1962). *Theory of Groundwater Movement.* Princeton University Press, Princeton, NJ.

Ponce, V. M. (2014). *Engineering Hydrology.* Macmillan Education, London.

PCA (Portland Cement Association) (1964). *Handbook of Concrete Culvert Pipe Hydraulics.* Portland Cement Association, Skokie, IL.

Rainville, E. D. (1964). *Elementary Differential Equations*, 3rd ed. Macmillan, New York.

Raudkivi, A. J. and R. A. Callander (1976). *Analysis of Groundwater Flow.* Arnold, London.

Rich, G. R. (1984a). Water hammer. In *Handbook of Applied Hydraulics*, 3rd ed., C. V. Davis and K. E. Sorensen (eds.). McGraw-Hill, New York, section 27.

Rich, G. R. (1984b). Surge tanks In *Handbook of Applied Hydraulics*, 3rd ed., C. V. Davis and K. E. Sorensen (eds.). McGraw-Hill, New York, section 28.

Richard, G. A., P. Y. Julien and D. C. Baird (2005). Case study: modeling the lateral mobility of the Rio Grande below Cochiti Dam, New Mexico. *J. Hydraul. Eng.*, 131, 931–941.

Roberson, J. A. and C. T. Crowe (1985). *Engineering Fluid Mechanics*, 3rd ed. Houghton Mifflin, Boston, MA.

Roberson, J. A., J. J. Cassidy and M. H. Chaudhry (1997). *Hydraulic Engineering*, 2nd ed. Wiley and Sons, New York.

Robinson, S., R. Dixon, C. Preece and K. Moodley (2007). *Engineering, Business and Professional Ethics.* Routledge, London.

Salas, J. D. and J. Obeysekera (2014). Revisiting the concepts of return period and risk under non-stationary conditions. *J. Hydrol. Eng.*, 19, 554–568.

Salas, J. D., G. Gavilan, F. R. Salas, P. Y. Julien and J. Abdullah (2014). Uncertainty of the PMP and PMF. In *Handbook of Engineering Hydrology: Modeling Climate Change and Variability*, ed. S. Eslamian. CRC Press, Boca Raton, FL, 575–603.

Salas, J. D., J. Obeysekera and R. M. Vogel (2018). Techniques for assessing water infrastructure for nonstationary extreme events: a review. *Hydrol. Sc. J.*, 63, 325–352.

Salas, J. D., M. L. Anderson, S. M. Papalexiou and F. Frances (2020). PMP and climate variability and change. *J. Hydrol. Eng.*, 25, doi: https://doi.org/10.1061/(ASCE)HE.1943-5584.0002003.

SCS (1986). Tour brochure: Mississippi field trip. In Proceedings of the 3rd International Symposium on River Sedimentation, Jackson, MI, March 31–April 5.

Senturk, F. (1994). *Hydraulics of Dams and Reservoirs, Water Resources Publications*, Highlands Ranch, CO.

Shearman, J. O. (1990), *HY-7 - User's Manual for WSPRO - A Computer Model for Water Surface Profile Computations: Federal Highway Administration Report FHWA-IP-89-027.* Washington, DC.

Shimizu, Y., M. W. Schmeeckle and J. M. Nelson (2000). Three-dimensional calculation of flow over two-dimensional dunes. *Ann. J. Hyd. Eng., Japan Soc. Civ. Eng.*, 43, 623–628.

Shin, Y. H. and P. Y. Julien (2011). Case study: effect of flow pulses on degradation downstream of Hapcheon Dam, South Korea. *J. Hydraul. Eng.*, 137, 100–111.

Simon, A. L. and S. F. Korom (1997). *Hydraulics*, 4th ed. Prentice-Hall, Englewood Cliffs, NJ.

Singh, V. P. (1997). *Kinematic Wave Modeling in Water Resources*. Wiley, New York.

Smith, C. D. (1985). *Hydraulic Structures*. University of Saskatchewan Press, Saskatoon.

Spangler, M. G. (1962). Culverts and conduits. In *Foundation Engineering*, G. A. Leonards (ed.). McGraw Hill, New York, 965–999.

Stedinger, J. R., R. M. Vogel and E. Foufoula-Georgiou (1992). Frequency analysis of extreme events. In *Handbook of Hydrology*, D. R. Maidment (ed.). McGraw-Hill, New York, ch.18.

Stein, O. R., P. Y. Julien and C. V. Alonso (1993). Mechanics of jet scour downstream of a headcut. *J. Hydraul. Res.*, 31, 723–738.

Steininger, A. (2014). *Dam Overtopping and Flood Routing with the TREX Watershed Model.* M.S. thesis, Colorado State University, Fort Collins, CO.

Streeter, V. L. (1971). *Fluid Mechanics*, 5th ed. McGraw Hill, New York.

Sturm, T. W. (2001). *Open Channel Hydraulics*. McGraw-Hill, New York.

Swamee, P. K. and A. K. Jain (1976). Explicit equations for pipe-flow problems, *J. Hydraul. Div.*, 102, 657–664.

Terzaghi, K. and R. B. Peck (1967). *Soil Mechanics in Engineering Practice*, 2nd ed. Wiley, New York.

Todd, D. K. (1964). Groundwater. In *Handbook of Applied Hydrology*, V. T. Chow (ed.). McGraw-Hill, New York, section 13.

USACE (US Army Corps of Engineers) (1985). *Hydropower. EM-1110-2-1701.* Office of the Chief of Engineers, Department of the Army, Washington, DC.

USBR (US Bureau of Reclamation) (1976). *Design of Gravity Dams.* US Department of the Interior, US Government Printing Office, Washington, DC.

USBR (1977). *Design of Small Dams.* 2nd ed. US Department of the Interior, US Government Printing Office, Washington, DC.

USBR (1983). *Design of Small Canal Structures.* US Department of the Interior, US Government Printing Office, Washington, DC.

USBR (1997). *Water Measurement Manual*, 3rd ed., US Department of the Interior, US Government Printing Office, Washington, DC.

USBR (2006). *Fish Protection at Water Diversions: A Guide for Planning and Designing Fish Exclusion Facilities.* US Department of the Interior, Denver, CO.

USBR (2007). *Rock Ramp Design Guidelines.* US Department of the Interior, Denver, CO.

USBR (2011). *Seismic Loads on Spillway Gates, Phase I: Literature review. Report DSO-11-06.* Dam Safety Technology Development Program. Denver, CO.

USBR (2015). *Bank Stabilization Design Guidelines. Report SRH 2015-25.* US Department of the Interior, Denver, CO.

USBR–USACE (2015). *Large Wood National Manual: Assessment, Planning, Design and Maintenance of Large Wood Influvial Ecosystems: Restoring Process, Function and Structure.* Available from www.usbr.gov/pn/.

USDA (US Department of Agriculture) (2007). *Stream Restoration Design. Part 654 National Engineering Handbook.* US Department of Agriculture, Washington, DC.

US Navy (1971). *Design Manual: Soil Mechanics, Foundations and Earth Structures. NAVDOCKS DM-7.* Department of the Navy, Naval Facilities Engineering Command, Washington, DC.

Velleux, M., P. Y. Julien, R. Rojas-Sanchez, W. Clements and J. England (2006). Simulation of metals transport and toxicity at a mine-impacted watershed: California Gulch, Colorado. *Env. Sci. Technol*, 40, 6996–7004.

Velleux, M., J. England and P. Y. Julien (2008). TREX: spatially distributed model to assess watershed contaminant transport and fate. *Sci. Total Environ.*, 404, 113–128.

Velleux, M., A. Redman, P. Paquin, R. Santore, J. F. England Jr. and P. Y. Julien (2012). Exposure Assessment for Potential Risks from Antimicrobial Copper in Urbanized Areas *Env. Sci. Technol.*, 46, 6723–6732.

Vennard, J. K. and R. L. Street (1975). *Elementary Fluid Mechanics*, 5th ed. Wiley, New York.

Vischer, D. and W. H. Hager, eds. (1995). *Energy Dissipators*. Balkema, Rotterdam.

Vischer, D. and R. Sinniger (1999). *Hydropower in Switzerland*. Society for the Art of Civil Engineering, Zurich.

Walton, W. C. (1970). *Groundwater Resource Evaluation*. McGraw Hill, New York.

Ward J. P. (2002). *Hydraulic Design of Stepped Spillways*. Ph.D. Dissertation, Colorado State University, Fort Collins, CO.

Warren, H. J. (1983). Water Measurement Structures. *Design of Small Canal Structures*, US Department of the Interior, Denver Federal Center, Denver, CO, 243–297.

Woo, H. S., W. Kim and U. Ji (2015). *River Hydraulics*, 2nd ed. Paju, Geonggi Do, Cheong Mun Gak, South Korea.

Woolhiser, C. A. (1975). Simulation of unsteady overland flow. In *Unsteady Flow in Open Channels*. Water Resources Publications, Washington, DC, 485–507.

WMO (1986). *Manual for Estimation of Probable Maximum Precipitation. 2nd Edition, Operational Hydrology Report No. 1, WMO No. 332*. World Meteorological Organization, Geneva.

WMO (2010). *Manual on Stream Gauging, Vol. I Fieldwork. WMO No. 519*. World Meteorological Organization, Geneva.

Wylie and Streeter (1978). *Fluid Transients*. McGraw-Hill, New York.

Yang, C. T. and F. J. M. Simoes (2008). GSTARS computer models and their applications, part I: theoretical development. *Intl. J. Sediment Res*, 232, 197–211.

Yang, C. Y., K. LaForge, S. Doidge, T. Beckwith, C. Fogarty and P.Y. Julien (2020). *Middle Rio Grande Isleta Reach: Isleta Diversion Dam to Rio Puerco Morphodynamic Processes (1918–2016) and Silvery Minnow Habitat (1992–012)*. US Bureau of Reclamation, Colorado State University, Fort Collins, CO.

Young, R. B. (1983a). Baffled outlets. In *Design of Small Canal Structures*. US Department of the Interior, Denver Federal Center, Denver, CO, 308–333.

Young, R. B. (1983b). Transitions and erosion protection. In *Design of Small Canal Structures*. Department of the Interior, Denver Federal Center, Denver, CO, 335–347.

Index